NATO ASI Series

Advanced Science Institutes Series

A series presenting the results of activities sponsored by the NATO Science Committee, which aims at the dissemination of advanced scientific and technological knowledge, with a view to strengthening links between scientific communities.

The Series is published by an international board of publishers in conjunction with the NATO Scientific Affairs Division

A Life Sciences B Physics	Plenum Publishing Corporation London and New York
C Mathematical and Physical Sciences D Behavioural and Social Sciences E Applied Sciences	Kluwer Academic Publishers Dordrecht, Boston and London
F Computer and Systems Sciences G Ecological Sciences H Cell Biology I Global Environmental Change	Springer-Verlag Berlin Heidelberg New York London Paris Tokyo Hong Kong Barcelona Budapest

PARTNERSHIP SUB-SERIES

1. Disarmament Technologies	Kluwer Academic Publishers
2. Environment	Springer-Verlag
3. High Technology	Kluwer Academic Publishers
4. Science and Technology Policy	Kluwer Academic Publishers
5. Computer Networking	Kluwer Academic Publishers

The Partnership Sub-Series incorporates activities undertaken in collaboration with NATO's Cooperation Partners, the countries of the CIS and Central and Eastern Europe, in Priority Areas of concern to those countries.

NATO-PCO DATABASE

The electronic index to the NATO ASI Series provides full bibliographical references (with keywords and/or abstracts) to about 50000 contributions from international scientists published in all sections of the NATO ASI Series. Access to the NATO-PCO DATABASE compiled by the NATO Publication Coordination Office is possible in two ways:

- via online FILE 128 (NATO-PCO DATABASE) hosted by ESRIN, Via Galileo Galilei, I-00044 Frascati, Italy.

- via CD-ROM "NATO Science & Technology Disk" with user-friendly retrieval software in English, French and German (© WTV GmbH and DATAWARE Technologies Inc. 1992).

The CD-ROM can be ordered through any member of the Board of Publishers or through NATO-PCO, Overijse, Belgium.

2. Environment – Vol. 7

Springer
Berlin
Heidelberg
New York
Barcelona
Budapest
Hong Kong
London
Milan
Paris
Santa Clara
Singapore
Tokyo

Transboundary Water Resources Management

Institutional and Engineering Approaches

Edited by

Jacques Ganoulis

Aristotle University of Thessaloniki
Department of Civil Engineering
Division of Hydraulics and Environmental Engineering
Thessaloniki 540 06, Greece

Lucien Duckstein

University of Arizona
Systems and Industrial Engineering Department
Tucson, AZ 87521, USA

Peter Literathy

Institute for Water Pollution Control
Water Resources Research Centre
H-1095 Budapest, Kvassay J. u. 1, Hungary

Istvan Bogardi

University of Nebraska-Lincoln
Department of Civil Engineering
Lincoln, NE 68588-0531, USA

Springer

Published in cooperation with NATO Scientific Affairs Division

Proceedings of the NATO Advanced Study Institute "Transboundary Water Resources Management: Technical and Institutional Issues", held on Skopelos Island, Greece, May 15–25, 1994

Cataloging-in-Publication Data applied for

Die Deutsche Bibliothek - CIP-Einheitsaufnahme

Transboundary water resources management : institutional and engineering approaches ; [proceedings of the NATO Advanced Study Institute "Transboundary Water Resources Management: Technical and Institutional Issues" held on Skopelos Island, Greece, May 15 - 25, 1994] / ed. by Jacques Ganoulis ... Publ. in cooperation with NATO Scientific Affairs Division. - Berlin ; Heidelberg ; New York ; Barcelona ; Budapest ; Hong Kong ; London ; Milan ; Paris ; Santa Clara ; Singapore ; Tokyo : Springer, 1996

(NATO ASI series : Sub-ser. 2, Environment ; Vol. 7)

ISBN 3-540-60714-5

NE: Ganoulis, Jacques [Hrsg.]; Advanced Study Institute Transboundary Water Resources Management: Technical and Institutional Issues <1994, Skopelos>; NATO: NATO ASI series / 02

ISBN 3-540-60714-5 Springer-Verlag Berlin Heidelberg New York

Printed in Germany

Typesetting: Camera-ready by authors/editors

SPIN: 10469808 31/3137 – 5 4 3 2 1 0 – Printed on acid-free paper

PREFACE

On May 15-25, 1994, an Advanced Study Institute (ASI) sponsored by the NATO / Scientific Affairs Division (SAD) was convened on Skopelos island, Greece. The Institute had three main objectives.

(1) To present and discuss alternative approaches for sustainable water resources management in transboundary river basins

(2) To assess the state of knowledge and recent advances on the subject, and

(3) To review institutional issues and co-operative programs for effective transboundary river management.

Alternating among lectures, computer workshops and short presentations from about 80 scientists (15 invited lecturers and 65 participants), the following three topics were presented and discussed, corresponding to the *three parts* of this book

(1) *Institutional Considerations*: institutional analysis, hydrodiplomacy and negotiation strategies

(2) *Engineering Planning and Management*: risk and reliability in water quantity and quality data, compromise decision-making under risk and integrated transboundary water resources management

(3) *Current Practices and Operational Examples*: case studies from international river basins and regional seas: Danube, Rhine, Elbe, Black Sea, Aral Sea, US-Mexico, Balkan and Baltic Countries.

Apart from NATO/SAD, funding was also provided by the Aristotle University of Thessaloniki Research Committee and the Hydraulics Laboratory, Division of Hydraulics and Environmental Engineering. The success of the Institute was due to the contributions of the attendees and especially to those of the Institute's organizing committee: Lucien Duckstein and Istvan Bogardi and co-directors Jacques Ganoulis (Greece) and Peter Literathy (Hungary). Special acknowledgment is addressed to Dr. Luis Veiga da Cunha, Director of NATO/ASI programs for his assistance in organizing the institute. Thanks are also extended to staff and graduate students from the Aristotle University of Thessaloniki A. Papalopoulou, A. Zorba, G. Iliadis, F. Kouvoukliotis, S. Seitanis, G. Mavidou and Ol. Sortikou, who handled efficiently the local arrangements and to Mitzi Austin, Tucson, Arizona for her assistance in the technical editing of the final manuscript. The inspiring atmosphere of the Prince Stafylos Hotel and the natural beauty of Skopelos island may also have contributed greatly to the success of the Institute.

Jacques Ganoulis
Director
Hydraulics Laboratory
Aristotle University of Thessaloniki
Thessaloniki, Greece

TABLE OF CONTENTS

PART 2: ENGINEERING PLANNING AND MANAGEMENT

PART 3: CURRENT PRACTICES AND OPERATIONAL EXAMPLES

Chapter VI: Transboundary Water Management in Regional River Basins

INTRODUCTION

When river basin boundaries do not match national political borders, issues and problems of transboundary water management emerge. Historically, rivers and lakes have been used to determine frontiers between countries and due to this fact they have been the origin of numerous conflicts throughout history (the Rhine between France and Germany, the Rio Grande between the USA and Mexico, the Oder and Neisse between Germany and Poland, the Amur and Ussuri between Russia and China, etc.). In many cases, several countries occupy parts of the upstream or downstream area of the river catchment. This makes the issue of water resource sharing even more complicated (the Nile between Egypt and Sudan, the Middle East conflict over the Jordan river, the Danube between many European countries, the Elbe between the Czech Republic and Germany, etc.).

Within national borders, management of water as a resource involves a number of *internal issues,* the most important of which may be listed as follows

- Disparities between regions
- Fluctuations in seasonal and longer time scales
- Inequality between needs and supply
- Conflicts in use between different sectors (water supply, agriculture, industry)
- Institutional, legal, economic and social issues.

When one deals with sharing transborder water resources between riparian countries, a number of *external issues* should be added to the above, such as

- Differences in political, social, and institutional structures
- Different objectives, benefits, and economic instruments
- International relations, regulation, and conflicts.

On a global scale, the importance of transboundary water resources is far from negligible: according to reports submitted to the UN, about 50% of the land of our planet (excluding Antarctica) is located in internationally shared water catchments. In this area about 40% of the world's population lives, extending over more than 200 international river basins.

The basic question at hand, as in the case of this book, is how and through what kind of processes water in transborder regions may unify rather than divide the nations and how stakeholders in international water catchments may increase their benefits without making the others lose benefits. The issue is complex because from technical to ecological considerations, issues of domestic and external political policy are involved.

In order to analyze and understand the origin of water-related conflicts and provide the "optimal" or "acceptable" or "most beneficial" solution for all involved parties, various approaches have been developed so far by different scientific and interested communities, such as the engineering, law, economic, political and social sciences communities, administrators and policy makers. For managing transboundary water resources, the different theories and models may be categorized in two groups

- Descriptive or process models
- Quantitative or outcome models.

The *first approach* comes mainly from law experts and political analysts, who focus on describing the anatomy of a given situation of conflict or cooperation. They insist on determining the function of different parameters and factors influencing the behavior of each country, such as the political perception of the importance of water, the international image and status of the country and also economic and institutional issues. Such models, including the behavior of institutional structures, international negotiation strategies, alternative dispute resolutions and political models are presented in *Part 1* of this book. They are mainly prescriptive and not predictive. They do not necessarily give a quantitative output (such as costs and benefits), but they are extremely important for understanding the processes and for analyzing the origin and the evolution of conflicts or cooperation.

The *second approach* has been developed mainly by engineers and management experts. Depending on the number of objectives and decision-makers and the combination between them, models may be formulated as *optimization, multi-objective trade-off* computerized codes or on the basis of the *team and game theories*. Most of these models are based on the fundamental economic notion of Pareto optimality and they are predictive in the sense that they suggest a quantitative "optimal" situation, which should be to terminate a conflict by an equitable resolution between the interested countries. Recent advances and related theoretical developments in this area are presented in *Part 2* of this book, including the application of the fuzzy set theory. However, the success in practice of this kind of *engineering* or *rational modeling* is mainly dependent upon the acceptance between interested actors and countries of the model assumptions, which rely on a set of prescribed objectives and the relative weights or preferences between conflicting goals. In the real world this is not usually the case, and therefore, there is a need to develop better, easier-to-use, interactive and reliable predictive models for transboundary water resources management.

In order to do that, a valuable source of information comes from the lessons we can learn from current practices, international conflicts and agreements that take place throughout the world in international river basins and regional seas. A summary of the operational examples reported in *Part 3* of this book is given in the map in Fig. 1. Both international freshwater resources in rivers and lakes, such as the Danube, the Rio Grande and the Great Lakes and coastal waters in regional seas, such as the Baltic Sea and the Black Sea are included.

We hope that by combining the expertise and the state-of-the-art knowledge of different scientific communities and disciplines, such as engineering, geography, environmental and social sciences, this book will contribute to the development of new methods, theories and models in order to resolve more efficiently conflicts and controversial issues in the management of transboundary water resources.

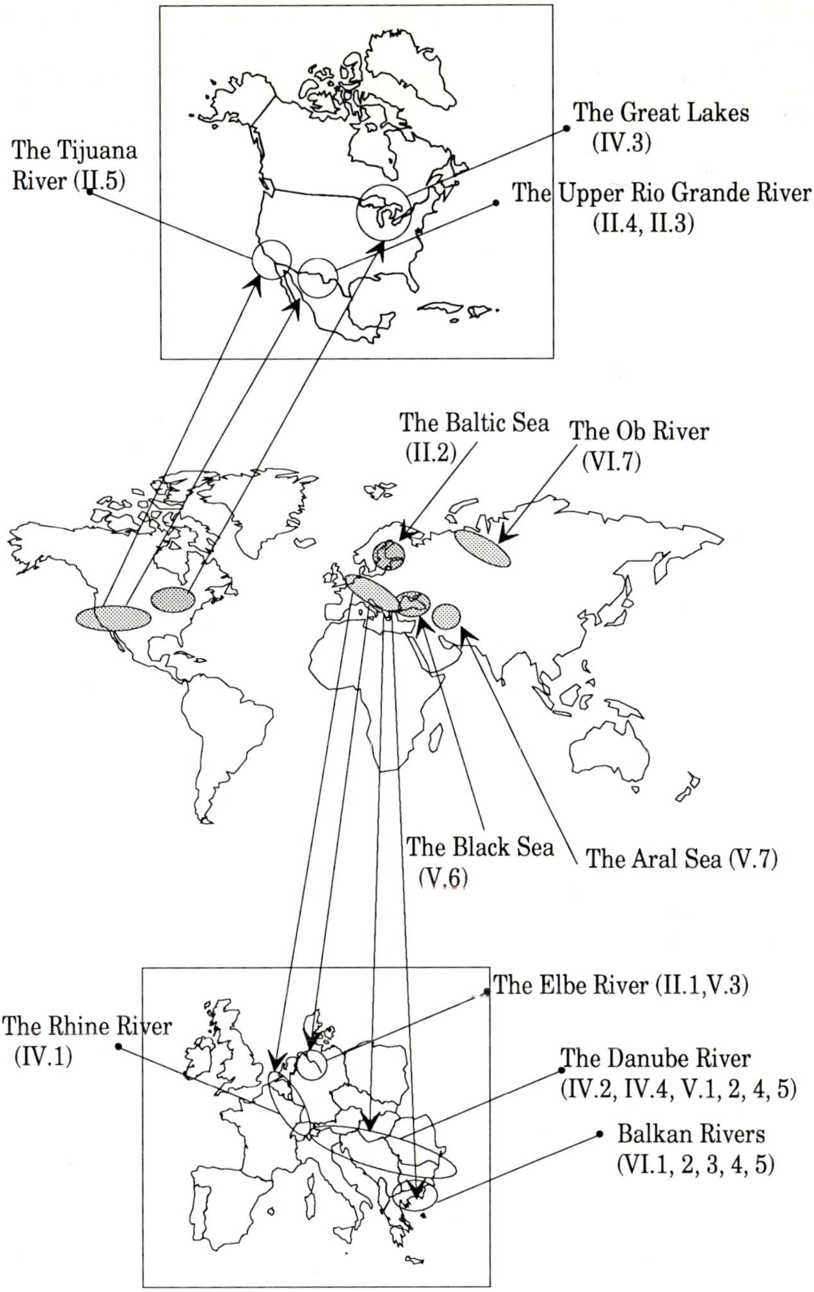

Fig. 1 Case studies of transboundary water resources management reported in this book.

PART 1

INSTITUTIONAL CONSIDERATIONS

INTRODUCTION TO PART 1

Part 1 of this book deals with policy issues and corresponding institutional analysis so as to be able to generate, implement and follow up water-related agreements, treaties, conventions and contracts. The first chapter of Part 1 presents conceptual tools and the second chapter poses a variety of sample issues and problems.

We dare recommend to the reader to begin reading with Waterstone's paper I.1, as it provides a framework not only for Part 1, but also for the rest of the book. The companion paper II.4 (also by Waterstone), which illustrates the concepts, definitions and interrelationships between the elements of a transboundary water problem, could be read shortly after I.1.

Specifically, Waterstone (paper I.1) defines institutions as a set of seven explicit and interrelated rules, which orders human actions; it further distinguishes three levels of decision systems (or institutional levels) namely:

- a policy level, where policy decisions are made
- an implementation level, where policy decisions are such as a given resource allocation is implemented
- an operational level, where explicit conditions are set (for example, allowable procedures for users).

Vlachos (paper I.2) introduces the concept of a second level of diplomacy to resolve transboundary water resources conflicts; his paper deals mostly with policy level institutions, in relation, however, to implementation level ones. Vlachos points out that any kind of international agreement is, at best, difficult; hence, all available tools should be utilized. Furthermore, particular attention should be paid to the relationship between institutional decision makers and the public; Vlachos distinguishes public awareness, involvement, participation and mobilization and briefly mentions the process of complexification, a problem which has been treated in greater detail in Casti (1994).

Eaton and Eaton (paper I.3) describe six possible options for negotiators that may be combined into a single scheme; these options constitute policy level decisions. Essentially, their approach covers the spectrum of game theory, from non-cooperative games (so called "lose-lose") to fully cooperative ones ("win-win"). Naturally, the latter option is recommended, which parallels Vlachos' advice (I.2).

Schumann and Simon (II.1) present a precise and specific description of the institution (at policy level) that has been set up to protect the river Elbe. For example, the "scope" rules (see I.1) are described precisely by specifying the tasks of each of the eight working groups of the International Commission.

Transboundary water resources issues between Baltic States are raised by Paukstys (II.2). Here, still using the three institutional levels described in Waterstone (I.1), it appears that the water resources policy has been set at least in part by a convention; Paukstys thus deals mostly with implementation level decision making, yet it appears that the convention or treaty has practically neglected an important pollution vector, namely, karst water transport. This problem is a good example of an implementation task (to control water pollution) that cannot be done properly because the policy decision of jointly considering karst ground water pollutant transport has not been made by the concerned actors (the Baltic States).

Gunning (II.3) dwells on the impact of the North American Free Trade Agreement (NAFTA) on water resources along the more than 3,300 km long U.S.-Mexican border. She points out that newly created policy level and then implementation level institutions should help reduce the considerable ecosystem strain on border region development, which is expected to increase even more rapidly than recently with NAFTA. Gunning also points out the necessity of public involvement, if not mobilization, as Vlachos (I.2) has done.

Part 2 of Waterstone's contribution (II.4) essentially shows how a simulation approach can integrate the three levels of decision making, using the actual case study of the Upper Rio Grande river basin. It should be noted that the term "boundary" denotes the separation between various water jurisdictions, albeit in the same state (New Mexico), yet similar problems occur as along international boundaries.

Finally, Brown (II.5) deals with the California-Mexico border and proposes to use a watershed level approach to attempt to preserve the biodiversity in the extremely pollutant-loaded San Diego-Tijuana area. Brown's approach is at the policy/implementation level; he uses the term "political ecology" in roughly a parallel sense with Vlachos' hydrodiplomacy.

In view of the imprecision of many quantities and concepts in most of the papers, it seems that further modeling and research in this field could use a fuzzy logic approach under risk. For example, collecting and sharing transboundary data on water quality may be based on a risk analysis framework, using fuzzy arithmetic (Ganoulis, 1994); institutions may be defined as a set of fuzzy rules (Bardossy and Duckstein, 1995), especially since any actor or stakeholder may belong to more than one of the seven institutional types. Fuzzy rule-based analysis allows several rules to be fulfilled simultaneously to varying degrees.

It also appears that most papers in this Part 1 (Chapters I and II) deal (implicitly) with policy/implementation levels of institutions. The next two parts, on the other hand, will place stronger emphasis on the implementation and operational levels.

References

Bardossy, A. and L. Duckstein (1995) *Fuzzy Rule-based Modeling in Geophysical, Economic, Biological and Engineering Systems.* CRC Press, Boca Raton, FLA.

Ganoulis, J. (1994) *Engineering Risk Analysis of Water Pollution: Probabilities and Fuzzy Sets.* VCH, Weinheim, Oxford, New York, Tokyo, 306 pp.

Chapter I: Institutional Analysis, Hydrodiplomacy, and Negotiation Strategies

I.1 A Conceptual Framework for the Institutional Analysis of Transboundary Water Resources Management: Theoretical Perspectives

Marvin Waterstone
Department of Geography and Regional Development
University of Arizona, Tucson, AZ 85721
USA

Introduction

It has become increasingly evident over the past several decades that natural resource and environmental issues transcend the maps that humans have used to divide up the globe. Enhanced greenhouse warming, depletion of stratospheric ozone, acid rain, as well as conventional air and water pollution clearly do not heed the artificial lines on maps. The same can be said about natural resources, which are the shared heritage of the citizens of more than one jurisdiction. The emplacement of political lines often disadvantages one group of resource users while benefiting or privileging another.

In such cases, mechanisms must be found for the efficient and equitable allocation of such resources. While engineered works often play a significant role in such circumstances, an appropriate institutional framework must be in place to allow the physical solutions to move forward. This paper begins with a brief overview of the special challenges presented by transboundary situations. It then presents an examination of some of the problems which have impeded the attainment of more effective institutional arrangements in resource management, and offers a remedy for several of these problems. Paper (II.4) then applies this approach to a specific case of transboundary water resource management, the Upper Basin of the Rio Grande in the USA.

Managing Transboundary Resources

Since late 1991, 46 new international boundaries have come into existence (Udall and Varady, 1993). Most of these are the result of the dissolution of the former Soviet Union, and the disintegration of Yugoslavia. New nations have coalesced along ethnic and cultural lines. Their boundaries represent points of contact and contest between social systems with differing histories, values, beliefs and political and legal structures. Such boundaries often are overlaid upon shared resource systems like water and frequently dissect common aquifers and rivers. Resource problems that may be difficult enough to resolve when they occur within one jurisdiction become much more intractable when they exist within several. Issues of national or state sovereignty, differential wealth, fragmented political and legal power, and ethnic and cultural differences all become much more salient when common resources and shared biophysical systems are divided by lines on maps and on the ground. According to Ingram, "[a]lmost 40 percent of the world's population lives within river basins shared by two or more nations" (1993: 1).

In many cases, such issues might be resolved by technological or structural solutions. For example, it may be possible to construct bi- or multi-national treatment facilities to ensure adequate water quality for a region. Likewise, it is possible to develop water storage projects which serve entire watersheds and provide flood control, water supply, hydropower and other purposes. However, for such solutions to proceed, an appropriate institutional framework must be in place. Many authors recognize this as a significant need (see, e.g., Mumme, 1993; Ingram and White, 1993; Szekely, 1993). For example, in a recent introduction to a two volume special issue of the Natural Resources Journal devoted to the management of North American transboundary water resources, Dworsky and Utton use the terms "institution" or "institutional" sixteen times within the space of eight pages (1993). Interestingly, and not uniquely, however, they never define these terms. If transboundary problems require changes in institutional arrangements for solution, clearly one essential step is the development of an appropriate framework for defining and analyzing institutions. In the remainder of this paper, I turn to these tasks. More detailed discussions can be found in Gregg et al. (1991) and Waterstone (1994), from which the following sections draw freely with the assumptions that the discussion is necessary and that the present volume will reach a different, largely non-overlapping audience.

Institutional Analysis and Resource Policy

Calls for greater emphasis on the institutional aspects of resource management have been frequent and persistent. However, severe terminological and methodological problems have hampered advances in institutional analysis generally and have been prevalent and problematic in institutional aspects of resource management research as well.

Over twenty years ago, O'Riordan indicated that:

> One of the least touched upon, but possibly one of the most fundamental, research needs in resource management is the analysis of how institutional arrangements are formed, and how they evolve in response to changing needs and the existence of internal and external stress. There is growing evidence to suggest that the form, structure and operational guidelines by which resource management institutions are formed and evolve clearly affect the implementation of resource policy, both as to the range of choice adopted and the decision attitudes of the personnel involved (O'Riordan 1971, 135).

This call for more research in institutional analysis built on the work of many others (including Kristjanson 1954; Ostrom 1962; Fox and Craine 1962; Sewell 1969) who had also identified institutional factors as important considerations in resource management. However, by 1975, Mitchell, in a canvass of the literature on institutional analysis, documented a continuing lack of progress (even in defining the problem).

More recently, others have noted the ongoing failure to come to grips with these issues (see, e.g., Cortner and Marsh 1987; Ingram et al. 1984). While there have been a number of recent attempts (for typical examples of this work, see Larkin 1986; Wandschneider 1986) to utilize institutional approaches to resource management and policy design problems, many still are problematic in the terms Mitchell used in 1975:

> ...due to lack of comparable terms of reference, studies about institutional arrangements in water management have been unable to build up a verified body of theory or knowledge. Instead, there is a large number of essentially idiosyncratic studies (1975, 255).

Mitchell goes on to recommend a resolution:

> If investigators could pursue a series of studies, using essentially the same terms of reference, an opportunity would emerge for checks upon reliability and validity as well as verification (Mitchell 1975, 257).

In the research described here, the intention was to produce just such a framework for analyzing institutions that would yield a consistent set of concepts and vocabulary, that would be flexible and that would be useful for comparative analyses to produce generalizable insights suitable for guiding policy design.

Institutional Analysis--Basic Concepts

In the following sections, I define and discuss the central constructs which are employed in the empirical research. The vocabulary I am employing here is not particularly "user friendly", but I believe it is necessary for addressing some of the long-standing confusion in this area of research. In fact, the development and elaboration of this consistent set of terms and concepts is one of the most significant achievements of this work. Without such a vocabulary and the concepts which underlie the terms, it is unlikely that progress will be made in understanding this important component of social scientific research.

Each of these constructs has been taken from some portion of the social science literature. The approach was eclectic and freely mixed constructs from a variety of paradigms and perspectives. Substantively, the research drew most directly on the work of public choice theorist Elinor Ostrom and the institutional economics tradition of John R. Common, upon which Ostrom builds.

Institutions

Institution is a term which has many meanings in both ordinary and scholarly discourse, and the ambiguity surrounding the term has slowed progress in many areas of research which could benefit from more rigorous institutional analysis. Many authors continue to use the term institution to mean organization (e.g., a governmental agency). However, many other researchers have used the term to indicate a broad (and usually unspecified) array of laws, regulations, policies, administrative rules and formal and informal guides to action and decision making (see, e.g., Riker, 1980; Ingram et al., 1984; Shepsle, 1986; North, 1986; Matthews, 1986; Young, 1986).

For the purposes discussed here, Ostrom's (1986) definition proved to be a useful one, and it is the meaning which should be understood throughout the remainder of this paper. Following Commons (1957) and Ganz (1971), Ostrom equates institutions with sets of rules. She refers to rules as "prescriptions commonly known and used by a set of participants to order repetitive, interdependent relationships" (1986, 5). Rules are created by individuals (acting under yet other rules, as discussed below) to achieve order and predictability within defined situations. An institution is such a set of nested rules.

The rules which comprise an institution may be formal and codified, as they are in constitutions, statutes and regulations. Alternatively, they may be informal and implicit, as are many of the rules that govern relationships within a family or within an organization or firm. However, for the purposes described here, rules must be subject to human intervention and change, as contrasted, for example, with physical laws, which influence human behavior, but for the most part are unchangeable by human action; however, it must be noted that the discursive representations of physical phenomena which comprise such laws are themselves human constructs.

It should also be understood that from the perspective of an individual actor, institutions (i.e., the rules that influence most everyday behavior) are both instrumental and nested. That is, institutions are connected to different spheres of everyday life, and individuals operate (with greater or lesser power) within several institutions simultaneously. For example, the rules which apply to the family exist coincidentally with those which apply to the workplace.

Rules

By defining institutions as sets of rules which order human action, I propose the following universal set of rules to move beyond the vagueness and lack of specificity which has plagued institutional analysis. This generally applicable and comprehensive set of rules (whether formal or informal) includes seven rule categories, which, taken together, comprise any institution.

(1) Position rules that specify a set of positions (i.e., the actors) and how many participants hold each position.
(2) Boundary rules that specify how participants are chosen to hold these positions and how participants leave these positions.
(3) Scope rules that specify the set of outputs that may be affected by a position.
(4) Authority rules that specify the set of actions assigned to a position.
(5) Aggregation rules that specify the manner in which individual actions are to be translated into collective choices.
(6) Information rules that authorize channels of communication among participants in positions and specify the language and form in which communication will take place.
(7) Payoff rules that prescribe how benefits and costs are to be distributed to participants in positions. (Adapted from Ostrom, 1986.)

Institutional Levels

Ostrom observed that, "Most public choice theorists `know' that multiple levels of analysis are involved in understanding how rules affect behavior. But this tacit knowledge of the multiple levels of analysis and how they intertwine is not self-consciously built into the way we pursue our work" (1986, 7). The explicit treatment of multiple levels of analysis, I believe, is another significant aspect of the empirical research described here. Wantrup identified three levels of decision systems (or what I am calling here, institutions) which he called the operational, the organizational and the policy levels (1970, 36). In the development of the analytic framework and the empirical research described in this paper, Wantrup's concepts were retained, but the terminology was altered slightly by referring to the second level of institutions as the "implementation" rather than the "organizational" level. The reason for doing so is to eliminate possible confusion due to the fact that action at any level can, and commonly does, occur within organizations.

The rules (i.e., the seven categories) at the operational level are those which govern actors' behavior in the world. The rules (again the seven types) at the implementation level are those which determine the rules at the operational level. The same seven rule types, when employed at the policy level, determine the rules at the implementation level. Thus, the outputs of decision making at each level (above the operational level) are the rules which will govern action at the next lower level. This model of the decision system, including the relationships between the three levels, is shown in Figure 1. [Ostrom (1986) describes a similar scheme and terms the three levels operational choice, collective choice and constitutional choice.]

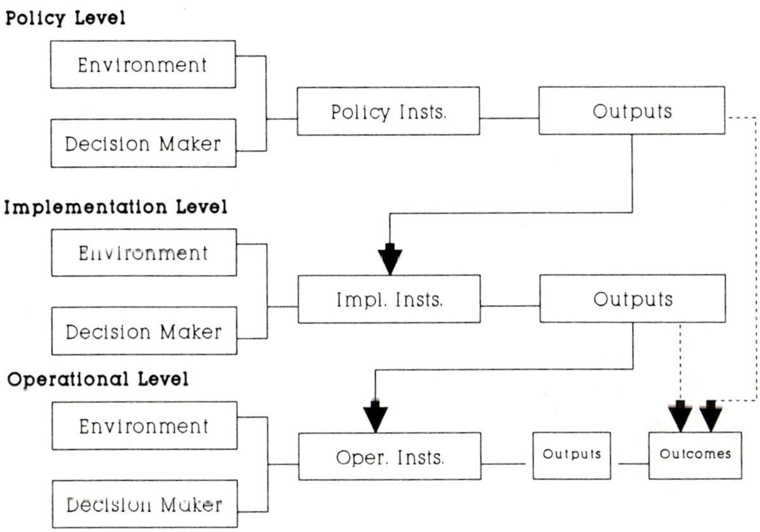

Figure 1 Institutional levels.

A simple example of this hierarchical arrangement would be the following: at the operational level, U.S. water users have permits which describe the conditions under which they can utilize water. The deployment of such a permit system (as well as the contents of particular permits) is usually determined by a water agency at the implementation level. The rules under which such a water agency operates usually are determined by a state legislature at the policy level.

This notion of institutional levels begins to get at several nagging questions of regarding power and collective action. Viewing the higher levels of institutions as sites for rule-making for lower levels provides a way to analyze differentiated institutional positions and therefore, differentiated power relationships. Those with positions and authorities at higher (i.e., implementation and policy) levels of institutions may not set the rules for operational level actors directly, but they often establish the procedures (i.e., the rules) for making such rules.

Institutional Analysis and Policy Design

With these basic concepts elucidated, I turn now to a generalized process model of institutional analysis and transformation. This model is intended to be descriptive rather than normative. It is intended to render a more or less faithful account of how institutional change actually occurs, not how I or anyone else think that it should occur. Because the empirical application which is described in the paper II.4 deals with a set of public policy issues, the following sections emphasize institutional analysis in such a relatively formalized framework. However, the basic conceptual model is equally applicable to less formalized institutions such as families, social groups and the like. In these cases, the kinds of analyses performed as well as the rules themselves are likely to be less formal than is described below.

Figure 2 depicts a process in which a problematic situation [the unspecified dissatisfaction with the status quo as described by Dewey (1938)], occurs at the operational level. This problematic situation then becomes diagnosed as a gap [defined by Livingston (1987; 1989) as a difference between what exists and what is thought to be desired]. The gap, once identified, may stimulate action to remove the dissatisfaction (through strategies to close the gap), action which may take the form of changing the rules at the operational level, at the implementation level and possibly, at the policy (or even higher) level.

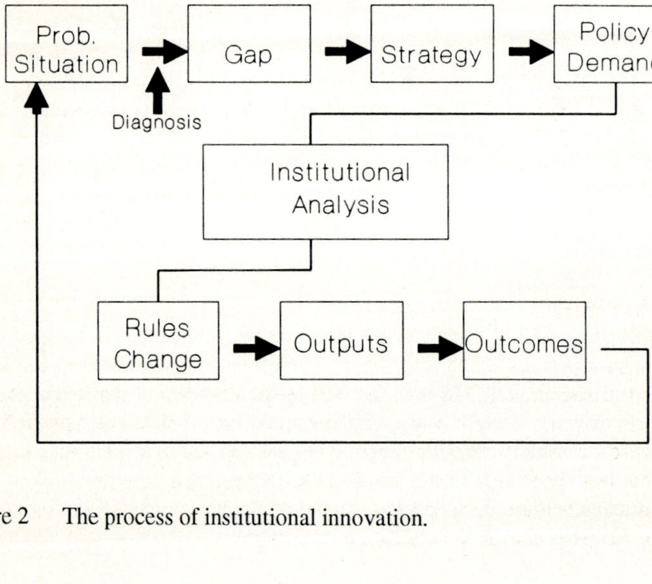

Figure 2 The process of institutional innovation.

From a behavioral perspective, one or more actors will have experienced gaps (perhaps from differing perspectives) which will have prompted them to adopt strategies [often through the formation of coalitions; see Knoke (1989); Laumann and Knoke (1987)] aimed at closing those gaps (refer to Figure 3). Of course, it is also possible (or in some cases, likely) that actors will seek remedies to problematic situations by going outside the rules altogether.

However, if actors attempt to work within the existing institutions, one element of such strategies is likely to be a call upon policy makers (i.e., usually actors at higher levels of the institution who have such authority) to take action in the form of changing rules. The actors may specify the rule changes which they want to see adopted (if these can be articulated), or they may emphasize the gaps which they hope to see closed. In either case, they place a demand upon higher-placed actors to act for them. These higher-placed actors, in responding to such demands, may pursue a process of institutional (i.e., rule) change.

The process of institutional analysis begins at this point, when policy makers (i.e., those in positions to change rules) call upon analysts to advise on the options available and their respective advantages and disadvantages. The task of the policy analyst is to prescribe specific rule changes (institutional innovations) which, if adopted and implemented, are likely to lead to closure of the gaps which gave rise to the policy demands.

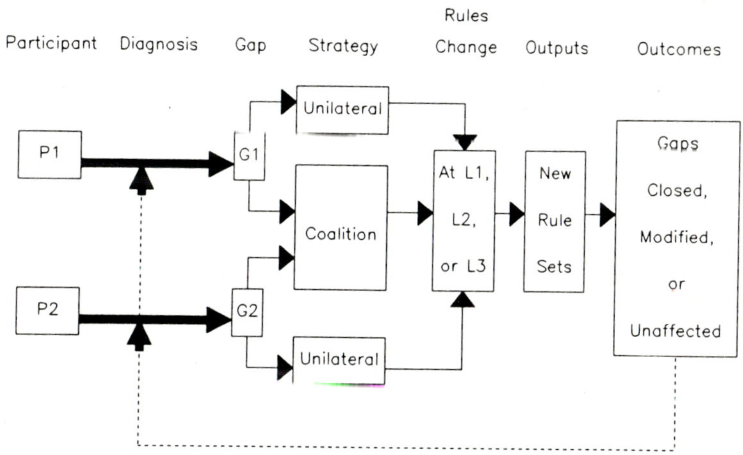

Figure 3 The behavioral model.

These rule changes are the outputs of the institutional change process. The institutional change will also have other consequences, some direct and some indirect, and some intended and others unintended. All are what many analysts (e.g., Easton 1979, 351) refer to as outcomes. Actors then weigh these outcomes and decide whether they are satisfactory or whether they leave the initial problematic situations unresolved or create new ones. In these latter cases, the process of institutional change becomes an iterative one.

This is conceptualized by a number of commentators as anti-structural activities by actors (Gregory, 1989; Lee, 1990; Mouzelis, 1989; Stones, 1991). For example, when wildlife enthusiasts perceive that species are endangered because of inadequate streamflow due to agricultural diversions, they may seek to change the rules which govern particular rivers.

It is actors' abilities to monitor these outputs and outcomes, their perceptions of the nature of these effects and their evaluation of them (i.e., their assessment of the relationship between outputs and outcomes) that determines (along with the rule sets in place at the relevant levels) participation in subsequent rounds of institutional transformation. If actors are satisfied, they may then work to preserve the new status quo. If they become aware of inadequately addressed gaps, new gaps or undesirable outcomes, they may (to the degree they are able) participate in a new round of rules changes and will oppose efforts to allow the new rules to be viewed as normative by asserting that they are counter-productive, damaging or needlessly costly.

One area of long-standing interest in which such problems arise with great regularity is water resource management. As described above, these issues take on added complexity in transboundary contexts. In the paper II.4, I turn to an application of the concepts just elaborated.

Conclusion

This paper provides a specific definition for the term institution. This analytic approach allows institutions to be specified and modeled with a degree of precision previously unattainable. It also allows institutional analyses to move forward in ways that will be necessary if the challenges of transboundary water management are to be met successfully.

References

Commons, JR (1957) Legal foundations of capitalism. Madison: The University of Wisconsin Press.

Cortner, HJ and FL Marsh (1987) Institutional analysis in community decision-making: A case example from Southern Arizona. Water Resources Bulletin 23: 317-324.

Dewey, J (1938) The public and its problems: An essay in political inquiry. Chicago: Gateway Books.

Dworsky, L and AE Utton (1993) Introduction. Natural Resources Journal 33: 1-8.

Easton, D (1979) A systems analysis of political life. Chicago: The University of Chicago Press.

Fox, IK, and Craine, LE (1962) Organizational arrangements for water development. Natural Resources Journal 2: 1-44.

Ganz, JR (1971) Rules: A systematic study. The Hague: Mouton.

Gregg, F et al (1991) Institutional response to a changing water policy environment. Available from the National Technical Information Service (Report Number PB 92-161405).

Gregory, D (1989) Presences and absences: Time-space relations and structuration theory, in Social Theory of Modern Societies: Anthony Giddens and His Critics, eds D Held, J Thompson. Cambridge: Cambridge University Press.

Ingram, H (1993). Transnational Water Resources Management: Learning from the U.S.-Mexico Example. The Fourth Annual Abel Wolman Distinguished Lecture. Washington, D.C.: National Research Council.

Ingram, H and White, DR (1993) International Boundary and Water Commission: An institutional mismatch for resolving transboundary water problems. Natural Resources Journal 33: 153-175.

Ingram, HE et al. (1984) Guidelines for improved institutional analysis in water resources planning. Water Resources Research 20: 232-334.

Knoke, D (1989) The political sociology approach to institutional analysis. Paper presented at the Institutional Analysis Seminar, School of Renewable Natural Resources and Department of Political Science, University of Arizona, February, 1989.

Kristjanson, K (1954) Institutional arrangements in water resources development. Land Economics 30: 347-362.

Larkin, A (1986) Environmental impact and institutional adjustment: Application of Foster's principles to solid waste disposal. Journal of Economic Issues 20: 43-61.

Laumann, EO, and Knoke, D (1987) The organizational state: A perspective on national energy and health domains. Madison: University of Wisconsin Press.

Lee, R (1990) The micro-macro problem in collective behavior: Reconciling agency and structure. Journal for the Social Theory of Behavior 20 213-233.

Livingston, ML (1987) Evaluating the performance of environmental policy: Contributions of neoclassical, public choice and institutional models. Journal of Economic Issues 21: 281-294.

-------------------- (1989) Normative and positive apsects of institutional economics: The implications for water policy. Paper presented at the Institutional Analysis Seminar, School of Renewable Natural Resources and Department of Political Science, University of Arizona, March, 1989.

Matthews, RCO (1986) The economics of institutions and the sources of growth. The Economic Journal 96: 903-918.

Mitchell, B (1975) An investigation of research barriers associated with institutional arrangements in water management. In Institutional arrangements for water management: Canadian experiences, ed., B Mitchell. Waterloo: University of Waterloo, Department of Geography.

Mouzelis, N (1989) Restructuring structuration theory. The Sociological Review 37 613-635.

Mumme, S (1993) Innovation and reform in transboundary resource management: A critical look at the International Boundary and Water Commission, United States and Mexico. Natural Resources Journal 33: 93-120.

North, D (1986) The new institutional economics. Journal of Institutional and Theoretical Economics 142: 230-237.

O'Riordan, T (1971) Perspectives on resource management. London: Pion Limited.

Ostrom, E (1986) An agenda for the study of institutions. Public Choice 48: 3-25.

Ostrom, V (1962) The political economy of water development. American Economic Review 52: 450-458.

Riker, WH (1980) Implications from the institutions. American Political Science Review 74: 432-446.

Sewell, WRD (1969) Geographical research in water management in Canada: Inventory and prospect. In Water management research: Social science priorities, WRD Sewell, RW Judy, and L Ouellet. Ottawa: Queen's Printer.

Shepsle, KA (1986) Institutional equilibrium and equilibrium institutions. In Political Science: The Science of Politics, ed. H. Weisburg. New York: Agathon.

Stones, R (1991) Strategic context analysis: A new research strategy for structuration theory. Sociology 25: 673-695.

Szekely, A (1993) How to accommodate an uncertain future into institutional responsiveness and planning: The case of Mexico and the United States. Natural Resources Journal 33: 397-403.

Udall, SL and RG Varady (1993) Environmental conflict and the world's new international borders. Transboundary Resource Report 7: 5-6.

Wandschneider, PR (1986) Neoclassical and institutionalist explanations of changes in Northwest water institutions. Journal of Economic Issues 20: 87-107.

Wantrup, SVC (1970) Resource conservation: Economics and policies. Berkeley: The Univesity of California Press.

Waterstone, M (1994) Institutional analysis and water resources management. In E Parent and L Duckstein, Editors, Engineering Risk in Natural Resource Management. NATO ASI Series E: Applied Sciences. Dordrecht: Kluwer Academic Publishers.

Young, O.R. 1986. International regimes: toward a new theory of institutions. World Politics 39: 104-122.

I.2 Hydrodiplomacy and Dispute Resolution in Private Water Resources Conflicts[1]

Evan Vlachos
Sociology and Civil Engineering
Colorado State University
Fort Collins, CO 80523
USA

Transboundary Water Changes

The last 20 to 30 years are characterized by significant changes in the planning, design, and management of water resources all over the planet. Mounting concerns about the environmental impacts of human activities, potential climatic shifts, expanding populations and demands as well as new knowledge are all expressions of the pressing need to develop alternative institutional schemes for managing in an integrated manner scarce natural resources. Many nations and regions have increasingly been turning attention to both streamlining existing administrative mechanisms and to introducing innovative institutional arrangements with regard to quantitative and qualitative aspects of their water resources. At the same time, the political significance of water becomes most important not only because of its scarcity in densely populated regions, but also because of its sharing across national boundaries.

In this context, geometrically expanded populations and increasing agricultural and industrial activities indicate the possibility of critical, global water shortages in the near future. As Mandel (1991) has indicated, in the twentieth century the world has witnessed substantial growth in the demand for fresh water along with noticeable declines in the uncontaminated supply of this resource. Even more, large scale population dislocations, socio-economic changes, and national and international events and upheavals have coalesced into an underlying apprehension and have focused attention to the need for more integrated, anticipatory and far-reaching water policies and strategies.

Experts disagree about how much water is available in given regions. International river basins have become arenas for competitive exploitation of shared water supplies by neighboring nations. However, awareness is growing that nations must cooperatively manage, engineer, and conserve available water resources. Indeed, as Gleick points out, no region of the world with shared international water is exempt from water-related controversies, though the most serious problems occur in water-scarce regions. Without cooperative management a zero-sum competition will emerge over water. Seasonal and regional water shortages may exacerbate social tensions and precipitate violence.

---

[1] This paper has been expanded and benefitted from revisions of two earlier communications: "Incentives for Regional Cooperation with Regard to Shared Water Resources," (UN Freshwater Consultative Forum, Geneva, Switzerland, December 1993) and from "Transboundary Water Conflicts and Alternative Dispute Resolution" (VIIIth International Water Association Congress, Cairo, Egypt, 1994).

Sharing and cooperation can provide benefits that exceed those achieved by attempts to maximize individual and national self-interest. Ideally, such cooperation requires a new "diplomacy", alternative institutional arrangements, large financial resources, and effective adjudication or conflict management mechanisms.

Today, for example, the major watercourses of the Middle East are increasingly becoming the focus of concern between neighboring countries. Increasing utilization of surface and underground water resources has raised awareness of their finite nature. Only continuous communication between the states concerned can be the central mechanism to peaceful solutions of festering disputes over shared waters. The Center for Strategic and International Studies predicted in 1988 that "by the year 2000 water, not oil, will be the dominant resource issue of the Middle East." As population grows and water uses multiply, the supply remains relatively constant, and the sources are often far from the ever-growing centers of demand. Yet, water resources management is not a simple matter of identifying reservoir sites, pipelines, agricultural irrigation schemes, and groundwater pumping. In the Middle East, religion, culture, politics and tradition greatly complicate what at best would be a difficult multinational resource-management problem. Some experts indicate that time is running short, that disaster vis-a-vis Middle eastern sources of water is around the corner. Others maintain that there is no general problem because nations simply can reallocate water from irrigated agriculture to municipal and industrial uses and utilize more efficient water technologies. Beyond the quantity of water, and the use of that water, is also the question of water quality. Deteriorating water quality may in effect be the trend that brings the Middle East to the limits of water availability.

Using the Balkans as a further illustrative example, similar water interdependencies can be brought forward. In the case of Evros/Maritza (between Bulgaria, Greece and Turkey) there are no major water supply problems as there are no other water uses besides irrigation. Between Greece and Albania, in the Aoos River there have been protests on the part of Albania due to the construction of a large dam on the Greek side. From 1965 on water resources problems in the Axios/Vardar River (Greece and Yugoslavia) have increased due to intensive irrigation and accelerating pollution. The greatest challenge in the region is the Nestos/Mesta Rivers between Greece and Bulgaria. Despite earlier agreements, Bulgaria has been withholding supplies for its increased agricultural and industrial needs. From 1975 on the Nestos flow declined from 1500 million CM to 600 million CM resulting to repeated Greek protests. A series of negotiations since 1965 have not resulted in any agreement and this fact poisons relations between the two countries. More recently, noticeable pollution from the Bulgarian heavy industry (including thermal pollution from a Bulgarian nuclear plant) have raised the level of tension in a region of Greece highly dependent on irrigated agriculture.

Thus, a series of trends and developments are creating an air of imminent crisis regarding water supplies. Factors underlying this context of urgency include:

a) the high variance of water supply in many parts of the planet, resulting in dramatic fluctuations, exacerbated by periodic droughts;

b) rapid population growth, coupled by dramatic shifts from rural to urban areas, the rise of major metropolitan agglomerations and related increasing urban water demands;

c) expanding agricultural uses and intensive irrigation developments;

d) deterioration of water quality, both the result of agricultural practices (irrigation, pesticides, fertilizers, herbicides) and of urban and industrial uses;

e) decreasing groundwater availability coupled with contamination of a large number of aquifers;

f) increasing environmental concerns and ecosystemic considerations, including questions of wetlands, migratory birds, concerns with potential climatic anomalies, the possibility of global warming and the entire gamut of anthropogenic disturbances in the surrounding environment;

g) transfrontier water dependencies and the international nature of water supplies. As Kolars (1991), for example, has pointed out, over 50 percent of the populations of the Middle East depend upon water from rivers that cross an international boundary.

There is no need to repeat the context of competing and conflicting demands all over the planet. What is particularly noticeable in many examples of transboundary water dependencies is that countries can come close to the brink of war over water schemes (e.g. between Hungary and Slovakia regarding the Gabcikovo, Nagymaros dams) for which no substantial, detailed impact assessments have been made (Rose 1993).

But what has further complicated the search for thoughtful, coordinated and long-range solutions has been the fact of rapid social changes, of challenges to centralized power, and of the realignment of ideological camps in kaleidoscopic combinations of resource policies, environmental awareness, developmental alternatives and increased pace of technological developments. Further aggravating the nature and rate of all such transformations are also the confluence of popular and scientific interest in environmental protection; data and information explosion, new understandings as to the structure and functioning of environmental systems; new institutional mechanisms for education and training and for the dissemination of knowledge; increased emphasis on globalization and interdependence; and, profound questions as to development and environment, especially as exemplified in North-South differentiations.

Recent history, rapid socio-economic changes and the transitions necessitated by the turbulent decades of the 80s and 90s underscore the increasing emphasis on the variety of environmental challenges, the search for sustainable development, the promotion of integrated planning and management, and the attempt to combine structural and non-structural solutions to persistent water resources problems. If we were to summarize four broad areas of on-going transformations in water resources, we should emphasize: a) conceptual breakthroughs, including shifting paradigms in terms of ecosystems, sustainability, heterarchization, complexity, uncertainty, turbulence and interdependence of surrounding environments; b) methodological advances, especially multi-purpose/multi-objective approaches, Decision Support Systems, Risk Analysis, and the implications of rapidly expanding computational prowess; c) organizational mobilization, in terms of new administrative mechanisms, institutional arrangements, renewed interest on river basin interdependencies, contingency planning, Alternative Dispute Resolution, etc.; and, d) contextual changes, signifying the entire gamut of on-going and future quantity and quality problems, new areas of concern, shifting priorities, potential intervention mechanisms, etc.

By now it should be apparent that the themes of interdependence, globalization, ecosystemic unity and sustainable development throw new light to traditional understandings of the geopolitical nature of water. River basins can, then, become an expression of "transnational regimes" in environmental politics (from global to bilateral scales). Such regimes can be defined either as a set of norms, rules, or decision-making procedures that produce some convergence in the actors' expectations in water related issues; or, as explicit rules specified in multilateral legal instruments regulating national actions (such as conventions or protocols). Historically, the depletion of national water resources, recurring droughts, expanding socio-economic demands have fueled confrontations and have forced international exchanges and cooperation. From the 805 AD deed of Charlemagne to a monastery for the navigation of the Rhine more than 3800 unilateral, bilateral, or multi-lateral declarations or conventions on water have been identified with 286 actual treaties of which 61 refer to 200 international river basins (Sironneau, 1993).

Looking back at the experiences of such efforts, it has been noted that the obstacles erected in the way of the efficient harnessing of these resources by the political boundaries that cut across them have often proved very difficult to surmount (Godana, 1985). But, recently, there has been not only increased awareness but also noticeable agreement among many States that the interconnection of water resources and the interdependence of developmental interests can be best affected by a commitment to principles of cooperation, effective international law, and a wide range of complimentary mechanisms of conflict management and consensus building.

The geopolitical significance of water is further accentuated by the variety of the conditions outlined above and especially by its scarcity, maldistribution and misuse. The mismatch between political boundaries and natural river basins, becomes, then, a focal point for the variety of difficulties reported in the literature vis-a-vis joint planning, allocation of costs, advantages of scale, exercise of power and coordination, and the whole range of issues associated with integrated, holistic management. Cooperation and conflict are, then, expressions of the same quest for improving effective planning and management, for promoting new ways for sustainable development, and for accommodating the realities of geography to the social context of shared water resources. Such international responsibilities tend to fall into three general categories. First, the "downstream responsibilities" of the water of one state which flows into another. Second, the "upstream responsibilities" of states whose activities may extend upstream and affect another state. And, finally, "cross-stream responsibilities" of countries whose common border is formed by a river, or even when they share underground reserves (Rose 1993).

Environmental Diplomacy and Water Conflicts

The central point of the previous remarks is that demands for water will expand as a result of both continued economic development and population growth. At the same time, broad questions of environmental preservation will persist and allocation patterns especially of shared water resources will continue being demanding and controversial. Present water management arrangements are not well equipped to deal with the new planning spaces. The practices which have evolved during the development to the present stage are no longer adequate to meet the demands of a desired larger framework for socioeconomic development as well as the search for environmental accountability. Indeed, as Rose (1993) has observed:

> "There is a severe water shortage in much of Europe and the Mediterranean region caused by chronic mismanagement of resources, political opportunism, war and drought. At the heart of the conflicts which characterize water resource allocation is governments' insistence on cutting up the water cycle into tiny conceptual or managerial parts. Water management is considered separately from land management, groundwater from surface water, human use from ecosystems and so on. In Greece no less than seven ministries govern the use and management of water. There is little or no coordination between their often competing objectives, a pattern repeated to varying degrees in most countries. The lack of horizontal coordination between user interests is mirrored in a similar lack of vertical collaboration between national and regional administrations over shared objectives. When water resources are shared between sovereign states, problems of water allocation, use, quality and conservation are exacerbated immeasurably, occasionally spilling over into conflicts in which sovereignty itself is an issue."

The comprehensiveness of water resource planning has been the subject of controversy and debate in the literature. It has been recognized, however, that in order to be able to maximize the benefits from any water resource project a much larger systemic analysis of the surrounding environment is needed, a broadening of the traditionally narrow planning and management approaches, and an increased sensitivity to decision-making problems associated with multi-objective and multi-purpose actions. In this vein a recent publication of the World Bank (1993) has elaborated the need for a comprehensive framework for analyzing policies and options where water scarcities exist, inefficiencies persist and environmental damages are becoming apparent. In such a context analyses at a river basin level become part of a national strategy for water resource management that includes elements for formulating public policies, regulations, incentives, public investment plans and environmental protection (as well as the interlinkages among them).

Geography suggests that, by virtue of physical unity and regardless of political divisions, a river basin should be developed and managed as a single indivisible whole. This is because moving water ties land together and interference with its movement has repercussions elsewhere in the basin. While geographic ties prescribe the unitary development of river basins, politics and history distort this process. The nation-state covets its sovereignty. Moreover, the desire to maximize benefits for itself provides the nation state with a powerful incentive to exploit resources unilaterally. This behavior, then, can generate international frictions, conflicts, and even war. We are, therefore, faced with a situation in which states, confronted with what they deem sub-optimal choices tend to adopt non-cooperative stances, although cooperation, on specific issues, would be in their mutual interest.

As Mandel (1991) has emphasized, the sources of dispute over international river basins encompass such diverse disciplines as ecology, economics, social psychology, religion, cultural anthropology, engineering, hydrology, and geopolitics. The author suggests a three-stage explanatory process starting from a non-cooperative setting (with pre-existing general antagonism); to a stage of environmental imbalance encompassing perceived growing scarcity of usable water and inequalities in distribution; to, finally, power asymmetries among riparian nations and little reciprocal interdependence.

The creation of a water management "regime" may facilitate the emergence of mutually beneficial agreements among the local or regional actors, ensuring that friction does not lead to conflict. An everyday definition of a regime is a "mode or system of rule or government." In contemporary international relations theory, however, regimes consist of procedural and normative guides to state behavior. Areas in which international regimes have grown most recently include outer space, the oceans, telecommunications, trade commodities, etc. Various types of international regimes have existed for centuries in such areas as the law of the sea, monetary affairs, trade, and communications. But the technological, political and economic forces which have brought nations into these more recent "webs of interdependence" have extended deeper than ever before into provinces that once had been the exclusive jurisdiction of national governmental policies.

The last fifty years has come about from nations diverting and developing international water resources, of using the resources as a conduit for waste disposal and assimilation, and of preserving certain environmental amenities beyond the historical uses of international rivers. These agreements reflect an interest to engage in comprehensive national planning, basin-wide management, and multi-sectoral development and water quality control. Thus, among professionals, the discussion of water resources' management has moved from a focus on efficiency, through multi-purpose planning, to comprehensive management, coupled with the increasing concern for ecosystem maintenance. This emerging trend has fostered an increased need for interdisciplinary management and the search for sustainable development along natural boundaries rather than artificial administrative divisions.

At the same time, as Susskind (1994) notes, the task of generating international agreement on anything is extremely difficult. The procedures currently used to formulate global agreements were not designed for complex ecosystemic and environmental problems, negotiators need to have unique interdisciplinary and technical skills, linkages among environmental concerns and other policy issues must be adequately established, and, effective monitoring and enforcement arrangements must also be devised and properly implemented. In such a demanding task Susskind ends with the pessimistic note as to the serious obstacles to global cooperation, including the increasing split between the developed nations of the North and the developing nations of the South; the persistence of national sovereignty; and the lack of incentives sufficient to bring nations to the table for serious and sustained negotiations.

In the same vein LeMarquand cites five important policy factors which influence a country's position about international rivers, namely: image, or the perception of national posture in foreign policy; international law, or the legal principles that enable countries to get involved in negotiations; linkage to other bilateral or multilateral issues; reciprocity and the desire for mutual commitment and obligation; and, sovereignty and the perception of loss of control over domestic resources implied by collective or bilateral constraints. The last is again part of a skepticism about the suitability of the state system with the search for a "rational" environmental management and with the desire for coping with "endangered-planet" type problems. To put it simply, questions are raised as to whether a potent mix of international bodies, scientific and other political groups, and nation states can effectively be mobilized to meet transboundary environmental challenges.

Rather than elaborating further such general principles, we may outline some discernible trends and developments which are affecting and will be affecting planning and management practices, especially with regard to shared water resources:

a) There is an increasing gap between development needs and available capital, leading to the need for efficient projects and mechanisms for equitable implementation (fairness).

b) Issues and actors are multiplying rapidly, leading to complexification, potential institutional gridlock and dissension. Planning and management must become, then, more anticipatory, participatory, multi-purpose and contingency oriented.

c) Climatic anomalies and attendant ecosystemic impacts accentuate the vulnerability of surrounding environments and reinforce risk assessment and risk management approaches.

d) Transboundary interdependencies enlarge the spatial envelope of concern and lead to efforts for cooperation as well as transnational conflict management mechanisms.

e) Appropriate responses must be developed in order to solve discontinuities between traditional practices and new institutional demands, including demographic changes, rapid urbanization, need for rehabilitation and infrastructure maintenance, conjunctive water use, multi-sectoral planning, as well as shifting water priorities.

There are currently a variety of multi-institutional approaches ranging from local institutions involving riverine institutions, national and international NGOs and research institutions, private sector participants (especially construction firms), bilateral and multi-lateral donors, riparian government institutions, all the way to transnational river basins (Scudder 1993). All such multiple actors and institutions face problems of vertical and horizontal integration not only in water resources projects, programs and policies; but, also, within and between other resource management entities and local, national, regional, and international organizations. These are principles from international law, bilateral and multilateral treaties, binding acts of international organizations, rules of customary international law, and, judgments of international courts and tribunals. All such components, provide for some general principles and rules with regard to environmental matters (Sands, 1994). Such principles and rules include sovereignty over natural resources, good neighborliness and international cooperation, the precautionary principle (especially when there is scientific uncertainty), the polluter pays principle, and, the common but differentiated responsibility.

Cross-cutting this complex legal and institutional plane are also theoretical problems of the legal doctrines governing the utilization by States of shared water resources. The four doctrines and theoretical approaches most usually encountered include:

a) the doctrine of absolute territorial sovereignty (the so-called Harmon doctrine), where a riparian may do what it will with the water;

b) the doctrine of absolute territorial integrity, regarding uninterrupted downstream flow;

c) the doctrine of limited territorial sovereignty or equitable utilization, which accords recognition to a riparian's jurisdiction, but, at the same time, ensures downstream states a reasonable share of water in reasonable condition; and

d) the doctrine of community co-riparian states, community of interests in the water or, integrated drainage basin development transcending national boundaries.

The question of scale, boundary, and geographic planning unit is of central concern for efforts of definition of the problem, identification of parties-at-interest, and potential implementation mechanisms. While many authors recognize the theoretical advantages of comprehensiveness, integration, and unified planning and management, successful practices seem to lag behind. The problems in implementation have been noted above but it is important to emphasize what we can summarily call as the on-going process of complexification creates a great diversity of interests, differences in resources accessibility, complex organizational arrangements, and competing and conflicting water policies. Finally, diachronic considerations complicate matters further by temporal shifting of goals and preferences, priorities, political consensus, leadership, sponsors, prices, etc.

What are the implications, then, of the above remarks for understanding potential water conflicts over shared waters? First, we need to recognize the difficulties associated with formal, legalistic approaches which tend to emphasize predominantly conduct rather than reasonable management (especially where there is no agreed-upon river regime). Second, existing legal approaches (necessary for international conduct and enforcement) can be supplemented by such flexible mechanisms as:

- second track diplomacy (environmental diplomacy or "hydrodiplomacy").

- Alternative Dispute Resolution through international bodies or in the spirit of Agenda 21 (1992 Rio Conference/UNCED).

- technical/professional or independent panels of experts or what have been called "epistemic communities."

- public awareness, participation and mobilization.

The operational terms of the above for international law and for other management mechanisms are complimentarity and implementation. In existing multilateral agencies there seems to be three vexing problems. One has to do with the historical and cultural inertia of past differences and of established practices. The second, revolves around the calculation of all costs involved in shared waters' development. And, the third, has to do with the incorporation of social and environmental concerns into planning. These broad problems emphasize once again the need for negotiation, for third party expertise and for dialogue based on factual information.

Yet, as Lynton Caldwell has also noted, there exists a notable paradox of international environmental cooperation, since national governments are characteristically reluctant to initiate cooperative efforts in this area. In essence, national political leaders are not that much interested in pursuing international environmental cooperation unless their own national (and often personal) interests are involved (Caldwell 1988). Plainly, internal, local issues tend to drive international decisions. The reasons that persuade governments to negotiate international environmental agreements and to make formal commitments to cooperate are not necessarily those that determine whether, how, or when governments will act.

But while these reservations should alert us as to the difficulties involved in managing shared water resources current national water policies and developmental plans, international organizations, joint planning/harmonization efforts, peace talks, monitoring on a continuous basis, conducting scientific studies, cooperation and exchanges are all parts of a supportive scene of "sharing" and of the spirit of integrated, holistic, unified planning and management. This spirit is also emphasized in such notable efforts as the Israel-Jordan Peace Treaty (Article 6 and Annex II), the Mekong River Agreement or the establishment of a joint Commission between Portugal and Spain to coordinate distribution of Iberian waters.

Water Conflict Management

By now the argument has become long and elaborate as to what integrated water resources development entails. Many of the items discussed above give rise to disputes and cause conflict. Conflicts arise out of scarcity, out of differences of goals, objectives, and priorities, out of social and historical differentiations (such as pre-existing antagonisms among riparian nations), out of competition and skewed power distribution among nations or stakeholder groups, and, often, out of cognitive differences (misunderstanding of facts and with credibility gaps as to the very nature, extent and quality of data). Other authors (such as Naff and Matson, 1984) prefer to use in their conflict/cooperation potential such considerations as riparian positions, degree of interest in water problems, perceptions of the water issues involved, internal power to influence decisions, and, external support to influence decision.

Two terms that are often encountered are "settlement of disputes" and "conflict resolution." The first usually connotes a more or less permanent institutional machinery for formal settling of disputes and of ironing out differences. The second can be more conveniently used for more informal ways of managing disagreements and for avoiding or minimizing conflict escalation. These include such principles of international law as equitable utilization, prevention of significant harm to other states, obligation to notify and inform, obligation to share data, cooperative management of international rivers, obligation to resolve disputes peacefully, as well as formal treaties.

As many authors have repeatedly pointed out, there is a long history of water-related disputes ranging all the way from conflicts over access to adequate water supplies to intentional attacks on water systems during wars (Gleick 1993). As stated previously, such conflicts, tensions and frictions are expected to increase as growing populations, urbanization and economic development will require more water for agricultural, municipal and industrial uses. At the same time, water availability may be coming up against what Falkenmark has labelled as the "water barrier," namely a level of supply below which serious constraints to development will arise (quoted in Gleick, 1993). Such limits will be further stretched by potential climatic anomalies which in turn could intensify regional conflicts between water-rich and water-poor nations.

In discussing water conflicts, we may want to recognize three distinct phases, i.e., conflict creation, conflict management, and conflict resolution. In the first (the genesis of conflicts), the emphasis is on diagnosis, anticipation and prevention, problem architecture, and joint fact-finding. The second phase denotes a stage for the development of confidence and trust among states through such mechanisms as mediation, arbitration, neutral expert fact-finding, etc. Finally, conflict resolution involves consensus building and depolarizing of conflicting interests through public finding processes or adjudication.

The search for a typology of conflicts and of appropriate responses has led to many speculative thrusts and interesting conceptual schemes. The implication of such attempts is that we should pay more attention to international river basins where the conflict potential is greatest, especially rivers forming a shared boundary between nations; where human action (e.g., construction of a dam) trigger disruptions; in cases of power asymmetries and non-cooperative settings; and, finally, in cases of noticeable environmental imbalances (as e.g., desertification, drought, water shortages, etc.). It is at this point that writers usually distinguish between "crisis politics" and "routine politics," or between "acute conflicts" (foreign policy, confrontation) and "low politics" (aspects of economic welfare) (See among others Frey, 1993).

International relations have become so complex that alternative dispute resolution means have become important in managing or resolving inter-societal conflicts. The search for alternatives to legal institutions to arbitrate disputes has been prompted not only by saturation of legal mandates, but also by increasing litigation and confrontation. Mediation, as a compromised discussion between disputants aided by a neutral third party whose judgment is respected, has become a viable alternative to adversarial processes. The gamut of adjudication, arbitration, mediation, conciliation and even "principled negotiation", expresses various alternative processes of dispute resolution. But criticisms have also risen as to whether such processes can compensate for inequitable power relations or can provide incentives for compliance or acknowledgement of the third party decision when there is no recourse to legal sanctions.

Supporting these mechanisms is also the role of public participation as an important means to reach project ends in development activities and in promoting environmental goals. Thus, in order to promote an effective and equitable implementation of an iterative process of management there is a need to accentuate the use of public involvement and participation as an important means for increasing awareness of and engagement with water resources planning as a way of minimizing conflict. Public involvement can vary all the way from simple informing to highly structured means of participatory processes in which the public takes an active part throughout all phases of planning and policy-making. Obviously, there are different degrees of control over the entire process, but the underlying assumption rests on the commitment to seek public input at progressive planning stages. By definition, participative planning is an iterative process beginning with problem identification all the way to evaluation of alternatives and with feedback loops linking publics, planners and decision-makers.

Public participation and negotiation become tools in maximizing agreement when there are disagreements not only as to the nature of the problem but also as to desirability of certain outcomes (goals). More important, "citizen participation," and "public involvement" have become not only the noble cries of an environmentally conscious society, but also the presumed imaginative tools for creative planning and for the achievement of a broad consensus for desired action. The last two decades have brought about not only a revolution in thinking, but also a renewed commitment to participatory decision making and a host of legal mandates for improving the quality of the surrounding environment or for arresting adverse consequences of proposed actions. The degrees of public engagement and mobilization of various publics can be better understood through a tripartite distinction between public awareness, public involvement and public participation. In other words, the creation of a supportive public climate of conflict management, increased consensus, and in international cases a reinforcement of the spirit of "transnational commons." Public awareness implies one-way information and alerting to community or national water issues. Public involvement implies two-way communication and a means of engaging community members in the exchange of information (dialogue). Finally, public participation is the most intense form of interaction between authorities, experts and citizens and implies more than anything else truly joint planning and democratic delegation of power and shared leadership.

The responsiveness of a given political and institutional culture and the creation of a supportive climate for negotiation include, among others, such considerations as: the nature of citizen's rights and their legal standing vis-a-vis environmental matters; the scope and political effectiveness of environmental policies, programs and projects; the role and effectiveness of mass media; the role of education (at all levels) in promoting environmental awareness, in encouraging forms of active participation, and in nourishing voluntarism; the presence and extent of citizen activists; and, the overall socio-political culture and historical traditions.

But there is an apparent dilemma in all such recent developments of participatory decision making and conflict management. As it has been noted within the last decade or so, societies have tended to advocate the simultaneous growth of participatory democracy and of expertise in decision making. It becomes difficult to maximize both of these value preferences and strains appear between the idealized conceptions of citizen participation and the harsh demands of public policy making and implementation, especially when they entail transfrontier considerations. Again, as many authors on the topic have also observed, participation is a double edged sword: planners and decision makers must be open to working with citizens, and citizens must be active and competent in planning. The basic proposition remains, however, namely that the broader the base of citizen involvement and participation, the more the potential influence resource managers and citizens can bring to bear on public policies and plans and the easier it can be to create a transnational climate of trust in transfrontier water dependencies.

Challenges and Opportunities

Levi in his seminal work Contemporary International Law has succinctly observed that international disputes are nowhere defined in conventions or treaties. But when an international dispute is found to exist states are obliged to settle it by peaceful means in such a manner that international peace and security and justice are not threatened (Levi, 1994). Some states have their own institutions for the settlement of regional disputes (as e.g. the European Union), while other international agencies have specialized procedures. There are a multitude of international organizations and agencies within which disputes can be settled peacefully. Other Accords (such as the Helsinki Accord of 1975) refer directly to disputes among participating states, while the U.N. Convention on the Law of the Sea (1982) foresees the creation of an international tribunal.

There are quite a number of methods available to disputing parties, but no standard legal structure as to how to proceed. Important here is the entire gamut of voluntary methods of no legally binding nature, but representing the wish to settle disputes without decisions by third parties (e.g. negotiation, mediation, inquire, and conciliation). In other cases, third parties can through arbitration or adjudication help diffuse and settle conflicts. All such efforts and similar voluntary mechanisms for helping settle conflicts from the backdrop for the practice of Alternative Dispute Resolution (ADR).

Considerable evidence for the practice of ADR from many parts of the earth creates a basis for some realistic optimism (Gray 1993). Theoretically, collaboration for addressing a wide range of environmental disputes involves three phases, i.e., problem setting (problem definition or problem architecture), direction setting (predominantly negotiations over substantive problems), and implementation (systematic management of inter-organizational relations and monitoring of agreements). All alternative processes proposed (direct negotiations, good officers, mediation, conciliation, etc.) could allow the parties to reach a faster, more timely, and hopefully, more appropriate resolution of disputes.

In this context, one can observe in on-going management practices the convergence of formal international legal practices, Alternative Dispute Resolution methods, and innovative water institutions. The search for comprehensive approaches is calling for more integrated water resources management across boundaries, jurisdictions and uses. Furthermore, cross-boundary, comprehensive, integrated river basin approaches become more feasible when international institutions can serve as magnifiers of public pressure and as sounding boards in competitive politics (as was the case in the Baltic and North Sea ministerial meeting and successive ozone talks after 1987 -- Haas et al. 1993); as springboards for the creation of what the same author has labeled "epistemic communities," i.e., experts in environmental issues sharing principled values regarding the enhancement of collective welfare; as enhancers of a contextual framework for credible commitments among states; and as vehicles for building national capacity through inter-organizational networks and transferring of water relevant information. Such catalytic functions can help promote a more global approach and strengthen any relevant effort for integrated water resources management.

Although analytical techniques alone are not sufficient for building better water allocation agreements or managing conflicts, they must be considered in any integrated water management considerations. The following include some of the more important lessons learned from the water resources field and experiences in negotiating international environmental regimes:

1. Integrated management across water uses and jurisdictional boundaries is the key to sustainability.

2. Water development and management must move from a project-by-project focus to programmatic planning and systems management. The use of the river basin as a unit of regional planning is reemerging as a focal point.

3. A variety of institutional models for managing and developing water resources are available. These range from authorities and bilateral agreements to multilateral compacts to river basin and regional authorities.

4. Data are rarely neutral. Data collection itself is driven by values and is ultimately negotiable.

5. Technical expertise is necessary, but not sufficient, for achieving integrated water management -- so too with political agreements. The critical water management skills and institutional arrangements increasingly are those designed to facilitate management in the gray area between the political and technical design and operation.

6. Changing demographic conditions in both arid and humid areas are requiring new agreements or institutions to reapportion old allocations and facilitate negotiations on new apportionments through relevant compacts.

7. Assisted negotiation approaches such as mediation by a third, neutral party are proving useful to meet such needs.

8. The key conditions for successful water sharing remain those of equity of any agreement and of successful control by affected parties of their own water resources.

In a recent article discussing Middle East negotiations Aaron Wolf (1993) has also proposed a series of specific guidelines and cautions for efforts in facilitating shared water negotiations. These include:

- identification of the actors, salience of interests and power;
- insistence on common criteria for analysis;
- inventing options for mutual gains and equitable division of existing resources;
- determination of feedback mechanisms for perpetuating agreement.

LeMarquand on the other hand emphasizes for successful international agreements concerning water the centrality of a requisite "climate for agreement" which includes such conditions as:

- Countries with the same technical perception of a problem.
- Similar tastes for consumption of goods and services.
- When water quality is an issue, the use of similar industrial production technologies.

- The existence of an extensive network of transnational and transgovernment contacts between countries.

- The participation of a small number of countries.

- The desire of one large country to have an agreement.

- The necessary development by one country of a good or service for its own use which may benefit other countries.

The common thread in any discussion of transboundary water conflicts emphasizes how new strategies are needed because water (and for that matter natural) resources problems are becoming both highly complex and globalized. Some even argue that the traditional spatial environmental envelope has collapsed and project boundaries - and their impacts and consequences - are now much more diffuse. Thus, there is a need for bringing back an environmental approach that requires drastic measures of ecological rehabilitation, innovative institutional mechanisms, and a balance between autonomy and cooperation. Such global approaches entail also improvement of environmental monitoring and information by expanding the factual basis of comprehensive river basin models. In addition, they also imply a framework for negotiations which stresses the importance of comprehensive institutional formats and clarity in national and international decision making processes.

Given such considerations and strong socio-political divisions (even centrifugal forces and fragmentations in many nations) there are three responses that we should consider. First improve efforts towards the utilization of "hydrodiplomacy" in terms of understanding alternative dispute resolution and conflict management efforts to transboundary water resources. Second, recognize again the river basin approach as a cooperative mechanism and authority, and as being much more sensitive to ecosystemic interdependencies. And, third, place particular emphasis on integrated water resources management (including the building of more robust water resources institutions).

As the international scene turns to questions of sustainable development, the restoration and rehabilitation of degraded environments, and to the creation of new cooperative arrangements centering around shared water resources, it becomes apparent that institution building, comprehensive management and alternative dispute resolution efforts will be central quests in the years to come. Diminishing or degraded water resources and their potential impacts on international security, provide unique opportunities for cooperative institutions and for cooperative transnational behavior.

The international conference in Dublin on Water and the Environment (1992) and the UN Conference on Environment and Development (UNCED) in Rio de Janeiro during the same year are but only two recent manifestations of the centrality of water, environment and development. The transition to the 21st century will require also an institutional order of cooperation, of comprehensive management principles, and of sharing of experiences gained in the practice of ecosystemic principles in water resources projects. Paths to effectiveness for some authors imply the boosting of governmental concern; the enhancement of a contractual and bargaining environment, and, finally, the building of national capacity (Levi et al. 1991). Others (e.g. Carbonneau, 1989) discuss larger educational and epistemological goals in ADR and the existing legal system including the building of decentralized alliances, provision of prenegotiation assistance to individual countries, new approaches to treaty drafting, expansion of the role for nongovernmental interests, balancing science and politics, or encouraging issue linkage. The question that should worry us, though, is to what extent sovereign states, multinational corporations, NGOs, or existing international bodies can respond in sharing long acquired power and in implementing action that promotes ecological interdependence and globality of increasing environmental challenges. At the same time, rapid population increases and expansion of activities are creating unprecedented situations requiring new economic paradigms of planning and action. The main problem will be how to achieve integrated planning and management within institutional frameworks which have evolved under different conditions and for needs which are incongruous with the present and certainly will differ from projected futures.

In the early 1990s there has been a flurry of activities in trying to articulate "principles" or "declarations", "model rules" or "non-binding agreements" for shared water resources (and for broader natural resources interdependencies). The Helsinki Rules on the Uses of the Waters of International Rivers (1966), the Delft Declaration ("A Strategy for Water Resources Capacity Building", 1991), Statement and Report of the Copenhagen Informal Consultation on Integrated Water Resources Development and Management (1991), the Dublin Statement on Water and Sustainable Development (1992), Mediterranean Charter for Water (1992), and many others with such summary titles as the Bangkok Statement, the New Delhi Statement, the Beijing Declaration, and so on, all have attempted to articulate concerns and potential codes of conduct in transboundary water practices, on river basin planning and on integrated approaches to shared water resources management.

Thus, all over the world the concern with water resources, increasing environmental awareness and a series of natural resources "crises" have coalesced in reinforcing a trend toward comprehensive management and interdisciplinary planning. The overall thrust of such efforts in water planning and management in the last 20 years (certainly in the U.S.) underscore among many other goals:

- the internalization of a wider range of costs and benefits of water management decisions;
- increased emphasis on anticipatory planning and an enlarged time horizon;
- commitment to participatory planning through the empowerment of many additional parties-at-interest;
- contingency planning and richer menu of planning and management options;
- use of risk assessment in order to define the range levels of acceptable and assumable risks in decision making;
- a discernable trend toward comprehensive or unified management along with special efforts to involve all levels of government and to attain support of the individual citizens;

- integrated ground- and surface-water management and the use of the basin as a management area;
- preference for "non-structural" solutions and for conservation measures that recognize that neither "availability" nor "requirements" should be treated as unalterable; and,
- the emerging coordination of water with other natural resources, especially land-use controls and planning.

Such a list of on-going developments can be further expanded but it is only indicative of the major shifts and transformations in water resources planning and management in the last two decades. They are the manifestations of the centrality of the plight of contemporary society with regard to expanding numbers, limitations of resources and the degradation of our living environment. A rapidly transforming environment has created a new environmental ethos concerning ecosystemic interdependencies and characterized by continuous socio-cultural changes; shifting roles of various institutions (notably the government); technological changes affecting the distribution and utilization of water resources; and, contextual changes involving new patterns of understanding the world around us and the knowledge as to critical interrelationships such as new ecological paradigms and environmental insights.

Reflecting back on many of the observations made as to integrated water resources planning and management, one is struck by the obvious need for normative or goal-oriented approaches that would accomplish two purposes: voluntary joint planning through common goals, and, second, justify third-party dispute resolution efforts animated by principles of collective welfare. We need, then, a combinatory approach between ex ante justifications (teleological planning, basin-wide principles, ecosystemic, proactive emphasis) and post hoc measures (legal regimes, adversarial mechanisms, reactive/corrective emphasis). Such a combination would be based on integrated management for efficient and equitable water use; on principles of regional cooperation; on functionally interrelated problems and projects; reduction in advance of points of friction and stress; understanding of contextual forces and conditions; and, elimination of competing and conflicting situations through risk management. The end result of such a posture would be a contingency preparedness for change, the ability to cope pragmatically with the challenges of transboundary interdependencies, the appreciation of long-term planning, and the capacity to manage peacefully and in a pro-active manner long-standing conflicts about shared water resources.

References

Amy, Douglas J. 1987. The Politics of Environmental Mediation. New York: Columbia University Press.

Bingham, Gail. 1986. Resolving Environmental Disputes: A Decade of Experience. Washington, DC: The Conservation Foundation.

Birnie, Patricia W., and A.E. Boyle. 1992. International Law and the Environment. Oxford: Clarendon Press.

Caldwell, Lynton K. 1988. "Beyond Environmental Diplomacy: The changing institutional structure of international cooperation" in John E. Carroll (ed.) International Environmental Diplomacy. Cambridge: Cambridge University Press.

Carbonneau, Thomas E. 1989. Alternative Dispute Resolution. Urbana: University of Illinois Press.

Carroll, John E. 1983. Environmental Diplomacy: An examination and a prospective of Canadian-U.S. transboundary environmental relations. Ann Arbor: The Univ. of Michigan Press.

Chauhan, D.R. 1981. Settlement of International Water Law Disputes in International River Basins. Berlin: Erich Schmitt Verlag.

Choucri, Nazl (ed.). 1993. Global Accord: Environmental Challenges and International Responses. Cambridge: The MIT Press.

Delli Priscoli, Jerome. 1992. "Collaboration, Participation and Alternative Dispute Resolution (ADR)". Technical Paper Prepared for the World Bank.

Falkenmark, Malin, Gunnar Lindh, Lenart De Mare, and Carl Widstrand. 1980. Water and Society: Conflicts in Development. Part 2. Water Conflicts and Research Priorities. Oxford: Pergamon Press.

Frey, Frederick. 1993. "The Political Context of Conflict and Cooperation Over International River Basin" Water International, 18(1):54-68.

Gleick, Peter H. 1993. "Water and Conflict: Fresh Water Resources and International Security." International Security 18(1):79-112.

Godana, Bonaya Adhi. 1985. Africa's Shared Water Resources: Legal and Institutional Aspects of the Nile, Niger and Senegal River Systems. London: Frances Pinter Publishers.

Gray, Barbara. 1993. "Negotiating Collaborative Solutions to Disputes Over Water Resources." Paper prepared for the Interamerican Dialogue on Water Management, Miami, October 28-30.

Haas, Peter M., et al. (ed.). 1993. Institutions for the Earth: Resources of Effective International Environmental Protection. Cambridge, Mass: The MIT Press.

Hurrell, Andrew, and B. Kingsbury (eds.). 1992. The International Politics of the Environment. Oxford: Clarendon Press.

Kolars, John F., and W.A. Mitchell. 1991. The Euphrates River and the Southeast Anatolia Development Project. Carbondale: Southern Illinois University Press.

Lang, Reg. 1986. "Achieving Integration in Resource Planning." In Reg Lang (ed.) Integrated Approaches to Resource Planning and Management. Calgary: The University of Calgary Press.

LeMarquand, David G. 1977. International Rivers: The Politics of Cooperation. Waterloo: University of British Columbia.

Levi, Werner. 1991. Contemporary International Law: A Concise Introduction. Boulder: Westview Press.

Mandel, Robert. 1991. "Sources of International River Basin," Paper presented at the Annual Meeting of the International Studies Assn., Vancouver, BC, March.

Mitchell, Bruce (ed.). 1990. Integrated Water Management: International experiences and perspectives. London: Belhaven Press.

Naff, Thomas, and Ruth C. Matson (eds.). 1984. Water in the Middle East: Conflict or Cooperation. Boulder: Westview Press.

Newson, Malcolm. 1992. Land, Water and Development: River basin systems and their sustainable management. London: Routledge.

Patchen, Martin. 1988. Resolving Disputes Between Nations: Coercion or Conciliation? Durham: Duke University Press.

Porter, Gareth, and J.W. Brown. 1991. Global Environmental Politics. Boulder: Westview Press.

Rose, Laurence. 1993. "Shared Water Resources and Sovereignty in Europe and the Mediterranean." IBRU Boundary and Security Bulletin, 1(3):62-67.

Sands, Philippe (ed.). 1994. Greening International Law. NY: The New Press.

Sirronneau, Jacques. 1993. "L'Eau Resource Strategique; Menaces et Enjeux de l'hydropolitique." Geopolitique, No. 43(Fall):45-69.

Susskind, Lawrence E. 1994. Environmental Diplomacy: Negotiating more effective global agreements. NY: Oxford University Press.

Trolldalen, J.M. 1992. International Environmental Conflict Resolution. Washington, DC: NIDR.

Westing, Arthur H. (ed.). 1986. Global Resources and International Conflict: Environmental Factors in Strategic Policy and Action. Oxford: Oxford University Press.

Wolf, Aaron. 1993. "The Jordan Watershed: Past Attempts at Cooperation and Lessons for the Future." Water International 18:5-17.

I.3 Negotiation Strategies in Transboundary Water Disputes

Joseph W. Eaton and David J. Eaton [1]
University of Pittsburgh
Forbes Quadrangle
Pittsburgh, Pennsylvania 15260
USA

The Universality of Negotiation Patterns

Individuals, groups and nations often try to negotiate their differences with competing stakeholders. The process is circumscribed by applicable laws, the readiness of law-enforcement agencies to enforce them and the respective goals and priorities of each of the negotiating parties. This model has to be modified in international disputes over the management of water resources. Each sovereign nation is a law unto itself. Even when an applicable treaty exists, each country can choose to ignore it or interpret its terms as it pleases, in ways that other stakeholders might regard as a violation. This paper reviews the use of universal alternate negotiation strategies to modify this anarchistic framework that has characterized transboundary water management in the Middle East for almost half a century. plus pertinent illustrations from North America and the Indian sub-continent. Decision maker and those who may negotiate on their behalf have a choice of six universal negotiations strategies:

1. Negative non-zero sum games, or "Lose-Lose" solutions
2. Unilateral creation of new facts
3. Zero-sum games or "Win-Lose" negotiations
4. Positive non-zero sum games or "Win-Win" negotiations
5. Conflict and threats of violence
6. No action, causing opportunity costs from neglect and/or delay

"World Government" advocates share as their goal the adoption of an international body of laws. Their dream of a better world includes the adjudication of disputes between states about environmental pollution or water rights by international courts which arbitrate or enforce

---

[1] Joseph W. Eaton is Professor Emeritus in the School of Public and International Affairs, the University of Pittsburgh, Pittsburgh, PA 15260. David Eaton is Bess Harris Jones Centennial Professor of Natural Resources Policy Studies, Lyndon B. Johnson School of Public Affairs, University of Texas, Austin, Texas. This research was support by a grant from the U.S. Institute of Peace in Washington, D.C., as well as administrative support from the University of Pittsburgh and from the University Research Institute and Policy Research Institute of the University of Texas at Austin.

international conventions and standards. But such a peace committed world does not yet exist. No member of the United Nations has as yet been willing to accept significant reduction in their sovereign powers, as did the 50 sovereign states which compose the United States.

A dispute over water rights between California and Arizona can be finally adjudicated by the United States Supreme Court. Mexico and the United States are not bound by such a World Federalist order. They have to negotiate over their differences politically without referral to a supra-national authority.

Distinguished international experts have adopted guidelines on how to manage transboundary water resources, the Helsinki Declaration (International Law Association, 1992) and the Seoul Rules (Utton, 1981). But they include some mutually exclusive general principles. For example, consider Turkey's unilateral decision to impound behind newly constructed dams within its sovereign territory a portion of the flow of the Euphrates River to irrigate its Southeast Anatolia region,. This so-called "GAP Project," will ultimately include a coordinated network of 22 dams and 17 hydroelectric power stations.

As a result, the volume of Euphrates River water available to the lower riparian nations , to Syria and Iraq, will be reduced permanently (Gruen, 1992). Turkey's strong defense force makes it unlikely that either Iraq or Syria can use military force to reverse this reduction of their previously available share of the river. Nor does any international body have jurisdiction. Disputes such as these can be dealt with only on the basis of the above mentioned negotiation strategies.

Option 1: Lose-Lose Solutions or Negative Non-Zero Sum Games

Groundwater management in the Rio Grande/Rio Bravo Basin along the Border of Texas and Mexico poses particularly difficult policy challenges because of the technical uncertainties and laissez faire water right standards within both nations.(Eaton et al. 1992). The quality and quantity of groundwater in the border area's high-growth centers are already diminishing because each land owner is free to pump as much water as he or she wants from a well on their property, although they share common aquifer. In the absence of joint planning and control of this limited natural resource, the population growth on both sides of the border will increasingly exacerbate water shortages on both sides of the border. Estimated withdrawals from the aquifers are twice the recharge rate, which consequently will deplete the recoverable storage over time.

Each country's legal inability to limit withdrawal of sub-surface waters by property owners is leaving them without domestic power to adopt a joint policy. Compounding the problem is that the two countries have different domestic laws governing groundwater use. Any agreement that is fair bilaterally would require one or both nations to strike down state laws and local precedent for allocating withdrawal rights, which would become politically controversial. The Texas Water Development Board predicts a sharp decline in the amount of water withdrawn from the El Paso aquifers—from 253 million cubic meters per year in 2020 to about 74 million cubic meters in 2030. While the United States and Mexico are able to cooperate in the multilateral negotiations to phase out the use of ozone - depleting chemicals, they are unable to address the need to assure their next generation a sustainable supply and quality of water in the aquifers under their shared border. They appear to be locked into a "lose - lose" negative non-zero sum game process.

Option 2: Creating New Facts

By creating facts a sovereign nation can act unilaterally in what it believes to be its interest. Turkey's GAP Project, as discussed above, is an illustration of unilateral action by a nation that creates a new hydrologic reality. Another illustration is Syria's construction of small dams along the tributaries of the Yarmuk River which originates within its boundaries. These dams have the combined capacity to withhold a major portion of the winter rains before they flow into Jordan, where this high quality source of fresh water is Jordan's principal source of irrigation water.

After many of these Syrian dams were completed, a treaty was negotiated between Syria and Jordan in which the already established dams were accepted as no longer subject to change. As the militarily weaker country, Jordan agreed to a permanent reduction of its previously available volume of Yarmuk River water. Jordan could not demand that the dams be destroyed or be kept empty, even though Syria now retains a larger portion of the Yarmuk River water than its prior entitlement which had been informal negotiated in 1953 under auspices of Ambassador Eric Johnston, an American mediator.

Creating Facts was a favored strategy during the Cold War. Both the Soviet Union and the United States took provocative actions short of war and then forced their adversary to adjust to unilateral change of the status quo, such as the construction of the Berlin Wall or the adoption of the Marshall plan to give economic development aid to the western portion of occupied Germany but not to the Eastern portion, the so called German Democratic Republic.

Option 3: "Win-Lose" Negotiations or Zero Sum Games

In some negotiations material concessions by one party result in an equivalent gain by other partners. This approach can be the preferred negotiation model when one side believes itself to be powerful relative to other stakeholders.

This "Win-Lose" strategy involves the transfer of assets from one party to another without compensation. It was the basic principle of the Versailles Peace treaty after World War I. The two defeated nations, Germany and Turkey, had to yield territory to the victors in return for peace. In addition, Germany was required to pay heavy reparations.

Ambassador Itamar Rubinstein, a participant-observer of secret peace explorations between Israel and its Arab neighbors between 1948 and 1952 (Rabinovich, 1991) attributes the failure of these efforts to the expectation by King Abdallah of Jordan, General Husni Zaim of Syria, and various Egyptian political leaders that Israel would be amenable to a zero-sum game strategy. They proposed several schemes under which Israel would have to surrender some of its limited territory and water resources in return for diplomatic recognition by its Arab neighbors. The Israelis thought the territorial concessions were too severe and the quality of "peace" that was being offered too circumscribed.

In the current post-Madrid peace process the zero-sum strategy continues to be preferred by the Palestinians. Their representatives have sought voluntary re-distribution of the existing regional water supply in their favor, based on the much higher per capita supply of water for Israelis and the Palestinians in the occupied territories. The Israeli government defends this status quo in part on the basis of prior use—rights acquired through governmental concessions during the period when both all of Palestine was ruled by Great Britain under a League of Nations Mandate. Since then Israel has invested much of its capital in technological hydrological innovations, while investment in the infra-structure of the West Bank and the Gaza strip were held back by the high priority assigned by Arab and Palestinians leaders to maintenance of a never-ending armed struggle strategy against the very existence of the Jewish State.

After King Hussein publicly announced the end of Jordan's state of war with Israel in Washington, D.C. on July 25, 1994, Prime Minister Rabin ordered a one time unilateral transfer of about 2 - 3% of its traditionally available volume of fresh water for the benefit of Jordan, where water shortages in the cities during the summer limit the running water supply to a few hours a week. In the subsequently signed peace treaty, positive non-zero sum game or "Win - Win" transactions were dominant.

There always is a normative dimension when people negotiate. Stakeholders rarely evaluate their respective concessions equally. For the Israelis, the transfer of water was a zero sum game concession. Jordan's increased supply required an equivalent reduction of Israel's volume. Arab factions opposed to the peace process viewed the agreement also as a zero sum game concession, but in Israel's favor. A one season reduction of 50 Million Cubic Meters of fresh water by Israel was dismissed by them as an inconsequential gesture in return for the political acceptance of Israel's right to exist. For the Arabs, this was a major concessions, while the Israelis took their right to statehood for granted. It was not dependent on Arab acknowledgment.

Option 4: "Win-Win" Negotiations or Positive Non-Zero Sum Games

"Win-Win" types of agreements can lead to better feelings among negotiating partners because each side is better off under the agreement than without it. Positive non-zero sum solutions, or "win-win" agreements, offer gains to both sides with no concessions which one of the sides would regard as a "loss." Changes in the relationship between the opposing parties are considered which could generate gains for all sides. It is the strategy most favored by the American Foreign Aid Program. Technical assistance and economic as well as military aid have been offered to countries in return for their support of American foreign or economic policies.

In the Arab - Israeli conflict, the positive non-zero sum strategy model was temporarily made operative in the Emir Feisal-Chaim Weizman agreement at the 1919 Paris Peace Conference. These heads of both the Zionist and the Pan-Arab national movements pledged to support each other's aspirations. Both leaders hoped this coordination would advance their respective demands for territorial control over parts of the Middle East which the Ottoman Turks had been forced to evacuate. The agreement became moot when the British and French governments failed to live up to what the Pan Arab movement believed to be its entitlement to sovereignty over most of the territories liberated from Ottoman rule.

In 1953, President Eisenhower appointed Mr. Eric Johnston as a special U.S. Ambassador to defuse Middle East tensions. He was able to broker a de facto water allocation agreement between technical personnel representing Jordan, Israel, Egypt and Syria. The Arab negotiators acquiesced to Israel's plan to pump Jordan River water into its National Water Carrier to provide drinking water to some of its cities and to irrigate the arid Negev (southern) region. Jordan was permitted to build the Ghor Canal to irrigate large areas of land and also to supply some of its hill district towns, including Amman, its capital. U.S. aid funds helped to finance both projects. The conflicting parties, officially at war with each other, agreed to cease interference with each other's water diversion projects.

Another case of "Win-Win" negotiations is the effort during the last few years of environmental non-governmental organizations in Israel, Jordan and Egypt to cooperate under auspices of the Environmental Law Institute in Washington, D.C. They have agreed to pool information to facilitate a joint approach to protect the sensitive Red Sea ecosystem against pollution, as well as to encourage trade, employment, and tourism. When King Hussein and Prime Minister Rabin (with the encouragement of Hosni Mubarrak, the Egyptian Prime Minister) agreed to construct a road linking Egypt with Jordan via Israel and to permit two way tourist and commercial traffic between the neighboring towns of Elath and Aquaba, they took a first step to implement this positive non-zero sum solution that leave all sides better off.

Terms of the PLO-Israeli interim agreement for peaceful co-existence concluded in September, 1993 to establish a Palestinian Interim Self Government Authority also includes a "win-win" provision, a protocol on Israeli-Palestinian cooperation that includes cooperation for enhancement of the available supply of water. At last year's conference between Palestinian and Israelis in Zurich, Switzerland participants identified a number of "Win-Win" water management suggestions which were to be further explored. A few are listed below:

> Recharge of ground water by collecting winter rains that now flow unused into the Mediterranean or the Dead Sea.
>
> Desalination of brackish water in the coastal regions of Gaza, Israel, Egypt, and Jordan to increase the region-wide supply of fresh water.
>
> Purchase of fresh water as a market commodity from any nearby country with a surplus.

The "Win - Win" approach was even more pronounced in the Peace Treaty between the State of Israel and the Hashemite Kingdom of Jordan signed on October 26th 1994. For instance, Israel agreed to acknowledge Jordanian sovereignty over near borderline wells, which Israeli farmers had drilled on Jordanian territory during their 46 years of war. But Jordan agreed to rent them back to leave the operations of the Israeli farmers undisturbed.. Under Article IV of their treaty, the formerly hostile countries adopted the following "Win - Win" regulations regarding their operation and maintenance:

> "Operation and maintenance of the wells and systems on Jordanian territory that supply Israel with water, and the electricity supply shall be Jordan's responsibility. The operation and maintenance of these wells will be contracted at Israel's expense to authorities or companies selected by Israel."

> " Jordan will guarantee easy unhindered access to personnel and equipment to such wells and systems for operation and maintenance. This subject will be further detailed in the agreements to be signed between Jordan and the authorities or companies selected by Israel."

Such cooperative arrangements were made within the context of a bi-national "Joint Water Committee" to develop plans "for purposes of increasing water supplies and improving water use efficiency, within the context of bilateral, regional or international co-operation" The two countries also agreed to minor rectification of their common border under circumstances to "eliminating or mitigating effects on their respective economies " and "giving each other enough time to make the necessary adjustments."

Option 5: Conflict or Threats of Violence

Breakdown of peaceful negotiations can be followed by a stalemate or acts of conflict, including armed struggle. An intransigent stakeholder or one unwilling to accept a concession or sharing a transboundary water resource might seek a militarily-imposed outcome. However, resort to force involves the risk of losing the war, and it can leave the war-initiator worse off. After Iraq invaded Kuwait because it was not willing to make the concessions which Iraq had demanded, the infra-structure of both nations were terribly damaged.

Armed struggle also dominated the Arab - Israeli negotiating process for over 45 years. In 1964, a little more than a decade after adoption of the above mentioned the Eric Johnston formula for the peaceful (and "Win - Win") distribution of the waters of the Jordan-Yarmuk rivershed had been informally adopted by area's stakeholders, the Arab League nations met in Cairo to cancel this arrangement.

Under strong Syrian pressure, the assembled heads of Arab states agreed to finance to diversion of Banias Springs, which originate in Syria about 1 kilometer above the Israel border. This source of about 25 percent of the upper Jordan River water would have been diverted in a new channel to empty into the Yarmuk River, far from Israel's control.

The Banias diversion project never got off the ground. Israel artillery destroyed the earth moving equipment shortly after it had been brought into the area. While the diversion channel was never completed, intermittent artillery fire in the area continued until Syria was evicted by Israeli troops from its position in the Golan Heights overlooking the Jordan Valley.

The defeat of the combined Arab League forces in the Six Day War produced no change in their "eternal war" strategy. The Arab political leaders knew that the Soviet Union would provide new weapons and might intervene militarily if Israeli troops moved towards one of their capitals to impose a peace agreement. The heads of the Arab League States at conference at Khartoum in 1967 felt sufficiently confident of these geopolitical realities to refuse to accept UN Security Council Resolution 242, replacing it with an Arab-Moslem nation commitment that there be "no negotiation, no recognition, no peace" with Israel.

The stalemate ended in 1991. The invasion of Kuwait by Iraq divided the previously unified international policy of the Arab League states. The most powerful and wealthiest Arab countries invited the United States to lead them in evicting the Iraqis by armed force, which was achieved in a 100-hour electronic war. But most Palestinian leaders had sided with Iraq's brutal dictator. One result is that Palestinians were perceived as a potentially subversive element in some of the more conservative Arab countries. Saudi Arabia, the Gulf states, and Kuwait ousted many persons of Palestinian origin, including some who had played a major and constructive role in the modernization and in the economic and social development of these oil rich nations. In the United States and the West, Palestinians are carefully watched by security services because of that among them are Palestinian and Moslem fundamentalist action factions, which conspire to engage in terrorism.

Option 6: No Action and Opportunity Costs of Delay

Failure to act can be a negotiation option, one that can cause heavy opportunity costs from delay. Water management in the Ganges / Brahmaputra / Meghna Rivers in South Asia (the Basin) is a pertinent case illustration. Unlike the Indus basin, which India divided with Pakistan in 1960, the Ganges-Brahmaputra/Meghna system are not regulated by a treaty. Nepal, Bhutan, India, and China upper riparians. India is a middle riparian and both Bangladesh and India are lower riparians. About 450 million people, or an eighth of mankind, live within the Basin where annual monsoons create huge rivers as they strike against the Himalayan mountains.

The drop in altitude from mountains to sea creates an hydroelectric energy potential of in the order of 200,000 megawatts. But the water and power potential has yet to be tapped. Monsoon rains drain destructively as floods carrying run-off and erosion of topsoil to the sea, while in the dry season the farmers clamor for water. Rapid population and livestock growth contribute to the destruction of forests and fields, leading to ecological degradation. Over-pumping of ground water leads to desertification, and receding spring levels in the hills.

Water resource specialists in Nepal, India, and Bangladesh believe that sustainable water resource development is possible and necessary in the Basin. Each country recognizes that alone it cannot fully control its rivers or resolve its water problems. Each agrees that political understanding must precede a definition of mutual interest, which in turn can be followed by technical planning. So why the delay in coordinated water management? One reason is the inadequate financial and technical resources, which constrain what the co-riparians can do. Nepal and Bangladesh are among the world's ten most impoverished countries, and half of India's 844 million people live in poverty. Political sensitivities also have been important factors limiting the scope and pace of water resource cooperation. Beyond the human constraints, the magnitude of nature's challenges—floods, droughts, erosion, siltation, and other physical processes—overshadows each nation's unilateral efforts.

As George Verghese has written (Rabinovich, 1991): "The Himalaya, a geologically young and fragile mountain chain, is being skinned alive. A vicious cycle of erosion, floods, loss of productivity, and migration threatens the future of scores of millions. This damage to the Himalaya is a common danger, because what happens in Nepal or India affects Bangladesh. Disparities in size, economic strength, and states of development of the countries within the Ganges-Brahmaputra basin have not made cooperation in water resource development any easier. There are risks in partnership for the poor newly-emerging nations with the proximate forbidding Indian giant.

Development problems on Nepal and Bhutan are accentuated by the fact that these Indian neighbors are landlocked. Bangladesh can hardly take action to control water resources when over 90 percent of the stream flows originate in India (Shajahan, 1989). India's rich northeast is also nearly land-locked, but for a narrow corridor between Nepal and Bangladesh which links this region with the Indian heartland. There is no escape from either or these geopolitical or hydrological realities. The Brahmaputra and Barak were the natural arteries of communication from Assam, Manipur, and Tripura to Calcutta, with Chittagong providing an outlet to the sea. Partition severed these links. They have yet to be meaningfully restored."

Conclusions

The choice of negotiation technique is always controversial and subject to political considerations. Preferences depend on the balance of power among transboundary stakeholders and the cost of concessions. The more powerful and wealthy stakeholders can resort to the creation of facts with minimal risks of counteraction by weaker and impoverished neighbors. They also can afford to make gestures of friendship through "Win-Lose" agreements in the interest of enhancing regional stability. "Win-Win" solutions may not be always be sufficient when naturally limited water resources are under consideration.

The armed struggle strategy has turned out to be counter-productive when it was applied by the Arab League to impose a significant reduction of the previously available fresh water resources on Israel. During the 1967, 1973 and 1982 wars, Israel was able to gain control of most of the disputed sources of fresh water to which Arab countries were upper riparians.

A significant proportion of the Palestinian nation and of Pan Arab countries have begun to rely on peaceful negotiations strategies, in response to which the World Bank and donor nations have pledged to provide investment capital for regional projects. If the Madrid peace process continues to reduce past hostilities, significant improvements the regional infra-structure and level of prosperity of the former belligerents are likely to occur. But such a shift in favor of negotiated solutions to transnational problems continues to be opposed by Muslim fundamentalist like the Hamas and Hezbollah factions and a number of Islamic nation ruled by uncompromising military cliques, including Syria, Lybia, Sudan, Iraq and Iran.

There also are the parliamentary opposition parties in Israel which take the position that the withdrawal of some Israeli settlements in the Golan Heights, the Westbank or Gaza strip in return for a full peace treaty and the end of economic boycott of Israel by most Arab League nations. would be dangerous Zero-sum game concession, They believe that future Arab leaders could abrogate their peace treaties - "a mere piece of paper", once Israel had completed such proposed territorial withdrawals.

International agreements are inherently unstable unless it remains in the interest of all negotiation parties to adhere to them. Long range security guarantees are not easily engineered, not in transnational conflict, not in labor negotiations or in business contracts. But even with this in-built element of uncertainty, conflict resolution by negotiation is generally preferred to reliance on war and armed struggle or on ignoring the problem - where negative consequences tend to be immediate and devastating. There are risks in every negotiated solution between sovereign powers whose rulers can abrogate agreements made by their predecessors, but such risks are widely believed to be reduced when agreements are made under the active sponsorship of the major world powers, as is the case in the current Madrid negotiation process between Israel and its Arab neighbors.

References

Gruen G.E., (1992) The Water Crisis: The Next Middle East Conflict, Los Angeles, California, The Simon Wiesenthal Center, Revised Edition.

International Law Association, (1992) "The Helsinki Rules On The Uses Of Waters Of International Rivers", Chapter 2, Article V, in Dante A. Caponera, Principles of Water Law and Administration, Rotterdam, Holland, A.A. Balkenna.

Rabinovich I., (1991) The Road Not Taken: Early Arab-Israeli Negotiations, New York, Oxford University Press.

Shajahan M., (1989) "Regional Cooperation for Flood Control", in International Seminar of Regional Cooperation for the Protections of the Environment of South Asia, New Delhi, Indian Council for South Asian Cooperation.

This section is composed of material from the text of Chapter 3 of David Eaton and David Hurlbut, (1992) Challenges in the Binational Management of Water Resources in the Rio Grande Bravo, Austin, Texas, The University of Texas.

Utton A.E., (1981) "The Development of International Groundwater Law", in Ludwick A. Teclaff and Albert E. Ulton, International Groundwater Law, London, Oceana Publications, Inc.

Verghese G., (1992) "Learning to Say: Open Sesame", in David Eaton, Editor, The Ganges/Brahmaputra Basin: Water Resource Cooperation Between Nepal, India and Bangladesh, Austin University of Texas.

Chapter II: Examples of Policy Schemes and International Treaties

II.1 A Transboundary Water Management Organization: The International Commission for the Protection of the Elbe River

A.H. Schumann
Ruhr-University Bochum, Germany

M. Simon
International Commission for the Protection of the Elbe River

The Elbe River - its Basin and its State

The Elbe River is the third largest stream (behind the Danube and the Rhine) in Western Europe. Its catchment covers an area of 148.268 sq. km (Fig.1). It belongs to the international river basins, since it is shared between the Federal Republic of Germany (FRG) (65.4 %) and the Czech Republic (33.8 %) (the area parts of Austria and Poland are less than 1 % and can be neglected). The population living within the river basin totals 25.7 million people. The average population density is 173 per sq. km. Within the catchment, three cities with more than one million inhabitants are situated: Berlin (3.4 mio), Hamburg (1.6 mio) and Prague (1.2 mio). The national importance of the Elbe river can be demonstrated by the percentage of citizens living in both countries within the river basin: 60.9 % of all Czechs and 24.6 % of all Germans. In the past, the water policy was orientated to waste water removal instead of treatment and purification. As a result, the load of the Elbe river with organic nitrogen, heavy metals, slowly biodegradable and toxic substances damaged the ecosystem significantly and reduced the possibility for drinking water supply and irrigation along the river. In 1989, the water quality was comparable to the quality of the Rhine river in the early seventies, during its maximum pollution. The Elbe River can be seen as a victim of socialist environmental policy. The extensive utilization of this river as a sewerage for non-treated wastewater by the former German Democratic Republic (GDR) and the Czechoslovakian Socialist Republic damaged the river seriously.

Before the institutional framework to overcome the transboundary problems in the Elbe river basin is characterized, the specific problems of water management, which have been inherited from different political developments in the past, should be discussed. Re-unified Germany mirrors the problems of the necessary rehabilitation of environmental conditions in a former socialist country. Present day water management issues in eastern and western European countries may differ considerably from those in the past, and efforts to develop common standards may differ much from past efforts to improve water management conditions. Between 1945 and 1990 in East Germany, decisions were based on the ideology of the socialist oligarchy. Under these conditions, decision makers were also interested in developing the water supply system, as water was one of the staple foods which should be available in sufficient quantity for a low price in order to demonstrate the welfare of the socialist state to its citizens. As a result, water supply systems were developed with the aim of meeting all quantitative needs of society.

borders of the river basin

borders of federal states within Germany

state borders

Figure 1 The basin of the Elbe river.

The water price was subventioned, and as a result, water consumption per inhabitant was much higher than in the West. By the political orientation to provide sufficient water for the people, the problems of water quality were neglected in East Germany. For example, between 1960 and 1989, the water supply capacity was increased by 240 percent in the German Democratic Republic (GDR), but in 1990 nearly 50 % of all inhabitants received drinking water of low quality (beyond the standard of the European Community) (Clausnitzer, 1990). Water quality data were viewed as a state secret, and this problem became politically non-existent. As public discussions of environmental problems were not possible in socialist countries, no pressure groups existed that could demand a solution of such problems. Administrative rules regarding the control of wastes and reduction of environmental damages were part of the governmental structure, but their enforcement had a low priority in comparison to the problems of the national economy. In its last years of its existence, the GDR only invested 0.5 % of the total social product in environmental protection. In the Federal Republic, this percentage was 1.7 % (Simon, 1991). The results of the different developments in East and West Germany can be demonstrated with the example of wastewater treatment (Fig. 2). In 1990, 87.5 % of the waste water in the western part of Germany was treated biologically; in East Germany, only 30 % was treated. In the former GDR, 26.1 % of the waste water was treated mechanically, but 14.8 % was not treated at all (in the Federal Republic only 2.8 %).

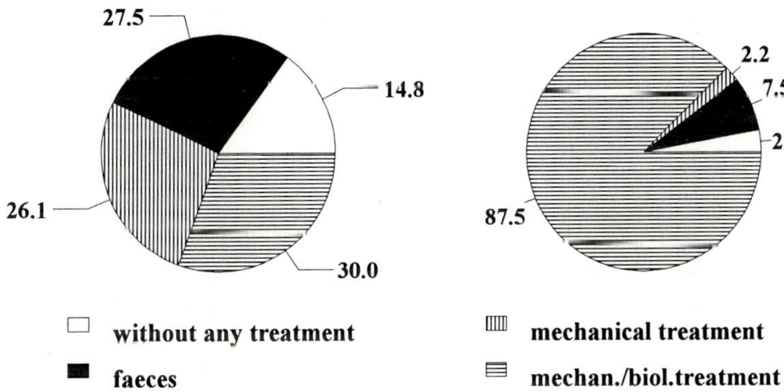

Figure 2 Wastewater treatment in the GDR and in the Federal Republic (after Stat. Bundesamt, 1992; Clausnitzer, 1990).

How will East Germany achieve the standards of the West Germany? In Fig. 3, the large financial demand is shown, which would be necessary for East Germany to reach the EC-standards by the year 2000. The total demand amounts to 211 billion DM. The main part of this is needed for the improvement of waster water treatment. The amount of 125 billion DM for this purpose is nearly 25 times as large as the money that was invested for waste water treatment in the West Germany in 1991.

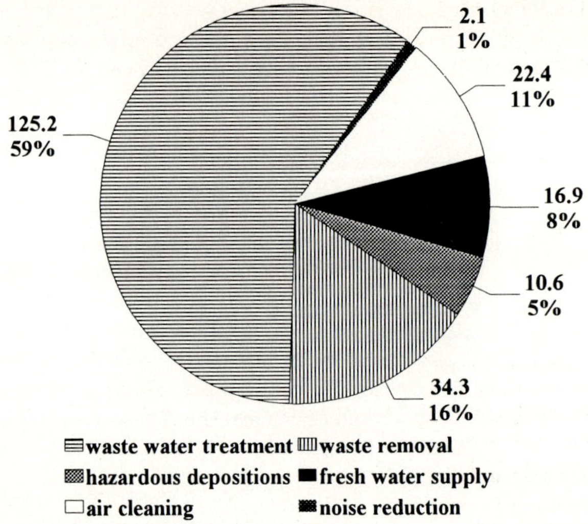

Figure 3 Demand of financial means to improve the environmental conditions in East Germany to the year 2000 (total demand: 211 000 000 000 DM).

After the political changes in Europe, improvement of water quality and restoration of the suppositions for the utilization of the Elbe river runoff became one of the most challenging problems of water policy in Western Europe. The foundation of the International Commission for the Protection of the Elbe River (ICPE) became the institutional framework to overcome the transboundary problems in the Elbe river basin. Once the international treaty of the foundation of the ICPE was signed on October 1990, the basis for a cooperation between Germany and the Czech Republic to improve the water quality of the Elbe river was formed.

The agreement about the foundation of the ICPE was the first international treaty signed by the re-unified Germany. The partners signing the contract were the Federal Republic of Germany (FRG), the Czechoslovakian Republic and the European Union (EU). The main targets of their cooperation are:

- to enable the utilization of the Elbe river for drinking water supply and agricultural water use
- to ensure the development of a natural ecosystem with a healthy diversity of species
- to reduce the waste load transported via the Elbe river into the North Sea.

To realize these targets, the improvement of the state of the Elbe river and its main tributaries in their physical, chemical and biological aspects and the increase of the ecological value of the Elbe river are the most important tasks.

The Organization of the ICPE

The ICPE is organized into international working groups in which all three partners are represented (Fig. 4).

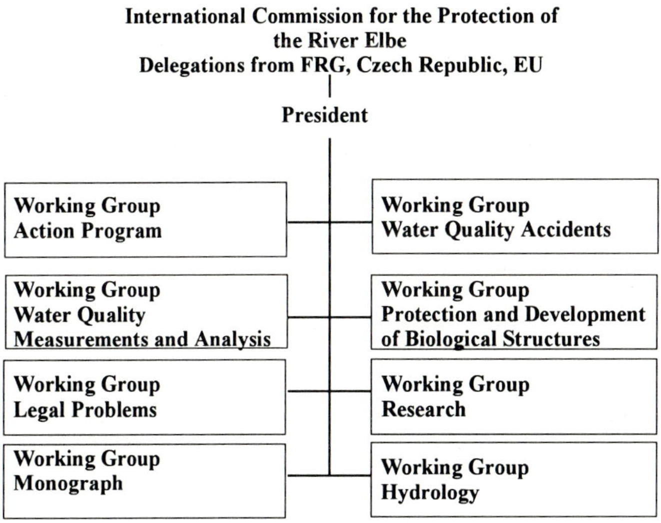

Figure 4 Structure of the ICPE.

The main tasks of these working groups are:

a) Working group Action Program:

· to work out programs for actions to improve the water quality and the ecological state in cooperation with other working groups,

· to work out an overview on the state of waste water treatment plants and the spill-off of waste water from communities and industries along the Elbe river and within the catchment,

· to set up a list of priority for harmful substances for which the spill-off should be stopped or at least reduced,

· to evaluate the waste loads originating from non-point sources and to develop a strategy for reducing them,

· to work out water quality targets for the Elbe river and its tributaries under consideration of water quality demands for water users, the protection of the North Sea and ecological needs.

This working group has a sub-group responsible for formulation of a common standard of the minimum requirements for waste water quality for communities and industries and also the step-by-step realization of a standard for monitoring waste water loads from point-sources.

b) The working group "Water Quality Measurements and Analysis" has the following tasks:

· to work out a yearly program of water quality measurements and analysis for the Elbe river and the estuaries of the main tributaries,

· to publish the measured values yearly,

· to publish reports on the water quality of the Elbe river and its tributaries and to evaluate the changes in a two year time step.

This group has three sub-groups; one is responsible for the organization of the international network of automatic stations for water quality measurements; the second for the hard- and software of these stations under consideration and for the need for data exchange, and the third is responsible for control of a common standard in water quality analysis with the aim of producing data that are directly comparable.

c) The working group "Water Quality Accidents" is responsible for the setup of an international warning and alarm system, a list of facilities with a potential risk for water quality accidents and for measurements to reduce these risks.

d) The working group "Research" is responsible for

· a proposal for a research program (1992-1995) to analyze the load of harmful substances and its evaluation,
· for the research program on ecomorphology and ecosystems of the Elbe river, of selected tributaries and the river banks

e) The working group "Protection and Development of the Biological Structures of the River and its Banks" has the tasks:

· to work out a plan of action for immediate protection and ecological improvements of the biotopes along the river,

· a proposal for the protection of actually endangered ecological reaches,

· an ecological study on the measures necessary for improvement of aquatic and terrestrial communities of live

f) The working group "Legal Problems" is responsible for all treaties and agreements relevant for the targets of ICPE.

g) The working group "Monograph" will document the collected information

h) The working group "Hydrology" is responsible for the hydrological service.

First Results of the ICPE

Starting at the end of 1990, the work of the ICPE was organized in three temporal levels (ICPE, 1992) (Tab.1).

Tab. 1 Time steps of ICPE- work between 1990 and 2000

Time steps	Main activities of ICPE
I. 1991	Analysis: -state of waste water treatment plants with a population equivalent > 50 000 p.e. -state of waste water treatment of chemical, pharmaceutical, cellulose, pulp, metal industries Result: Action Plan for Improvement of Water Quality 1992- 1995 (accepted 1. June 1992)
II. 1992- 1995	Planned activities: -completion of all treatment plants under construction with capacities of more than 20 000 p.e. -planning of treatment plants with an capacity of more than 50 000 p.e. and start of the construction by 1993 -reduction of industrial waste water load -installation of an water quality measurement network -warning and alarm system -realization of a research program Result: Long- term Program for the Development of Water Quality and Ecology
III. 1995- 2000	Planned activities: -Definition of a new standard of waste water treatment within the river basin -Analysis of non- point sources -set-up of new water quality targets

As the first step, in 1991 a program for immediate improvement of water quality between 1992 and 1995 was set up. This program includes a list of priorities for the construction of waste water treatment plants as well as an evaluation of the effects of these measures and an estimation of the costs.

The basis of this first action program was an analysis of the waste water treatment of communities with a total population equivalent of more than 50.000 inhabitants (which represents more than 50 % of the total load from communities) and of chemical, pharmaceutical, cellulose, pulp and metal working industries with regard to 15 different substances and parameters.

The realization of this program will reduce the waste water load significantly. The program was accepted 1 June, 1992 by the environmental ministers of both states and a representative of the EU. It is the basis for the work of the ICPE in the second phase (1992-1995).

In the program, the following measures are intended for the time period of 1992-1995:

1. For the improvement of communal waste water treatment:

a) All treatment plants under construction with a population equivalent of more than 20.000 (category 1) should be completed until 1995

b) The planning for construction of waste water treatment plants with a population equivalent of more than 50.000 (category 2) should be completed. Construction should start until 1995.

c) Considering the limitations of financial means in the first step, biological treatment plants should be build. The planning should consider the need for further waste water treatment (denitrification, phosphate elimination).

By June, 1993, the construction of 28 of the planned 61 treatment plants of the first category was completed (ICPE, 1993) (Tab. 2). The construction start for treatment plants of category 2 was planned for 62 plants by 1993. This aim was realized for 15 plants only. Especially in Germany, the planning was much more difficult than intended as a result of changes in administration structure after the re-unification and the new legal framework of planning.

2. For the reduction of industrial waste water loads, a list of priorities was set up in order to reduce the load of some substances with extremely harmful effects. As a result of the changes in national economics between 1989 and 1991, the industrial production as well as the waste water amount was reduced significantly (Tab. 4).

The most important result is expected not only from the reduced production of chemical and cellulose industries in Germany, but also from the improvement of waste water treatment for some industries in the Czech Republic.

Table 2 Realization of treatment plants with a population equivalent of more than 20.000.

State	planned waste water treatment plants	planned capacity (1000 p.e.)	realized until June 1993	realized capacity
Germany	31	9071	16	4846
Czech Republic	30	5025	12	1780
Sum	61	14096	28	6626

Table 3 Realization of the start of construction for treatment plants with a population equivalent of more than 50.000.

State	planned wastewater treatment plants	planned start of construction (1991- 1993)	start of construction realized by June 1993	planned start by end of 1993
Germany	65	55	10	22
Czech Republic	13	7	15	5
Sum	78	62	25	27

Table 4 Reduction of the waste water from industries between 1989 and 1992.

Substance	year	Chemical C.R.	industry F.R.G.	Cellulose C.R.	industry F.R.G.	sum (tons/ yr)	reduction to (percent)
COD	1989	43 400	358 055	84 000	358 561	844 381	-
	1991	30 785	61 503	37 700	66 372	197 175	23.4
	1992	26 720	35 665	19 100	17 429	98 929	11.7
Ammonia-Nitrogen	1989	2 000	25 559	-	-	27 559	-
	1991	1 215	6 778	-	-	7 993	51.1
	1992	980	4 441	-	-	5 421	27.4
Mercury	1989	2.3	17.1	-	-	19.4	-
	1991	1.84	3.1	-	-	4.5	23.2
	1992	1.76	0.2	-	-	1.96	10.1

The reduction of the loads of Biological Oxygen Demand (BOD), Chemical Oxygen Demand (COD), Total Nitrogen and Total Phosphate are shown in Figs. 5 and 6.

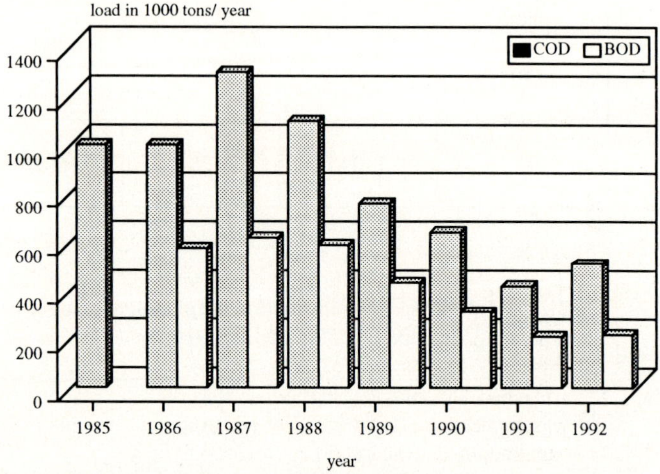

Figure 5 Development of the annual load of Biological and Chemical Oxygen Demand at the water quality station Schnackenburg / Elbe.

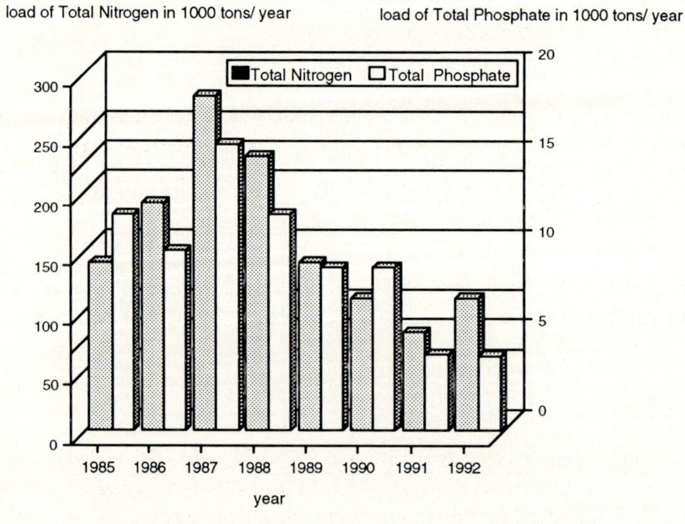

Figure 6 Development of the annual load of Nitrogen and Phosphate at the water quality station Schnackenburg / Elbe

Until the end of 1993 the water quality information network was improved significantly. It includes 18 stations for the automatic measurements of selected water quality parameters, two national data centers and the international data center in Magdeburg (Fig. 7).

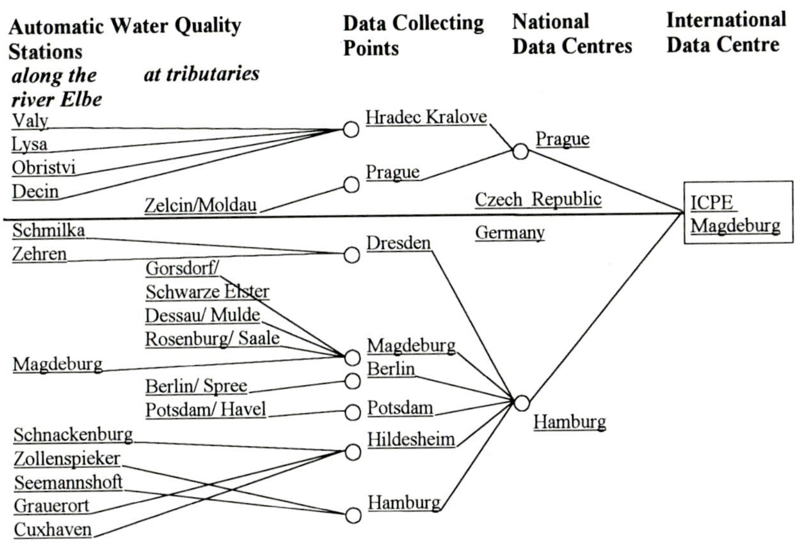

Figure 7 Structure of the water quality data network.

A warning system was set up ensuring the information flow during accidents and the warning of water users. In the field of research coordination, 82 research proposals were evaluated. In Germany, 39 research projects with a total financial demand of 22 Mio DM have been realized. In the Czech Republic, 10 projects have been sponsored with a financial demand of 5.8 Mio DM.

For the protection and development of ecological systems proposals were made on the basis of a program for immediate necessary action and a preliminary study for a long-term program.

Future Activities

In the second phase of the cooperation between 1992 and 1995, the realization of the above mentioned program will be analyzed yearly. The third phase of the program will be prepared. An agreement for the standard of waste water treatment under consideration in the EU regulatory framework and the need for temporal solutions for bridging the financial problems until the EU standard can be reached will be prepared. The list of chemical substances and biochemical parameters to be considered for improving the water quality by the year 2000 will be extended. The impact of non-point sources of harmful substances, especially from intensively used agricultural areas, will be evaluated. New water quality targets will be formulated. As a result, at the end of the second period, a long term program for improvement of water quality and development of ecology will be set up, based on the newest results of monitoring and research. The third phase between 1996 and 2000 is dedicated to the program implementation. Its realization will be evaluated at the end of the years 1997 and 2000.

Summary and Conclusions

The work of the ICPE shows that measures for the development of water quality in international rivers demand a strong cooperation in transboundary water management. The analysis of the state of the river and the evaluation of possible impacts of planned actions by all partners on the river in its entirety is a common task ensuring the transparency of planning for all sides. Objectivity is very important for the most effective solution of international environmental problems. This objectivity can only be reached by a free exchange of information and data as it becomes evident in the work of ICPE.

References

Clausnitzer, E. (1990): Die Wasserwirtschaft der DDR und die Lage in der Wasserversorgung gwf Wasser- Abwasser 131 (1990) Nr.11

ICPE (1992): Arbeitsplan der IKSE bis zum Jahre 2000, Spindleruv Mlyn, 21.09.1992

ICPE (1993): Bericht über den Stand der Durchführung der im "Ersten Aktionsprogramm (Sofortprogramm) zur Reduzierung der Schadstofffrachten der Elbe und in ihrem Einzugsgebiet" enthaltenen Maßnahmen, Stand: Juni 1993, Magdeburg, 2.11.1993

Simon, M. (1991): Die Belastung der Elbe und ihrer Hauptnebenflüsse auf dem Gebiet der ehemaligen DDR Wasser und Boden (1991) H.4

Statistisches Bundesamt (1992): Statistisches Jahrbuch 1992 für die Bundesrepublik Deutschland Statistisches Bundesamt Wiesbaden 1992

II.2 Transboundary Water Resources Management in the Baltic States

Bernardas Paukstys
Geological Survey of Lithuania
Konarskio 35
2600 Vilnius
Lithuania

Introduction

Integrated planning and economic development compatible with environmental sustainability at near-border territories should be agreed upon among the neighboring countries. As today often happens in post-soviet countries such as Lithuania, human activity in one of them (waste water discharge into surface streams, development of deposits of useful minerals, groundwater abstraction, design or construction of hazardous objects, etc.) is not coordinated with the environmental impact on the bordering country.

The European map has changed significantly during the last decade. After the fall of the Berlin wall - symbol of German division - on 9 November, 1989 and the collapse of the Soviet Union, the Commonwealth of Independent States appeared with new frontiers and new international environmental problems. Artificial political boundaries themselves do not influence natural environment; the dividing social, political and cultural differences of neighboring countries they create institutional and organizational difficulties in environmental relations between them. However, political boundaries between the neighbors, in an ideal case, can also become a place for meetings and negotiations.

The problem of transboundary water resources management has been an international problem for years. Sixty-three international treaties and conventions related to transboundary water resources management have been signed among European countries up till now; the first, on Danube River, was agreed upon in the seventeenth century. However, large quantities of political documents, even if they are very good, do not repeal the problems. The extent of the dilemma related to the sharing of international water resources is directly dependent on the number of frontiers. The scale of the quandary could be binational and multinational. Both multinational and binational transboundary water management problems will be discussed in this paper, using the example of Lithuania.

Helsinki Conventions - Multinational Efforts For Protection of the Baltic Sea

The Baltic Sea (Fig. 1) is one of the major brackish water basins in the world and also one of the most polluted seas. It is bordered by the nations of Russia, Estonia, Latvia, Lithuania, Poland, Germany, Denmark, Sweden and Finland. Four further countries, namely, Norway, the Czech Republic, Slovakia and Belorus form part of a catchment draining into the Baltic (Fig. 1). The area of the Baltic region totals 2.1 million km^2, including the drainage basin (1.7 million km^2) and the Baltic Sea itself (Carlsson 1992).

Figure 1 Location of the Baltic Sea and Baltic States.

Hundreds of surface water streams and groundwater from a catchment area discharge into the Baltic. It is directly influenced by such large cities as St.Petersburg, Kaliningrad, Riga, Tallinn, Klaipeda, Helsinki, Stockholm, Rostok, Gdansk and a great number of smaller ones situated on the seaside. The Sea is affected by the industrial, agricultural and energy production activities of almost 80 million people (20 years of International Cooperation 1994). In such a catchment shared by many nations, an approach to an integrated study of water flux to the Baltics requires combined efforts of all the relevant countries.

In 1960, scientists have expressed great concern over the increasing pollution of the Baltic Sea. In 1973, the first international experts' meeting was initiated by Finland to consider the possible subjects for the joint convention to protect the Baltic. The Convention on the Protection of the Marine Environment of the Baltic Sea Area, the Helsinki Convention, was signed by seven Baltic States (Lithuania, Latvia and Estonia were part of USSR then) in Helsinki, Finland on 22 March, 1974. It became effective on 3 May, 1980, after all the Baltic states ratified it. The Baltic Marine Environment Protection Commission was established through the Convention, also known as Helsinki Commission (HELCOM).

The duties and rights of the countries bordering the Sea are well defined in general terms in the Convention. Fundamental principles and obligations of the countries are to "individually or jointly take all appropriate legislative, administrative or other relevant measures to prevent and abate pollution and to protect and enhance the marine environment of the Baltic Sea Area without causing an increase of pollution outside the Baltic Sea Area" (20 years of International Cooperation 1994).

In April, 1992, the government of Finland convened a Diplomatic Conference on the Protection of the Marine Environment of the Baltic Sea Area. At this conference, a revised Convention was signed by the nine Baltic States, Norway, the Czech and Slovak Republics and the European Economic Community. Strategic approaches and principles for the Baltic Sea Joint Comprehensive Environmental Action Program were adopted. The program is expected to last at least 20 years, with the cost of the implementation project estimated at about 18 billion ECU. The long term program has 6 components:

-policy, legal and regulatory measures;

-institutional strengthening and human resources development;

-investment activities;

-management programs for coastal lagoons and wetlands;

-applied research;

-public awareness and environmental education.

Once fully implemented, the program is expected to have a strong beneficial impact on the water quality of the rivers in the Baltic Sea drainage area. The quality of coastal waters can be expected to improve relatively rapidly. Reducing the load of nutrients and heavy metals and strengthening of environmental management capacities throughout the Baltic Sea region will help to restore the ecological balance of the open sea and also of the coastal lagoons and wetlands.

Baltic Sea-Input from the Surface Streams

The Baltic Sea is a unique and highly vulnerable ecosystem. It is shallow sea (average depth is 55m) and isolated from the ocean. A large number of industrial towns and agricultural areas fringe the Baltic Sea, resulting in pollution of surface and groundwaters draining into the Baltic. Two hundred and fifty large and small rivers discharge 450 km^3 of waters (and pollutants) into the Sea , the largest being the river Neva with a mean annual flow of approximately 2400 m^3/second or 80 km^3 /year (Carlsson 1992).

The annual river runoff varies significantly. Fig. 2 shows the measured data on water flow for the period from 1950 to 1990. The rivers bring human-produced chemicals and organic substances, which cause major environmental problems in the Baltic sea, such as a decrease in the oxygen content of deep water layers, enlargement in concentrations of nitrate and phosphorous which influence significantly eutrophication processes and also augmentations in concentrations of heavy metals and organic compounds. It has been calculated that about 40 thousand tons of oil are discharged into the Baltic Sea (The environmental problems of the Baltic Sea region, 1991). 336 million m^3 of untreated waste water have been discharged annually into the Nemunas River, which flows through the Republics of Belorus, Lithuania and Russia. 73,5%, or 247 mln. m^3 of the total waste amount comes from Lithuania, 24,4% from

Russia (Kaliningrad region) and 2,1% from Belorus. The Neman and Sovetsk paper and pulp plants (Russia) spew 220 thousand m^3 of contaminated waste water into the Nemunas river (The environmental problems of the Baltic Sea region, 1991).

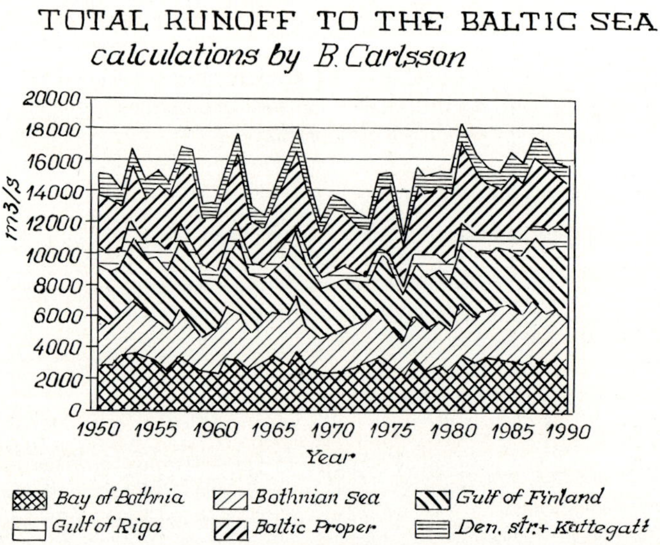

Figure 2 Total river runoff into the Baltic Sea during the period from 1950 to 1990.

Hydrological discharge is being observed and calculated in all the Baltic area countries. Hydrological data are normally collected, processed and stored by national hydrologic centers. Daily values of river runoff or data of higher resolution in time are available from national hydrological centers.

Assessment of discharge of contaminants to the Baltic Sea is also carried out by surrounding nations. Two Baltic Sea Pollution Load Compilations (PLC) based on data from all Baltic countries have been published by HELCOM, the first in mid 1980', the second in 1993 (Baltic Sea Environmental Proceedings 1993).

The ultimate goals of Pollution Load Compilations are:

-to compile the total load of important pollutants entering the Baltic Sea via rivers and from point and non-point sources on a basis of harmonized monitoring methods;

-to follow up the long-term changes in the pollution load from various sources;

-to provide information for periodic assessment work.

The main pollutants discharged into the different parts of the Baltic Sea by the rivers are presented in Table 1.

Table 1 Total pollution load entering the Baltic Sea via rivers, t/year (after Baltic Sea Environment Proceedings 1993)

Region	Phosphorus	BOD_7*	Nitrogen
Bothnian Bay	2 134.1500	79 792.9750	35 033.7000
Bothnian Sea	1 850.6400	88 536.1300	42 985.0000
Archipelago Sea	664.0000	7 780.0000	7 870.0000
Gulf of Finland	7 641.7000	201 934.9000	109 529.5000
Gulf of Riga	2 704.9000	101 806.9000	79 965.8000
Baltic Proper	>14 158.1300	>529 861.6200	182 136.1800
Belt Sea and W. Bay	1 699.2300	>4 528.2600	38 821.0000
The Sound	223.9800	> 488.9400	7 591.0000
The Kattegat	2 283.300	> 23 425.7650	63 806.000
Total	>33 360.0300	>1 038 155.4900	567 730.1800

*BOD_7 -Biological oxygen demand = total organic compounds x 0.145

To follow up the effects of pollution on the marine environment of the Baltic Sea, a joint monitoring program has been coordinated by the Helsinki Commission since 1979. For the program, special guidelines were published on the amount of stations, parameters, methods to be applied in sampling, analytical methods, frequency of sampling and the reporting of results. The program has been updated every five years. The data are distributed annually to all Contracting Parties of the Commission,and a common data base for data processing is established.

Groundwater - Important Component of the Water Budget of the Baltics

Precipitation and surface streams are not the only variables which form the water balance of the Baltic Sea. A well known picture on the water cycle in nature (Fig. 3) shows the third important element, which should not be neglected. This element is groundwater. It is more difficult to quantify the groundwater component of pollution transport to the Baltic. Although groundwater is traditionally regarded as being less vulnerable to pollution than surface waters, it is known that many of the Baltic States experience such a heavy pollution load that groundwater (particularly the upper aquifers) is extensively contaminated. Its discharge, either directly or via rivers, into the Baltic should also be taken into account.

Unfortunately, in none of the HELCOM reports is groundwater being considered. The major groundwater discharge (1.3 million m^3/day) comes from the zone of active water exchange. Thirty million m^3/day of groundwater reach the Baltic Sea via the rivers. Altogether, groundwater discharge accounts for 13% of total surface water runoff (Mokrik 1990). The directly dissolved ion discharge into the Baltic from Prequaternary (deep) aquifers reaches 200 thousand tons/year, via the rivers--about 400 thousand tons/year. In addition, 1.5 million tons/year of dissolved ions discharge into rivers from Quaternary (upper) aquifers (Mokrik 1990). The figures show that groundwater can also have quite an important influence on the quality and water budget of the Baltic Sea.

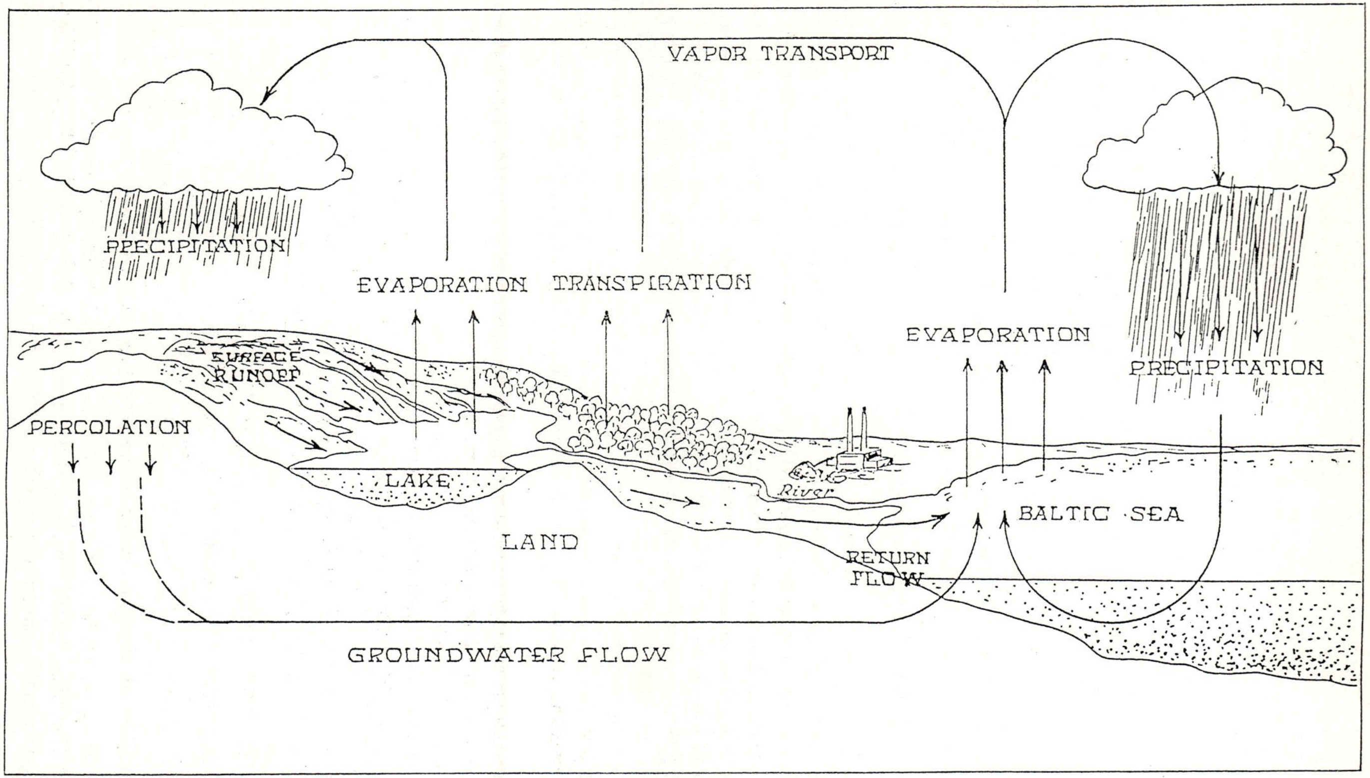

Figure 3 Natural water cycle in the Baltic Sea.

Human activity disturbs the natural equilibrium of ecosystems. It also influences water balance significantly. It is estimated that groundwater runoff in the future will decrease by 1,5 times due to abstraction (Mokrik, 1990). This creates another environmental problem--increase of return flow or sea water intrusion into aquifers (see Fig. 3). This phenomenon increases with expansion of water consumption. The problem of sea water intrusion is also important for the Baltic countries, because along almost all of the shore line, many water users are located. Groundwater is mainly being used for drinking water supply in the Baltic States. More than 40 thousand water supply wells are being used for groundwater abstraction, with a total yield of nearly 4 million m^3/day, causing changes to the groundwater regime in the neighborhood of 1000-4000 km^2 (Kiruchin, 1993). Mineralization of the main aquifer used for drinking water supply in Riga, for example, increased to 3.5 g/l due to intrusion of saline sea water. Water consumption is increasing with enlargement of population and development of industry in the Circum Baltic countries. It is necessary to observe interaction of fresh and saline water and to predict sea water intrusion in order to protect the aquifers from deterioration

Evaluation of Pollution Transport Via Groundwater

The idea of a joint project for the development of compatible groundwater monitoring methodologies, leading to determination of the contribution of groundwater-borne pollution to the water composition and balance in the Baltic Sea was first raised at the International Hydrogeological Symposium in Palanga, Lithuania, in 1992 (Juodkazis, 1992). All Circum-Baltic countries were invited to take part in the project. Although all the countries around the Baltic have a tradition of groundwater monitoring, mapping and research, the methods employed differ. The various nations use different mapping methodologies, sampling and analytical techniques and groundwater monitoring philosophies. Therefore, it was proposed to combine the intellectual forces of hydrogeologists around the Baltic and to create a compatible system of information collection and distribution. Groundwater monitoring activities carried out by the member countries may be considered as their contribution to the joint project.

The proposed project will be of an umbrella type and will consist of several different individual tasks, which could be solved in separate phases:

- development of compatible groundwater monitoring methodologies for the Circum-Baltic countries,
- elaboration of the joint mapping techniques in the basin of the Baltic Sea;
- unification of the systems for collection of hydrogeological data and compilation of compatible data banks;
- analysis of interaction between groundwater flow, sea water and bottom sediments;
- prognosis of effects of groundwater component on water balance and quality in the Baltic Sea and recommendations of necessary remediation actions.

Based on the results of the joint project, the new legal documents or supplements to existing HELCOM Guidelines leading to international groundwater resources management could be agreed upon among the nations of the Baltic.

The State Geological Survey of Lithuania and Vilnius University were authorized by the participants of the Conference to oversee the implementation of this project. In October 1992, at the meeting of the HELCOM Environmental Commission in Tallinn, Estonia, a proposal was made by a representative of Lithuania for organizing groundwater monitoring activities in circum-Baltic States and for establishment of a separate working group at HELCOM. The proposal was supported by the representatives of most Baltic countries but rejected by the chairman of the Environmental Commission The same question was raised again at the International Conference ECOBALT-93 in June, 1993 in St.Petersburg (Voronov, 1993).

The State Geological Survey of Lithuania and Vilnius University, executing the resolution of the Conference ECOBALT-93, contacted Geological Surveys, Universities, other scientific institutions and consulting companies of Circum Baltic countries involved in groundwater monitoring activities and invited them to participate in the joint program. Thirty-seven letters were mailed to 12 countries and 12 answers were received from Belorus, Estonia, Finland, Sweden, Norway, the Czech Republic, Germany, Latvia, Poland, Russia and Denmark. Almost all countries support the idea described in the program and agree that the main emphasis at the initial stage of the project should be put on standardization of joint methodical principles for groundwater monitoring.

The project was also discussed at the meeting of the directors of the Geological Surveys of the Nordic and Baltic States in Vilnius in April, 1994. It was decided to prepare a project proposal and to present it to different funding organizations. The Geological Surveys of Denmark and Lithuania accepted the responsibility of further implementation of project ideas, but implementation of new ideas usually takes time due to institutional misunderstandings. The specialists-hydrogeologists of the Baltic countries will do their best to try to convince decision makers (actors) of relevant institutions of the necessities and benefits of the proposed actions for groundwater monitoring.

Binational Lithuanian-Latvian Transboundary Water Problems

Lithuania and Latvia, the sister countries, are the only ones left in the world speaking a language from the same Baltic language group. The two countries share not only a common language but also water resources. Ten thousand five hundred km^2 of the Lithuanian territory belongs to the drainage area of rivers falling into the Baltic Sea through the Latvian territory. Therefore, binational transboundary water management problems appear between the countries. The problem is enhanced by the fact that small towns and industrial units in Lithuania are equipped with waste water treatment plants, whereas major cities of the country have no treatment facilities. The two largest cities, Vilnius and Kaunas, discharge approximately 50% of total pollutants into water bodies. It is estimated that after the construction of biological waste water treatment plants in the four biggest Lithuanian towns of Vilnius, Kaunas, Klaipeda and Siauliai, the water quality in the main rivers will improve by as much as 75%. Discharge of organic pollutants will be reduced by 67% or 40 thousand tons/year (The environmental problems...1991) Pollution of the rivers flowing through Latvia will also be reduced, but today waste water from the Northern part of Lithuania flows into the Baltic Sea through Latvian territory.

As was described in previous chapters, river water quality is being observed by Baltic Sea nations including Lithuania and Latvia, and information on river flux is published in HELCOM proceedings. Much less data is available about groundwater level and quality fluctuations. Meanwhile, due to concentrated groundwater withdrawal at near-border zones, wide and deep depression cones (level drawdown), some having a radius of hundreds of kilometers, develop in the aquifers shared by neighbor countries. On the other hand, the largest (and most polluted) rivers of Lithuania and Latvia, the Nemunas and the Daugava, are the sources of formation of groundwater yield of large river bank waterworks where the drinking water quality completely depends on the pollution extent of the rivers. This shows that even between small neighboring countries, large transboundary environmental problems exist. Several cases could be presented as typical examples of transboundary environmental relations, but most clearly the necessity of solving trans-frontier problems of shared groundwater resources could be shown on the karst region of Lithuania and Latvia. The roots of the problem are related to uncoordinated institutional activities when making decisions on the use of natural resources.

Karst Region - Need for Protection of Shared Aquifers

In the Northern part of Lithuania at a border with Latvia, in a territory of more than 1000 km^2, thin strata of glacial sediments (0-10m) are underlain by the Upper Devonian gypsic formation. Karst process is developing in this formation. On the surface, karst is expressed by a dense network of sinkholes, which differ in size and shape. In some areas, the amount of sinkholes exceeds 200 units per square kilometer.

During the last decades, land subsidence and formation of new sinkholes is noticeably increasing. According to the data of aerial mapping, on one of the sites of intensively karstified territory with an area of 3 km^2, 61 new sinkholes appeared in a 23-year period (from 1967 till 1990) (Paukstys, 1991).

Human activity, groundwater abstraction and intensive agricultural pollution are considered to be the main reasons of karst intensification. Exploitation of an hydraulically interconnected system of aquifers disturbs the hydrodynamic equilibrium in them and accelerates the removal of filling material from underground cavities, while regional groundwater pollution by agricultural activity increases its aggressivity with respect to karst rocks (gypsum).

From the geological point of view, the karst region is well investigated in Lithuania and Latvia. The territory is covered by the hydrogeological and engineering-geological mapping, with a scale of 1:50,000. Geological, hydrogeological, morphological and other features of surface and subsurface karst forms are investigated in detail.

A monitoring system has been created on the karstified territory of Lithuania. It consists of 130 observation points installed into different layers of the active water exchange zone: drilled and dug wells, natural springs and surface streams (Klimas and Paukstys, 1990). Observation data of groundwater level and quality fluctuations from 1977 are available in the databases of the Geological Survey of Lithuania. Results of groundwater monitoring show that the quality of karst waters changes significantly with time (Table 2).

Table 2 Changes of salinity (mg/l) and total hardness (meq/l) of groundwaters in observation wells of karst region

Observation time	Well No. 1397		Well No. 1401		Well No. 1386	
	salinity	hardness	salinity	hardness	salinity	hardness
1979	485	5.11	1307	20.09	-	-
1980	424	4.63	1425	18.07	-	-
1981	410	4.97	1145	15.12	2105	24.76
1982	323	3.18	785	8.96	1549	20.14
1984	280	2.67	-	-	1221	18.17
1985	279	2.69	249	3.30	1089	13.44
1986	272	2.44	555	6.04	899	10.88
1987	260	2.50	468	4.99	859	10.55
1988	259	2.46	336	2.78	1088	19.08

The main reason for such changes is groundwater abstraction, which decreases groundwater pressure and causes the drawdown in karst aquifers. Fresh precipitation and surface water can easily infiltrate through the thin soil layer into the karst aquifer, change groundwater composition (make it fresher) and more karst rocks are being dissolved in it. Therefore, groundwater withdrawal should be regulated in the region.

According to the vulnerability of groundwater (thickness of sedimentary cover, density of sinkholes, character of recharge-discharge zones, pollution extent), the karstified area was divided into four land groups, each with a different extent of limitation of agricultural activity. A protection zone for the region was also defined.

A special program of environmentally sound human activity in the karst region was initiated by the Geological Survey and approved by the Government of Lithuania. The main idea of the program is to regulate groundwater abstraction, establish the system of organic agriculture (not using chemical fertilizers) and organize economical activity in a way which will lead to the mitigation of groundwater pollution and human impacted karst development.

Karst prevention, as described above, and groundwater protection actions have been implemented in Lithuania, but just across the artificial political border, in Latvia, at a distance of several kilometers in a densely karstified territory with the same geological-hydrogeological conditions, quarrying of gypsum deposits is being planned. For this purpose, it is necessary to lower the ground water level by 30 meters. The mining project was prepared, and quarrying was planned to start in 1991, but through the actions of the green movement, gypsum exploitation was stopped then. Now, however, at an economically difficult time for Latvia, the possibilities of development of its own natural resources are under discussion again. It is obvious even for non-geologists that such a big groundwater drawdown will create a groundwater depression cone not only in Latvia but also in Lithuania. This will lead to unpredictable consequences of karst development in both countries. That also means that environmental protection measures taken in Lithuania won't achieve any results.

The presented examples show that on the basis of results of geological and other environmental investigations, international or binational treaties, conventions or other political documents for prevention of transboundary pollution, creation of a system for water resources management and regulation of economic activity on near-border territories could and should be initiated. A comprehensive international system for regulation of human activity that could be transferred successfully from one region to another should be created.

Transboundary water management relations are well described in the Convention on the Protection and Use of Transboundary Watercourses and International Lakes signed in Helsinki on 17 March, 1992 between the Baltic Sea countries. The Convention says that the countries "shall cooperate in the conduct of research into and development of effective techniques for the prevention, control and reduction of transboundary impact. To this effect, the Parties shall, on a bilateral and/or multilateral basis, taking into account research activities pursued in relevant international forums, endeavor to initiate or intensify specific research programs". Further "the Parties shall establish programs for monitoring the conditions of transboundary waters".

The only thing that is needed in relations between relevant institutions of Lithuania and Latvia is to ratify and to follow the requirements of the Convention.

References

Baltic Sea environment proceedings, 1993, No 45, 165 pp.

Carlsson, B. 1992: Fresh-water runoff to the Baltic Sea from surrounding land areas (in Swedish with English abstract). Proceedings the Nordic Hydrological Conference in Alta, August, 1992, Norway.

Juodkazis, V. (ed.) 1992: Hydrogeological mapping and ground water monitoring in the Baltic States.Reports of Intern.Symp., Vilnius, Lithuania, 90 pp.

Kiruchin,V.A. 1993: Underground waters seen as one of the factors of environment in the Baltic Sea basin.Ecological hydrogeology of Baltic Area. International Scientific Seminar. Abstracts of Reports, 18-19.

Klimas, A.J. and Paukstys, B.P. 1990: Lithomonitoring organization in karst region of North Lithuania.Moscow, Engineering geology, 4, 75-80 [in Russian].

Mokrik, R. V. 1990: Hydrogeology of the coastal part of the Baltic States. Authoref. of doctoral thesis. Moscow, VSEGINGEO, 55 pp. [in Russian].

Paukstys, B. 1991: Impact of human activity on the development of gypsum karst in Northern Lithuania. Authoref.of doct. thes., Moscow, VSEGINGEO, 25 pp. [in Russian].

Voronov, A. N. and Kryza, J. 1993: The tasks of ecological hydrogeology in the sphere of study of groundwater contamination and the forecast of the pollution results. Ecological Hydrogeology of the Baltic Area .International Scientific Seminar.Abstracts of Reports.3-5.

The environmental problems of the Baltic Sea Region. 1991: Abstracts collection, Vilnius, 56pp.

20 Years of International Cooperation for the Baltic Marine Environment 1974-1994. Helsinki Commission, 1994, 40 pp.

II.3 The Projected Impact of the North American Free Trade Agreement on Transboundary Water Management between Mexico and the U.S.A.

Marie J. Gunning
Yale School of Forestry and Environmental Studies
New Haven, Connecticut 06511
USA

Introduction

Rapid economic and population growth along the US - Mexican border has had numerous environmental implications. Chief among these is the impact of this growth on the two countries' shared water resources. Today, both sides of the border face problems with respect to both water quality and water availability (Emerson, 1991). Not only do they represent a potential threat to the heath to border inhabitants, but they are also likely to be a major constraint to the area's future economic growth.

The magnitude of the border's water problems highlight the need for change in the institutional structures which have guided the management of the two countries' water resources for the past decades. As described below, the recently enacted North American Free Trade Agreement (NAFTA) may provide an important platform from which these much needed changes can begin to occur. By virtue of the new institutions, standards and enforcement mechanisms it presents, as well as the funding which will accompany them, this treaty represents exciting new opportunities for resource management in both countries. As such, NAFTA promises to be a milestone in the Unites States' and Mexico's transboundary water relationship.

In support of this conclusion, this paper presents three principle sections. First, is an overview of the present status of the two countries' shared water resources, with a discussion of the management issues each side currently faces. Following this is a brief review of the organizational and legal structures which, here to date, have played a major role in the management of those water resources. Finally, the paper explores the specific provisions of NAFTA which may influence those structures, and in doing so, shape the future of the border area's transboundary water management regime.

Regional Setting: A Brief Overview of the US - Mexican Border Area

The importance of improved transboundary water management is centered on the extensive binational frontier shared by the US and Mexico. Ranging 3,326 km in length (WFB, 1994), the "border area" is defined by treaty as consisting of a territory 100 km on either side of the two countries' inland and maritime boundaries. (La Paz Treaty, 1984). The US-Mexican border crosses six Mexican and four US states, while covering nearly a quarter million square miles of land (Border Plan, 1992). By historical agreement, the border coincides with the beds of the Rio Grande and the Colorado River for nearly half of its length. Its remaining sections transverse mostly high altitude, desert-like landscapes which receive approximately 10 inches or less per year (WFB, 1994). Freshwater resources shared by the two countries include two of North America's longest inland waterways, the Rio Grande/Rio Bravo and Colorado rivers, as well as numerous smaller rivers and tributaries (i.e., the New, the San Pedro, and the Tijuana rivers). In addition, several large, and numerous smaller aquifers are shared, as well. Until now, however, border area groundwater usage has generally been limited by the extremely poor quality of the water in many of those aquifers. Consequently, surface waterways, particularly the Rio Grande and its tributaries, serve as the border region's main sources of water (US Council Report, 1993).

Land Use and Population Changes

In the early part of this century, the primary economic activities in the border area were mining, agriculture and ranching. Because of the region's low average annual precipitation, the latter activities were able to maintain economic viability mainly through the intensive use of irrigation (Rudolph, 1988). However, by 1965 the economic mix along the border region began to change. At that time, the Mexican government initiated a national industrialization program. A central point of this program was the encouragement of more intensive economic development at the border, via the so called Maquiladora Development program.

The core concept of the maquiladora model is that businesses outside of Mexico can import both capital goods and raw materials, on a duty free basis, into a designated 20 km zone on the Mexican side of the border (Miller, 1986). Parts assembled in the border area with lower wage Mexican labor are then exported out of the country, at a reduced tariff rate applied solely against the value added component of the product generated in Mexico.

The maquiladora program has been extremely successful in attracting US and foreign capital investment into the Mexican border region. The number of maquiladora manufacturing and assembly plants has grown rapidly, particularly within the past ten years. By the early 1990s there were an estimated 2,000 maquiladora factories operating in Mexico (Vaznangh, 1992). At that time these businesses were producing over $4.1 billion in domestic value added and generating more than 8% of the country's total foreign exchange earnings (Hufbauer and Schott, 1992). More importantly, the program has created over a half a million new jobs along the border. As a result, the area now has one of the lowest unemployment rates in all of Mexico (US Council Report, 1993).

Not surprisingly, a concentration of low skill employment opportunities, (in the new maquiladora factories, as well as, in traditional agriculture), acts as a magnet by drawing labor migrants into the border area. As a result, in recent years the area's population has increased at a blistering pace, particularly within the seven sets of large "sister" cities which have developed along the border. Estimates prepared for the US's Integrated Border Environmental Plan indicate that, in just these seven areas, population grew at a rate of nearly 3% per year throughout all of the 1980s (Border Plan, 1992). If similar growth rates are maintained over the next decade, the projected population for these cities for the year 2000 is over 10.3 million. Further, since there are approximately 2 million border inhabitants living outside these seven urban areas, the actual population of the border region even today is close to 9.5 million. Thus, at prevailing rates of growth the border region's population could top 13 million by the end of this decade. (US Council Report, 1993).

US - MEXICAN BORDER REGION: POPULATION STATISTICS

METROPOLITAN AREA	1990 POP.	1980-90 COMPOUND ANNUAL GROWTH RATE	PROJECTED POP. BY 2003
Tijuana	747,381	4.94%	1,399,654
San Diego	2,498,016	2.98%	3,660,511
Mexicali	601,938	1.67%	746,129
Calexico	109,303	2.60%	152,636
Nogales (Mexico)	107,119	4.64%	193,106
Nogales (US)	29,676	2.20%	39,371
Ciudad Juarez	797,679	3.47%	1,242,171
El Paso	591,610	2.11%	776,574
Nuevo Laredo	219,468	0.77%	242,445
Laredo	133,239	2.99%	195,300
Reynosa	282,667	2.48%	388,500
McAllen	383,545	3.08%	568,848
Matamoros	303,392	2.42%	414,063
Brownsville	260,120	2.18%	344,151
TOTALS	7,065,153		10,363,461
AVERAGES	504,654	2.94%	740,247

Source: US Council of the Mexico-US Business Committee

Implications for the Border Area's Water Resources

Rapid industrialization along the border, and its associated population growth have had a serious effect on the supply and quality of the region's water. As a result, water issues are today widely considered to be the most critical border environmental problem (Williams, 1993). In particular, the pollution of the region's surface waters is a major concern.

Border water pollution has a number of different sources. The most acute of these is the region's lack of adequate sewage treatment, particularly within Mexico. Each day, millions of gallons of polluted water are released into surface waterways at the border. For example, the major Mexican city of Juarez has, for many years, had no sewer treatment system. Consequently, by late 1992, it was discharging over 22 million gallons of untreated wastewater each day into both local agricultural lands and the Rio Grande river (Exec. Office, 1993). Similarly, waste inflows have greatly outstripped treatment capacity in the growing city of Nuevo Laredo, Mexico. There, as much as 25 million gallons of raw sewage is dumped into the Rio Grande each day (Villa, Testimony 1991).

There are numerous other examples of either inadequate, or nonexistent wastewater treatment facilities along both sides of the border. The serious implications of this shortfall in treatment capacity are increasingly obvious. In fact, the governments of both countries have had to publicly acknowledged that the drinking water supplies in many communities along the border now pose a human health concern, due to the potential of hepatitis and other waterborne diseases (Border Plan, 1992).

Also, while scientific evidence remains unclear, there is great concern that the dumping of hazardous materials by nearby maquiladora businesses may be creating even more dangerous health problems along the border. For instance, scientists are continuing to investigate the possible connection between unusually high rates of a rare infant brain cancer (anencephaly) in certain border communities, and the dumping of industrial wastes (IER, 1993). As one independent group of US medical professionals group has reported, the accumulation of municipal and industrial wastes along the border interface has reduced many areas of the border to a "virtual cesspool" of pollution (NRDC, 1991).

As quality issues have grown, so too have concerns about adequate water availability. Although Mexico and the US have enjoyed a long history of relatively successful shared water use agreements, throughout much of this period area population and industrial densities were much lower. Accordingly, in the past water allocation, rather than water availability, was often the most important factor in determining issues such as land use development. Now, following a period of intensified development, maintaining adequate water supplies is a serious concern for many border communities.

For example, in the United States the Rio Grande provides the vast majority of the border region's water supply. At present its annual use is fully allocated; approximately 300,000 acre feet of its waters are designated for use by municipal and industrial users, and 2.8 million acre feet apportioned for use by agriculture (Marstan, 1991). However, during most years there is not enough water available to fully satisfy these agreed upon allocations. In fact, it is estimated that the total consumption of both the Rio Grande and Colorado Rivers is presently more than 40% of renewable supply (Szekely, 1993). With these water supply limitations, further development will almost certainly require the difficult task of shifting water use rights among the various classes of existing users. Restrictions on water availability may potentially pose stiff challenges to maintaining the region's economic vigor.

Historical Responsibility for Managing US - Mexican Transboundary Water Resources

The impact of these issues on living conditions at the border has placed great pressure on the organizations with responsibility to manage the region's transboundary water issues. Although there are a number of local, state and federal agencies within both countries involved in various aspects of border water use in the region, the organization which historically has had lead responsibility for transboundary water management is the International Boundary Water Commission, United States and Mexico ("IBWC"). The IBWC is the only binational commission with permanent institutional status established by treaty (Mumme, 1984). Despite this unique distinction, by pursuing a relatively narrow agenda of issues (Mumme, 1993), and forming policy in mostly closed door sessions the Commission for many years enjoyed a low public profile.

Today, however, with the manifestations of serious water problems so apparent, the IBWC has come under persistent, widespread public criticism (Ingram and White, 1993). The severity of pollution along the border, as well as the absence of an effective action plan with which to begin to manage the problem, has led many to conclude that the IBWC has not coped well with the region's rapidly shifting socio-economic condition. (see among others, Mumme, 1993; Szekely, 1993; Dworsky and Utton, 1993; Emerson 1991).

It appears that the IBWC's ineffectiveness with regard to the border's current transboundary water challenges is as much a function of its legal structure, as its management approach. With regard to the former, the structure and modern mandate of the IBWC can be traced back to the "Treaty Regarding Utilization of Waters of Colorado and Tijuana Rivers and of the Rio Grande", or more commonly, the 1944 Water Treaty. As its name suggests, the treaty is strongly oriented toward conflict resolution. The areas of appropriation and delivery of the two countries' transboundary waters are its primary focus (Water Treaty, 1944). The agreement not only establishes the basic organizational structure of the Commission (two separate national sections, each responsible to their respective governments), it also specifies its broad functional responsibilities. These include boundary demarcation, channel rectification, construction and maintenance of water storage, and sanitation and sewage facilities (Water Treaty, 1944).

By the 1970s it became clear that the issues affecting the two countries's transboundary water relationship had become more complex than envisioned in the original 1944 Treaty. As a result, in 1979 the two parties agreed to what is known as Minute* 261. Under this agreement, the IBWC was given specific authority to address transboundary water pollution problems, including sanitation and sewage problems, salinity and industrial and pesticide contamination (Mumme, 1984).

Despite this authority, the IBWC has been slow to respond to those issues. In fact, rather than taking the lead, some contend that the Commission has consistently chosen a very narrow interpretation of these formal powers (Mumme, 1984). It appears that one reason for this seemingly very conservative management approach has been the growing overlap of agencies (national, state and local) which are involved in various border water management issues. In particular, in the period when the border's water problems began to intensify, the two countries were establishing new, national environmental authorities (in the United States, the Environmental Protection Agency, or "EPA", in 1971; and in Mexico the Secretaria de Desarrollo Urbano y Ecologia ("SEDUE") several years later). Arguably, the creation of these new agencies resulted in a much more complex, and competitive, political environment for organizations such as the IBWC. Rather than risk competing for a leadership role with those other agencies, the IBWC attempted to insulate itself by maintaining a narrow technical focus (Mumme, 1993).

This issue of overlapping institutional jurisdictions was made even more apparent in 1984. At that time the countries entered into the "Agreement Between the United States of America and the United Mexican States On Cooperation for the Protection and Improvement of the Environment in the Border Area", also known as the La Paz Agreement. This agreement provides an explicit framework for the two countries to consult, and attempt to resolve transboundary environmental problems (Mumme, 1994). The lead roles in implementing this plan were given to each party's national level environmental agencies (the EPA and in Mexico, SEDESOL, the successor organization to SEDUE). This was done by designating each of those agencies as national coordinators of the agreement. The primary role of these coordinators is to manage and monitor the implementation of the agreement's agenda - which explicitly includes transboundary pollution (La Paz Treaty, 1984).

---

*An IBWC "minute" records the formal agreement of the two parties to a cooperative action (Water Treaty, 1944). Much of the implementing work of the 1944 Treaty is handled via the Commission's minutes.

Yet, Mexico and the United States also agreed in the La Paz Agreement that, "Nothing in this agreement shall prejudice or otherwise affect the functions entrusted to the (IBWC) under the Water Treaty of 1944. (La Paz Treaty, 1984).

A common criticism of the La Paz Agreement is that it lays out a number of general promises regarding cooperative action on the border environment, but does not offer specific mechanisms with which to implement these goals (CBO, 1993). More subtly, it can also be argued that it impedes improvement on border water issues by failing to specifically delineate IBWC's role in the execution of the La Paz agenda. Instead, the agreement may have increased the complexity of the political environment which the IBWC had to work within. It formally communicated that the responsibility for transboundary pollution issues resided with the countries' national environmental agencies, without addressing potential jurisdictional issues created by virtue of the 1944 Water Treaty's delineation of the IBWC's functional responsibilities.

Unfortunately, the issue of overlapping institutional authority was also not clarified in the two countries' next subsequent agreement on border environmental issues. That agreement, the Integrated Environmental Plan for the Mexican-US Border Area (the "Border Plan"), was entered into in 1992 . The explicit objective of the Border Plan was to strengthen the principles of the La Paz agreement. It was negotiated with the purpose of creating a comprehensive border management plan that included more targeted program goals, with funding available for their implementation (Border Plan, 1992). Particular emphasis was given to water management. For instance, the Plan proposed USD $731 million dollars in contract funding for just the EPA's Office of Water; USD$199.5 million dollars in US wastewater treatment and drinking water construction grants; and USD$ 60 million in Mexican sewer and wastewater treatment funding.

Yet, like the agreement it was designed to supplement, the Border Plan has been criticized for containing more promises of improvements than actual specific action plans, as well as for lacking enforcement authority with which to achieve these goals (Villa, 1991). The effectiveness of the Border Plan has been furthered hindered by the United States' failure to fully fund its monetary commitment under the plan (Pastor, 1994). Perhaps most importantly, as with the La Paz Agreement, the EPA and SEDESOL are delegated lead responsibility for its implementation. Although the executive summary of the first stage draft of the Plan makes mention of the long history of the IBWC, like the La Paz, the Plan itself does not specify the IBWC's specific responsibilities for advancing its goals.

The absence of a definite, articulated role for the IBWC within the La Paz agreement may reflect that there are no institutional analogs to the Commission for other transboundary resources (i.e., air and land pollution). As well as that the national coordinators, the EPA and SEDESOL, are themselves operating within much different legal, political and cultural settings.*

---

* One very important example is that the EPA has been established as an independent Federal agency, and works within a strong, common law regime. In contrast, SEDESOL, which was only given its environmental responsibilities in 1992, works within a civil law system.

As a result, it may be that these two organizations have not yet developed enough of a working rapport to tackle the difficult task of incorporating in the activities of a unique, bilateral commission such as the IBWC. Or, perhaps, it may be that the IBWC is, in fact, a "social artifact" as some have suggested (Mumme, 1993), and should not have a lead role in the Border Plan's agenda.

Whichever the reasons, the net result has been a lack of strong leadership on transboundary water management issues. By either design or default, in recent years the IBWC has responded cautiously to the area's changing water resource issues. And, at the same time, the countries' national environmental agencies have failed to commit the resources and the finances to the area's water problems.

The North American Free Trade Agreement

Into this void has been passed what some have called the "greenest" trade agreement in history (Reilly, 1992). Adopted into law on January 1, 1994, the North American Free Trade Agreement (NAFTA), may be one of the most significant multi-lateral trade treaties in history. By fostering the phase-out of trade tariffs and trade barriers, NAFTA links the economies of Canada, Mexico and the United States. The result is the creation of one of the world's largest markets, with over 365 million people and an estimated $6.5 trillion of market production (Gantz, 1993).

Its sheer economic magnitude, and the likelihood that the treaty will serve as a standard of measurement for future trade agreements, helped to generate a huge amount of public interest and debate throughout the NAFTA negotiations. This climate, in conjunction with the increasingly obvious manifestations of environmental problems along the border, allowed environmental groups to have an unprecedented level of involvement in the crafting of NAFTA (Esty, 1994).

The results of this involvement take several forms. Not only were a number of environmentally related trade provisions included in the treaty, but the governments of Mexico and the United States also agreed to a special supplemental agreement, called the "North American Agreement on Environmental Cooperation" ("NAAEC"), as well as the establishment of the Border Environmental Cooperation Agreement ("BECA"). Combined, these agreements represent a significant milestone in the history of natural resource management between the US and Mexico.

More specifically, the influences of these changes on transboundary water management can be broadly categorized into two areas: the creation of new organizations and mechanisms to respond to environmental disputes; and the introduction of explicit policies and standards which could benefit both new and existing organizations involved in water resource management.

New Organizations:

The Commission for Environmental Cooperation ("CEC")

With regard to the first of these categories, the NAAEC establishes a new, central organization for managing environmental cooperation. Called the Commission for Environmental Cooperation ("CEC"), its primary role is to create an atmosphere of mutual cooperation on environmental issues (NAFTA, NAAEC: Article 8). Although the actual activities to meet this rather broad mandate have not yet been specified, the Commission has the broad responsibility of promoting the objectives and policies of the NAAEC agreement. These objectives are outlined in the beginning of the environmental supplement and include: promoting "sustainable" development based upon cooperation and mutually supportive economic policies; increasing the cooperation between the parties to better conserve, protect, and enhance the environment; and promoting pollution prevention policies and practices (NAFTA, NAAEC: Article 1).

To complete these various tasks the Commission comprises three separate sections. These are: (1) its Council, (2) a Secretariat, and (3) a Joint Public Advisory Committee. The Council of the CEC represents a cabinet-level (top governmental) position. It is chaired by the respective heads of each of the countries' federal environmental protection agencies (or their equivalent). Importantly, the Council has been given a central role in the governance of environmental matters between the three countries. With specific regard to the Mexican-US border region, the CEC has direct responsibility for developing recommendations on transboundary and border environmental issues, such as long range transport of air and marine pollutants (NAFTA, NAAEC: Article 10).

Whereas the Council has a central policy making role, the Secretariat of the CEC has responsibility for many of the operational functions of the Commission. It handles reporting requirements, budgets, and requests on enforcement matters. Notably, this last area is likely to give the Secretariat important influence on the policies of the NAFTA and the NAACE dealing with enforcement. As currently established, the Secretariat will review all submissions from non-governmental organizations or individuals asserting that one of the parties has failed to meet the requirements of the agreement. Upon such review the Secretariat decides whether such submission ".. merits a response from the Party (so accused)" (NAFTA, NAACE: Article 14). Hence, as the effective "gatekeeper" on non-party complaints, the Secretariat is likely to have a significant role in translating the written policies of both NAFTA and NAACE into action.

The Joint Public Advisory Committee is the third section of the CEC. This group was created to provide advice to the Council on scientific and technical matters. Although not required under the NAAEC, the agreement also specifically allows each party to convene either national advisory and/or governmental committees (NAFTA, NAAEC: Articles 18 & 19). These would provide for the participation of non-governmental organizations, as well as federal, state or provincial governments in advising on the on-going implementation of the agreements.

New Policies, Procedures and Enforcement Mechanisms:

In addition to creating a centralized body responsible for environmental cooperation, the NAAEC environmental side agreement puts forth several new policies and procedures which may affect transboundary water resource management. The most significant of these is Article 5 within the NAAEC. This segment deals with enforcement of environmental regulations. For the first time in an international document, the parties have pledged to ensure that their respective environmental laws and regulations will be enforced through appropriate government action (Browner, 1993). Such action will include training inspectors, monitoring compliance, and publicly releasing non-compliance information. Record keeping and reporting are required, as well as, a system of licenses and permits (NAFTA, NAAEC: Article 5). This section of the agreement also stipulates that each party will have under their respective laws, judicial, quasi-judicial or administrative enforcement proceedings to remedy such violations (NAFTA, NAAEC: Article 7).

Also, Articles 22-36 of this agreement go on to specify the exact methods of response if a party has what is considered to be a persistent pattern of failure to enforce its environmental laws and regulations. In such a case the two parties would first have consultations on the matter. If not so resolved, either party could then request that a special session of the Council review the issue. Although the Council has been granted the authority to attempt to resolve such matters with such review, under its current charter the group is limited to making recommendations to the parties only (NAFTA, NAAEC: Article 23).

If these recommendations do not resolve the issue, the Council can then request that an arbitral panel be convened. Should this arbitral panel find that a persistent pattern of failure does in fact exist, the two parties in disagreement must agree on an action plan for resolving the failure. After such a plan has been established, the party complained against must implement this action plan, under threat of monetary penalty.

The length and sequencing of the procedures, and the use of the open-ended "persistent pattern" enforcement criteria suggest that such penalties might be rare. Nevertheless, the process spelled out by the NAAEC is potentially an important new mechanism for dealing with border water resource issues. This reflects the fact that it represents the first intergovernmental dispute resolution mechanism which links environmental enforcement with trade and other monetary benefits (Browner, 1993). Arguably, such linkage will be an effective motivator for both the US and Mexico to effectuate the water resource management goals outlined in bilateral agreements like the 1992 Border Plan.

Another change introduced by NAFTA and the NAAEC is the establishment of a number of general, bilateral environmental obligations. These include increasing scientific research and technology in environmental matters; completing ("where appropriate") environmental impact assessments; and the periodic preparation and publication of reports on the state of the environment (NAFTA, NAAEC: Article 2:1). Also, under NAFTA Mexico and the US have agreed to a broad set of new sanitary and phytosanitary standards for the products in which they trade (NAFTA, Chapter 7 and 9). These chapters begin to address the difficult issue of the two countries' divergent product and processes standards. While not directly linked to water, these new standards will likely, over time, affect the management of transboundary water problems. For instance, over time, the agricultural standards advanced may influence the character of non-point source pollution of pesticides. Given the present heavy concentrations of pesticides in agriculture run-off, and the problem they pose in the two countries' shared waterways, standards which direct producers towards the use of less toxic compounds would also obviously benefit water management.

Finally, also under NAFTA, the parties have specifically affirmed that they will not encourage capital investment in their respective countries by relaxing health, safety or environmental standards (NAFTA, Chapter 11:1114). The enforcement mechanism for this particular issue is quite limited (if violations take place, remedies are limited merely to consultations). Nevertheless, the inclusion of this point signals a recognition by the parties of the need to protect against the so-called "race to the bottom" on environmental standards (in which countries or regions compete for business investment by relaxing environmental requirements). By attempting to address environmentalists' concerns about this issue the parties have created what might be considered a "moral" commitment among themselves on general environmental standards.

Other NAFTA Related Effects on Transboundary Water Resources

In addition to the new organizations and policies flowing from NAFTA and the supplemental NAAEC, their passage has brought other, less direct developments to border water resource management. Most significant of these is, by far, the passage of the Border Environmental Cooperation Agreement ("BECA"). The agreement to enter into the BECA was ratified along with the passage of the NAFTA. Its purpose is to help finance the massive environmental improvements needed along the border, via two new organizations: the Border Environment Cooperation Commission (BECC), and the North American Development Bank (NADBank) (Exec. Office, 1993). The BECC's function is to provide technical and planning assistance for environmental infrastructure projects which have significant transboundary effects. It will not own or manage any projects, but will function essentially as a certifying agent for projects financed by the new NADBank (IER, 1994). Importantly, the US State Department has overall responsibility for US involvement in the BECC. This same agency also oversees the IBWC. Given this shared oversight and IBWC's proven expertise on technical matters, one possible scenario is that the IBWC would take a lead role in the project certification process, and provide general technical expertise to the CEC. This would leave the newly formed NAAEC (which is chaired by the heads of the EPA and SEDESOL), with the lead on managing the overall border clean-up effort, as well as addressing emerging transboundary issues, such as water availability.

The NADBank was established solely for the purpose of accessing money for border environmental and health related improvements. It is expected to provide over $2 billion in new project financing, both as a supplement to existing funding sources, and as an inducement to the participation of private capital (IER, 1994). Much of this project financing is expected to occur on the Mexican side of the border, where infrastructure needs are greatest, private financing options are often limited and tax exempt financing is not available (US Council, 1993).

The financing support which the BECA organizations hope to deliver will be further leveraged by the separate border clean-up funding package agreed to by the US and Mexican governments. Direct spending by the two governments on border environmental projects over the next several years is projected to total close to $1.5 billion dollars (Esty, 1994). Those funds will be in addition to a $1.8 million dollar loan from the World Bank Group to the Mexican government for the remediation of pollution problems along the border (IER, 1994). It appears clear that as a result of the scrutiny which the NAFTA negotiations generated, funding for border environmental improvement projects became a much higher political priority. Compared to the funding history of both the La Paz Agreement and the 1992 Border

Plan, NAFTA represents a tremendous new level of financial commitment by both governments. Thus, it may well be that the objectives of the La Paz and Border Plan agreements - namely the implementation of a comprehensive border management plan - will receive new emphasis in the near future.

Concluding Analysis

As a result of rapid population and industrial growth there are obvious and growing transboundary water problems along the Mexican-US border. These recent problems have not been adequately addressed by the institutions which have the mandate to do so. In particular, the IBWC, which for many years successfully managed the two countries' shared water resources, has been unable to provide much needed leadership in directing a response to these challenges. This failure can be traced to a complex array of factors. The most predominant of these has been an increasingly competitive operating environment, in which boundaries of jurisdictional authority have become increasingly blurred. This environment may have exacerbated the IBWC's already conservative interpretation of its powers. Also, despite official recognition of the border's water problems there has been a general lack of funding for needed water infrastructure projects. Although the two countries have designated their respective national level environmental agencies to take the lead on implementing recent bilateral border agreements, the breath of problems along the border and the very different legal and political structures of the EPA and SEDESOL has, to date, limited their effectiveness on transboundary water issues.

The newly enacted NAFTA, its environmental side agreement and the two new environmentally related organizations created at the time of its passage (the BEEC, and the NADBank), offer new, and very specific opportunities to address this lack of definition on institutional authority. By virtue of the attention which the treaty's passage has generated, as well as the financial resources the treaty has generated there is a greater likelihood that clearer, more effective lines of jurisdictional authority will in fact be established. From their position as heads of the new BECC, the EPA and SEDESOL have been given the clear opportunity to coordinate the activities of the various agencies acting at the border, and potentially, to help define a more active and effective role for the IBWC.

Yet, it is important to also note that it will be the implementation of these programs which will define the success of NAFTA and its side agreements. The passage of these agreements clearly represent an additional overlay of institutional responsibilities in the border arena. Absent strong leadership, NAFTA will not improve, and could worsen, the management of the border's water problems.

The NAFTA treaty also represents a significant increase in funding commitments for border infrastructure projects, particularly those involving wastewater treatment. This is an important, much needed change, and it should effect a marked improvement in water quality in many border areas. It must also be recognized, however, that based on recent estimates the total cost of the environmental clean-up at the border may range between $5 and $12 billion dollars (McMinn, 1993). Since this is obviously well beyond the scope of the funding allocated with the treaty's passage, it is clear that NAFTA is not a complete answer to the funding issues associated with the border's water problems. The participation of private industry in financing the border's water infrastructure needs, via either a production based tax and/or user fees, is seen as absolutely critical.

Finally, it should be noted that despite their laudable environmental objectives, the institutions, policy mechanisms and funding sources which NAFTA and its environmental side agreements introduce are all still primarily reactive in their design. In spite of the new ground which they break, the mechanisms they represent do not, per se, advance the type of proactive, more integrated ecosystem management approach which many deem as essential for successful future transboundary resource management (Dworsky and Utton, 1993). Because of the magnitude of the problems already present at the border, this may be unavoidable at the current time. Yet, it is important to note that the purpose of NAFTA is to facilitate additional economic growth. In order to make a <u>real</u> difference at the border, the new organizations and policies it creates need to incorporate future concerns into their management strategies.

Clearly, the role of NAFTA in responding to Mexico and the United States' transboundary water problems will depend heavily on the implementation of the new organizational structures and environmental policy mechanisms which it introduces. While probably not "the" answer for the border's complex array of water issues, by virtue of the breath of changes it introduces, NAFTA is certain to have a dramatic impact on transboundary water resource management between Mexico and the US.

References

Agreement Between the United States of America and the United Mexican States on Cooperation for the Protection and Improvement of the Environment in the Border Area, August 14, 1983 United States-Mexico, entered into force February 1984. (The "La Paz Agreement"). BNA publication 31:1401.

CBO (United States Congressional Budget Office), 1993. "A Budgetary and Economic Analysis of NAFTA". July 1993.

Dworsky, L. and Utton, A., 1993. "Assessing North America's Management of Its Transboundary Waters". Natural Resources Journal, Winter 1993, Volume 33.

Esty, D. 1994. "Making Trade and Environmental Policies Work Together:Lessons from NAFTA". Aussenwirtschaft, 49.Jahrgang (Spring 1994), Heft 1, Zurich.

Gantz, D. 1993. "Principle Feature of the North American Free Trade Agreement", in <u>Making Free Trade Work in the Americas</u> (B.Kozolchyk, ed., 1993) pp. 34-51.

Hufbauer, G. and Schott, J. 1992. <u>"Trade Effects of NAFTA: A survey", in North American Free Trade: Issues and Recommendations</u>. Institute for International Economics, Washington, D.C. 1992.

Ingram, H. and White, D., 1993. "International Boundary and Water Commission: An Institutional Mismatch for Resolving Transboundary Water Problems". Natural Resources Journal, Winter 1993, Volume 33.

International Environment Reporter ("IER"), Bureau of National Affairs. Washington D.C.
-April 7, 1993. "Texas Border Families Sue 88 Firms Alleging Toxics Caused Birth Defects".

-October 6, 1993. "Water Pollution Biggest Problem Facing Border Region, Former Official Says".

-April 18, 1994. "Implications of International Environmental Agreements Between U.S.- Mexico".

Miller, Robert, 1986. Mexico: A History. University of Oklahoma Press, Norman, Oklahoma. Third Edition.

Mumme, S., 1984. "Regional Power in National Diplomacy: The Case of the U.S. Section of the International Boundary and Water Commission". Publis, Fall 1984, Volume 14.

----- 1993. "Innovation and Reform in Transboundary Resource Management: A Critical Look at the International Boundary and Water Commission, United States and Mexico". Natural Resources Journal, Winter 1993, Volume 33.

----- 1994. "Enforcing International Environmental Agreements: Lessons from the U.S.-Mexico Border". Journal of Environment and Development, Winter 1994, Volume 3:1.

Natural Resources Defense Council (NRDC), 1991. Report on Environmental Protection within the North American Free Trade Agreement, as referenced in the Congressional Record on September 19,1992, during the testimony of William Reilly.

North American Free Trade Agreement, 1994. Final Version,including the supplemental agreements, as published by CCH, Incorporated.

Pastor, R., 1993. "NAFTA's Green Opportunity". Issues in Science and Technology, Summer 1993, Volume 9:47-54.

Rudolph, J. (Ed), 1985. Mexico: A Country Study. American University, Third edition, Washington, D.C..

Szekely, A., 1993. "Emerging Boundary Environmental Challenges and Institutional Issues: Mexico and the United States". Natural Resources Journal, Winter 1993, Volume 33.

Treaty Regarding Utilization of Waters of Colorado and Tijuana Rivers and of the Rio Grande., Mexico and the United States (The Water Treaty), 1944. As found in U.S. Statutes at Large, 79th Congress, 1st Session 1945. Vol. 59, Part2. U.S.. Government Printing Office, Washington, D.C.

United States Congressional Hearing on September 16, 1992, "Environmental Impact of the Proposed North American Free Trade Agreement" A Hearing before the Senate Committee on Finance. U.S. Confessional Report Serial No. 102-2/1039. Including the testimony of: William Reilly, U.S. Environmental Protection Agency.

United States Congressional Hearing on September 30, 1991, "Protecting the Environment in NAFTA Negotiations". A Hearing before the Sub-committee on Regulating Business Opportunities and Energy". U.S. Confessional Report Serial No. 102-144. Including the testimony of: Peter Emerson, Environmental Defense Fund; James Marstan, Environmental Defense Fund; Michael McCloskey, Sierra Club; Hector Villa, Texas Water Commission District 15 Manager.

United States Congressional Hearing on March 16, 1993. A Hearing before the U.S. Senate Committee on Environment and Public Works. U.S. Confessional Report Serial No.103-1/75. Including the testimony of Lawrence Williams, Sierra Club; Raymond McMinn, Asociacion de Maquiladores de Sonara.

United States Congressional Hearing on November 11, 1993. A Hearing before the House Committee on Merchant Marine and Fisheries. U.S. Congressional Report Serial No.103-1/329. Including the testimony of Carol Browner, U.S. Environmental Protection Agency.

United States Council of the Mexico-U.S. Business Committee, 1993. "Analysis of Environmental Infrastructure Requirements and Financing Gaps on the U.S./Mexican Border". July, 1993.

United States Environmental Protection Agency, 1992. "Environmental Plan for the Mexican-U.S. Border Area: First Stage (1992-1994)", (The Border Plan). Issued in February, 1992. Reproduced by the U.S. Department of Commerce National Technical Information Service, Springfield, Va.

United States, Executive Office of the President, "The North American Free Trade Agreement, Expanding U.S.Exports, Jobs and Growth. A Report on Environmental Issues". November, 1993.

Vaznaugh, R. 1993. "Extraterritorial Jurisdiction and NAFTA". Hastings International and Comparative Law Review, Fall 1993, Volume 17:207.

WFB - World Factbook 1994. Prepared by the Central Intelligence Agency, Washington D.C., U.S.A. 1994.

II.4 Transboundary Water Resources Mangement in the Upper Rio Grande Basin

Marvin Waterstone
Department of Geography and Regional Development
University of Arizona, Tucson, AZ 85721, USA

Introduction

The Rio Grande flows southward nearly 2,000 miles (about 3,000 km) from its headwaters in southern Colorado through the state of New Mexico and then on into Texas, forming the international boundary between the United States and the Republic of Mexico, and finally into the Gulf of Mexico. It drains an area of over 350,000 square miles (roughly 920,000 square km), with over half of this area in Mexico.

It is the lifeblood of this semi-arid region, supplying drinking water for more than 3.5 million people, as well as for agricultural, recreational, hydropower and industrial purposes. Population (and related water use) has been growing rapidly throughout the area, especially along the US/Mexico border, and is likely to continue well into the next century. In addition, much of this population growth has been, and is likely to continue to be, predominantly in urban areas. One implication of this trend is the shift from the historically dominant agricultural use of water to emerging needs related to urban development and manufacturing and to increasingly important environmental and recreational uses.

The first third of the river's length ends at Ft. Quitman, Texas, where by international agreement the upper basin is distinguished from the lower. From Ft. Quitman, the Rio Grande (or Rio Bravo, as it is known in Mexico) forms the international boundary between the two countries. The study that is reported here focuses on the Upper Basin, from the headwaters to Ft. Quitman, Texas, and therefore concentrates on the transboundary issues between the three US basin states of Colorado, New Mexico and Texas. However, some comments will be offered on the particular issues concerning the US/Mexico border portion of the Rio Grande basin as well.

Study Objective

The principal purpose for this research is to investigate the potential impacts of future, uncertain conditions (e.g., climate change) on the management and use of the Rio Grande. In a general sense, this is a study of physical/structural and institutional resilience. We are interested in determining the ability of the Rio Grande system to accommodate stresses induced by a variety of possible future circumstances.

This research is guided by several assumptions. The first is that the system has a significant degree of flexibility and resilience under the existing operating rules. This resilience will allow managers to handle additional burdens on the system, whether these arise simply from extrapolations of present conditions, or from new (and currently unforeseen) circumstances.

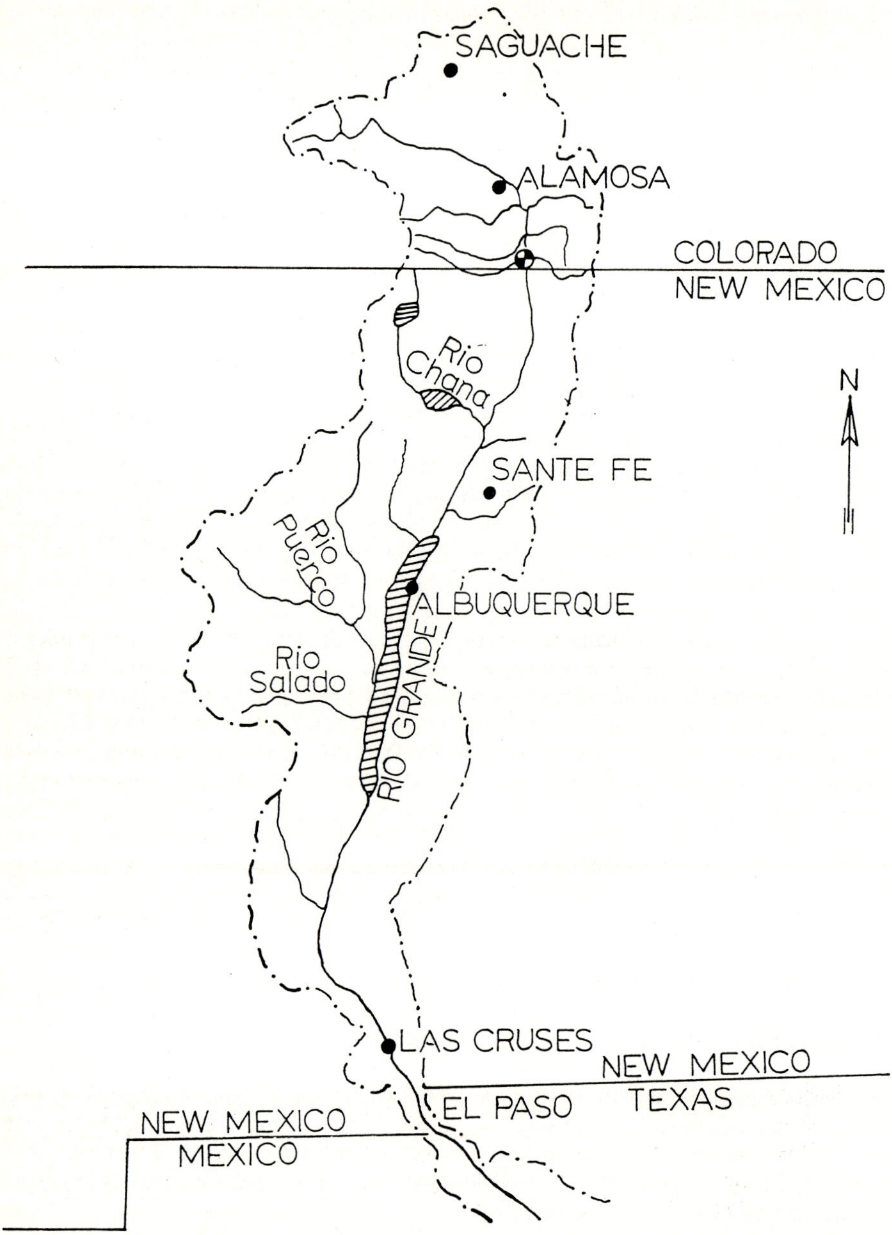

Figure 1 Upper Rio Grande basin.

However, a second assumption is that this resilience is finite and that at some point it will not be possible to accommodate additional stress on the system under the present operating rules. In other words, it is possible to conceive of future changes (whether they stem, for example, from additional population and water demand or from altered climate scenarios and altered water supplies) that will be beyond the ability of managers to cope, while still complying with all of the operating requirements presently in place. This, in turn, leads to the final assumption that if and when such conditions arise, the rules (i.e., the institution that governs the Rio Grande system) will have to be changed at levels above the operational level in order to accommodate the new demands on the Rio Grande. The procedures for assessing these assumptions, and their implications, are described in a later section of the paper.

The Physical System of the Upper Rio Grande

At present, the physical system of the upper Rio Grande basin comprises a series of eight major linked reservoirs operated principally by the US Army Corps of Engineers (USACE, Department of Defense) and the US Bureau of Reclamation (USBR, Department of Interior). The reservoirs are managed to fulfill a variety of purposes including flood control, municipal and industrial water supply, irrigation supply, hydropower production, sediment control, environmental enhancement for water quality and wildlife (i.e., through the maintenance of minimum streamflows) and recreation.

As with any set of multi-purpose reservoirs, these are managed to strike an appropriate balance among varying, and sometimes competing, uses. For example, in this semi-arid region, a continual concern is to maximize water storage for irrigation supplies. However, there is also a continual need to preserve space in these reservoirs for accommodating flood waters and preventing downstream flood losses. Attaining these multiple objectives is also complicated, in this case, by the fact that the reservoirs are also managed in a multi-state, multi-jurisdictional, transboundary context.

In order to accomplish these various purposes and to satisfy the transboundary conditions, the reservoirs are managed on a day-to-day basis under a series of complex rules. In terms of the framework described in Part I of this paper, these rules constitute the operational level of the water management institution for the Rio Grande basin. These rules, in turn, are nested within (and are compliant with) a set of implementation level rules, and a set of policy level rules.

The Institutional System of the Upper Rio Grande--Overview

It is now possible to characterize, in a general way, the water management system of the Rio Grande. This characterization enables us to understand how operation level rules come into being, how they articulate with implementation-level and policy-level rules and how they might be changed if they become deficient. The detailed description of the institution is presented in the following section.

Policy-Level Rules -- At this highest level (at least of those included in the present analysis) of the institution are several overarching sets of formal requirements which mandate how implementation-level (and eventually operational-level) activities should operate. (Of course these rules are themselves formulated under yet other rules, at what some call the constitutional level of the institution, which includes the rules under which the US Congress and Executive Branch agencies operate.) One of these policy-level rules includes the set of water delivery requirements mandated under the Mexican Water Treaties of 1906 and 1944 (which, among other purposes, allocate the water supply of the Rio Grande between the US and Mexico). Another is the Rio Grande Compact of 1938, which allocates water among the three basin states (Colorado, New Mexico and Texas), and is the primary mechanism for dealing with interstate, transboundary water problems (by specifying the ways in which the reservoirs must be operated to assure equitable distribution of the flow of the Rio Grande among the basin states).

In addition, at the policy level are rules specified by the Reclamation Act of 1902 (which govern irrigation and other water uses in the reservoirs operated by USBR), the multiple US statutes which authorized the construction and operation of the reservoirs operated by the USACE (principally for flood control and navigation purposes), the Appropriation Doctrine of "first in time, first in right" (which determines the priority dates and quantities that individual users are allowed to withdraw and use from the Rio Grande or its tributaries), the system of Federal Reserved Water Rights for Indian tribes (and other purposes), various environmental and water quality laws and the water laws of the three basin states themselves. Finally, there are a series of informal rules which attempt to accommodate the traditional water-using activities of the Rio Grande basin, but which are not codified into formal statute.

These policy-level rules specify the implementation-level actors, delineate the scope of their activities (including allowable and proscribed actions) and impose responsibilities and confer authorities upon them. In other words, these policy level rules indicate the kinds of goals and objectives to be achieved through water management, as well as those who will be responsible for devising the means to meet these goals. In general, specification of these means is left to the implementation level actors of the institution.

For issues internal to the US, the actors at the policy level are primarily members of either federal or state legislatures (i.e., either the US Congress or the legislatures of the basin states). Conflicts at this level are generally resolved through the political process (in which, it should be noted, ordinary citizens have actual positions and authorities) or through the courts.

For international issues, additional actors will include the heads of state of the US and Mexico, the US Department of State and its Mexican counterpart, the US Environmental Protection Agency and the Mexican environmental agency (SEDOSOL), as well as the state-level agencies in the US and Mexico. The rules are embodied in Treaties and other international instruments, and conflicts are generally resolved through diplomatic procedures.

Implementation-Level Rules -- These are the rules which specify precisely how, and under what conditions the reservoirs (and other facilities) are to be operated. The actors at this level, who are mostly members of executive branch agencies (principally the USACE, the USBR, the US EPA, the basin states' water agencies, and for international issues the International Boundary and Waters Commission), as well as the major water users, working within the rules from the policy level, develop the strategies and mechanisms (i.e., the operational-level rules) for day-to-day operation of the structural and non-structural features of the Rio Grande system. (It should be stated that many actors may hold positions at all three institutional levels, but will have different authorities and scopes of activity at each level.) Conflicts at this level are often worked out through inter-agency bargaining and agreements, memoranda of understanding, and the like. Informal rules also operate at this level to accommodate implementation level actors' expertise and judgments. These informal rules also provide an added measure of flexibility and resilience to the system. However, unlike the formal rules, these latter algorithms tend to be associated with individuals' expertise and experience, and are therefore somewhat ephemeral.

Operation-Level Rules--On a day-to-day basis, the features of the Rio Grande are thus operated according to this diverse set of formal and informal rules from the policy and implementation levels. The policy-level rules tend to be goal- and objective-setting, while the implementation-level rules tend to be mechanism-specifying. These rules include the Standard Operating Procedures (SOPs) for the reservoirs under various conditions. These SOPs determine the gate settings, and in turn, the flow of the river on a daily (or sometimes, hourly) basis. The operational-level rules also include the system of water rights for users, water quality permits and many others. Taken together, these rules determine who may use water and in what ways.

The Institutional System of the Upper Rio Grande--Detailed Analysis

In this section, I want to move from a general description of the institutional framework in the Upper Rio Grande to a detailed examination of the rules in place. At this point, the assessment focuses primarily on the formal rules and attempts to draw out their implications in terms of the study's goal of examining flexibility and resilience in the Rio Grande water management system.

Policy Level Rules (formal) -- As stated, these rules designate the implementation level actors, indicate their goals and objectives, and specify their authorities and scope of activities. These rules also identify the mechanisms for changing the implementation-level rules. The principal policy-level rules are enumerated hereafter:

1848-Treaty of Guadalupe Hidalgo: established the present US/Mexico border, and gave rise to the first border commission.

1902-Reclamation Act: organic act for authorizing and constructing Federal reclamation (irrigation) projects; sets goals for controlling reservoir levels throughout the region; regulates the use of water along the river; controls the ability to shift water to other uses (e.g., prohibits use of reclamation water for non-irrigation purposes); requires reclamation water to be used only in the confines of specified districts; specifies the maximum number of acres any one person may irrigate); it is administered by USBR.

1905-Elephant Butte dam and reservoir authorized, completed 1916; original purpose was irrigation supply for New Mexico and Texas (now includes hydropower); operated by the USBR, with water distributed (except for treaty-required deliveries to Mexico) by the Elephant Butte Irrigation District (EBID) and the El Paso County (Texas) Water Improvement District No. 1.

1906-US Mexico Treaty: requires the states of the Rio Grande basin above Ft. Quitman, Texas to deliver 60,000 af / year to Mexico and is administered by the International Boundary and

Water Commission. This is a fixed amount and must be delivered each year as a first priority of the water management system.

1925-Middle Rio Grande Conservancy District (MRGCD) formed.

1935 - El Vado dam and reservoir completed on the Rio Chama (tributary to the Rio Grande); owned by MRGCD, but operation turned over to USBR in 1956; purpose was to provide irrigation supply for MRGCD, and by agreement with the Department of Interior, to six pueblo Indian tribes in the region; is a "post-Rio Grande compact" reservoir, and has restrictions on the release of "native water" (as differentiated from water diverted from the Colorado River basin through the San Juan Chama diversion project, described below) when New Mexico is in debit status under the Rio Grande Compact; also stores San Juan Chama water (described below), which can be released for beneficial use (this water is not governed by the Rio Grande Compact).

1939-Rio Grande Compact signed by states and congress (effective May 31, 1939): federal law that supersedes state laws; imposes delivery requirements on Colorado to New Mexico and on New Mexico to Texas; amount of water to be delivered in any year varies depending on snowfall and consequent streamflow (i. e., deliveries to downstream states are not fixed, but are indexed to annual water supply conditions); administered by the Rio Grande Compact Commission (made up of representatives of the three basin states and a presidential appointee representing the US government. This is the most important priority (after satisfying the treaty requirements with Mexico) for the water managers of the Rio Grande system.

1940-Platoro reservoir authorized, completed in 1951; purposes were irrigation supply (for agriculture on the Conejos river in Colorado) and flood control; operated by the USBR in consultation with the State of Colorado for water rights issues and the U. S. Army Corps of Engineers (USACE) for flood control matters; it is a "post-Rio Grande compact" reservoir and has restrictions on the release of native water when Colorado is in debit status under the Rio Grande Compact.

1944 -- US-Mexico Treaty: divides the Rio Grande/Rio Bravo below Ft. Quitman, Texas between the US and Mexico; administered by the International Boundary and Water Commission (IBWC).

1948 -- Jemez Canyon Dam and reservoir authorized, completed 1954; operated by the USACE for flood control and sediment control. 1950-Abiquiu Dam and reservoir authorized, completed in 1963; operated by the USACE originally only for flood and sediment control; in 1981 was authorized to store 200,000 af of San Juan Chama (SJC) water (described below), and in 1988 was further authorized to store Rio Grande water in the SJC space. 1960-Cochiti reservoir authorized, completed in 1975; operated by the USACE originally for flood and sediment control; in 1964 was authorized for recreation, fish and wildlife purposes. 1962-Heron reservoir authorized, completed in 1971; operated by USBR for the sole purpose of storing water from the San Juan basin.

1963 -- San Juan-Chama (SJC) Project authorized to import San Juan River water (a tributary of the Colorado River) into the Rio Grande system via the Rio Chama, completed in 1971; currently about 110,000 acre feet/year (with a maximum possible diversion of 270,000 acre feet in any year (an acre foot is approximately 250,000 gallons); water can be used for municipal, irrigation, domestic and industrial uses, as well as for recreation and fish and wildlife; it is operated by the Department of Interior; water is contracted by the City of

Albuquerque, New Mexico (48,200 af), MRGCD (20,900 af), Santa Fe (New Mexico) Metropolitan Water Board (5,605 af) and several other entities using about 1,000 acre-feet/year or less.

1964 -- Recreation pool authorized for Cochiti lake.

1968 --National Wild and Scenic Rivers Act; has implications for minimum flow requirements and other water management-related actions.

1972 -- Closed Basin Project in Colorado authorized; project consists of 170 salvage wells which pump groundwater from an unconfined aquifer and discharge it into the Rio Grande; purpose is to help Colorado meet its Rio Grande Compact obligations and to help all 3 basin states meet their delivery obligations to Mexico under the 1906 treaty.

1973 -- Minute 242 (of the Rio Grande Compact), "Permanent and Definitive Solution to the International Problem of the Salinity of the Colorado River" adopted and gives the IBWC some limited jurisdiction over international groundwater and water quality issues.

1974 -- recreation pool of 50,000 af authorized for Elephant Butte reservoir.

1979 -- Minute 261 (of the Rio Grande Compact) "Recommendations for the Solution of Border Sanitation Problems" adopted by the IBWC, expanding its authority in areas of water quality.

1981 -- 200,000 af of SJC water storage authorized at Abiquiu reservoir.

1988 -- 200,000 af of Rio Grande water storage authorized at Abiquiu reservoir.

Implementation Level Rules (formal) -- Based on the foregoing policy-level rules, which set a variety of goals and objectives but do not specify how these should be met; the rules at the implementation level designate the operational level actors, provide their goals and objectives and specify their authorities and scope of activity. These rules also identify the mechanisms for changing the operational-level rules. For this assessment, I will only focus on these rules for reservoir operations.

Heron Reservoir -- The operation of this reservoir is guided by two major principles: 1) irrigators relying on SJC water create temporary, additional depletions from the Rio Grande, and these must be made up quickly by releases of SJC water from Heron reservoir so that "no residual effect is felt in the natural waters of the Rio Grande; and 2) no carryover storage of contracted SJC water is permitted from one diversion season to another in Heron reservoir. Therefore, contractors release water from Heron reservoir and seek storage in downstream reservoirs; this has resulted in very heavy releases from Heron reservoir in the fall and winter months. Heron reservoir is operated in compliance with the project authorization, the Reclamation Act and the Rio Grande Compact. Because there is no storage provision for native water, all natural flow is passed through Heron reservoir.

El Vado Reservoir -- Because the reservoir was constructed after 1929 it is considered a "post-compact" reservoir, and therefore is operated under the provisions of the Rio Grande Compact. However, these conditions apply only to the native water stored in El Vado, and not to any SJC water which may be in storage. The principal operating goal for El Vado is to meet the annual irrigation needs of the MRGCD; water is stored during the spring runoff season, and then released to meet needs during the irrigation season. However, these operations are constrained by Rio Grande Compact rules: 1) no storage may be held in post-compact reservoirs upstream of Elephant Butte reservoir when the amount of native water (exclusive of "credit water" and "imported water" which are defined by the Rio Grande Compact) in both Elephant Butte and Caballo reservoirs falls below 400,000 acre-feet; 2) any native water (i.e., non-SJC water) stored in El Vado must be held in storage to the extent of any debit accrued by New Mexico under the Rio Grande Compact. The reservoir must also be operated to meet the water needs of six Indian pueblos and must release water when the natural flow of the Rio Grande is insufficient to supply water for irrigating the combined 8,847 acres of these indigenous people. This determination is made by the Bureau of Indian Affairs and the USBR. Finally, El Vado must be operated to meet the irrigation needs of the downstream users on the Rio Chama between El Vado and Abiquiu dam.

Abiquiu Reservoir -- Abiquiu dam is operated principally for flood and sediment control (in accordance with the Flood Control Act of 1960, PL 86-645), and operations for these purposes are coordinated with the reservoirs at Jemez Canyon and Cochiti Lake. In this regard, Abiquiu dam is operated to limit flow in the Rio Chama to downstream channel capacity. However, irrigation releases from El Vado reservoir are also passed through Abiquiu, and so releases are contingent upon flow conditions on the mainstem of the Rio Grande as well. Generally, water is stored in Abiquiu during April and May and released during June and early July. However, after July 1, if flow on the mainstem drops below 1,500 cubic feet/second (cfs), any storage remaining in Abiquiu is carried over until November to prevent irrigators from diverting water illegally.

Platoro Reservoir -- Platoro reservoir is operated principally for irrigation storage and flood control. The reservoir is operated in accordance with procedures laid out in its authorization and the follow-up regulations. These flood control operations are coordinated between the Division of Water Resources (State of Colorado), the USACE and the USBR. The State of Colorado has responsibility for administering the water rights on the Conejos river (a headwaters tributary of the Rio Grande); the USACE has responsibility for determining the flood control operations at Platoro, and the USBR has the responsibility for the operation and maintenance of Platoro. Because Platoro is a "post-compact" reservoir, it must store any native water when Colorado is in debit status under the Rio Grande Compact. This has been the case many times since the reservoir was completed, and therefore it has been of limited utility for irrigation purposes.

Jemez Canyon Dam and Reservoir -- This dam, on the Jemez River (a tributary of the Rio Grande) is operated by the USACE for flood and sediment control. Normally, Jemez River water is passed through the reservoir with little, if any, regulation. Flood flows are retained briefly and then released as rapidly as downstream conditions permit. Flood regulation at Jemez is coordinated with operations and conditions in Cochiti Lake. The sediment pool is maintained by capturing Jemez River inflow, and this amount is made up to the rest of the system by releases of SJC water elsewhere in the system by exchange.

Cochiti Lake Dam and Reservoir -- This is the only flood control reservoir for snowmelt runoff on the mainstem of the Rio Grande. It is a multi-purpose reservoir and is operated by the USACE to: 1) regulate damaging floods; 2) retain sediment and 3) provide recreation and fishery opportunities with SJC water. The flood control operations are intended to protect the middle valley of New Mexico, and the dam is operated to control flows between Cochiti and Elephant Butte reservoir to maximum channel capacity. Snowmelt runoff is stored in April and

May and released in June and early July. However, under PL 86-645 (which contains the operating criteria for all of the Middle Rio Grande Project flood control reservoirs--Jemez Canyon, Abiquiu and Cochiti) if during the months of July, August, September and October there is more than 212,000 acre-feet of storage available for regulation of summer floods, and if the inflow into Cochiti is less than 1500 cfs, no water will be withdrawn from storage in Cochiti reservoir, and the inflow from upstream flood-control storage will be retained in Cochiti until after November 1. Again, this is to prevent irrigators from withdrawing water illegally during the irrigation season.

Elephant Butte Reservoir -- The primary purpose for this reservoir is to store all inflow and release only Project demands for irrigation. Since 1985, a reserve capacity of 50,000 acre-feet in summer and 25,000 acre-feet in winter has been maintained to store flood inflow between Cochiti and Elephant Butte. Other releases are made for power production through the Elephant Butte hydropower facility.

Caballo Reservoir -- This reservoir principally provides a re-regulation function for the hydropower releases at Elephant Butte. It is located only 25 miles downstream of Elephant Butte, and the connecting channel has a design capacity of 5,000 cfs. The flood control operations at Caballo are directed by the IBWC, while the USBR operates the dam generally and controls releases for irrigation demand. The water discharged from Elephant Butte for power operations during winter is stored at Caballo for release for irrigation use during the next summer.

Operational Level Rules (formal) -- It should be clear that the foregoing sets of rules at the policy and implementation levels give rise to the actual requirements for day-to-day operation of the reservoirs themselves. The rules designate the reservoir operators and specify their activities (e.g., gate settings, etc.). It should also be clear that if these operational-level rules should prove deficient for any reason, it is to these other levels that one must turn to develop new rules at the operational level.

Summary -- I want to stress that the institution (i.e., the sets of rules) just elaborated is by no means comprehensive. Various environmental quality, fish and wildlife, and other federal rules come into play, as well as the rules embodied in state and local regulation of water rights and water use. Some of these rules appear at the policy level, some at the implementation level, but all influence the operational-level rules, and the actual functioning of the system. Finally, all of these water-related rules are embedded within a larger set of rules governing other (although inter-connected) spheres of social and political life. It is possible, however, as has hopefully just been demonstrated, to extract a portion of these rules for critical analysis. In the next (final) section, I examine how this is being done.

Bringing it all Together: The Simulation Model

In order to attain the goals of the research, we have constructed and calibrated a simulation model (using object-oriented simulation software) of the Upper Rio Grande basin, which includes both the physical components of the system and the institutional (i.e., rules at the various levels) parameters just described. In this work we articulate these rules in such a way that they can be included in the simulation model of the Rio Grande system in the form of decision-tree structures which determine particular actions under specified conditions. In this way, as we begin to project future changes in the Rio Grande basin, the model is already equipped to assess situations which will require changes in the operational or implementation level rules in order to allow additional, unanticipated stress to be accommodated. While the institutional side is in many ways much more complex than the physical system, we believe that this analytic framework makes it manageable.

Most of the work, to date, has involved building the simulation model, collecting the necessary data on both the physical and institutional parameters and calibrating the model (described below). Forecast simulations (which are just commencing at the time of this writing) will begin with base case scenarios, which project current supply and demand conditions (including their underlying causal parameters) and seek to identify situations of water over- and under-supply in the future. Analyses of these projections will also seek to determine if these plausible deviations from past and present conditions can be accommodated under the current operating rules of the Rio Grande system, and if so, through what kinds of alterations. If the magnitude of these changes exceeds current operational flexibility, we will analyze what changes would be necessary at the operational level and how such changes could be achieved given the requirements of the rules at the policy and implementation levels. Finally, simulations including changes in supply and/or demand attributable to future perturbations (e.g., climatic changes) will be constructed to identify the incremental contribution of such altered conditions. Again, the ability of the system to accommodate such stresses will be analyzed, and the necessary changes evaluated. The general framework for the analysis (using climate change as an illustrative forcing function) is shown in Fig. 2. The major components of the system (physical and institutional) and their integration are presented in the next sections.

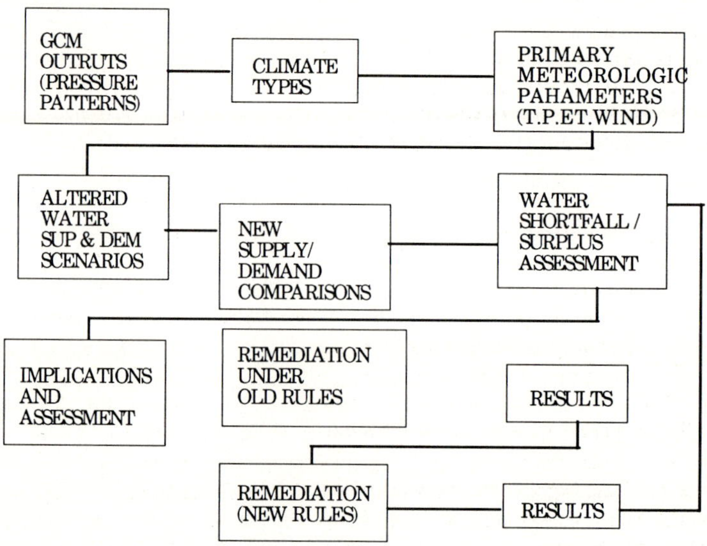

Figure 2 Analytical framework.

Simulating the Physical System -- In this part of the model, we have simulated the operations of the eight major reservoirs of the Upper Rio Grande basin (see Table I) and the linkages among them, in terms of storage capacity, inflows and outflows, evapotranspiration and channel losses, channel capacity and water demand. The simulation model is built up from the operational-level rules (e. g., for reservoir releases) which govern the day-to-day activities of the reservoir operators. Embedded within these rules (either explicitly or implicitly) are the multiple purposes for which the reservoirs are operated and the sets of priorities among these purposes (which actually play out the implementation- and policy-level rule specifications). The model also includes historical water demands of all major water users in the upper basin for the period 1970-1989. The model's performance was calibrated by "back-casting" these historical data.

Table I: Major Reservoirs in the Upper Rio Grande Basin.

RIO CHAMA
HERON RESERVOIR (CONSERVATION)
EL VADO (CONSERVATION)
ABIQUIU DAM AND RESERVOIR (FLOOD CONTROL, CONSERVATION, SEDIMENT)

CONEJOS RIVER
PLATORO RESERVOIR (FLOOD CONTROL, CONSERVATION)

JEMEZ RIVER
JEMEZ CANYON DAM AND RESERVOIR (FLOOD CONTROL, SEDIMENT)

RIO GRANDE
COCHITI LAKE (FLOOD CONTROL, RECREATION, SEDIMENT)
ELEPHANT BUTTE DAM & RESERVOIR
(CONSERVATION, RECREATION)
CABALLO RESERVOIR (FLOOD CONTROL, CONSERVATION)

Simulating the Institutional System -- We are interested in exploring both the resilience of the system and the consequences of exceeding this resilience for the many users of the system (see Table II). In this latter regard, we are therefore concerned to identify the ways in which the operating rules for the Upper Rio Grande basin can be modified if future conditions (whether driven by physical forcing functions such as climate change, or by socio-economic forcing functions such as population change, or both) make these rules ineffective. Clearly, this means incorporating the higher level (i.e., the implementation- and policy-level) rules into the simulation model. These higher-level rules operate variously as constraints or objectives in the simulation model. When new conditions force changes in operational-level rules, these changes are frequently constrained by the requirements of implementation-level rules (which are themselves frequently constrained by the policy-level rules). In many cases, however, it is possible (by appealing to the higher levels of the institution) to alter the nature of the constraints in such a way that the operational-level rules can be changed to meet new conditions. We are currently working on this task.

Table II: Entities in the Upper Rio Grande Basin.

STATES	MAJOR CITIES	OTHER ENTITIES
COLORADO	ALAMOSA	CONSERVANCY AND
NEW MEXICO	LOS ALAMOS	IRRIGATION DISTRICTS
TEXAS	SANTA FE	NDIAN PUEBLOS
	ALBUQUERQUE	ACEQUIAS
	SOCORRO	RECREATION
	LAS CRUCES	INTERESTS
	EL PASO	STATE AGENCIES
	CIUDAD JUAREZ	FEDERAL AGENCIES
		REGIONAL AGENCIES

Using this rules-based approach for building the physical component of the model has allowed us to identify periods in the record when the system has not been operated according to the formal rules and to investigate (through extensive interviews with the USACE and USBR system managers) why operations have diverged periodically from the formal requirements. This has led to some interesting findings regarding the ways in which the system's built-in flexibility has been utilized to accommodate new demands and uses.

As just one example, we identified higher than normal summer-season releases from one of the reservoirs beginning in about 1985. It turned out that the USBR operators had determined that it would be possible, under certain well-specified conditions, to release water from this reservoir in high quantity over short time periods and to allow for a new demand: white water recreation. It is quite clear that while these releases do not follow the formal rules governing the operation of the system, the managers are still clearly engaged in rules-governed behavior. Our approach allows us to identify such deviations from normal operating procedures. Where these deviations are systematic, we are now attempting to capture the rules-governed behavior of the operators as an expert system and to incorporate these informal rules into the institutional component of the simulation model.

Conclusion

While the application identified here is not yet complete, it does highlight the utility of defining institutions as rules which can be analyzed in consistent and comparative ways. By including these rules along with the specifications of the physical system in the simulation model, we have the potential to analyze not only the current operational status of the Upper Rio Grande basin, but also the mechanisms by which operations can be changed as new and problematic conditions arise in the future.

II.5 A Watershed and Ecosystem Approach for Transboundary Wastewater Management in the Tijuana River Catchment

Christopher Brown
Departments of Geography, San Diego State University
San Diego CA 92182-0381 and
The University of California, Santa Barbara
Santa Barbara, CA 93106, USA

Introduction to the Region and Problem

The Tijuana River Watershed (TRW) lies in the westernmost section of the United States (US)-Mexico border, encompassing portions of the City and County of San Diego in the United States and the Municipio de Tijuana in Mexico. The drainage basin of the Tijuana River is 1,735 square miles in size and contains four main tributaries, Cottonwood Creek, Tecate Creek, Rio de las Palmas, and Rio de Alamar; approximately 30% of the basin lies in the US and 70% lies in the Republic of Mexico. (Fig. 1 provides a locational map of the region, (Pryde, 1986)).

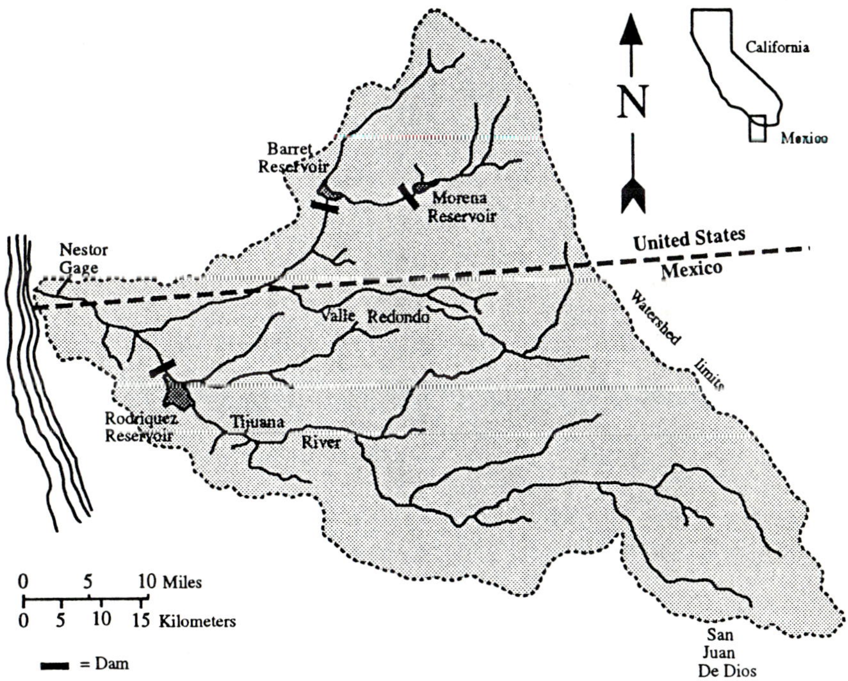

Figure 1 Map of the Tijuana River Watershed depicting local dams and surface water reservoirs. Compiled by David Van Mowerick from original in Pryde, 1986 and Zedler et al., 1992.

As such, the watershed is binational in nature, providing numerous challenges for policy development and implementation in the management of water resources; adequate management of wastewater flows is a particularly difficult challenge the region faces. Three dams and reservoirs lie within the watershed, providing varying degrees of flood control and local water supply for Tijuana and San Diego. La Presa Rodriguez and Reservoir lie on the Rio de las Palmas in Mexico and provide Tijuana a measure of local water storage and flood control; the Tijuana River Flood Control Channel in the urban area of Tijuana aids in this flood control. Barrett and Morena Dams and Reservoirs lie along portions of Cottonwood Creek in the US, providing a limited degree of flood control and water supply, although problems with infrastructure in the US limit Barrett's ability to provide local water supply and require large amounts of fresh water to be spilled during the rainy winter season. These dams and reservoirs also add complexity to the hydrologic regime in the basin; attempts to provide flood control and water supply upstream interact to complicate each objective and also impact water quality in the lower basin.

Of particular interest in the hydrology of the watershed is the Tijuana River Estuary, an area of approximately 2,900 acres in the lowest reaches of the basin. This estuarine ecosystem exists due to freshwater drainage in the river basin mixing with the normal tidal flows and pulses of the Pacific Ocean; a rich diversity of floral and faunal life contributes to the natural diversity of the area (Zedler, Nordby, and Kus, 1992). In 1982, the Tijuana River National Estuarine Research Reserve (TRNERR) was established to preserve the ecosystem; with this came a rich opportunity for preservation, as well as significant management challenges owing to the proximity of large urban areas upstream. The region has experienced extensive industrialization, economic development and population growth in the post World War II period (Huntley, et al; 1993), causing numerous impacts on the quality of the physical environment and human health.

One of the most intractable water resource challenges that has long faced the San Diego-Tijuana region is the management, treatment, and ultimate disposal of municipal wastewater within the lower reaches of the Tijuana River. These wastewaters originate in the Municipio de Tijuana, cross the international border between the United States (US) and Mexico and generate a myriad of negative impacts on human and environmental health that effect residents of both Mexico and the US (Meyer, 1983). At various times, the existing infrastructure within the US and Mexico is unable to handle these flows, causing in-stream releases of raw and treated sewage into the Tijuana River upstream from the border in Mexico. In addition, renegade flows enter the US through canyons at various points along the border (International Boundary and Water Commission (IBWC), 1993; Herzog, 1990), as indicated in Fig. 2.

This paper argues that the wastewater challenges in the TRW be approached via a watershed and ecosystem approach, specifically focusing on the manner in which human activities have major impacts on the viability of ecosystems within the basin. Implicit in this analysis is an examination of the location and spatial configuration of the sources of wastewater being generated, the hydrologic regime and the land uses and population that drive water use and generation of wastewater. By such a watershed approach, I explore various treatment technologies, the spatial configuration of these options and the policy options that attempt to implement these technologies, all within the regional context of the watershed. In doing so, I examine the basin and related ecosystems in an holistic manner that others have pursued in addressing basin wide management challenges (Gardiner, 1991).

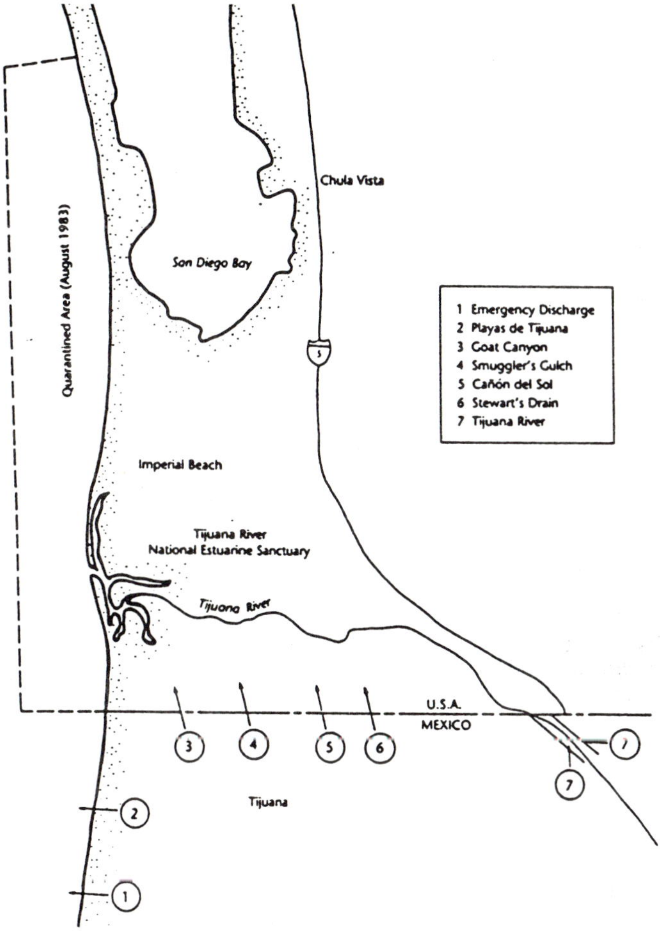

Figure 2 Renegade sewage discharges in the canyons that flow into the Tijuana River Watershed from areas in western Tijuana. (Source: Herzog, 1990).

I also seek to expand the range of options that may help solve the problems evident in the basin in a manner similar to that of previous watershed research done in other areas (National Academy of Sciences (NAS), 1968; Cleary, 1967; and Davis, 1968). Early work within this approach examines suitable spatial configurations of appropriate treatment technologies; latter stages of work explore how these treatment schemes and the attendant policy options can be implemented via some form of cohesive regional framework. This latter stage of work poses questions concerning how these options can be advanced within the regional context examined earlier. Past attempts by various levels of government on both sides of the border --primarily unilateral efforts -- have yielded little progress in advancing solutions to these problems. Given that many of the processes and elements that interact to cause water quality problems in the lower reaches of the Tijuana River Valley --particularly those associated with wastewater--

exist and occur within this watershed, unilateral "stop gap" measures to solve these issues are problematic at best and doomed at worst. Accordingly, a regional, ecosystemic and watershed approach to policy implementation within the Tijuana River basin is an approach that is promising as well as challenging for problem resolution.

Questions concerning implementation involve research into institutional arrangements that both presently exist or need to be newly created or enhanced. The importance of institutional issues has long been recognized by researchers examining resource use and environmental quality questions (Mitchell, 1989; Sanchez, 1993; Mumme, 1992, and 1993a; and Szekely, 1993).

Mitchell (1989) argues that geographers clearly have a role in the examination of institutions, as we study the spatial allocations of resources (and distribution of residuals of resource use), relationships between humans and the environment, and the complexity of regions. Accordingly, I argue that questions concerning implementation and institutional reform, *with respect to the spatial and environmental implications of these activities*, are clearly important to research into viable transboundary water resource management and well within the realm of valid geographic research.

Review of the Literature and Justification of a Watershed Approach

The Regional Issue in Detail - As indicated in the introduction of this paper, the wastewater issues facing the region are due to complex interactions of human activities, the manner in which these activities change over time, the naturally occurring (and human-altered) hydrology of the basin and the manner in which these elements all interact across space. The lower reaches of the basin clearly receive a significant share of these impacts (Angel, 1992; Cleeland, 1992; Dewyze, 1988; IRSC, 1989a and 1989b, 1992; and McNeil-Lehrer, 1992). Risks to environmental and human health that result from these interactions include various disease agents present in the wastewater flows (Devine, 1993), degradation of the quality of surface water flows that enter the estuary (Zedler, 1993), and degradation of coastal waters in the Pacific Ocean offshore from the estuary (Devine, 1993). Considerable efforts by both the US and Mexican governments over several decades have been unable to effect a solution to this problem. A detailed history of these efforts is outlined in a Regional Water Quality Control Board (RWQCB) paper (Meyer, 1983) and the Environmental Impact Statement for the International Wastewater Treatment Plant (IWTP) (Regional Environmental Consultants (RECON), 1991); the key points of this history are briefly summarized below.

Over the last 75 years, Tijuana has grown from a small town of about 500 people in the 1920's to a metropolis that presently ranges from 750,000 to over one million people. The history of wastewater management, treatment, and disposal challenges in the region is one of the above mentioned dramatic population growth in the Municipio de Tijuana, attempts on behalf of both the US and Mexican governments to manage wastewater flows, and the resultant impacts on natural resources in the Tijuana River valley. Early small septic systems were expanded to larger systems, and the International Boundary and Water Commission (IBWC) has aided these efforts by helping design and build a series of treatment facilities. (The IBWC is a joint US-Mexican federal agency with formal responsibility to deal with boundary and water resource issues along the US-Mexico border, (IBWC, 1981)). Despite these efforts, wastewater flows have continued to negatively impact the basin, providing the management challenges this paper addresses. These problems have been well recognized by residents of the Tijuana River Valley within the US and Mexico, as well as policy makers, researchers, and government officials at local, regional, and national levels (Dewyze, 1988; IBWC, 1993; Meyer, 1983; United States Environmental Protection Agency (USEPA), 1992; and USEPA/SEDESOL, 1991). A viable solution must be able to examine these issues at the appropriate scale; as introduced earlier in this paper, a watershed approach within the TRW is such an approach well grounded in the literature of both water resource management and geographic research.

Ecosystem and Watershed Approaches - Early geographic research introduced ecological concepts into the debate of both human-environment interactions and geographic thought. Barrows (1923) stated "geography is the science of human ecology," where the latter term is defined as a unifying concept between humans and their surroundings, focusing on human adaptation to the environment. Sauer (1923) further explores the role of humans as modifiers of the environment or as beings wholly influenced by this environment; he also introduces the *region* as a key unit of study, firmly addressing spatial variation and regional studies as major themes in geography. Both authors wrestled with the manner by which we examine human environment interactions, and the inclusion of ecological concepts in the discussion laid a foundation for latter work examining these issues in ecosystemic and watershed frameworks.

The ecosystem, as defined by Fosberg (1963), is "a functioning interacting system composed of one or more living organisms and their effective environment, both physical and biological, *and may include its spatial relations*" (emphasis added). Stoddard (1965) embraces this ecological viewpoint and the ecosystem as vehicles by which a systems approach of human-environment interactions is brought to geography. In addition, the different levels of complexity found in ecosystems parallel the geographic concept of scale; we can build conceptual models within this framework to seek an understanding of complex human-environmental systems at varying spatial scales and conceptual levels of complexity. Stoddard further argues that ecosystems are monistic and unified systems with a clear and present structure; by studying this structure or morphology, we can better focus on the relationships and functions that exist within this morphology and drive the specific system (Stoddard, 1965). Clearly, the ecosystem approach to human-environment interactions (and resulting issues of environmental quality) offers utility in the issues this paper addresses.

A watershed is defined as "a geographic region within which water, sediments and dissolved materials drain into a common outlet -- a point on a larger stream, a lake, an underlying aquifer, or an ocean" (USEPA, 1991). If we extend this definition to include the flora, fauna and humans that exist and interact within this region, *including certain spatial relations*, we can link ecosystemic and watershed perspectives. Rivers determine riparian and coastal wetland ecosystems; these subsystems in turn exist within the region we have defined as a watershed. This type of regionalization based on systematic analyses of physical resources has been advanced as a framework for managing environmental resources by the USEPA (Gallant et al., 1989). This theme has also been extended to the concept of bio-regionalism, which recognizes that resource issues and management challenges occur in transboundary settings, with successful management necessary within the ecological regions involved (Foster, 1984). Given the above, I argue that a watershed and ecosystemic approach is actually a specific example of regional approaches that have drawn on the work cited above, with considerable success in resolving water quality problems in other river basins (Caldwell, 1988).

Research by the NAS (1968) examined water resource issues in the Colorado River Basin by this approach and aggressively examined how to expand the range of resource use options involving large scale water projects and economic development pursued by government agencies in the US. Davis (1968) pursued a similar tack in his analysis of excessive dissolved oxygen in the Potomac Estuary, explicitly exploring the spatial configuration of various treatment techniques. Cleary (1967) aggressively pursued a similar approach in developing a basin wide plan for controlling sewage discharges in the Ohio River Valley. He firmly grounded his work in a watershed framework, with special attention given to the spatial context and configuration of sources of sewage and a management regime to control them.

More contemporary work has also used this approach, both domestically and internationally. The International Joint Commission, a binational Canadian-US advisory body, has used this approach for resolution of water quality issues in the international context of the Great Lakes Basin that lies along the US-Canadian border. The Commission actually elevated this position to a policy statement when they argued "restoration and enhancement of the boundary waters cannot be achieved independently of other parts of the Great Lakes Ecosystem with which these waters interact" (Great Lakes Water Quality Agreement of 1978).

Research Issues

Spatial Configuration Issues - I argue that an ecosystemic-watershed framework offers a useful perspective to examine wastewater issues in the Tijuana River basin, specifically addressing the spatial configuration of wastewater sources and treatment technologies to manage wastewater flows. This involves an examination of the hydrologic regime of the basin, focusing on both natural hydrology, as well as anthropogenic modifications to this regime. The manner in which precipitation, runoff, flood flows and groundwater linkages occur in the basin is important in this work, especially the spatial configuration of this hydrology. With these hydrologic data in place, present sources of wastewater negatively impacting the basin can be examined; of particular interest are critical areas where wastewater is either escaping from the existing collection network in Tijuana or forming renegade flows due to a lack of collection infrastructure.

In order to determine where future problem areas are likely to occur, land use, urbanization and population data for the watershed can be examined, specifically with respect to areas of new industrial and residential development, and areas likely to lack needed collection infrastructure in the future can be identified. Present and projected population distributions in the basin, especially areas of high present and future population, impact both finished water demand and waste water generation. The following questions can be explored by such an approach:

1) How does surface water hydrology impact the generation and transmission of wastewater flows? What role does stormwater runoff play in this dynamic?
2) What are the present major sources of wastewater contamination in the basin, and where are these located? How does wastewater quality vary across the basin?
3) What are present and future distributions of population? How will these in turn drive finished water demand and wastewater treatment load?

4) Based on existing and future land uses and population distributions, where are existing and future critical areas, particularly with respect to new industrial and related residential development?

Treatment Technologies - With a comprehensive representation of the basin's morphology in place, existing and emerging means of wastewater treatment can be examined, focusing on both conventional technologies (activated sludge, oxidation ponds, and bio-solids), as well as less commonly used techniques involving water reclamation (trickle filtration, biologically activated lagoons, and scalping plants). This examination can determine what techniques are being used and where as well as how well they treat pollutants in current wastewater flows. Also of interest is the manner by which existing techniques can be enhanced or new techniques can be developed to expand the range of water resource options involved, specifically with respect to two areas of water resource management -- provision of water supply and flood control.

If wastewater treatment is enhanced to provide reclaimed water, conventional treatment options can be expanded to provide supplies of non-potable water, especially if treatment plants are located near areas that have demand for this type of water. Flood control is also an integral part of water resource challenges in the basin (Pryde, 1986), especially as flood waters have major impacts on the downstream areas of the basin, particularly the estuarine ecosystem. Conjunctive use strategies whereby the management of wastewater could be coupled with aquifer recharge and mitigation of flood flows can play a positive role in the management of the estuary. This analysis can examine these options with respect to the morphology of the basin to answer a specific set of questions:

1) Given the spatial configuration of the basin, where would be suitable locations for various treatment technologies, given their variable ability to deal with total dissolved solids (TDS), metals, and toxic materials? How might they contribute to meeting water supply and flood control needs?
2) What areas of the watershed have wastewater with a low enough TDS load to be reclaimed? What areas or regions have a demand for reclaimed water?
3) Given the above, what role can alternative treatment technologies such as reclamation, biologically activated lagoon technologies, and small scale scalping plants play?
4) What are the relative costs of various techniques, and how do these costs impact their potential to be applied in an overall treatment plan?

The overall goal of this stage of work is to explore how these techniques can be located and linked together in a basin-wide plan to expand wastewater treatment options to address the broader set of water resource management questions outlined above, taking into account the entire range of activities in the watershed. Given a knowledge of the spatial configuration of the present and future sources of wastewater in the watershed, and insight into where and how various treatment regimes could be suitably located, how would a basin wide plan for sustainable wastewater treatment infrastructure be configured? *Suitability* in this sense refers to the ability of various technologies to adequately treat various pollutant loads, as well as how suitable these technologies are with respect to available human resources and the socio-cultural and economic fabric of the study area as encompassing portions of a developing country. Techniques well suited to developed countries are not necessarily appropriate to lesser developed countries, owing to socio-cultural and economic differences (Okun and Ponghis,

1975; Reid, 1982; and Schulz and Okun, 1984). *Sustainable infrastructure* refers to infrastructure that is effective after the initial period of investment, through the expected life of the project (Ostrum, Schroeder, and Wynne; 1993). I address these concerns in more detail later in this paper, yet I introduce them now to punctuate the intent to avoid constraining the analysis of treatment options to those appropriate only in modernized developed regions. By doing so, I seek to provide a framework flexible enough to truly allow a wider range of options to be examined.

Implementation Issues - The manner in which a basin wide wastewater management plan may be put into place is the final element to be addressed within this framework. As stated earlier in this paper, geographers are concerned with how institutions function to make decisions concerning resource use and environmental quality (Mitchell, 1989). The manner by which agencies make these policy decisions directly influences the spatial configuration of our built environment and infrastructure, and impacts environmental quality. Questions concerning these policy and institutional issues, posed both in the US and Mexican portions of the region, can explore cross-border dynamics within the watershed. An approach that shows much promise for this type of analysis is political ecology.

Blaikie and Brookfield (1987) state that "political ecology combines the concerns of ecology with a broadly defined political economy. Together, this encompasses the constantly shifting dialectic between society and land based resources and also within and among classes and groups within society itself." Bryant (1992) sees political ecology as "an attempt to understand the political sources, conditions, and ramifications of environmental change." In doing so, he examines the contextual sources of environmental change, the conflict that differential access to power and resources generates and the political impacts of environmental change (Bryant, 1992). The entire process of policy implementation involves agencies and individuals making decisions and wielding political power and resources to bring about a change in present resource use, with resulting environmental impacts. Herein lies the very nature of the political ecology of the issues at hand. Namely, how do socio-cultural, political and economic factors interact in the unfolding dynamic involving regional development and the development of related infrastructure?

Political ecology allows us to examine how political power and relevant socio-cultural and economic factors and processes involved in infrastructure development vary and interact along and across the border in this binational basin. By exploring how the related factors have caused existing policy to be crafted and implemented, we can investigate specific elements of institutional reform that would allow new policy to be put in place to advance wastewater treatment and solution of related problems. Cost issues are certainly of importance and need to be explored within this analysis as important economic factors. Yet this perspective argues that the analysis moves past the limits of traditional cost benefit work, as this latter analysis tends to overlook the non-economic facets of policy implementation. Questions important to examining these policy and institutional issues include the following:

1) What changes or enhancements to existing institutions need to occur to allow a treatment and management regime to be implemented?
2) What role can newly emerging institutions play in effective policy formulation and implementation?

3) What barriers or impediments to policy implementation exist within this institutional framework?
4) What processes that occur within this institutional framework (i.e., delegation and sharing of political power and primacy and means of funding infrastructure) may need to be altered to remove or lessen these impediments and advance this policy ?

A final issue that can be examined in this framework concerns a model of policy formulation and implementation that is suited to this binational region. Previous research to be discussed in greater detail later in this document can be explored as a starting point, but a larger goal is to craft a conceptual model that examines the range of political ecological elements identified above, generates a framework for posing and answering the related questions and fits the political ecology of this binational region.

Methods and Data

In Country and Collaborative Dimensions of the Research - Of major importance to this research approach is collaboration with researchers and officials in governmental agencies and research institutions in Tijuana, San Diego and other portions of Mexico and the US. Specifically, El Colegio de la Frontera Norte in Tijuana (COLEF, the northern border institute), the Office of Planning and Urban Development in the Municipio de Tijuana and La Comision Internacional de Limites y Agua (CILA, the Mexican Section of the International Boundary and Water Commission) are particularly important in this work in Mexico. (A preliminary list of agencies involved is included in Table 1 in the Appendix). In conjunction with a current collaborative research project San Diego State University (SDSU) and COLEF are pursuing, the research questions posed above are being explored in this collaborative manner to insure that the analysis does indeed occur within a watershed context, examining pertinent elements and attendant interactions on both sides of the border.

Spatial Configuration of Variables in the Basin - The first step in this research is to obtain data on the spatial configuration of the basin, specifically the basin's hydrology, the distribution of present population levels and land uses, and locations and nature of critical areas of wastewater generation. This provides a benchmark upon which projections of future changes in land use, increases of population, and potential increases in wastewater generation can be examined. Implicit in this early work is an examination of digital and analog maps of these variables as they occur in the region. The above referenced project within SDSU and COLEF is gathering these data to create a seamless spatial database for the basin, consisting of geographic information system (GIS) and analog map coverages of these variables. Specific data being collected include topography, hydrography, climate and precipitation, SPOT satellite imagery, air photos, vegetation, land use, and land cover; specific details of these data are provided in a written report on the project (IRSC, 1995).

An integral part of this work is a comprehensive air photograph interpretation of the basin involving a time series of air photos for the region to capture land use, land cover and extent of urbanization. These data are input into Arc/Info (a commercial GIS software) and linked to Mexican census and other data obtained from researchers at COLEF to model urbanization and population growth trends. By linking these census and urbanization data across time and space, we can identify past trends of development and urban growth and extrapolate these trends into the future. Thereby, we seek to determine how population is presently distributed across space and is likely to increase in the future, and how changes in land use and urban development will interact with this population growth. Wastewater generation is a direct effect of water use by humans in residential, industrial and commercial activities; this early work highlights areas where wastewater generation is presently occurring and likely to occur in the future.

Related to this work is archival research within COLEF, the City and County of San Diego, the IBWC and CILA, offices of the Municipio de Tijuana and other Mexican and US agencies to determine the configuration of the existing network for collection and treatment of wastewater flows in the basin. Specific items sought in this archival work are maps, plans, and projected infrastructure development schemes that indicate where existing and future wastewater management infrastructure will be located, especially relative to areas of increasing population and water intensive land uses likely to generate increasing levels of wastewater. Within Arc/Info, cartographic modeling is performed on the population, urbanization, hydrology and infrastructure data layers to determine how population distribution and urbanization trends are or are not co-occurring with present and planned infrastructure and what role surface hydrology may play. This analysis will yield maps of the critical areas where future development and urbanization will likely generate unmet treatment demand.

In conjunction with the above work, a water and wastewater sampling effort can help determine the spatial distribution of wastewater flows and surface water quality, thereby further focusing on critical areas in the basin. Analysis of these samples also reveals how the quality of wastewater flows contained in the collection and treatment network varies across space. Information is sought on the presence of toxics, the nature of spatial variations of TDS, quantities of renegade and collected sewage and impacts of renegade flows on surface water quality. Owing to the steep coastal terraces in parts of the basin and the manner by which this topography interacts with the region's hydrology, locations of sample sites are likely to be point sources of contaminated surface water runoff, located downstream from urbanized areas. As surface flows in the river are intermittent in nature owing to a winter precipitation regime, this will be a wet season sampling scheme. This sampling and analytical work is not designed to provide a comprehensive and long term set of surface water quality data, but rather a more limited set of data that can provide some general trends of spatial variability of surface water quality. Data on water quality and wastewater parameters are also sought with the various offices of the IBWC and CILA and will continue via archival efforts.

Given adequate surface hydrology data are available, possible means of modeling stormwater runoff can be explored, as this is a major component to surface water flows in the basin. The GIS and cartographic modelling work will indicate general areas for future wastewater treatment infrastructure; stormwater modeling can provide additional detail as to what areas in the basin are receiving significant stormwater flows and how the quality and quantity of these flows vary spatially. Overton and Meadows (1976) provide several concepts worth exploring in greater detail; an EPA stormwater management model examines stormwater runoff and sewage overflows in a combined system approach, and the pollutograph details urban sources of pollution within the vector of stormwater as a function of time. Subject to data availability, these concepts and others can be explored as a means of determining general patterns of stormwater runoff and the contribution this makes to degradation in surface water quality. Again, the intent in examining stormwater dynamics is to obtain a general idea of spatial variability of these flows to aid in focusing on critical areas of surface water pollution, not perform a long term and comprehensive analysis of flood flows.

Treatment Infrastructure and Technologies - The above analysis provides insight into the morphology of wastewater generation in the basin and an indication of what areas require enhanced wastewater treatment infrastructure. An engineering examination of existing and emerging treatment technologies builds on this to determine how these techniques meet treatment needs within the basin. Technologies to be examined include existing activated sludge and bio-solids techniques as proposed for currently planned treatment works, as well as lower technology techniques discussed earlier in the proposal. Before discussing this engineering analysis, some detail is provided concerning some of these alternative treatment technologies, especially those already being explored in the region.

SIDETRAN (Sistema Descentralizado de Tratamiento y Reuso de Aguas Negras en Zonas Urbanas) is a recently implemented example of a small scale, decentralized, biologically-based treatment technology designed to treat wastewater in Tijuana to standards such that effluent could be reused as an irrigation source (La Rue, S., 1993). To the degree that treated effluent can be reclaimed, this technology eliminates the need for expensive large scale outfall facilities, a major cost item in treatment plants. A variation on this theme is the development of upstream, small scale scalping plants located to take advantage of the variability of the level of total dissolved solids (TDS) in wastewater. Overall, these plants would route all of the solids removed in treatment to larger downstream treatment plants, but they would "scalp off" secondary or tertiary treated effluent in upstream portions of the valley that can be reclaimed for irrigation and other non potable uses. By optimally locating these plants (close to lower TDS wastewater sources and near areas where the need for reclaimed water is highest), provision of non-potable water can be coupled with reducing the load on traditional treatment plants. This option could make significant contributions to the basin and is also being argued as an option for treatment of wastewater elsewhere in the greater San Diego region by Elmer Keen (1993a and b).

A final less technology intensive option that has been proposed for the basin is a system of deep lagoons that uses natural biological processes to remove the bio-chemical oxygen demand (BOD) in wastewater. This technology has been successful on a smaller scale in Humbolt County and the Napa Valley of California and combines non-traditional treatment with provision of recreational areas and generation of tertiary treated effluent for non-potable purposes. This concept was proposed to the IBWC, has enjoyed some support by various interest groups and is being considered as a trial technology by the City of Imperial Beach (1993). Questions of scale demand attention (as this technology is fairly land intensive), and the ability of this technique to handle possible slugs of toxic wastewater is also of concern to regulatory officials, but this technique offers potential for wastewater treatment in the basin.

As introduced previously, suitability of these techniques involves two concepts, the first being how well they remove and handle solids, bio-chemical oxygen demand (BOD) and other pollutants found in wastestreams. This level of suitability can be explored in a traditional engineering framework as outlined by experts in the field of wastewater treatment (Fair, Geyer, and Okun; 1968 and 1971), with the intent being a technical determination of the degree of pollutant removal, relative cost of processes involved, and related issues. General cost data are available for various types and scales of plants (Okun and Ponghis, 1975; Reid, 1982; and Schulz and Okun, 1984) that allow compilation of relative cost figures for various technologies to aid in gauging their appropriateness (i.e., large activated sludge plants may cost approximately $10 million for each million gallons per day of treatment capacity, whereas SIDETRAN may cost $100,000 for the same amount of treatment capacity).

Secondly, these techniques can be examined with respect to how appropriate they are for use in the Mexican portion of the watershed, paying special attention to the socio-cultural and economic fabric of Mexico as a developing country. As introduced earlier, techniques well suited to developed countries tend to be highly capital and technology intensive, raising operation and maintenance (O&M) issues that pose problems in developing countries that may lack technological and managerial expertise (Okun and Ponghis, 1975; Reid, 1982; and Schulz and Okun, 1984). Large scale activated sludge plants requiring expensive and extensive O&M may work well north of the border, but they may be wholly inappropriate in Mexico. A related concept is *sustainable infrastructure,* infrastructure that is effective after the initial period of investment, with reasonable levels of O&M investment and lasting the expected life of the project (Ostrum, Schroeder, and Wynne; 1993). Since the goal of this research is to craft a wastewater management plan that is suited to both the physical parameters of the basin and related waste flows, as well as the relevant socio-cultural and economic context, this facet of suitability is as important as the more technically oriented one. The cost issue addressed earlier is an important consideration of suitability and one included in this analysis of appropriate technologies.

This latter examination of appropriateness of treatment technologies can then be linked with locations of crucial areas of wastewater generation and treatment via two possible methods for treatment plant location research. First, GIS maps of the basin showing population, urbanization and treatment need can be examined intuitively or heuristically to determine how treatment techniques fit the various areas. Areas with high levels of industrialization and potential for toxics, heavy metals and similar contaminants will require more advanced treatment regimes, most likely with pre-treatment processes. Conversely, less populated areas with no industry may be adequately served by lower technology plants employing some form of reclamation technique that could generate non-potable water supply. By visual inspection of the above referenced maps and consideration of how various treatment techniques fit various treatment needs, the nature of these treatment needs in various areas of the basin can be matched to techniques that handle related pollutants in an appropriate manner.

A second possibility is to utilize location/allocation algorithms with the various layers of data previously entered and processed to generate various configurations of treatment plants. The simplest level of analysis can explore this possibility with respect to minimizing distances that effluent and influent would need to be pumped (and by extension pumping costs). Previous discussions advanced the concept of the "appropriateness" of treatment technologies; this concept could be incorporated via various weighting schemes or alternate heuristics or algorithms. Since the goal is a treatment plan that fits all facets of the basin, moving past simple cost issues towards those of sustainability as outlined above can insure that the options explored fit the human dimensions of the basin and problems at hand, as well as meeting distance and cost minimization criteria.

Basin-wide Planning and Management Efforts - Given knowledge of the spatial configuration of wastewater treatment demands, socio-economic variables that effect this configuration and the appropriateness of various treatment regimes for meeting these demands, a basin-wide plan can be crafted for these treatment options, specifically with respect to location and allocation questions. Similar to work done by White (1977), Downs, et al. (1991), Okun (1977), and the USEPA (1991), this plan can expand the range of treatment options by examining wastewater treatment as one part of a larger riddle of prudent water resource management, specifically within a watershed context. Within this regional framework, attention is focused on the location of relevant elements and the spatial processes that link them together. Location of reclamation facilities must deal with the spatial variation of wastewater quality and demand for reclaimed water. Attempts to link wastewater treatment with conjunctive use schemes in the lower reaches of the watershed must do so with respect to the spatial configuration of the basin's hydrology and human attempts to modify this via flood control works. Lastly, proposed works must take into account the location and operation of existing and proposed wastewater treatment facilities (pump stations, collectors, interceptors, and outfalls). The initial mapping of the basin, the two pronged engineering analysis of treatment technologies and the location work described above is combined to form a basin wide plan, firmly established in a ecosystemic and watershed context and framework.

Examination of Implementation Options - The latter part of this research approach explores how this scheme may be implemented in the basin; a political ecology approach as outlined above serves well as the paradigm for this work. Using this paradigm, I seek to identify the significant *"political ecological"* dimensions of implementation issues, in an attempt to address the questions concerning policy and institutional reform posed earlier. Political ecology has been posed as an extension of earlier ecosystem work that allows us to examine ecosystemic, socio-cultural, and political dimensions of human-environment interactions by examining how power and related elements interact with the environment as we craft and carry out development policy. I briefly introduce a useful heuristic tool as a starting point for this stage of the research.

Work being done by Campbell and Olson (1991a and b) builds on earlier political ecology research and provides a specific heuristic framework by which we can approach development issues. The Kite seeks to bring the environmental facet of resource use to the core of the analyses (an ecological influence), to open up the spatial scale to include extra-local effects (a regional influence), and to gauge and incorporate the varying interest groups and the power they wield (a political influence). To do the above, The Kite analysis identifies the factors involved in development policies as socio-cultural, political, economic or environmental and places them in the framework depicted in Figure 3. The Kite framework then provides dynamic linkages to connect these variables, incorporating processes across different spatial and temporal scales. This framework deals with scale, space, power and time and provides a means by which researchers can examine the complex set of interactions that occur between society and the environment. Given the vastly different ways in which political and economic power are exercised across different levels of government and space in the US and Mexican portions of the border region, the Kite offers a viable way to approach implementation issues. As such, this heuristic is an extension of political ecology, firmly building on earlier research of water resource issues done by geographers.

Using the Kite as a starting point and blueprint, the socio-cultural, political, economic and environmental elements involved in regional and infrastructure development in the region can be identified and explored to determine how they interact as relevant policies are crafted and put in place. Archival research can uncover laws, regulations and infrastructure plans and schemes in both the US and Mexico concerning wastewater and water treatment in the basin to identify and explore the legal and political elements involved. (A list of initial items to be reviewed is included in Table 2 in the Appendix). Extensive and comprehensive newspaper clipping collections held within the libraries of the San Diego County Water Authority and the Tia Juana Valley County Water Authority can provide additional data concerning how these political, socio-cultural, and economic elements interact. Interviews with officials, staff, and researchers that work within the relevant agencies and organizations can uncover the history of economic and infrastructure development in the region. The goal of this interview work is to follow up on elements identified in the earlier archival work and clearly focus on how the political, economic, socio-cultural, and environmental elements within the basin have interacted to yield the present configuration of wastewater infrastructure. Enabling legislation, financing and bonding instruments and processes, political elements and processes, and socio-cultural influences can be explored in an attempt to find the elements specific to the basin that fit the Kite framework.

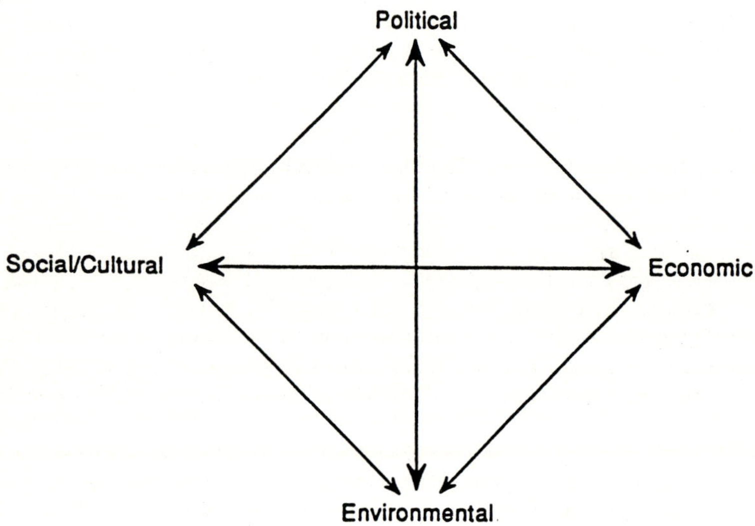

Figure 3 The Kite model of people-environmental interaction. Source: Campbell and Olson, 1991 (used with permission of the authors).

Once the elements in this framework have been identified, further work can examine the elements involved to identify institutions, laws and regulations, related processes and other elements that act as barriers or impediments to developing and implementing policy that would advance a basin wide plan. In a manner parallel to the way environmental impact reports and statements (EIR's and EIS's) identify *significant environmental impacts* of proposed land and water use activities, this approach explores the *political ecological impediments* involved in infrastructure development and the relative resistance these impediments pose to policy actualization. Once these impediments are identified, attention focuses on what changes within the institutional framework would need to be realized to lessen or remove these impediments, in a similar manner to how EIR's and EIS's attempt to mitigate the relative impacts of various projects and actions.

As a means of clarification, I offer a few examples of impediments identified in early research, as well as some suggested actions to eliminate or lessen them. Fiscal and monetary problems related to the recent peso devaluation in Mexico persist in the public and private sector, impeding infrastructure development. Pursuing funding for border infrastructure with newly emerging organizations arising from the side accords for the North America Free Trade Agreement may provide a means of dealing with this. A persistent, overly centralized political economy in Mexico impairs the abilities of municipalities to exercise "home rule" and funding ability (Crane, 1995). Recent attempts to decentralize La Comision Estatal Servicios Publicos de Tijuana (the state water commission serving Tijuana) and generate and retain local funding for infrastructure in Tijuana show promise for lessening this barrier (Dibble, 1994). Lastly, a very real strained social fabric exists along the border, both reflecting and in part due to a recent anti-immigration ballot initiative (Propostion 187) and immigration enforcement activities (Operation Gatekeeper). This tension genuinely impairs US-Mexico cooperation on all levels. Continued efforts (such as the SDSU/COLEF project mentioned earlier and the research approach advanced in this paper) to reach out across the border and cooperate on contemporary problems may help to lessen this impediment.

An important goal of this approach is to explore how the institutional framework identified in this latter work may be enhanced and changed to allow the treatment and management regime to be put in place. The importance of institutional changes has long been recognized by researchers examining these issues (Mitchell, 1989; Sanchez, 1993; Mumme, 1992, and 1993a; and Szekely, 1993), and a goal of this work is to seek region-specific opportunities for these changes. Several options concerning changes to existing institutions in the area appear fruitful at all levels of government; locally, regionally, nationally, and internationally. This stage of the research is an attempt to get past the broad based citations of "interjuristictional barriers, cultural differences, and sovereignty issues" that have characterized much of the past dialogue on these issues.

The last research issue to be addressed in this approach concerns the development of a model for policy formulation and implementation that is well suited to this binational region. The Kite offers a useful heuristic tool to begin the analysis, yet this tool only opens the dialogue and does not appear to be adequate to deal with the unique binational nature of the border region. The approach argued in this paper can yield a framework for the elements involved, a means to identify the relevant barriers to policy implementation, and some suggested means for overcoming these impediments. With these pieces of the puzzle in place and drawing on the work of previous policy researchers, a model of policy implementation can be synthesized that reflects the processes and elements that impact the infrastructure issues of the region. Waterstone (1994) offers the concept of institutions as sets of rules by which various shareholders operate as development and resource use is realized. Mazmanian and Sabatier (1989) offer a framework for policy formulation and implementation that seeks to strengthen the links between these two actions, paying special attention to the perspective of those involved and the role of program evaluation. The results of analyses performed within this framework can be linked with these other models to craft a model and graphic representation of the related elements that can help explain how effective policy in bi-national river basins may be effected.

Outcome and Utility of This Approach

The framework and perspective argued in this paper can provide several possible outcomes. Early work can yield a series of GIS map products showing existing and future population distributions, urbanization, and land uses, and a description of how these elements interact with the natural and human-altered hydrology of the basin across time and space to drive the generation of wastewater. These maps would also indicate present and planned wastewater infrastructure. Data concerning water and wastewater quality and stormwater flows would provide general trends to the spatial variability of these flows and how this variability may impact future infrastructure development. Combining these sets of data can drive a gap analysis that identifies critical areas where wastewater generated in the future is not likely to be adequately collected, treated and disposed of.

With this background in place, a range of existing and proposed treatment technologies can be examined to explore their suitability and appropriateness for various sub-regions of the watershed; this suitability is with respect to both technical and human resource issues as detailed above. These two early components drives the development of a basin wide management and treatment plan for wastewater that aggressively moves past single purpose treatment schemes that treat wastewater as an isolated facet of water resource management. The goal of this plan is to expand the range of treatment options within the basin to include linkages to provision of reclaimed water to meet non-potable water demand and conjunctive use schemes that incorporate flood control and aquifer recharge. As such, this plan is an holistic one treating the riverine bio-region and the relevant physical and human elements as the inter-connected elements they are.

The last stages of this research approach advance a political ecology examination into the manner by which this scheme may be put in place. Economic, socio-cultural, political and environmental facets of this plan and its implementation are examined in order to explore institutional changes needed to bring this plan about. Political ecological impediments to policy options involved are explored, as the analysis seeks to find means to overcome these impediments and allow this plan to be advanced within the regional context of the watershed. The importance of these institutional changes has long been recognized and this later stage of research seeks region-specific opportunities for these changes.

Conclusion

This paper discusses and advances an ecosystem and watershed approach that seeks to expand the range of treatment and management options for wastewater flows within the Tijuana River basin, specifically with respect to the spatial configuration of the basin and the manner in which relevant elements interact across space. This paper also explores new ground involving the application of regional political ecology to the problems at hand, expanding this paradigm to address water resource management issues in a binational and cross cultural watershed.

Political ecological impediments are explored in this effort, providing a means to grapple with the barriers to policy implementation that have historically impaired problem resolution. Taken as a whole, this conceptual model can provide utility in resolving wastewater management challenges in this basin, as well as applicability to other binational, cross-cultural basins.

References

Angel, S.M.. 1992. Peace Between the Borders: The San Diego-Tijuana Border Sewage Problem. Unpublished Masters Thesis. San Diego State University: San Diego, CA.

Barrows, H.H.. 1923. "Geography as Human Ecology." The Annals of the Association of American Geographers. 13(1):1-14.

Blaikie, P. and H. Brookfield. 1987. Land Degradation and Society. Methuen: London, England.

Bryant, R.L.. 1992. "Political Ecology: An Emerging Research Agenda in Third-World Studies." Political Geography Quarterly. 11(1):12-36.

Caldwell, L. K. 1988. Perspectives on Ecosystem Management for the Great Lakes. State University of New York Press: Albany, NY.

Campbell, D.J. and J.M. Olson. 1991a. "Environment and Development in Kenya: Flying the Kite in Kajiado District." The Centennial Review. 35(2): 295-314.

Campbell, D.J. and J.M. Olson. 1991b. Framework for Environment and Development: The Kite. CASID Occasional Paper No. 10. CASID: East Lansing, MI.

City of Imperial Beach. 1993. Resolution No. 93-4289. Concerning Interim Emergency Treatment and Disposal Facilities for Tijuana Sewage. 20 October, 1993. Imperial Beach, CA.

Cleary, E. J.. 1967. The ORSANCO Story: Water Quality Management in the Ohio River Valley Under an Interstate Compact. Johns Hopkins University Press for Resources for the Future: Baltimore, MD.

Cleeland, N. 1992. "Birth Defect is a Border Woe". November 22, 1992. San Diego Union Tribune: San Diego, CA.

Crane, R. "Federalism and Decentralization at the Municipal Level: The Case of Tijuana." Paper Presented at the 1995 Annual Meeting of the Association of Borderlands Scholars, 28 April,, 1995. Oakland, CA.

Davis, R. K. 1968. The Range of Choice in Water Management: A Study of Dissolved Oxygen in the Potomac Estuary. Johns Hopkins University Press for Resources for the Future: Baltimore, MD.

Devine, M. 1993. Personal Conversation, 10 September, 1993. San Diego County Environmental Health Department, San Diego, CA.

Dewyze, J. 1988. "Get Ready to Learn the Shocking Facts and the Horrifying History of the Runaway Tijuana Sewage that Threatens to Close our Beaches and Poison Our Beaches Unless Somebody Does Something Pretty Damned Quick." San Diego Reader, 23 November, 1988.

Dibble, S. 1994. "Tijuana Public Works Vote a First." San Diego Union Tribune. April, 1994.

Downs, P.W., K.J. Gregory, and A. Brookes. 1991. "How Integrated is River Basin Management?" Environmental Protection. 15(3): 299-309.

Fair, G.M., J.C. Geyer, and D.A. Okun. 1968. Water and Wastewater Engineering. John Wiley and Sons: New York, NY.

Fair, G.M., J.C. Geyer, and D.A. Okun. 1971. Elements of Water Supply and Wastewater Disposal. John Wiley and Sons: New York, NY.

Fosberg, F.R.. 1963. "The Island Ecosystem." Man's Place in the Island Ecosystem, a Symposium. Honolulu, Hawaii.

Foster, C.H.W. 1984. Experiments in Bio-regionalism. University Press of New England: Hanover, CN.

Gallant, A.L., T.R. Whittier, D.P. Larsen, J.M. Omernik, and R.M. Hughes. Regionalization as a Tool for Managing Environmental Resources. United States Environmental Protection Agency. Office of Research and Development: Corvallis, OR.

Gardiner, J.L. (ed). 1991. River Projects and Conservation: A Manual for Holistic Appraisal. John Wiley and Sons: New York, NY.

Great Lakes Water Quality Agreement. 1978. Governments of the United States of America and Canada. Ottawa, Canada and Washington D.C.

Herzog, L. 1990. Where North Meets South: Cities, Space, and Politics along the US-Mexico Border. The University of Texas Press:Austin, TX.

Hundley, N., J. Bortz, D. Lorey, and A.M. Pahissa. 1993. Transformation and Integration: The Borderlands, 1940-1990s. Unpublished Book Chapter.

Institute for Regional Studies of the Californias (IRSC), 1989a. Water Quality Issues of the California-Baja California Border Region. Ed. by Clifton Metzner. San Diego, CA: IRSC.

__________ (IRSC), 1989b. Water Quality Issues of the San Diego-Tijuana Border Region. Ed. by Clifton Metzner. San Diego, CA: IRSC.

__________ (IRSC), 1992. Environmental Issues of the Mexican-US Border Region. Ed. by Clifton Metzner. San Diego, CA: IRSC.

__________ (IRSC), 1993. Workshop on Water Quality and Provision in the Tijuana/San Diego and Mexicali/Imperial Valley Regions. 6-8 July, 1993. San Diego, CA: IRSC.

__________ (IRSC), 1995. Identifying Priorities for a GIS for the Tijuana River Watershed: Applications for Land Use, Planning, and Education. Ed. by R. Wright, K. Ries, and A, Winkel. IRSC: San Diego, CA.

International Boundary and Water Commission (IBWC). Assorted Treaty Minutes. Nos. 222, 261, 263, 266, 270, 283. Signed Between the US and Mexico. El Paso, TX.

__________ (IBWC). 1981. Joint Projects, United States and Mexico Through the IBWC. IBWC: El Paso, TX.

__________ (IBWC). 1992. Correspondence from C. W. Ruth to James Giannopoulos of the State Water Resources Control Board. Dated October 8, 1992. IBWC: El Paso, TX.

__________ (IBWC). 1993. Public Meeting to Address the Proposed Action for Additional Interim Works to Capture and Return Tijuana Sewage to Mexico. 6 October, 1993. Joint IBWC/Imperial Beach City Council Meeting.

Keen, E.A. 1993a. San Diego's North City Water Reclamation Plant, a Critical Assessment. Unpublished Manuscript. May, 1993. San Diego, CA.

Keen, E. A. 1993b. Personal Conversation. 19 October, 1993. San Diego State University, San Diego, CA.

La Rue, S.. 1993. "Low-tech Facility in Tijuana Hailed as One Sewage Solution." The San Diego Union Tribune. 20, October, 1993.

Mazmanian, D.A. and P.A. Sabatier. 1989. Implementation and Public Policy. University Press of America: Lanham, MD.

Metzner, C. 1993. Comments made at the Workshop on Water Quality and Provision in the Tijuana/San Diego and Mexicali/Imperial Valley Regions.7 July, 1993. San Diego, CA.

Meyer, E.L. 1983. Staff Report: History of Sewerage Facilities Serving the City of Tijuana, Baja California, Mexico. California Regional Water Quality Control Board, San Diego Region: San Diego, CA.

McNeil Lehrer Newshour. 1992. Story on Anencephaly Occurrences and Border Industrialization. November 19, 1992. Public Broadcasting System: Washington, DC.

Mitchell, B.. 1989. Geography and Resource Analysis. John Wiley and Sons, Inc: New York, NY.

Mumme, S. 1992. "New Directions in United State-Mexican Transboundary Environmental Management: A Critique of Current Proposals." Natural Resource Journal. 32:539-562.

Mumme, S. 1993a. "Innovation and Reform in Transboundary Resource Management: A Critical Look at the International Boundary and Water Commission, United States and Mexico." Natural Resource Journal. 33:93-120.

Mumme, S. 1993b. "Mexico's New Environmental Policy: An Assessment". Manuscript in Process of Publication. Colorado State University: Fort Collins, CO.

Mumme, S. and R. Sanchez. 1992. "New Directions in Mexican Environmental Policy." Environmental Management. 16(4): 465-474.

National Academy of Sciences (NAS), 1968. Water and Choice in the Colorado River Basin: An Example of Alternatives in Water Management. NAS: Washington DC.

National Science Foundation (NSF). 1995. Comments of anonymous review of Christopher Brown's dissertation enrichment proposal, Ref # : SBF-95-08432, issued 18 April, 1995. NSF: Washington, DC.

North American Free Trade Agreement Supplemental Agreements and Additional Documents. 1993. Washington DC.

Okun, D.A. 1977 Regionalization of Water Management: A Revolution in Wales. Applied Science Publishers, Ltd: London, UK.

Okun, D.A. and G. Ponghis. 1975. Community Wastewater Collection and Disposal. World Health Organization: Geneva, Switzerland.

Ostrum, E., L. Schroeder, and S. Wynne. Institutional Incentives and Sustainable Development: Infrastructure Policies in Perspective. Westview Press: Boulder, CO.

Pryde, P. 1986. "A Geography of Water Supply and Management in the San Diego-Tijuana Border Zone." In L. Herzog, ed., Planning the International Border Metropolis, pp. 45-54. Monograph no. 19. Center for US-Mexican Studies: La Jolla, CA.

Regional Environmental Consultants (RECON). 1991. Draft Environmental Impact Statement for the International Boundary and Water Commission International Treatment Plant and Outfall Facilities. RECON: San Diego, CA.

Reid, G.W. 1982. Appropriate Methods of Treating Water and Wastewater in Developing Countries. Ann Arbor Science Publishers: Ann Arbor, MI.

Sanchez, R. 1992. "US-Mexico Economic Integration and Environmental Effects." Proceedings of San Diego-Tijuana Borderlands: Problems and Prospects. 17-18 April, 1992. San Diego Mesa College: San Diego, CA

Sanchez, R. 1993. Personal Conversation at El Colegio de la Frontera Norte, Tijuana, BC, MEXICO. 5 March, 1993.

Sauer, C. 1923. "The Morphology of Landscape." Life and Land: A Selection of the Writing of Carl Sauer. Ed. by John Leighly. UC Press: Berkeley. CA

Schlessinger, D. 1993. City of San Diego Memorandum Concerning Results of City of San Diego/EPA/IBWC/SWRCB Policy Meeting to Discuss Acceleràtion of the IWTP. 5 November, 1993. San Francisco, CA.

Schulz, C.R. and D.A. Okun. 1984. Surface Water Treatment for Communities in Developing Countries. Intermediate Technology Publications: London, UK.

Stoddard, D.R.. 1965. "Geography and the Ecological Approach: The Ecosystem as a Geographic Principle and Method." Geography. 50:242-251.

Szekely, A. 1992. "Establishing a Region for Ecological Cooperation In North America." Natural Resource Journal. 32:563-622.

Szekely, A. 1993. "Emerging Boundary Environmental Challenges and Institutional Issues: Mexico and the United States." Natural Resource Journal. 33:33-46.

United States Department of the Treasury. 1993. Fact Sheet on the US-Mexican Agreement on the Border Environment Cooperation Commission (BECC) and the North American Development Bank (NADBank). Washington, DC.

United States Environmental Protection Agency (USEPA). 1991. The Watershed Approach: An Overview. USEPA, Office of Water: Washington, DC.

United States Environmental Protection Agency (USEPA). 1992. Minutes of the National Coordinators's Meeting, June 22-25, 1992. USEPA: Sante Fe, NM.

United States Environmental Protection Agency (USEPA) and Secretaria de Desarrollo Urbano y Ecologia (SEDUE). 1991. Integrated Environmental Plan for the Mexican-US Border Area, First Stage, 1992-1994. USEPA: Washington, DC

Waterstone, M. 1994. "Transboundary Water Resources Management: An Institutional Perspective." Paper presented at NATO Advanced Study Institute, 23 May, 1994. Skopelos Island, Greece.

White, G.F. 1977. Environmental Effects of Complex River Development. Westview Press: Boulder, CO.

Zedler, J. 1993. Personal Conversation, Department of Biology, San Diego State University. 24 May, 1993.

Zedler, J., C. Nordby, and B. Kus. 1992. The Ecology of Tijuana Estuary: A National Estuarine Research Reserve. NOAA Office of Coastal Management, Sanctuary and Reserve Division: Washington, DC.

Appendix

Table 1 Listing of Agencies to be Contacted for Interviews and Access to Infrastructure Plans.

US Agencies; officials and staff at the following agencies:

- City of San Diego Clean Water Program staff,
- City of San Diego Water Utilities Division staff,
- SDAWMD officers and staff,
- IBWC staff and officers,
- University faculty at SDSU, UCSB, and UCSD,
- San Diego County Water Authority staff,
- City of San Diego elected officials (council members and Mayor's Office) and staff,
- County of San Diego Board of Supervisors and staff,
- County of San Diego staff (Depts. of Public Health, Planning, and Land Use),
- TRNERR staff,
- Members of special interest groups (Citizens Revolting Against Pollution (CRAP), Citizen's Coordinate for Century Three, and others),
- Staff at consultants and contractors (RECON, Malcom Pirnie, Swanson Oswald, and others), and
- Officials at the USEPA, RWQCB, SWRCB, and USACE.

Mexican agencies; officials and staff at the following agencies:

- CNA --Comision Nacional de Aguas, Mexicali,
- INEGI --Instituto Nacional de Estadistica, Geografia, y Informacion, Tijuana, Mexicali, y Aquascaliente,
- SEDESOL --Secretaria de Desarrollo Social, Mexicali,
- CILA --Comision Internacional de Limites y Aguas, Tijuana y Ciudad Juarez,
- SAHOPE --Secretaria de Asentamientos Humanos y Orbas Publicas del Estado, Mexicali,
- CESPTT --Comision Estatal de Servicios Publicos de Tijuana y Tecate, Tijuana,
- CESPM Comision Estatal de Servicios Publicos de Mexicali,
- COLEF --El Colegio de la Frontera Norte, Tijuana y Mexicali,
- COSAE --Comision de Servicios Aguas Estado (Sistema de Agua Residuales y Negras, Punta Bandera), and
- DPDUE --Direccion de Planeacion del Desarrollo Urbano y Ecologico, Municipio de Tijuana

Table 2 Laws, Regulations and Treaty Documents Dealing with Water Provision and Wastewater Treatment in the US and Mexico.

US Laws and Regulations

- Federal Water Pollution Control Act (FWPCA), 1948 and FWPCA (1972) - P.L. 92-500,
- Federal Clean Water Act (FCWA) of 1977 (P.L. 95-217), the most recent amendment to FWPCA,
- Federal Safe Drinking Water Act (SDWA), 1974, (P.L. 93-253),
- Department of Health Services (DHS) Title 22 Regulations for non-potable water reclamation,
- The "Framework Document for Potable Reuse of Reclaimed Water for Surface Water Augmentation, Departments of Health Services (DHS) and Water Resources (DWR), and
- California Safe Drinking Water Act, Chapter 7, Section 4010, California Health and Safety Code

Mexican Laws and Regulations

- Ley Federal Para Prevenir y Control de la Contaminición Ambiental (Federal Law for the Prevention and Control of Environmental Contamination),
- Reglamento Para la Prevención y Control de la Contaminación de Aguas, Regulation for the Prevention and Control of the Contamination of Water,
- Ley Nacional de Agua (National Water Law),
- Plan de Ordenamiento Ecologico del Estado, Baja California, Norte (General Ecological Inventory of the State of Baja Norte,
- Integrated Project for Potable Water and Sanitary Sewer of Tijuana, and
- Ley General de Equilibrio Ecologico y Proteccion Ambiental (General Law on Ecological Balance and Environmental Protection).

US-Mexico Environmental Treaty Documents

- Treaty Regarding Utilization of the Waters of the Colorado and Tijuana Rivers and of the Rio Grande, (1944 Water Treaty),
- Cooperation for the Protection and Improvement of the Environment in the Border Area, 1983 (La Paz Agreement); Annex/Article 2 is of particular interest, and
- IBWC Treaty Minutes #'s 261, 263, 270, 283 dealing with border sanitation and the Tijuana River Basin in Particular.

PART 2

ENGINEERING PLANNING AND MANAGEMENT

INTRODUCTION TO PART 2

This part comprises two chapters. Chapter III provides a theoretical background and also engineering models and schemes for managing transboundary water-related data and for compromise decision-making at mostly the implementation level. Chapter IV focuses on applications at the operational level.

Ganoulis (III.1) provides a *risk analysis framework*, which may be important in two areas

(1) Data collection and sharing between riparian countries

(2) Decision-making process for settling an agreement between partners.

As explained in two examples (river pollution, fuzzy regression), stochastic techniques and fuzzy arithmmetic may be suitable for quantifying risks in transboundary data collection and management.

Bogardi, Bardossy and Duckstein (III.2) develop a multicriterion algorithm that incorporates fuzzy logic; this approach, labeled *fuzzy composite programming*, makes it possible to address simultaneously two fundamental issues in transboundary water management:

(1) Conflict situations at two levels

- (a) conflict among goals, objectives and criteria, in particular, economic, environmental and social criteria
- (b) conflict among groups of actors.

(2) Uncertainties

- (a) in criterion values
- (b) in the preference functions of the actors or interest groups.

It should be noted that such a model is preferably to be applied only after the policy level institutions have been firmed up.

Duckstein, Bella, Stakhiv and Waterstone (III.3) provide another application of fuzzy composite programming dealing with the implementation level decisions of the same upper Rio Grande case, as used earlier by Waterstone (II.4).

Buck (IV.1) shows how existing implementation institutions at the border between France and Germany along a 180 km stretch of the Rhine River lead to good operational level decisions. Here composite programming (non-fuzzy) seems to provide good compromise schemes to manage the multiobjective system.

Next, Nachtnebel (IV.2) presents the operational level case study of conflict between the Czech Republic and Austria along a reach of the Danube River, where economies (hydropower, flood control) are in conflict with ecology (wetland forests, ground water, water quality).

Eberhardt, Rassam and Duckstein (IV.3) deal with the weekly operational level decisions of

Lake Ontario, which forms part of the U. S.- Canadian border. The policy and implementation decisions have been agreed upon, so that decision support systems for operational purposes can be designed. Two such systems described in the paper are tested, including a simple linear multiattribute value model that seems to perform well. This paper, which corresponds to present practice, shows that government agencies, such as the U. S. Corps of Engineers, proceed with caution when implementing new concepts, here, multicriterion decision making. The use of more sophisticated techniques, such as the fuzzy composite programming algorithm used in both Bogardi et al. (III.2) and Duckstein et al. (III.3) may be introduced later and slowly.

Similarly, the so-called multicriterion Q-Analysis Technique presented by Eder, Nachtnebel and Duckstein (IV.4) will take a certain time before being considered for use by action agencies. Yet this is one of the rare approaches that can rank not only a set of discrete alternatives evaluated by several or many conflicting criteria, but, as illustrated in the paper by a lower Austrian Danube case study, the approach can be used to help reduce an overly voluminous criterion set.

In summary, it appears that risk analysis and multicriterion decision-making approaches may help in sharing data, as well as modeling and solving conflicts at the implementation and operational levels in transboundary and international river basins. This presumes that policy level decisions have been agreed upon; unfortunately, we do not have as good a grasp on modeling those types of decisions, as we do in specifying implementation-operational level decision-making.

Chapter III: Risk Analysis and Compromise Decision Making in Water Disputes

III.1 Engineering Risk Analysis in Transboundary Water Resources Management

Jacques Ganoulis
Hydraulics Laboratory, School of Engineering
Aristotle University of Thessaloniki
54006 Thessaloniki
Greece

Introduction

This paper emphasizes the role of engineering risk assessment and management in transboundary water resources conflicts. This task may be important in two domains

(1) Data collection and sharing between riparian countries

(2) Decision-making process for settling an agreement between partners.

Concerning the first domain, different hydrological and water quality characteristics, such as water resources potential, water requirements, and water pollution data are needed in bilateral and international agreements on transboundary water resources management. These quantities are usually documented as the result of long-term monitoring and data collection in transborder regions (Wright et al., 1993). Some important issues related to transboundary data acquisition and management are

- Development of transboundary hydrologic data bases
- Monitoring of transborder environmental data
- Sharing transboundary data bases.

Point estimates of the above quantities are often subject to disagreement and suspicion between riparian actors. This is mainly due to the low reliability of such estimates, the origin of which is related to various uncertainties in obtaining data by different technological tools of the riparian institutions. A better common basis for cooperation and agreement may be reached when quantification of hydrologic data and water quality parameters is accompanied by a reliability analysis of their estimates. This may be based on stochastic methods, interval estimation, and the fuzzy set theory (Ganoulis, 1994).

The second domain of application of engineering risk analysis, which is the decision-making process under uncertainty, is largely developed in the papers which are included in this chapter. Having estimated the risk for several alternatives, it is important to define the consequences to the riparian partners by deciding a particular agreement under risk. This is very useful in order to estimate the range of levels of acceptable or assumable risks.

This paper focuses on the issue of risk and reliability in data collection and analysis. Engineering risk and reliability assessment and management may be considered as a general framework with applications to the process of resolving transboundary water-related conflicts.

Risk and reliability have different meanings and may be used differently in various disciplines such as engineering, mathematics, economics, medical and social sciences. The situation is sometimes confusing because terminologies and notions are transferred from one discipline to another without modification or adjustment. *Engineering Risk Analysis* is mainly based on the quantification of various uncertainties which may occur in the processes. *Ecosystem risk* and *Health Risk* should be distinguished from *Engineering Risk* on water quality. The latter refers to exceedances of pollutant concentrations on the abiotic water quality, as far as the quality specifications and regulation are concerned (Ganoulis, 1994).

In this paper, after recalling some basic definitions of risk analysis in water management problems, hydrologic and water quality uncertainties are first analyzed. Then, methods and tools for quantifying risks in hydrologic data are briefly reviewed. Some examples of application by use of stochastic simulation and the fuzzy logic approach (Ganoulis and Duckstein, 1995) illustrate the underlying concepts.

Basic Elements in Transboundary Water Resources Management

Management of transboundary water resources involves addressing not only physical and technical issues but also social actors, institutions and administrative procedures (Fig. 1). According to LeMarquant (1990), 5 foreign-policy factors influence international water situations: (1) international posture of each country, (2) international law, (3) linkage between water and other issues, (4) mutual commitment (reciprocity), and (5) national sovereignty.

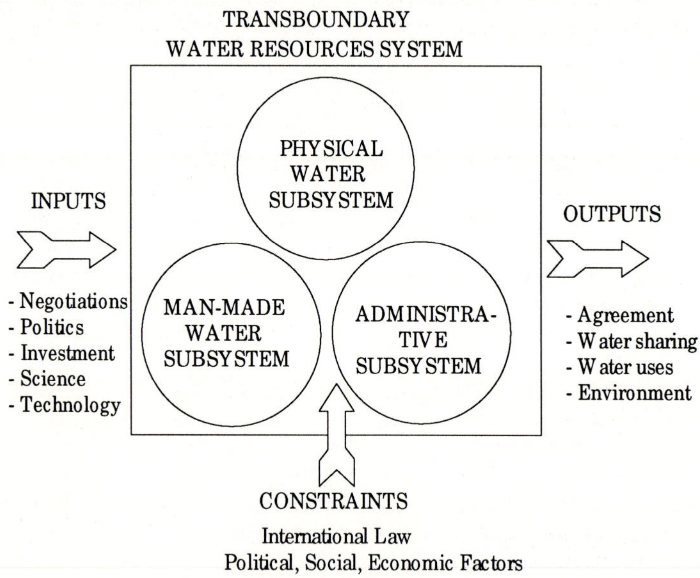

Figure 1 Elements in transboundary water resources system.

The main objective of an effective management of transborder water resources is to satisfy the demand of all partners, given the possibilities and limitations of water supply (Fig. 2). This balance between supply and demand should take into consideration both water quantity and quality aspects and the protection of the environment. Water quantity and quality problems are very much inter-related and should be studied in an integrated framework. According to Frey (1984), for understanding the origin of serious conflicts over international water systems, three main factors should be considered: (1) the importance of water (both in quantity and quality), (2) the relative power of the actors, and (3) the respective riparian position of the countries.

Definition of Risk

Let us consider the case of water quality. Water pollution may be detected by means of a characteristic variable expressing the existing state of water quality. This should reflect the quality of the water body, such as river, lake, coastal area or a groundwater aquifer, in terms of concentration of a specific polluting substance. For example, the maximum pollutant concentration of coliform bacteria at a specific site may be selected as the characteristic variable. This is a function of various conditions, such as wastewater loadings, flow characteristics and kinetic transformations controlling the water quality.

Consider the case of an international river discharging pollutants on a coastal area. As shown in Fig. 3, at the sea surface, a two-dimensional field of pollutant concentration may be measured, corresponding to the chemical composition of the pollutants. The latter could be nitrates, phosphates or coliform bacteria, found in variable concentration in municipal wastewaters with or without preliminary treatment. The maximum concentration of a specific pollutant C_m may be chosen to indicate some of the adverse effects produced by the river to the water environment.

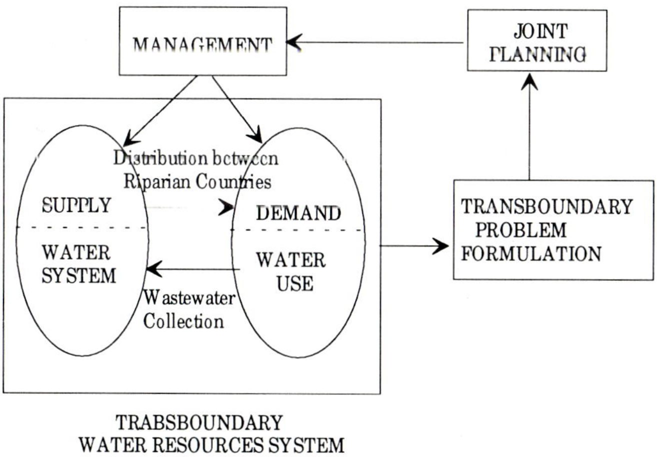

Figure 2 Transboundary water resources management system.

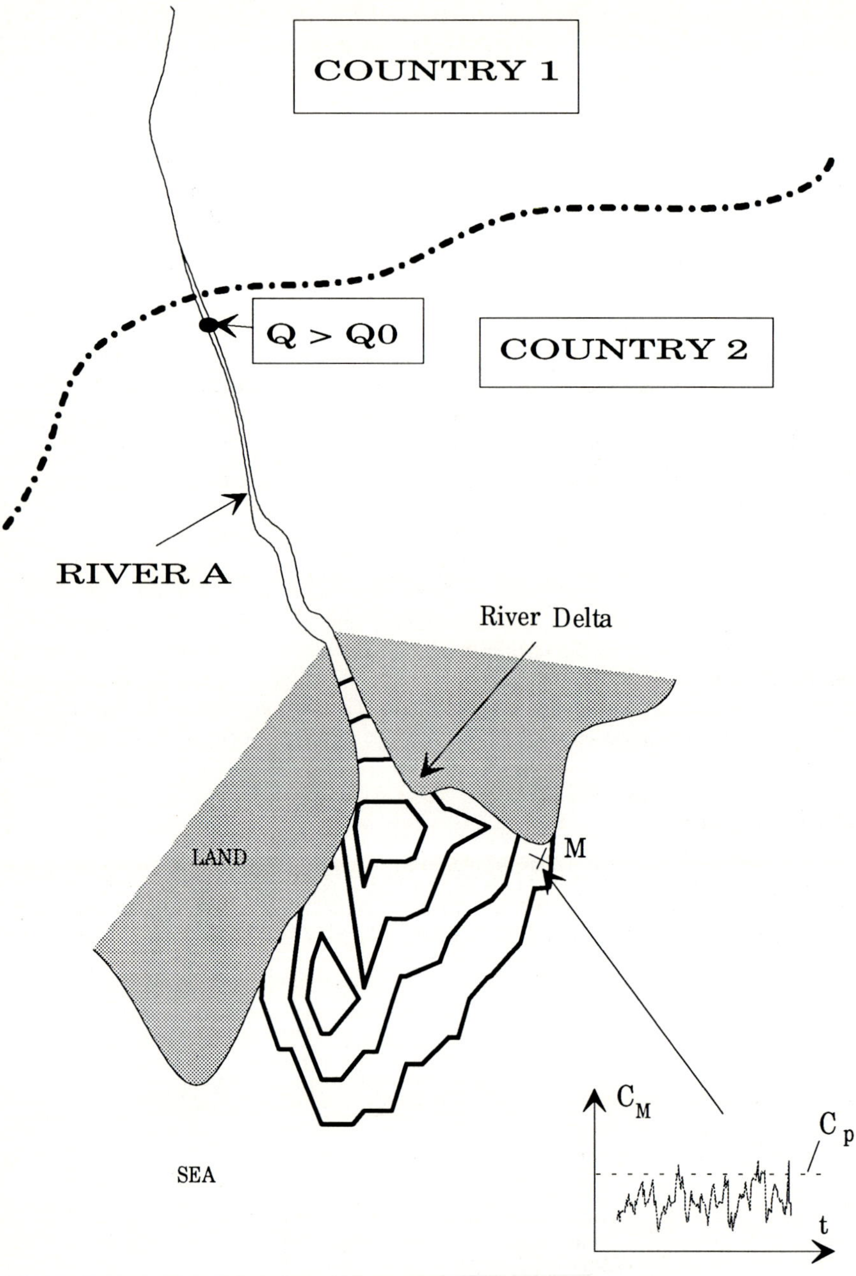

Figure 3 Two-dimensional pollution concentration field at the sea surface from a river.

Depending on the specific use of seawater, e.g. bathing, oyster farming, fishing, etc., environmental quality standards specify the *maximum allowable pollutant concentration* in seawater C_p (Fig. 3). This concentration should be considered as the *capacity* of the marine system to sustain pollution. A critical condition producing adverse effects and risk for seawater pollution occurs when the maximum pollutant concentration observed C_m exceeds the receiving capacity of the system, i.e., when $C_m > C_p$. Otherwise the system is safe, as far as pollution is concerned. So we have

$$\text{POLLUTION :} \quad C_m > C_p$$

$$\text{NO POLLUTION :} \quad C_m \leq C_p$$

As explained in Ganoulis (1994) the above definition may be extended if we consider as *load* ℓ a variable reflecting the behavior of the system under certain external conditions of stress or loading. There is a characteristic variable describing the capacity of the system to overcome an external load. We should call this system variable *resistance* r. There should be a *failure* or an *incident* when the load exceeds this resistance, i.e.,

$$\text{FAILURE or INCIDENT} \quad : \ell > r \tag{1}$$

$$\text{SAFETY or RELIABILITY} : \ell \leq r \tag{2}$$

Environmental Risk Assessment and Management

As shown in Fig. 4, application of environmental risk analysis consists of two main phases

(1) the assessment of risk, and
(2) risk management.

The assessment of risk is mainly based on modeling of the physical system, including forecasting of its evolution under risk. Although the main objective of risk analysis is the management of the system, it is not possible to do that if risk has not been quantified beforehand.

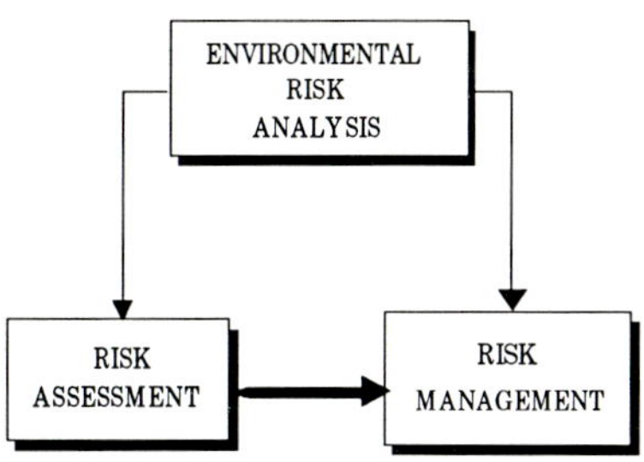

Figure 4 Risk assessment and risk management as elements of environmental risk analysis.

The phase of risk assessment involves the following steps

Step 1 : Risk or hazard identification

Step 2 : Assessment of loads and resistances

Step 3 : Uncertainty Analysis

Step 4 : Risk Quantification.

When it is possible to assess the risk under a given set of assumptions, then, the process of risk management may begin. The various steps in risk management are

Step 1 : Identification of alternatives and associated risks

Step 2 : Assessment of costs in various risk levels

Step 3 : Technical feasibility of alternative solutions

Step 4 : Selection of acceptable options according to the public perception of risk, government policy and social factors

Step 5 : Implementation of the optimal choice.

Because of the human and social questions involved, risk management is the part of the whole process which is the most important and most difficult to develop. From the engineering point of view, theories and algorithms of optimization under uncertainty, multi-criterion optimization and decision-making under risk are all applicable.

Uncertainties in Transboundary Water Problems

Several authors in the literature have analyzed different types of uncertainties and made various distinctions, such as between objective and subjective, basic and secondary, natural and technological uncertainties. Uncertainties are due to our lack of knowledge about the structure of various physical and biochemical processes and also to the limited amount of data available. Distinction should be made between

(1) aleatory or natural uncertainties or randomness and,

(2) epistemic or man-induced or technological uncertainties.

Aleatory Uncertainties or Randomness

It is postulated that natural uncertainties are inherent to the specific process and they can not be reduced by use of an improved method or more sophisticated model. Uncertainties due to natural randomness or aleatory uncertainties may be taken into account by using the stochastic or fuzzy methodologies, which are able to quantify uncertainties.

Epistemic or Man-induced Uncertainties

Man-induced uncertainties are of different kinds: (a) data uncertainties, due to sampling methods (statistical characteristics), measurement errors and methods of analyzing the data (b) modeling uncertainties, due to the inadequate mathematical models in use and to errors in parameter estimation, and (c) operational uncertainties, which are related generally to the construction, maintenance, and operation of engineering works. Contrary to natural randomness, man-induced uncertainties may be reduced by collecting more information or by improving the mathematical model.

Fig. 5 shows time series of a the river flow rate and nitrate concentration. These curves are based on collected data. They include both natural randomness and imprecision due to data collection, sampling and subsequent laboratory analyses.

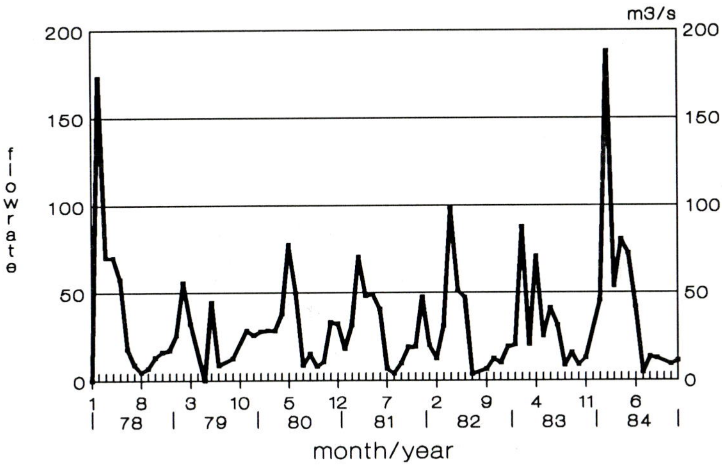

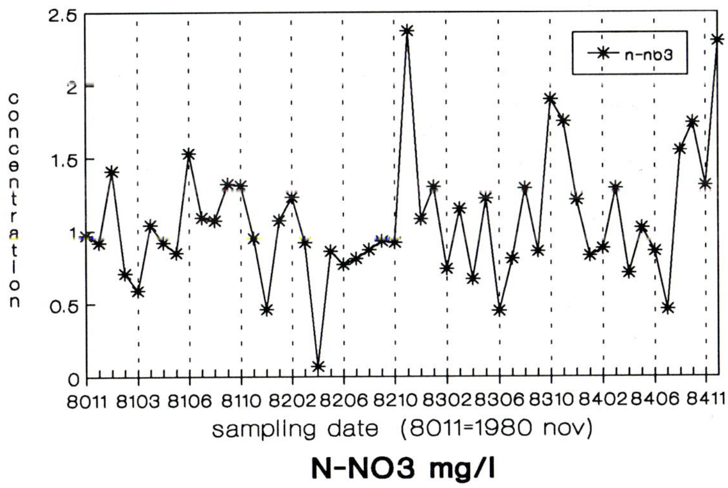

Figure 5 Typical time series of flow rate (m^3/s) and concentration of nitrate-nitrogen N-NO_3 (mg/l) measured in a river (Ministry of Agriculture, Greece).

An incident or failure is produced if the concentration C of a characteristic substance exceeds the limits given by water quality standards. For example, as shown in Fig. 6, dissolved oxygen (DO) may fall below 5 mg/l, which is generally accepted as the minimum concentration of DO necessary for the survival of different marine organisms.

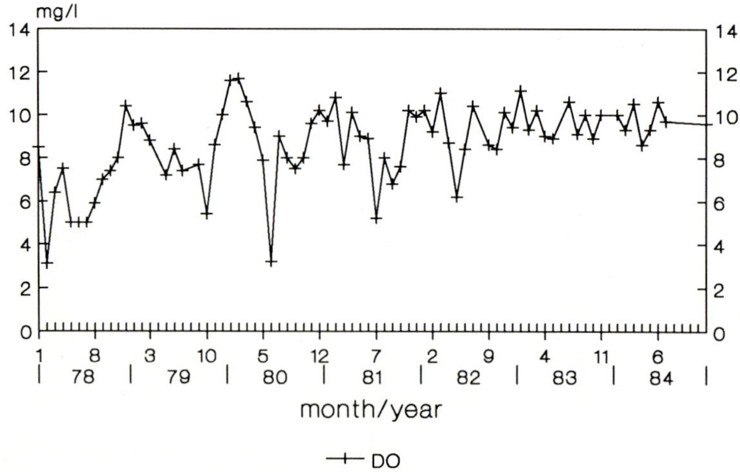

Figure 6 Typical time series of dissolved oxygen (DO) measured in a river (Ministry of Agriculture, Greece).

Risk Quantification

Because, by definition, risk analysis is related to uncertainties, quantification of risks should be based on methodologies that may take into account uncertainties. Two main theoretical approaches are available for doing this (1) the probabilistic approach, and (2) the fuzzy set theory. A review of such methods to quantify risks is given in Ganoulis (1994).

Although the stochastic approach is relatively well established, fitting of probability laws and analyzing the dependencies between random variables need large quantities of data which are not always available. If load and resistance are assumed to be independent, direct integration may be applied to quantify risk and reliability. Available data may be used to determine extreme values and the risk of exceedance, such as the hydrologic risk. Another possibility for quantifying risks is the formulation of stochastic differential equations. Monte Carlo simulation is a powerful technique for numerical representation of the system and subsequent risk quantification.

When very small samples of data are available or when variables are not easy to measure, fuzzy sets and fuzzy calculus may be utilized for analyzing imprecision. This approach is relatively new in engineering, but it has been proven very useful so far in informatics and control theory.

Direct Stochastic Evaluation

Loads and resistances are considered here to be positive scalars, applied on a single component of a system and are assumed to be time-independent. The latter assumption is necessary to enable direct integration for risk quantification. The above approach is generally known as static reliability analysis. The results to be presented below may be generalized when loads and resistances are vectors with several components.

Let us consider the *load* or *exposure* ℓ as a random variable L. In this case uncertainties associated with the estimation of L are quantified by means of probabilistic methods. Alternatively, fuzzy set theory may be used to calculate uncertainties in the load or exposure by considering ℓ as a fuzzy number $\tilde{L}$.

The *resistance* or *capacity* r is expressed in the same units as the exposure. Because in many cases it is also uncertain, probabilistic methods or fuzzy set theory may be used to describe resistance either as a random variable R or a fuzzy number $\tilde{R}$.

Suppose now that both ℓ and r are positive random variables and that probabilistic methods are utilized to quantify risk. In the general case, the risk p_F may be computed as follows (Ganoulis, 1994)

$$p_F = P(L > R) = \int_0^\infty \left\{ \int_0^\ell f_{LR}(\ell, r) dr \right\} d\ell \tag{3}$$

where $F_{L,R}$ is the joint probability density distribution function.

Introducing the conditional probability $f_{LR}(\ell/r)=f_{LR}(\ell, r)/ f_R(r)$ into Eq. 3 we obtain

$$p_F = \int_0^\infty f_{LR}(\ell / r) \left\{ \int_0^\ell f_R(r) dr \right\} d\ell . \tag{4}$$

Let L and R be independent random variables, i.e. $f_{LR}(\ell/r) = f_L(\ell)$. In this case, as shown in Fig. 7, the integral takes non-zero values in the overlap between the two curves $f_L(\ell)$ and $f_R(r)$, and Eq. 4 yields

$$p_F = \int_0^\infty f_L(\ell) \{ \int_0^\ell f_R(r) dr \} d\ell . \tag{5}$$

Introducing the distribution function $F_R(r)$ into Eq. 5, where $F_R(\ell) = \int_0^\ell f_R(r)\, dr$, the following equation is obtained

$$p_F = P(L > R) = \int_0^\infty F_R(\ell)\, f_L(\ell) d\ell \tag{6}$$

The successive steps for quantifying the risk by means of Eq. 6 are shown in Fig. 7.

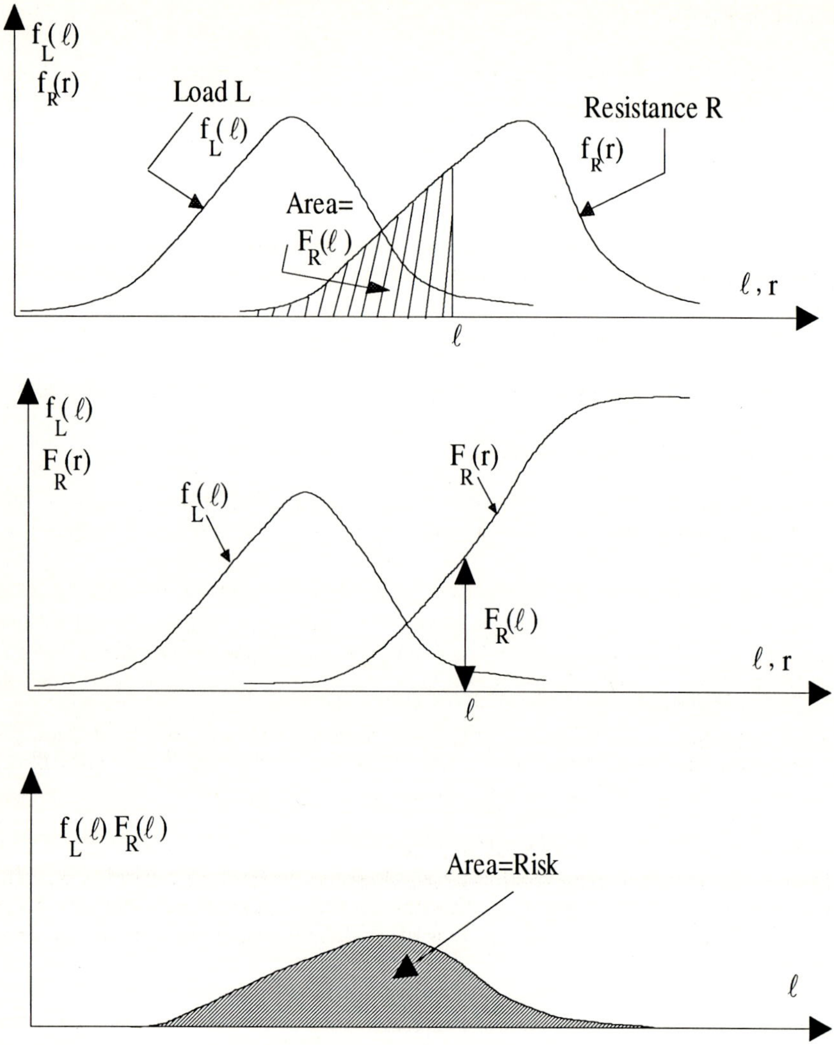

Figure 7 Steps for direct quantification of risk.

Fuzzy Regression

Obtaining functional relationships between available data, which are realizations of two or more variables, is very useful in order to compile available information by means of analytical expressions and make predictions using the available data.

In a stochastic framework, statistical regression between two or more variables is a very popular technique. It is a statistical method for smoothing available data by means of linear or non-linear functional relationships between variables. However, uncertainties related to the number of available data may create bias on the form and value of coefficients of the regression functions. This is the most serious limitation of classical regression analysis based on three basic assumptions concerning the deviations or approximation errors, which should

- (i) form a sequence of *independent* random variables,
- (ii) have a *zero* mean value,
- (iii) have a *constant* variance.

To check the above assumptions a sufficient number of data (usually more than 30) is necessary in order to obtain meaningful statistical parameters, such as the correlation coefficient. If the available sample of observations is small, the estimation of the regression function coefficients and the analysis of variance are biased and there is no guaranty that the basic assumptions on approximation errors are valid. On the other hand, if the statistical hypothesis for validation of a given regression is rejected on the basis of the analysis of variance, there is no alternative from rejecting the data, even if they were of good quality.

Most of the time, data for some of the required variables are scarce, mainly because of the high cost required to obtain them. In this case, an alternative solution to the classical regression analysis is the application of fuzzy regression.

Fuzzy regression may be considered as an extension of interval regression. Interval arithmetics (Moore, 1979) may be interpreted as a special case of fuzzy number calculus (Kaufmann and Gupta, 1985). It is very useful to start with arithmetic operations on intervals and then, to generalize on multiple intervals with variable confidence levels. These are, in fact, fuzzy numbers.

Let x_i be the vector of i independent observations and y_i the data vector of the dependent variable, with i = 1, 2, . . ., n. The n pairs of observations are given in the following form

$$(x_i, y_i) \qquad i=1, 2, 3, \ldots, n \tag{7}$$

Interval Regression

Instead of considering X an Y as random variables it may be assumed that both or one of them are the interval variables $\overline{X}$ and $\overline{Y}$.
Instead of the lower and upper limits c_l and c_2, an interval $\overline{C}$ may be defined by its center c_k and radius c_r. In this case we use the notation

$$\overline{C} = < c_k,\ c_r > \text{ where } c_k = \frac{1}{2}(c_1 + c_2) \text{ and } c_r = \frac{1}{2}(c_2 - c_1) \tag{8}$$

Let us consider now the variables $\overline{Y}_i$, $\overline{A}$ and $\overline{B}$ as the following intervals

$$\overline{Y}_i = < y_{ki}, y_{ri} >,\ \overline{A} = < a_k, a_r >,\ \overline{B} = < b_k, b_r > \tag{9}$$

For a given set of observation pairs (y_i, x_i) the following linear regression equation is defined

$$\overline{Y}_i = \overline{A} + \overline{B}\ x_i \tag{10}$$

Introducing the expressions (9) into Eq. 10, we obtain

$$< y_{ki}, y_{ri} > = < a_k,\ a_r > + < b_k,\ b_r >\ x_i \tag{11}$$

or

$$\left.\begin{aligned} y_{ki} &= a_k + x_i b_k \\ y_{ri} &= a_r + x_i b_r \end{aligned}\right\} \tag{12}$$

The coefficients a_k, a_r and b_k, b_r determine two regression lines which are lines of interval regression. They are subject to two conditions

(i) the "total" width between the two lines should be minimum
(ii) data points must be located between these two lines.

The above two conditions are sufficient to formulate mathematically the problem of finding the coefficients a_k, a_r and b_k, b_r. In fact, a linear programming problem may be written as follows

$$\sum_i (y_{2i} - y_{1i}) = 2 \sum_i y_{ri} \rightarrow \min$$

or

$$\sum_i y_{ri} = \sum_i (a_r + x_i\ b_r) = na_r + (\sum_i x_i)\ b_r) \rightarrow \min \tag{13}$$

where y_{1i}, y_{2i} are the lower and upper coordinates of the regression lines. The above condition (i) for minimization of the total width between the regression lines is mathematically expressed by the Eq. 13.

Conditions (ii) take the following mathematical form

$$y_i \leq y_{2i} = y_{ki} + y_{ri} = (a_k + x_i b_k) + (a_r + x_i b_r) \tag{14}$$

$$y_i \geq y_{1i} = y_{ki} - y_{ri} = (a_k + x_i b_k) - (a_r + x_i b_r) \tag{15}$$

Fuzzy Regression

The interval regression Eq. 10 is now replaced by the fuzzy linear regression equation

$$\tilde{Y}_i = \tilde{A} + \tilde{B}\ x_i \tag{16}$$

where $\tilde{Y}_i$, $\tilde{A}$ and $\tilde{B}$ are triangular and symmetrical fuzzy numbers (see appendix). Introducing the h-level interval (see Fig. 8), the Eq. 11 is now replaced by the following

$$[y_{ki}, y_{ri}]_h = [a_k, a_r]_h + [b_k, b_r]_h \; x_i \tag{17}$$

It is easy to realize that, in order to replace the interval regression lines by the h-level intervals, the fuzzy numbers $\tilde{A}$ (see Fig. 8) and $\tilde{B}$ should be defined as

$$a_k' = a_k \qquad b_k' = b_k \tag{18}$$

$$a_r' = \frac{a_r}{1-h} \qquad b_r' = \frac{b_r}{1-h} \tag{19}$$

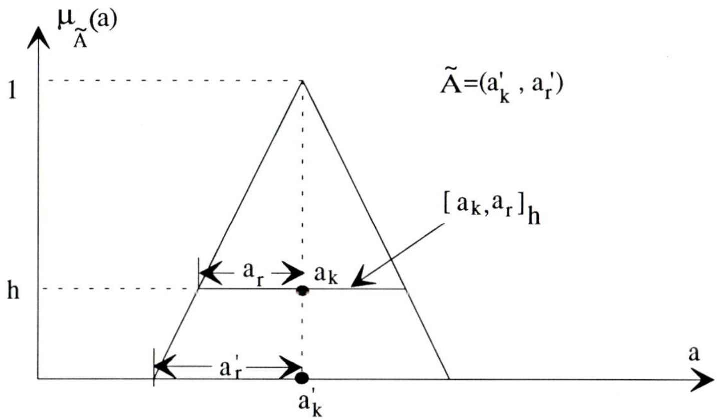

Figure 8 Definition of a triangular and symmetrical fuzzy number and the h-level interval.

where the coefficients a_k, a_r and b_k, b_r are the same as these defined by Eqs. 13, 14 and 15. For example the value h=0.5 may be chosen for the h-level interval corresponding to the regression lines which contain all data points. The fuzzy numbers $\tilde{A}$ and $\tilde{B}$ can be computed from Eqs. 18 and 19.

Examples of Application

Example (1): Direct Evaluation

To illustrate the meaning of the theoretical results, a numerical application is given for the case of river pollution. According to the environmental quality standards, which have been recommended by the environmental protection agency, there is pollution in a river when the pollutant concentration C(t) exceeds a critical concentration C_p. The permitted risk of exceeding the standards, i.e., the risk which does not affect the ecosystem, is supposed to be 10%. This means that we should have

$$p_F = P(C(t) > C_p) \leq 0.10 \ . \qquad (20)$$

If we multiply both terms in the above inequality by the river flow rate Q(t) we obtain

$$p_F = P(M_L(t) > M_R(t)) \leq 0.10 \qquad (21)$$

where

$M_L(t)=C(t)\,Q(t)$ is the pollutant load (mass per time)

$M_R(t)=C_p\,Q(t)$ is the allowed pollutant mass rate or the resistant load (mass per time).

Suppose now that both loads $M_L(t)$ and $M_R(t)$ are normally distributed. The resistant load is $M_R = N(45, 7.5^2)$ and the pollutant load $M_L = N(35, 5^2)$. It may be shown (Ganoulis, 1994), that Eq. 21 may be transformed as

$$p_F = 1 - \Phi\left(\frac{M_{R0} - M_{L0}}{\sqrt{\sigma_{ML}^2 + \sigma_{MR}^2}}\right) = 1 - \Phi\left(\frac{45-35}{\sqrt{7.5^2 + 5^2}}\right) = 1 - \Phi(0.864) \approx 0.19 > 10\%$$

where Φ is the cumulative error function.
We conclude that there is risk of pollution.

Example (2): Fuzzy Regression

Temperature data are given in Tab. 1 from 5 measurements taken at two different stations 1 and 2 in a coastal area.
Find a fuzzy regression between the two series of data.

Tab. 1 Temperature data at stations 1 and 2.

	$T_1(^0C)$	$T_2(^0C)$
1	11.2	9.1
2	13.7	10.2
3	15.0	9.3
4	21.4	7.4
5	22.4	18.2

Because the amount of data is small, application of the statistical regression appears to be problematic. The fuzzy linear regression (Eq. 16) was applied as an alternative, and the obtained results are shown in Fig. 9. The regression lines obtained as solution of the linear programming problem formulated by the Eqs. (13), (4), and (15) are taken as 0.5-level

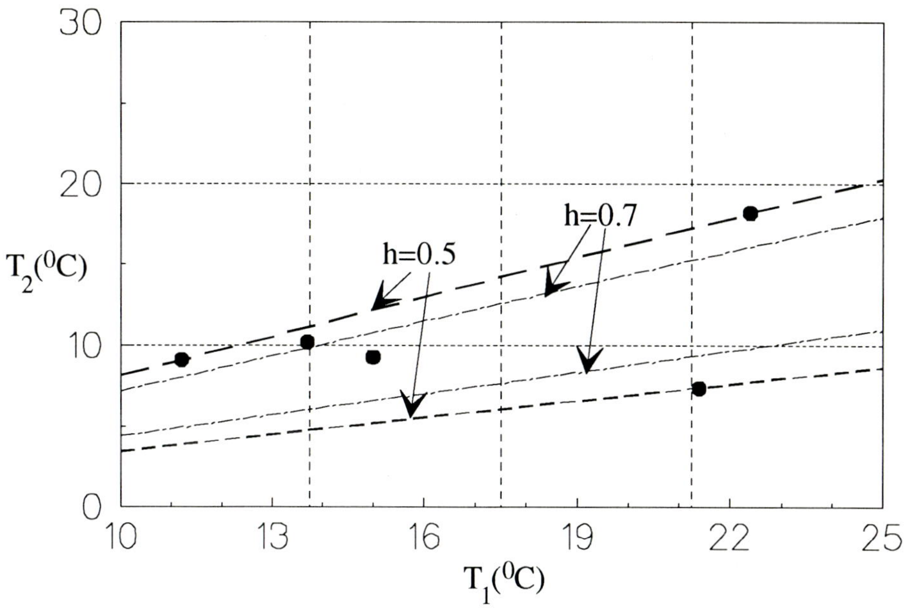

Figure 9 Fuzzy linear regression of a small sample of 5 data.

confidence intervals. It was found that

$$[a_k, a_r]_{h=0.5} = [0.0, 0.0] \text{ and } [b_k, b_r]_{h=0.5} = [0.579, 0.233].$$

Using the Eqs. 18 and 19 we have

$$\tilde{A} = (a'_k, a'_r) = (0.0,\ 0.0) \text{ and } \tilde{B} = (b'_k, b'_r) = (0.579,\ 0.466).$$

Conclusion

In this paper is shown how Engineering Risk and Reliability may be helpful for assessing uncertainties which may occur in transboundary data bases and international agreements relative to water sharing disputes. Methods and tools based on stochastic simulation and the fuzzy set theory are reviewed and their use for quantifying risks is illustrated in two simple and practical examples. These are a case of river pollution under risk and a fuzzy regression in a sample containing limited amount of data.

References

Frey, F.W. (1984) Middle East water: the potential for conflict or cooperation. In Naff T. and Matson, R.C. (eds.) *Water in the Middle East,* pp. 180-198, Westview Press, Boulder, Co

Ganoulis, J. (1994) *Engineering Risk Analysis of Water Pollution: Probabilities and Fuzzy Sets.* VCH, Weinheim, Oxford, New York, Tokyo, 306 pp.

Ganoulis, J. and L. Duckstein (1995) Fuzzy Regression between Atmospheric and Coastal Water Temperature for ecological Modeling, *PU.M.A.,* Ser. C, **3**, (1-4), 59-76

LeMarquant, D.G. (1990) International Development of the Senegal River. *Water International,* **15** (4): 223-230

Kaufmann, A. and M.M. Gupta (1991) *Introduction to Fuzzy Arithmetic: Theory and applications.* Van Nostrand Rheinhold, New York

Moore, R.E. (1979) *Methods and Applications of Interval Analysis.* SIAM, Philadelphia

Wright, R., Hepner, G. and Brown C. (1993) *Geographic Information Systems for Mexico-United States Border Environmental Research and Management,* Institute for Regional Studies of the Californias, San Diego State University, San Diego, USA, 31 pp.

Appendix

FUZZY NUMBERS

A fuzzy number may be characterized by three real numbers: two values of x where the membership function h reaches zero and at least, one where it reaches the value of 1. The simplest type of fuzzy number is a triangular one that is linear on either side of the peak (Ganoulis, 1994).

Let us define two triangular fuzzy numbers A and B by the triplets $A=(a_1,a_2,a_3)$ and $B=(b_1,b_2,b_3)$. We have

i) Addition: $A+B=(a_1+b_1,a_2+b_2,a_3+b_3)$

ii) Subtraction: $A-B=(a_1-b_3,a_2-b_2,a_3-b_1)$

Multiplication or division between two fuzzy numbers is not giving always a fuzzy number. However we can approximate them as follows

iii) Multiplication

$$A*B = [\min(a_1*b_1,\ a_1*b_3,\ a_3*b_1,\ a_3*b_3),\ a_2*b_2,\ \max(a_1*b_1,\ a_1*b_3,\ a_3*b_1,\ a_3*b_3)]$$

iv) Division

$$A/B=[\min(\frac{a_1}{b_1},\frac{a_1}{b_3},\frac{a_3}{b_1},\frac{a_3}{b_3}),\frac{a_2}{b_2},\ \max(\frac{a_1}{b_1},\frac{a_1}{b_3},\frac{a_3}{b_1},\frac{a_3}{b_3})]$$

III.2 Conflict Analysis Using Multiple Criterion Decision Making under Uncertainty

Istvan Bogardi
Ministry of Environment, Budapest, Hungary
on leave from Civil Engineering Department
University of Nebraska, Lincoln, USA

Andras Bardossy
Institute for Hydrology and Water Resources
University of Karlsruhe, Germany

Lucien Duckstein
Systems and Industrial Engineering Department
University of Arizona, Tucson, Arizona, USA

Introduction

The purpose of this paper is to present and apply conflict resolution techniques to evaluation and ranking of transboundary water resources development. It is a common endeavor to solve water-related conflicts through negotiations; the techniques presented herein are meant to be decision aids for such negotiations.

The dispute over shared water resources among nations has a long history. According to Gleick (1991, 1992), four classes of water-related disputes may be distinguished, i. e., water resources as 1) strategic goals, 2) targets during war, 3) tools or weapons of conflict and 4) roots of conflict. Conflict situations in transboundary water resources management occur on at least two levels: conflicts among criteria related to water management and conflicts of interest between the countries involved.

The first level of conflict, related to water management criteria, seeks a management or development scheme that is economically attractive, ecologically sound, socially acceptable and technically feasible. Generally, there is no scheme that satisfies all these criteria completely. Quite often in the history of water resources development, only the first and last criteria have been considered. This has led to a number of water projects, mostly of a hydraulic engineering nature such as dams, which resulted in environmental disruption and/or became socially unacceptable. Multiple-criterion decision making (MCDM) techniques can be used to address the conflict situation at this level. The main idea is to find the best compromise among the four types of criteria. It is likely that no criterion will reach its ideal value, but an acceptable lower value may be identified in order to work out a good water resources scheme. MCDM can be applied to help reach such a solution. There are many real-life applications of MCDM at this first level, including water resources (David and Duckstein 1976; Goicoechea et al., 1982; Tecle et al., 1988; Bogardi and Duckstein 1988; Eder et al., 1994).

The second level-of-conflict situation, differing interests of the countries involved, is typical of transboundary water management. As a consequence of this type of conflict, the "best" scheme obtained after addressing the first level may not be acceptable to any of the stake-holding countries. MCDM techniques have rarely been used to address this second level of conflict situation, let alone both levels, in a given decision making situation (Cordeiro Netto et al., 1993).

We present MCDM techniques which are capable of finding in two steps water management schemes under both levels-of-conflict situations. In the first step, we select the best scheme from the technically feasible alternatives without considering the different interests of the countries involved. In the second step, we address conflicts among those countries. In brief, we seek alternatives which are the closest to the best solution obtained in the first step and acceptable to all stakeholders.

In addition to difficulties posed by conflict situations, transboundary water resources management is aggravated by uncertainty. Uncertainty may stem from different sources, including estimates of various economic, ecologic or social criteria. Even within one country, preferences among criteria are often defined imprecisely. When several countries are involved, this uncertainty increases substantially. We believe that fuzzy logic is the most appropriate technique to account for these uncertainties.

This paper is organized together with a related paper (Duckstein et al., 1994a). Specifically, the present paper contains the methodological basis and formulates the two-step approach for addressing the conflict situations with an application example. Duckstein et al. (1994) present in detail a two-level conflict case study in the upper Rio Grande Basin, New Mexico, U. S. A, using Fuzzy Composite Programming (FCP). In the next section, Composite Programming (CP) is described as the selected MCDM technique. The elements of fuzzy logic are then presented, and FCP is applied to both conflict levels that face transboundary water management.

Composite Programming (CP)

Several techniques for MCDM analysis are available, including multiattribute utility methods and ranking methods such as ELECTRE (Goicoechea et al., 1982). The methodology selected in this study is composite programming (CP) [Bogardi et al., 1983; Bardossy et al., 1985; Duckstein et al., 1994b], which is an extension of compromise programming (Duckstein and Opricovic 1979). CP has been selected because of its ability to integrate fully the different elements of transboundary water management into a common framework. It also represents a compromise between mathematical sophistication and ease of use. CP is a distance-based MCDM technique according to the definition of Szidarovszky et al. (1986). Distance-based methods assume that the decision maker visualizes the decision making process within the framework of a spatial coordinate system.

After the problem has been properly posed in terms of system definition and statement of objectives, CP organizes a water management problem into the following format:

(1) definition of basic indicators;

(2) definition of management alternatives;

(3) progressive grouping of basic indicators into fewer and more general groups;
(4) definition of weights, balancing factors, and worst and best values for the indicators;
(5) evaluation of the alternatives;
(6) ranking of the alternatives;
(7) performance of a sensitivity analysis.

Each management alternative is associated with economic, ecologic, social and political impacts expressed as basic indicators or criteria. The basic indicators are grouped into smaller, more general groups called first-level indicators. This grouping continues until only a few indicators such as socioeconomy and ecology remain.

After the indicators have been grouped into such a structure, weights are assigned to the indicators representing their relative importance to the decision maker. For example, a moderate risk to humans may be considered more important than a more severe risk to a non-endangered species of mammals.

Next, balancing factors are assigned to each group of indicators. Balancing factors set the maximal deviations of the indicators and limit the ability of one indicator to be substituted for another one. In other words, selecting an appropriately high balancing factor ensures that an indicator must not be exchanged. For example, increased risk to an endangered species will not be offset by lowered risk to an abundant, non-endangered species.

The *best* and *worst* possible values of the basic indicators for the particular study are then assessed. For instance, the *best* value for management cost would be the cost associated with most inexpensive control, while the *worst* cost would be that associated with the most expensive alternative. Likewise, for human carcinogenic risk, the *best* values would correspond to zero risk or zero incremental risk and the *worst* value would be the risk level reached without any control action.

Evaluation of the various management alternatives proceeds by computing composite distances using the CP algorithm. The first step is normalization (mapping the basic indicators' values Z_i onto the 0-1 interval). This is necessary since the units and ranges of the various basic indicators can be quite different (e.g., probability, mortality, dollars) and difficult to compare directly. The ability to compare and evaluate variables having different units (eliminating the need to transform all units into a common unit) is one of the primary benefits of using CP. Using the maximum Z_{i+} and the minimum Z_{i-},, the normalized value S_i of Z_i can be computed as:

$$S_i = \left| \frac{Z_i - Z_{i-}}{Z_{i+} - Z_{i-}} \right| \tag{1}$$

This formula maps the normalized value S_i onto the 0-1 interval at the appropriate distance from the *best* and *worst* values. Next, the first-level composite distances are computed for each first-level group of basic indicators using the equation:

$$L_j = \left[\sum_{i=1}^{n_j} a_{ij}^{p_j} \right]^{1/p_j} \tag{2}$$

where:

S_{ij} = normalized value of basic indicator *i* in the first-level group *j* of basic indicators,
L_j = composite distance for first-level group *j* of basic indicators,
n_j = number of basic indicators in group *j*,
a_{ij} = weight expressing the relative importance of basic indicator in group *j* such that *n*,å,*i*=1 ?*j* a_{ij} = 1
p_j = the balancing factor among indicators in group *j*.

As an example, Fig. 1 shows the first-level composite distance (tradeoff) of two basic indicators, human carcinogenic risk and human noncarcinogenic risk, for five waste disposal alternatives (Stansbury et al., 1991). These two basic indicators are used to determine the composite distance for the first-level indicator, human health risk. The numbered points represent the location (distance from the ideal point) of each management alternative with regard to the indicators being composited. The ideal point (graph origin) represents the point on the trade-off curve where a management alternative would be placed if the criteria under consideration (carcinogenic and noncarcinogenic risks) were both at their best possible level.

It should be noted that the term "tradeoff" is used to mean composite distance and that the variables being traded-off (composited) do not necessarily need to be antagonistic to one another. For instance, one might trade off (determine the composite distance for) human carcinogenic risk and human noncarcinogenic risk even though the two factors are not antagonistic to one another.

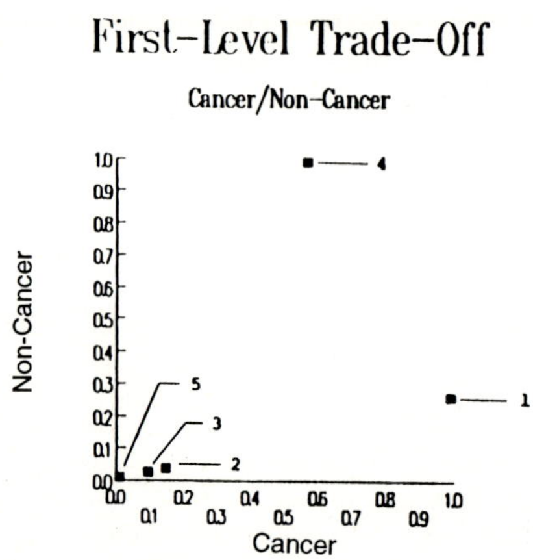

Figure 1 First-level tradeoff of two basic indicators for five waste-disposal alternatives.

The second-level composite distances are calculated using the first-level distances in the formula:

$$L_j = \left[\sum_{j=1}^{m_k} a_{jk} L_{jk}^{p_j} \right]^{1/p_k} \qquad (3)$$

where:

L_j = composite distance for second-level group *k*,
m_k = number of elements in second-level group *k*,
L_{jk} = first-level composite distance between group *j* and *k*,
a_{jk} = relative importance among elements in second-level group *k*, and
p_k = balancing factor for second-level group *k*.

Fig. 2 shows the second-level composite distance of the first-level indicators, human risks and ecological risks. These risks form the second-level indicator (overall risk). The process of computing successively higher levels of composite distances is repeated until the final composition between the highest level indicators is reached. At this point, in the case of two highest-level indicators, the final composite distance is computed as:

$$L = \left[a_1^p L_1^p + a_2^p L_2^p\right]^{1/p} \qquad (4)$$

where:

L = composite distance characterizing the actual state of the systems (e.g., ecology vs socioeconomy),
L_1 – composite distance for one group (e.g., ecology),
L_2 = composite distance for the second group (e.g., socio-economy),
a_1, a_2 = weights indicating relative importance between the two indicators,
p = balancing factor for the final composite distance.

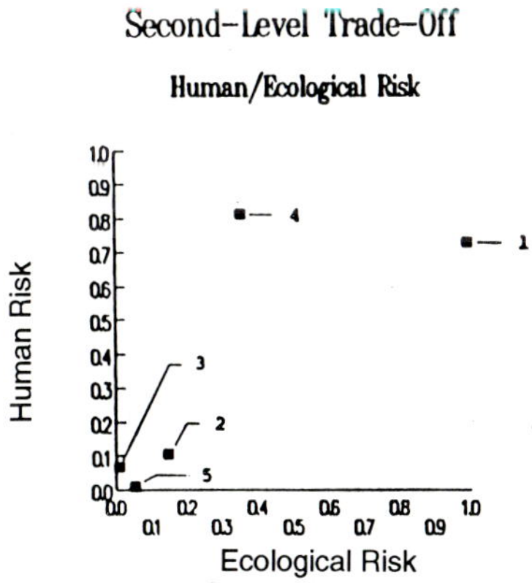

Figure 2 Second-level composite distance of two first-level indicators (overall risk).

Fig. 3 shows the final composite distances (tradeoffs) for this system. Uncertainties are introduced next, under the form of fuzzy logic.

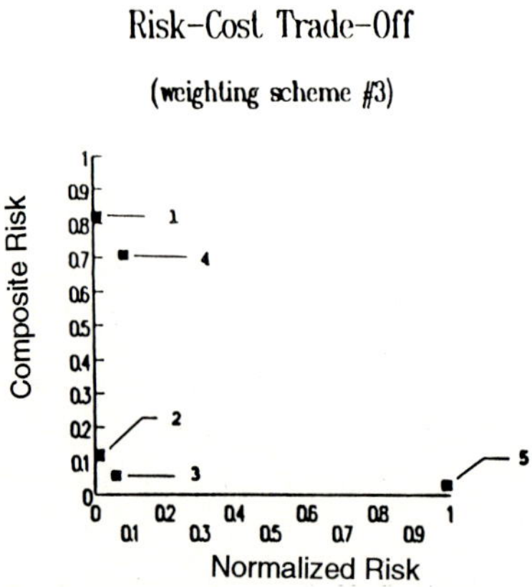

Figure 3 Final tradeoff (discrete format).

The Fuzzy Set Approach

Fuzzy Set Concepts

Fuzzy set theory (Zadeh, 1965; Zimmerman, 1985) is a mathematical method used to characterize and propagate uncertainty and imprecision in data and functional relationships. Fuzzy sets may be especially useful when insufficient data exist to characterize uncertainty using standard statistical measures (e.g., mean, standard deviation and distribution type).

Fuzziness represents situations where membership in sets cannot be defined on a yes/no basis because boundaries of the sets are vague. The central concept of fuzzy set theory is the *membership function* which represents numerically the degree to which an element belongs to a set. In a classic set, a sharp or unambiguous distinction exists between members and non-members of the set such that the value of the membership function of each element is either 1 for members or 0 for non-members.

However, many of the categories we commonly employ, such as "highly contaminated water," "high water table" or "numbers much greater than 1," are defined in linguistic terms and do not exhibit the distinct boundaries of the classical set. Even if the variables have precise numerical values, uncertainty comes from the extent to which, say, a four-feet-deep water table belongs to the fuzzy set "high water tables."

Fuzzy set theory introduces vagueness in order to simplify the handling of indistinct boundaries between classes (Klir and Folger, 1988). Thus, if an element is a member of a fuzzy set to some degree, the value of its membership function will be between 0 and 1. When the membership function of an element can only take on the values 0 or 1, fuzzy set theory reduces to classical set theory.

Fuzzy Sets as an Extension of Interval Analysis

The simplest method of considering uncertainty is to perform an interval analysis. An uncertain parameter, such as depth-to-water table, is assigned any value within an interval. Obtaining more information on the parameter allows us to "sharpen" the interval model, that is, we calculate with increased possibility that the parameter will take on certain value(s) within the interval. If the axioms and hypotheses of probability theory are verified, then probability theory is an extension of the interval analysis. If they are not verified, weaker hypotheses than those of probability theory are in order. In that case, we would use fuzzy set analysis, which allows few and weak prerequisite assumptions as an extension of interval analysis.

The Propagation of Fuzzy Sets

The propagation of uncertainty characteristics measured as intervals is the basis of traditional sensitivity analysis. The arithmetic of intervals is straightforward, and functions are easy to calculate (Dong and Wong, 1986). Multi-intervals can be defined in order to utilize additional information on an uncertain model parameter. For example, if we believe that the depth-to-water table is definitely greater than 2 feet and definitely less than 6 feet, most likely about 5 feet, but possibly somewhere else in the interval 2 to 6, then multi-intervals such as the ones sketched in Fig. 4 for $h=0$ and $h=0.5$ may be appropriate.

Fuzzy Numbers

A special class of fuzzy sets defined as *fuzzy numbers* describes the strength or degree of acceptance of the possible values of a number or a parameter. A fuzzy number has the following properties:

1. it is defined on the set of real numbers, rather than on a set of linguistic values;
2. its membership function always has a maximum value of 1;
3. its membership function consists of increasing and decreasing parts; and
4. a complete arithmetic is available to operate, rank or combine fuzzy numbers (Kaufmann and Gupta, 1991).

Furthermore, multidimensional functions of fuzzy numbers can be defined and transformed. The Appendix gives further details.

Referring to Fig. 4, the width of the membership function of a fuzzy number "about 5 but no less than 2 and not over 6" is $6 - 2 = 4$; this width is an interval length. If a so-called credibility level h = 0.5 is defined corresponding to a value for the strength of acceptance, then the level set h = 0.5 is defined as the interval [3, 5; 5, 5]. For h = 0, the level set is [2, 6].

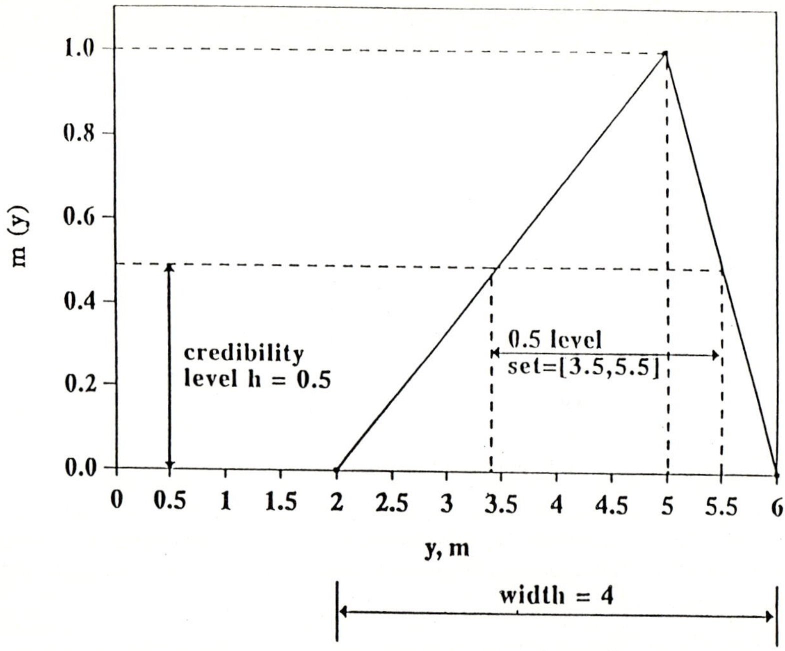

Figure 4 Membership function for a soil property of about 5, no less than 2, and not over 6.

Fuzzy Composite Programming (FCP)

The purpose of FCP is to provide a technique for embedding uncertainty/imprecision into the CP algorithm. Multicriterion decision-making tools such as CP, ELECTRE, multi-attribute utility theory and SEMOPS have been used to assist decision makers in dealing with problems involving multiple and conflicting objectives (Goichoechea et al., 1982; Keeney and Wood, 1977; Bogardi and Duckstein, 1992). However, uncertainties inherent in these problems are often treated separately from the decision process itself. That is, the decision analysis is often carried out using point estimates (no uncertainty) of decision variables, and uncertainty is then characterized by "plausible intervals" (Cothern et al., 1987). The decision maker must then either deal with point estimates, which provide no characterization of uncertainty, or with intervals showing bands of uncertainty, which are often large and difficult to use in a decision framework.

Fig. 5 illustrates the importance of considering uncertainty in decision-making, and the difficulty this causes. For example, if either management option A or B in Fig. 5 must be selected only on the basis of human health risk, then using point estimates of the risks leads to preference for management option B. However, when uncertainties spanning several orders of magnitude are associated with risk assessment (Dourson et al., 1983), it becomes difficult to determine which option is better.

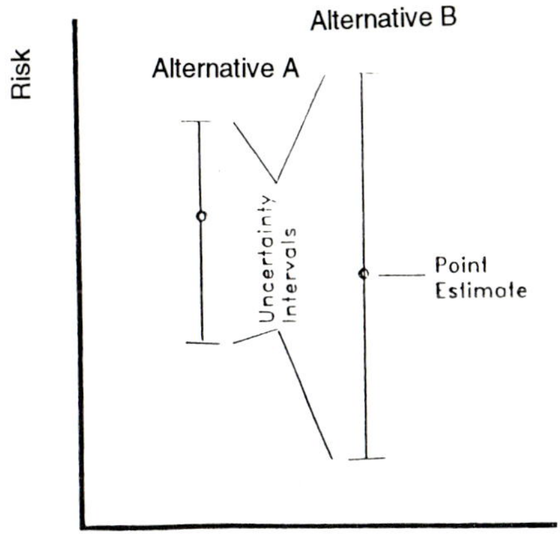

Figure 5 Point and interval estimates of risk.

Further, option B may possess other negative aspects (greater cost, loss of jobs), despite its acceptability on the basis of health risk. The FCP method formally ranks uncertain numbers such as ecological and economic estimates, and so allows the decision-making process to account for uncertainties.

Fuzzy Membership Values of Uncertain Criteria

Low, high and central values of basic water management indicators can be assessed and used to construct triangular membership functions, with the base level set between low and high values and the peak level set at central values. For example, the uncertainty in contaminant concentration is characterized by the highest, lowest and average recorded concentrations in groundwater. Other parameters such as population densities and consumption rates are also characterized in this manner. As a result, the uncertainty associated with the criteria is characterized by intervals surrounding the 0.0 level and a peak at the 1.0-level membership. Trapezoidal membership functions may also be used (Duckstein et al., 1994a). Several procedures have been developed to assess membership functions of fuzzy numbers (Chameau and Santamarina, 1987; Turksen, 1991).

FCP Algorithm

Evaluation of the various management alternatives under uncertainty proceeds as follows. Let $Z_i(k)$ be a fuzzy number for the *i*th basic indicator with a trapezoidal membership function $\mu(Z_i(k))$ [Fig. 6], where *k* is one element of the discrete set of management alternatives. A level cut or credibility level concept (Dong and Shaw, 1987) is used to decide the interval of each indicator at various credibility levels. As shown in Fig. 6, $Z_{i\alpha}(k)$ is the interval value of the *i*th basic indicator at level cut α . Since units of basic indicators such as ecology and economy are different and thus difficult to compare directly, the actual value of each basic indicator must be transformed into an index. Using the best value $BESZ_i$ of Z_i and the worst value $WORZ_i$ of Z_i for the *i*th basic indicator, the index function $S_{i,\alpha}(k)$, used to produce the index value of $Z_{i,\alpha}(k)$, can be represented by the piecewise linear function:

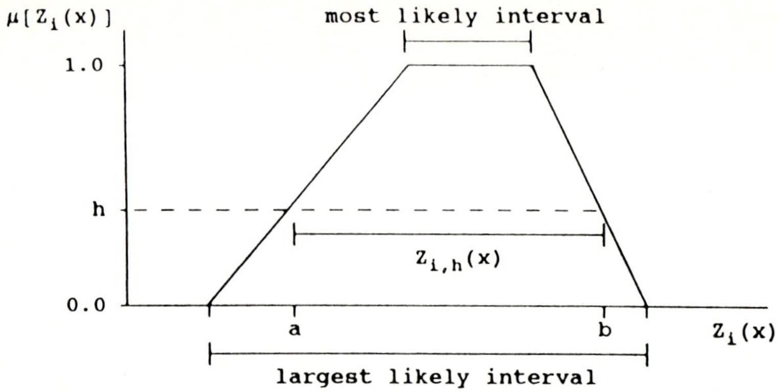

Figure 6 Fuzzy estimate of risk (trapezoidal membership function).

If $BESZ_i > WORZ_i$, the index function $S_{i,\alpha}(k)$ is defined by:

$$S_{i,\alpha}(k) = \begin{cases} 1, & z_{i,\alpha}(k) \geq BESZ_i \\ \dfrac{z_{i,\alpha}(k) - WORZ_i}{BESZ_i - WORZ_i}, & WORZ_i < Z_{i,\alpha}(k) < BESZ_i \\ 0, & z_{i,\alpha}(k) \leq WORZ_i \end{cases} \tag{5}$$

If $BESZ_i < WORZ_i$, the index function $S_{i,\alpha}(k)$ is defined by:

$$S_{i,\alpha}(k) = \begin{cases} 1, & z_{i,\alpha}(k) \leq BESZ_i \\ \dfrac{z_{i,\alpha}(k) - WORZ_i}{BESZ_i - WORZ_i}, & WORZ_i > Z_{i,\alpha}(k) > BESZ_i \\ 0, & z_{i,\alpha}(k) \geq WORZ_i \end{cases} \tag{6}$$

This index function takes on values between 0 and 1. As shown in Fig. 7, since variable $Z_{i,a}(k)$ is an interval with lower bound *a* and upper bound *b* (i. e., $Z_{i,a}(k) \in [a,b]$, the index functions, $S_{i,\alpha}(k)$ resulting from $Z_{i,a}(k) \in [a,b]$ is also an interval (i.e., $S_{i,\alpha}(k) \in [c,d]$.).

Symbolically, the interval $S_{i,\alpha}(k)$ of the index function for the *i*th basic indicator can be expressed as:

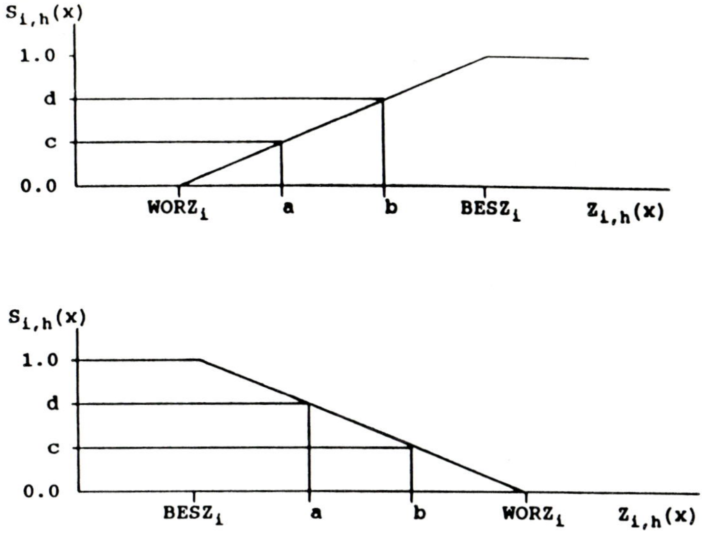

Figure 7 Function used to produce index of $z_{i,}(x)$.

$$S_{i,a}(k) = [\min S_{i,a}(k), \max S_{i,a}(k)] \tag{7}$$

where if $c \leq S_{i,a}(k) \leq d,\ \min S_{i,a}(k) = c$ and $S_{i,a}(k) = d$.

Next, index functions $L_{j,a}(k)$ for second-level composite indicators are defined as:

$$L_{j,a}(k) = \left[\sum_{i=1}^{n_j} w_{ij} \left(S_{i,a,j}(k)^{p_j} \right) \right]^{1/p_j} \tag{8}$$

where

n_j = number of elements in the second level group j,

$S_{i,a,j}(k)$ = fuzzy interval of the index function for the ith basic indicator in the second-level group j of basic indicator

a_{ij} = weight reflecting the importance of each of basic indicators in group j; $\Sigma\, w_{ij} = 1$ and

p_j = balancing factor among indicators for group j.

Then the fuzzy interval $L_{j,a}(x)$ of the index function for the second-level group j is given as:

$$L_{j,a}(k) = [\min L_{j,a}(k), \max L_{j,a}(k)] \tag{9}$$

Tradeoffs at other levels are calculated in a similar fashion until a final tradeoff distance representing composite ecology versus composite socioeconomy (in the example case considered herein) is determined for each alternative.

Ranking the Alternatives

If there are *K* management alternatives, the *K* -cuts $(L_a(k), k = 1,\ldots, K$ define *K* fuzzy numbers that correspond to the ecology versus socio-economy trade-offs. To determine the ranking of these *K* fuzzy numbers, the ranking method developed by Chen (1985) is applied. The method decides the ranking of fuzzy numbers by using a maximizing set and a minimizing set. The maximizing set *M* is a fuzzy subset with membership function μM given as:

$$\mu M(L) = \begin{cases} \dfrac{L - L_{min}}{L_{max} - L_{min}}, & L_{min} \leq L \leq L_{max} \\ 0, & \text{otherwise} \end{cases} \tag{10}$$

where for $k = 1,\ldots,K$, $L_{min} = \min[\min L_{a=0}(k)]$ and $L_{max} = \max[\max L_{a=0}(k)]$. Then the right utility value, $U_R(k)$ for alternative *k* is defined as:

$$U_R(k) = \max\{\min[\mu M(L), \mu L(k)] \tag{11}$$

The minimizing set *G* is a fuzzy subset with membership function μG given as:

$$\mu G(L) = \begin{cases} \dfrac{L - L_{max}}{L_{min} - L_{max}}, & L_{min} \leq L \leq L_{max} \\ 0, & \text{otherwise} \end{cases} \tag{12}$$

The left utility value, $U_L(k)$, for alternative *k* is defined as:

$$U_L(k) = \max\{\min[\mu G(L), \mu L(k)] \tag{13}$$

The total utility or ordering value for alternative *k* is then defined as:

$$U(k) = (U_R(k) + 1 - U_L(k))/2 \tag{14}$$

As a result, the final tradeoff distances of the disposal alternatives can be compared and ranked, even though they are uncertain (fuzzy). The alternative that has the highest ordering value in the discrete set of management alternatives is selected as the superior alternative. As an example after Stansbury et al. (1991), Fig. 8 shows the final trade-off between composite risk and cost considering five waste disposal alternatives. Note that there is much greater uncertainty in the risk estimates than in the cost estimates. The so-called UAD alternative provides the lowest-cost disposal alternative, but it does so at a relatively high risk. The Upland Secure site provides the lowest risk, but it requires a very large cost. The so-called CAD site provides the best trade-off between risk and cost. Table 1 shows the final ranking and the fuzzy trade-off distance L of the five disposal alternatives. If each country has a preferred alternative, then the CP procedure should be applied again to reach a compromise alternative acceptable across boundaries (Cordeiro-Netto et al.,1993; Duckstein et al., 1994a).

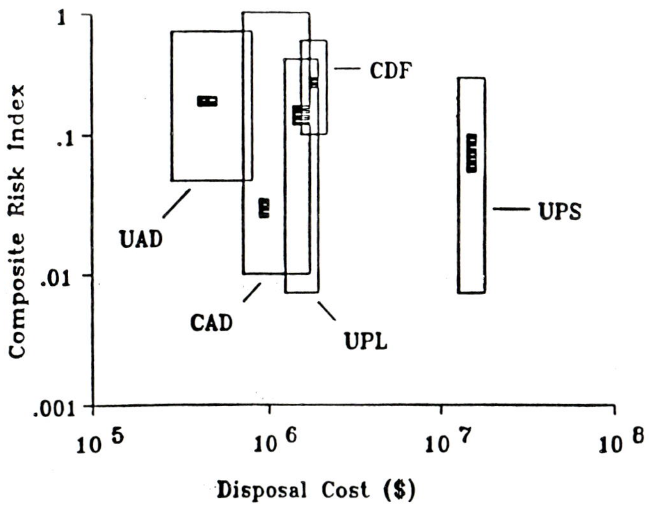

Figure 8 Final tradeoff between composite risk and cost.

Table 1. Final ranking of disposal alternatives.

Rank	Disposal Site	*Ranking Value
1	CAD	0.866
2	UAD	0.816
3	Upland	0.449
4	Upland Secure	0.408
5	CDF	0.315

*Ranking Value indicates relative acceptability of alternative where a value of 1.0 indicates an ideal alternative.

Illustrative Example

The conflict situation is as follows. Two countries disagree on problems of power production and related ecological consequences. The system consists of a watershed shared by the two countries. An upstream control point is available to allocate inflow for power generation and for ecological purposes. Water resources over the watershed include a lake with parts in both countries, a storage reservoir in Country A and river sections in both countries. Power production is shared in a five-to-one ratio between Countries A and B. Country A is mainly interested in the increase of power production, regardless of ecological or other consequences. Country B only marginally benefits from increased power production. Therefore, Country B's goals are improvement of the ecological, recreational and aesthetic state of the region.

Eight allocation alternatives are considered. The first water management alternative is maintenance of the status quo. Other alternatives are increased power and decreased power, both using the present production scheme, and choices of several different production schemes.

Conflict Resolution Method 1 - Separate Ranking of Criteria by Countries

As a first step, 14 basic indicators are agreed upon by both countries. Because the countries assign different importance to the same indicators, the indicators are considered twice. This results in a total of 28 indicators. The basic indicator values are the same, with the exception of recreation and aesthetics, which are assigned different values by each country. In this step each country forms its own groups of indicators, its own weighting and its own compensation values, resulting in a separate evaluation for each country. The structure of conflict resolution method 1 is given in Tab. 2. Indicator values pertaining to the alternatives are shown in Tab. 3, and the rankings of alternatives according to the two countries are given in Figs. 9, 10 and 11.

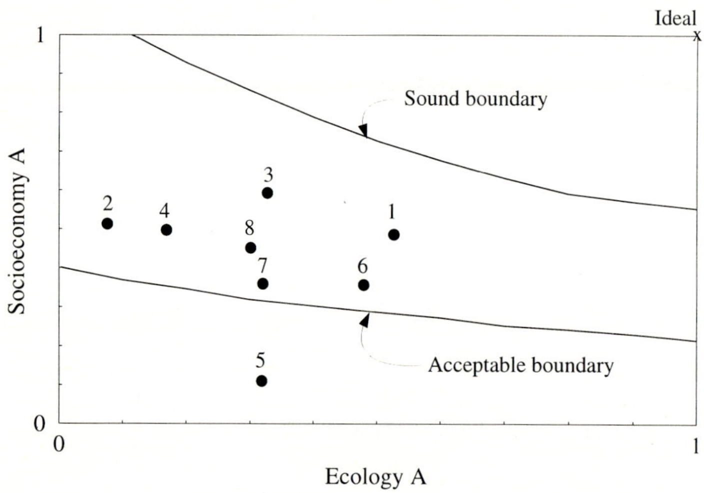

Figure 9 Conflict Resolution 1: Country A, No Uncertainty.

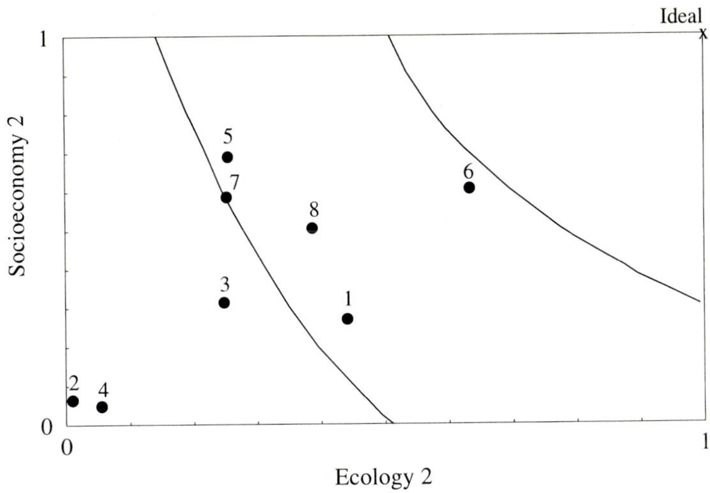

Figure 10 Conflict Resolution 1: Country B, no uncertainty.

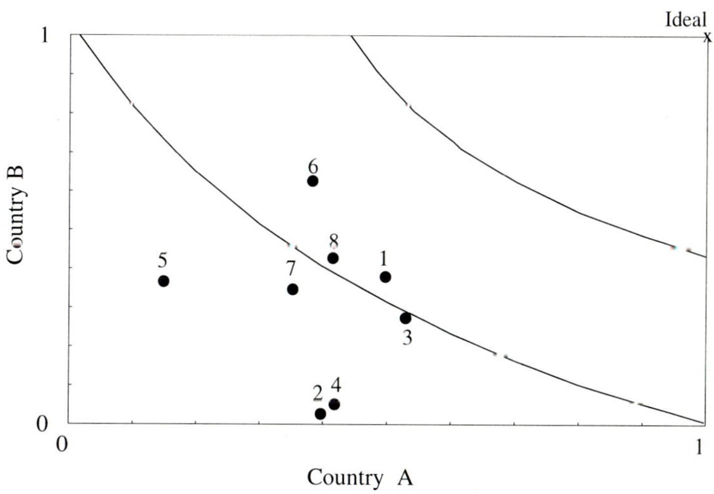

Figure 11 Conflict Resolution 1: Country A-Country B, no uncertainty.

Table 2. Structure of Conflict Resolution Method 1.

Country A				**Country B**			
LEVEL 1							
Group	Indicator	Weight	Compensation	Group	Indicator	Weight	Compensation
Biology	Land	0.333	3.0	Biology	Land	0.333	3.0
	Bird	0.333			Bird	0.333	
	Water	0.333			Water	0.333	
Water Quality	Lake	0.300	4.0	Water Quality	Lake	0.400	5.0
	Reservoir	0.500			Reservoir	0.100	
	River	0.200			River	0.500	
Economics	Tourism	0.150	2.0	Economics	Tourism	0.714	2.0
	Power Production	0.850			Power Production	0.286	
Recreation	Valley	0.167	2.0	Recreation	Valley	0.333	2.0
	Reservoir	0.500			Reservoir	0.167	
	Lakes	0.333			Lakes	0.500	
Aesthetics	Valley	0.400	3.0	Aesthetics	Valley	0.400	3.0
	Reservoir	0.400			Reservoir	0.100	
	Lakes	0.200			Lakes	0.500	
LEVEL 2							
Ecology	Biology	0.500	3.0	Ecology	Biology	0.500	5.0
	Water Quality	0.500			Water Quality	0.500	5.0
Socioeconomy	Economics	0.800	2.0	Socioeconomy	Economics	0.200	2.0
	Recreation	0.100			Recreation	0.400	
	Aesthetics	0.100			Aesthetics	0.400	
LEVEL 3							
Country A	Ecology A	0.200	2.0	Country B	Ecology B	0.667	2.0
	Socioeconomy A	0.800			Socioeconomy	0.333	
LEVEL 4							
Overall	Country A	0.500		Country B		0.50	2.0

Table 3. Conflict Resolution Method 1: Indicators vs. Alternatives.

	Management Alternatives							
Indicator - Country	1	2	3	4	5	6	7	8
Land biology-A	1300.	1103.	1200.	1100.	1500.	1400.0	1500.0	1350.0
Bird biology-A	170.0	140.0	155.0	140.0	200.0	185.00	200.00	176.00
Water biology-A	400.0	300.0	350.0	300.0	500.0	450.00	500.00	403.00
Water qual./lake-A	2.00	3.20	2.60	1.00	3.00	2.50	3.00	2.10
Water qual./resvr.-A	8.50	8.00	8.25	8.00	9.00	8.75	9.00	9.10
Water qual./river-A	7.20	8.70	7.70	8.20	6.20	6.70	6.20	7.20
Tourism-A	1000.	800.0	900.0	950.0	1200.	1100.0	1100.0	1050.0
Power production-A	500.0	650.0	575.0	575.0	350.0	425.00	425.00	500.00
Recreation/valley-A	3.00	2.00	3.00	2.00	4.00	4.00	4.00	3.00
Recreation/resvr.-A	3.00	4.00	3.00	3.00	2.00	3.00	3.00	2.00
Recreation/lake-A	2.00	1.00	2.00	1.00	4.00	3.00	3.00	3.00
Aesthetics/valley-A	2.00	1.00	2.00	1.00	3.00	3.00	3.00	3.00
Aesthetics/resvr.-A	3.00	4.00	3.00	2.00	2.00	4.00	4.00	2.00
Aesthetics/lake-A	2.00	1.00	3.00	1.00	4.00	3.00	3.00	3.00
Land biology-B	1200.	1100.	1200.	1100.	1500.	1400.	1500.0	1350.0
Bird biology-B	170.0	140.0	155.0	140.0	200.0	185.00	200.00	176.00
Water biology-B	400.0	300.0	350.0	300.0	500.0	450.00	500.00	403.00
Water qual./lake-B	2.00	3.00	2.00	1.00	3.00	2.00	3.00	2.10
Water qual./res-B	8.00	8.00	8.00	8.00	9.00	8.00	9.00	9.10
Water qual.//river-B	7.00	8.00	7.70	8.00	6.00	6.00	6.00	7.20
Tourism-B	500.0	300.0	400.0	450.0	1000.	800.00	700.00	600.00
Power production-B	100.0	1300.	115.0	115.0	70.00	85.00	85.00	100.00
Recreation/valley-B	2.00	1.00	2.00	1.00	4.00	3.00	3.00	3.00
Recreation/resvr.-B	1.00	2.00	2.00	1.00	2.00	3.00	3.00	2.00
Recreation/lake-B	2.00	1.00	2.00	1.00	4.00	3.00	3.00	3.00
Aesthetics/valley-B	2.00	1.00	2.00	1.00	4.00	3.00	3.00	3.00
Aesthetics/resvr.-B	1.00	2.00	2.00	1.00	2.00	3.00	3.00	1.00
Aesthetics/lake-B	2.00	1.00	2.00	1.00	5.00	3.00	3.00	3.00

Conflict Resolution by International Expert Commission

To look at the problem from the perspective of an international expert commission, one would group the 28 basic indicators by type regardless of the country by which they were nominated. Each group would contain the basic variables twice - once for each country. The countries would be free to assign percentage preferences to their own basic indicators so that the sum of preferences for each country would equal 50 percent.

This approach seeks a compromise at the lowest possible level, eliminating the need to rank and evaluate criteria within each country prior to considering international preferences. Table 4 shows the modified structure of indicators for conflict resolution. The indicator values will not change (Table 3). However, the different structure leads to a different overall ranking as compared to Conflict Resolution Method 1 (Fig. 12).

Table 4. Structure of Conflict Resolution 2.

<table>
<tr><td>Land A, Bird A, Water
Land B, Bird B, Water B</td><td>0.5
0.5</td><td>Biology</td><td rowspan="2">Ecology</td></tr>
<tr><td>Lake A, Reservoir A, River A
Lake B, Reservoir B, River B</td><td>0.5
0.5</td><td>Water Quality</td></tr>
<tr><td>Tourism A, Power A, Production A
Tourism B, Power B, Production B</td><td>0.5
0.5</td><td>Economy</td><td rowspan="3">Socioeconomy</td></tr>
<tr><td>...
...</td><td>0.5
0.5</td><td>Recreation</td></tr>
<tr><td>...
...</td><td>0.5
0.5</td><td>Aesthetics</td></tr>
</table>

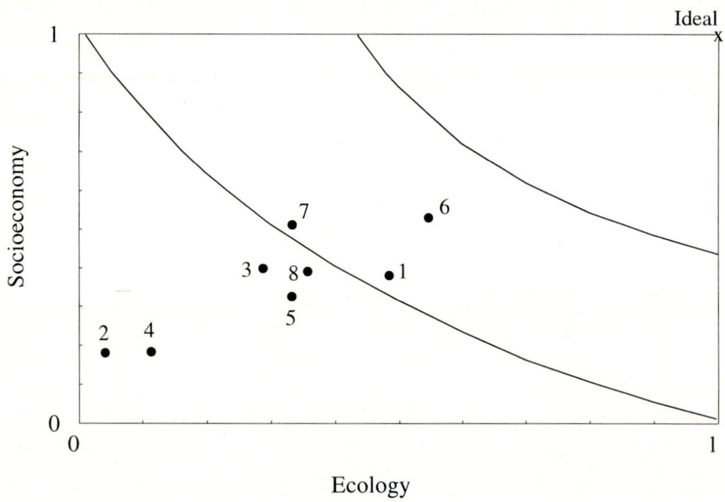

Figure 12 Conflict Resolution 2: No Uncertainty

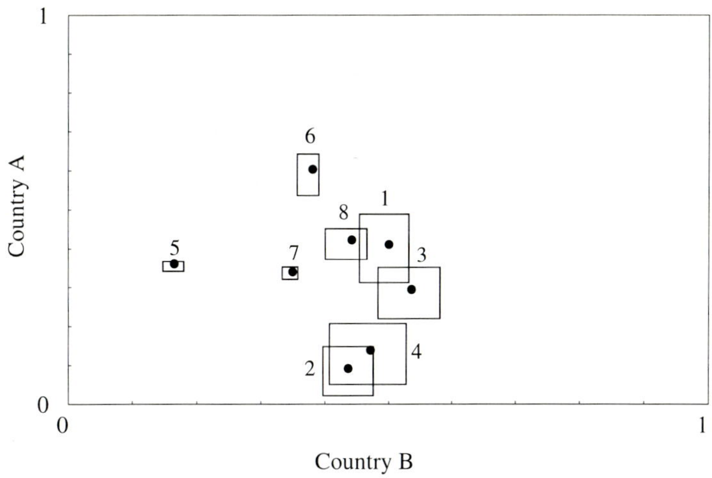

Figure 13 Conflict Resolution 1: Country A-Country B, With Uncertainty.

Consideration of Uncertainties

In this case some of the indicators can be estimated only imprecisely. Triangular fuzzy numbers are used to account for these imprecise indicator values. Then fuzzy composite programming is used with Conflict Resolution 1 (countries separately) and Conflict Resolution 2 (international expert commission). The results are shown in Figs. 13 and 14.

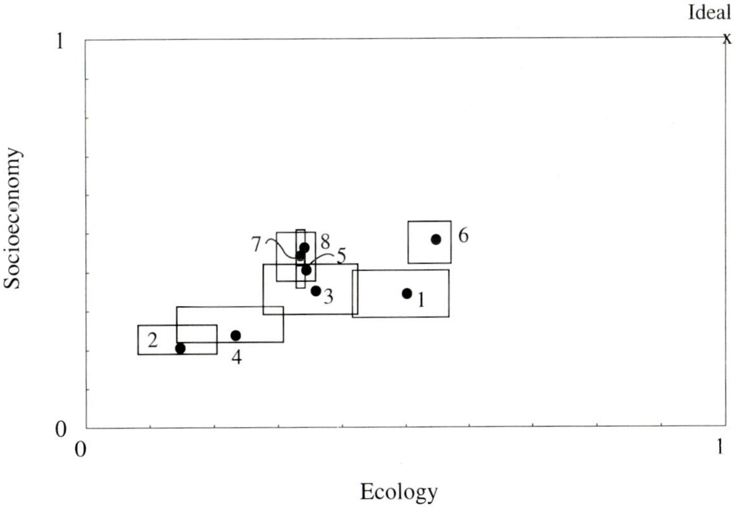

Figure 14 Conflict Resolution 2: With Uncertainty.

Conclusions

The following conclusions may be made:

1. There is no established user-friendly methodology to address double-conflict situations that face countries with a transboundary water management problem.

2. Composite programming, an available MCDM technique, has the capability to evaluate existing transboundary water management alternatives and select from among them.

3. To apply composite programming to transboundary water management, the steps are consideration first of the intercriterial conflicts, selection of the "best" water management scheme on that basis, consideration of conflict between the countries and finally development of a scheme that is acceptable to each party and as close as possible to the "best" scheme.

4. Transboundary water management involves uncertainties, such as the ones present in estimations of economic, ecological, social and political criteria.

5. Fuzzy set theory, often called fuzzy logic, may be practically applied to describe such uncertainties.

6. Composite programming can be combined with fuzzy logic in order to address both intercriterial conflicts and uncertainty in transboundary water management.

7. The results of these procedures may be applied to solve problems of international water management.

Acknowledgments

Research leading to this paper has been partially supported by funds from the University of Bodenkultur, Vienna, Austria under a grant to the first author, the Center for Infrastructure Research of the University of Nebraska, the U.S. Army Corps of Engineers and the U.S. Information Agency.

Appendix

Elements of fuzzy set theory

Encoding Imprecise Information

Many familiar examples can be found where information is expressed in imprecise form. Flooding as an attribute is often given in the form "tendency towards an increase" or "slightly below critical level". Such terms may be formalized as fuzzy sets, which are non-binary sets where the degree of membership of a variable or attribute x in this imprecisely defined set A is measured by a membership grade $m\ A(x)$. The grade ranges between 0 and 1. The value of $m\ A(x)$ is close to 1 if x "is believed to belong to" A and close to 0 if it is not "likely to belong to A. Since by definition no limitation is placed on the sets under consideration, one may use fuzzy sets for one or more of the following purposes:

(1) The first use may be to encode the <u>description</u> of the situation; thus, the linguistic

statement; "the day is quite hot" may be expressed as "The day *x* belongs to the set *B* of hot days with membership value $mB(x) = 0.8$. On the other hand, the statement "Tomorrow will be quite hot" may be expressed as a fuzzy set membership value or else as a prior probability distribution that the temperature *Y* will reach *y* degrees: $p(Y < y) = F(y)$.

(2) The second use of fuzzy set analysis may be to express a *preference* over an attribute value *x*, such as aquifer recharge dissolved oxygen (DO) and other water quality indicators, or the aesthetic (or recreational) value of a forest management scheme. For example, the membership function of DO in the set *A* of desirable value increases as an S-shaped fashion until 100 percent saturation concentration is reached and then decreases rapidly as sursaturation occurs.

(3) The third use of a fuzzy representation may be to express further *imprecision* in interpolation, control policy, regression equation specification (Bardossy et al., 1990) or failure phenomena (Duckstein et al., 1988). For example, if *h* is a water supply deficit: *h* = (demand) - (supply), then a value *h* belongs to the set of major failure *F* with a membership function $mF(h)$ increasing from h_{min} to h_{max}; namely, if *h* is a small fraction of the demand, then $m\ F(h)$ is close to zero; as deficit *h* becomes large, this deficit "incident" becomes a major failure, and $mF(h)$ approaches unity - similar to a "disvalue" function.

Fuzzy sets, fuzzy numbers, the extension principle

In this part, we summarize the basic definition of fuzzy sets, fuzzy numbers and fuzzy operations. Some of the definitions are restricted to the form used in this paper.

Definition 1: Let *X* be a set (universe). *D* is called a fuzzy subset of *X* if *D* is a set of ordered pairs:

$$D = \{x, mD(x),\ x \in X\}$$

where $mD(x)$ is the grade of membership of *X* in *D*. $mD(x)$ takes its values in the closed interval [0,1]. The closer $mD(x)$ is to 1, the more *x* belongs to *D*; the closer it is to 0, the less it belongs to *D*. If [0,1] is replaced by the two element set {0,1}, then *D* can be regarded as a subset of *X*.

Definition 2: The *h* level set of the fuzzy subset *D* is the set of those elements which have at least *h* membership:

$$D(h) = \{x: mD(x) \geq h\}$$

Definition 3: A fuzzy subset n^* is called normal if there is at lest one *z* such that $mn^*(z) = 1$.

Definition 4: A fuzzy subset of the set of real numbers is called convex if any three real numbers *x,y,z* such that $x \leq z \leq y$, can be defined as:

$$mn^*(z) \geq \min(mn^*(x), mn^*(y))$$

Definition 5: A fuzzy subset n^* is called a fuzzy number if n^* is a normal convex fuzzy subset of the set of real numbers. The membership value of a real number reflects the "likeliness" of the occurrence of that number; the level sets (intervals in this case) reflect different sets of numbers with a given minimum likeliness (Kaufmann and Gupta, 1991). Any real number can be regarded as a fuzzy number and is often called a crisp number in fuzzy mathematics.

Definition 6: If $X(1),\ldots,X(K)$ are sets and $D(1),\ldots,D(K)$ are fuzzy subsets of $X(1),\ldots,X(K)$ respectively, then the Cartesian product $D(1)x,\ldots,xD(K)$ has a membership function:

$$m(x(1),\ldots,x(K) \geq \min\{mD(k)(x(k)); k = 1,\ldots,K)$$

Zadeh, (1973) provides basic tools, in particular the extension principle, for extending point to point operations to fuzzy set operations, as shown next.

Definition 7: If *X* and *Y* are two sets, and *f* is a point to point mapping from *X* to *Y*

$$f: x \to y \;\; \forall x \in X, f(x) = y \in Y$$

then *f* can be extended to operate on fuzzy subsets of *X* in the following way:
Let *D* be a fuzzy subset of *X* with membership function *mD*, then the image or "value" of *D* if *Y* is the fuzzy subset *G* with the membership function

$$mG(y) = \max\{mD(x); y = f(x), x \in X\}$$

and $mG(y) = 0$ if there is no $x \in X$ such that $f(x) = y$.

Usually Definitions 6 and 7 are combined to yield the function of a Cartesian product, as given for example in Bardossy and Duckstein (1992).

Note that max (maximum) should, strictly speaking, be replaced by sup (supremum), but for computational purposes, all sets have a finite range. Applying the above principle to functions like $f(x,y) = x + y$ or $f(x,y) = xy$, the basic operations on fuzzy sets can be defined.

Definition 8: The mean or center of gravity of a fuzzy number a^* is defined as:

$$FM(a^*) = \frac{\int_{-\infty}^{+\infty} x \cdot ma^*(x)dx}{\int_{-\infty}^{+\infty} x \cdot ma^*(x)dx}$$

Definition 9: The width or support of a fuzzy number *a* is defined as:

$$FW(a^*) = \max\{x; ma^*(x) > 0\} - \min\{x; ma^*(x) > 0\}$$

Note that neither the mean nor the width of fuzzy numbers is necessarily finite. Further details and properties of fuzzy numbers can be found in Dubois and Prade (1980), Kaufmann and Gupta (1991) and other papers cited in the reference section.

References

Bardossy A, Bogardi I and Duckstein L, Serafini P. (eds) (1985) Composite programming as an extension of compromise programming. In: Mathematics of multi-objective optimization, CSIM, Udine, Italy; Springer-Verlag, Wien, pp. 375-408

Bardossy A and Duckstein L (1992) Analysis of a karstic aquifer management problem by fuzzy composite programming. Water Resources Bulletin, 28(1): 63-74, January

Bogardi I, Bardossy A and Duckstein L (1983) Regional management of an aquifer for mining under fuzzy environmental objectives. Water Resources Research, 19(6) 1394-1402

Bogardi I and Duckstein L (1988) Multiobjective approaches to river basin planning. Handbook of Civil Engineering, Technomic Publishing, Hasbrouck Heights, New Jersey, pp.415-452

Bogardi JJ and Duckstein L (1992) Interactive multiobjective analysis embedding the decision maker's implicit preference function. Water Resources Bulletin, 28(1): 78-88, January

Chameau J and Santamarina JC (1987) Membership functions: Comparing Methods of Measurement. International Journal of Approximate Reasoning, 1:287-301

Chen SH (1985) Ranking fuzzy numbers with maximizing set and minimizing set. Fuzzy Sets and Systems, 17, 113-129.

Cordeiro Netto O, Parent E and Duckstein L (1993) Water resources planning with multiple objectives and multiple actors: A case study in southern France. IFORS 93, XIII. World Conference on Operations Research, Lisbon, Portugal, 12-16 July

Cothern CR, Coniglio WA and Marcus WL (1987) Uncertainty in population risk estimates for environmental contaminants. Proceedings of the Society of Risk Analysis International Workshop on Uncertainty in Risk Assessment, Risk Management and Decision Making. September-October, Knoxville, TN

David L and Duckstein L (1976) Multi-criterion ranking of alternative long-range water resources systems. Water Reources Bulletin, Vol. 12, No. 4, pp.731-754

Dong W and Shaw HC (1987) Vertex method for computing functions of fuzzy variables. Fuzzy Sets and Systems, 24,65-78

Dong WM and Wong FS (1986) From uncertainty to approximate reasoning. Parts 1,2,3, Civil Engineering Systems, September, December, 1986, March 1987

Dourson ML and Stara FF (1983) Regulatory history and experimental support of uncertainty (safety) factors. Regulatory Toxicology and Pharmacology, 3:224-238

Dubois D and Prade CH (1980) Fuzzy sets and systems: Theory and applications. Academic Press, New York, p. 393

Duckstein L and Opricovic S (1979) Multiobjective optimization in river basin development. Water Resources Research, Vol. 16, No. 1, pp. 14-20

Duckstein L, Bella A, Stakhiv E and Waterstone M (1994a) Strategic and tactical trans-boundary water allocation under climatic uncertainty: Case of upper Rio Grande basin. To be presented at the NATO ASI Transboundary Water Resources: Management and Environmental Issues Conference, Skopelos, Greece, May 15-25

Duckstein L, Bardossy A, Tecle A and Bogardi I., Delgado M, Kacprzyk J, Verdegay JL and Vila MA (eds) to appear (1994) Fuzzy composite programming with application to wastewater management under changing physical conditions. In: Fuzzy Optimization, Recent Advances, Physica-Verlag, Heidelberg, Germany

Eder G, Duckstein L and Nachtnebel HP (1994) Ordering criteria by MCQA with application to an Austrian water resources planning case. To be presented at the EURO XIII Conference, July 19-22, United Kingdom

Gleick PH (1991) The vulnerability of runoff in the Nile Basin to climatic changes. The Environment Professional, 13, 66-73

Gleick PH and Irving M., Mintzer (eds) (1992) Effects of climate change on shared fresh water resources. Confronting Climate Change: Risk, Implications and Responses, Cambridge University Press, Chapter 9

Goicoechea A, Hansen DR and Duckstein L (1982) Multiobjective decision analysis with engineering and business applications. John Wiley and Sons, New York, New York

Kaufmann A and Gupta MM (1991) Introduction to fuzzy arithmetic: Theory and applications, Van Nostrand Rheinhold, New York

Keeney RL and Wood EF (1977) An illustrative example of the use of multiattribute utility theory for water resources planning. Water Resources Research, V.13, No.4

Klir GJ and Folger TA (1988) Fuzzy sets, uncertainty and information. Prentice Hall, New Jersey

Stansbury J, Bogardi I and Kelly WE, Ganoulis J (eds) (1991) Risk-cost analysis under uncertainty in water resources. In: Risk and Reliability in Water Resources and Environmental Engineering, Springer Verlag

Szidarovszky F, Gershon ME and Duckstein L (1986) Techniques for multiobjective decision making in systems management, Elsevier Scientific Publishing Co., New York

Tecle A, Fogel M and Duckstein L (1988) Multicriterion selection of wastewater management alternatives. ASCE J.Water Resources Planning and Management, Vol. 114, No. 4, pp. 383-398, July

Turksen IB (1991) Measurement of membership functions and their acquisition. Fuzzy Set and Systems, Vol. 40, No.1, pp. 5-34

Zadeh LA (1965) Fuzzy sets. Information and Control, 8, 338-353

Zahed LA (1973) Outline of a new approach to the analysis of complex systems and decision processes. In: IEEE Trans. Vol, SMC 3, pp. 28-44

Zimmerman HJ (1985) Fuzzy set theory and its applications. Kluwer-Nijhoff Publishing, Hingham, MA.

III.3 Strategic and Tactical Transboundary Water Allocation under Climatic Uncertainty: Case of Upper Rio Grande Basin, USA

Lucien Duckstein, Aimee Bella
Systems and Industrial Engineering Department
University of Arizona, Tucson, AZ 85721

Eugene Z. Stakhiv
IWR Research Division, U.S. Corps of Engineers
Casey Building, Fort Belvoir, VA 22060-5586

Marvin Waterstone
Dept. of Geography and Regional Development
University of Arizona, Tucson, AZ 85721

Introduction

The purpose of this paper is to examine how a multiple objective decision making technique may be used as an aid in analyzing the past developments of the Rio Grande management system and the present water use strategies employed by the decision makers. Then, by assuming that these developments of the past will propagate into the future, the most preferable operational strategies that could be used by these decision makers under climate fluctuation coupled with future population growth scenarios are derived. As pointed out in Duckstein et al. (1982), Szidarovszky et al. (1986) and Bardossy and Duckstein (1992), common problems in formulating a multiple objective model within a crisp (non-fuzzy) framework include the following:

1./ diversity of objectives,

2./ different importance and compensation among objectives,

3./ undefined numerical scales for some objectives.

As per Tecle and Duckstein (1992), current multicriterion decision making (MCDM) techniques may be classified into five categories: utility-type, outranking relations, distance-based, direction-based and mixed techniques. Simulation techniques such as river basin simulation schemes constitute additional tools that can be used to model decision making processes in conjunction with an MCDM analysis (Bogardi et al, 1991).

In this paper, we examine the suitability of an MCDM technique, namely fuzzy compromise programming, to model the upper Rio Grande basin conflict under climatic uncertainties using a generated discrete representation of the non-dominated set of alternatives. The paper is organized as follows. In the next section, the case study example dealing with the upper Rio Grande river basin management problem is presented and formulated. It is then analyzed using fuzzy composite programming. The last section consists of a discussion and set of conclusions.

Problem Formulation

The Rio Grande, the fifth largest river in North America, is an interstate and international river basin in the western United States (Fig. 1), where all waters are legally administered through the appropriation doctrine, international treaties and in interstate compacts. It originates in the snow covered mountains of Colorado (Fig. 2), flows primarily through the arid and semi-arid lands of New Mexico and Texas and discharges into the Gulf of Mexico. Because the Rio Grande flows through primarily arid and semi-arid regions, this river has had considerable influence on the region's inhabitants, more so than any other physical element of the environment.

One of the primary problems of the Rio Grande Basin has been to provide water to accommodate increasing populations in the region and to deal with the lack of a legal framework for recognizing the hydrological interconnection of surface and ground water in view of the water treaty of 1944 between the US and Mexico (Thomson and Thompson, 1991). Increasing populations have caused unregulated pumping of ground water which affects surface water flow (McQuillan, 1984) and may indirectly affect surface water treaty allocation in the future. The lack of surface water has been made up through interbasin diversions and ground water mining, but extensive ground water withdrawal from stream-aquifer systems will inevitably result in a reduction of available surface water supplies of the Rio Grande (Wilkens et al., 1980).

Figure 1 Major river basins in the United States.

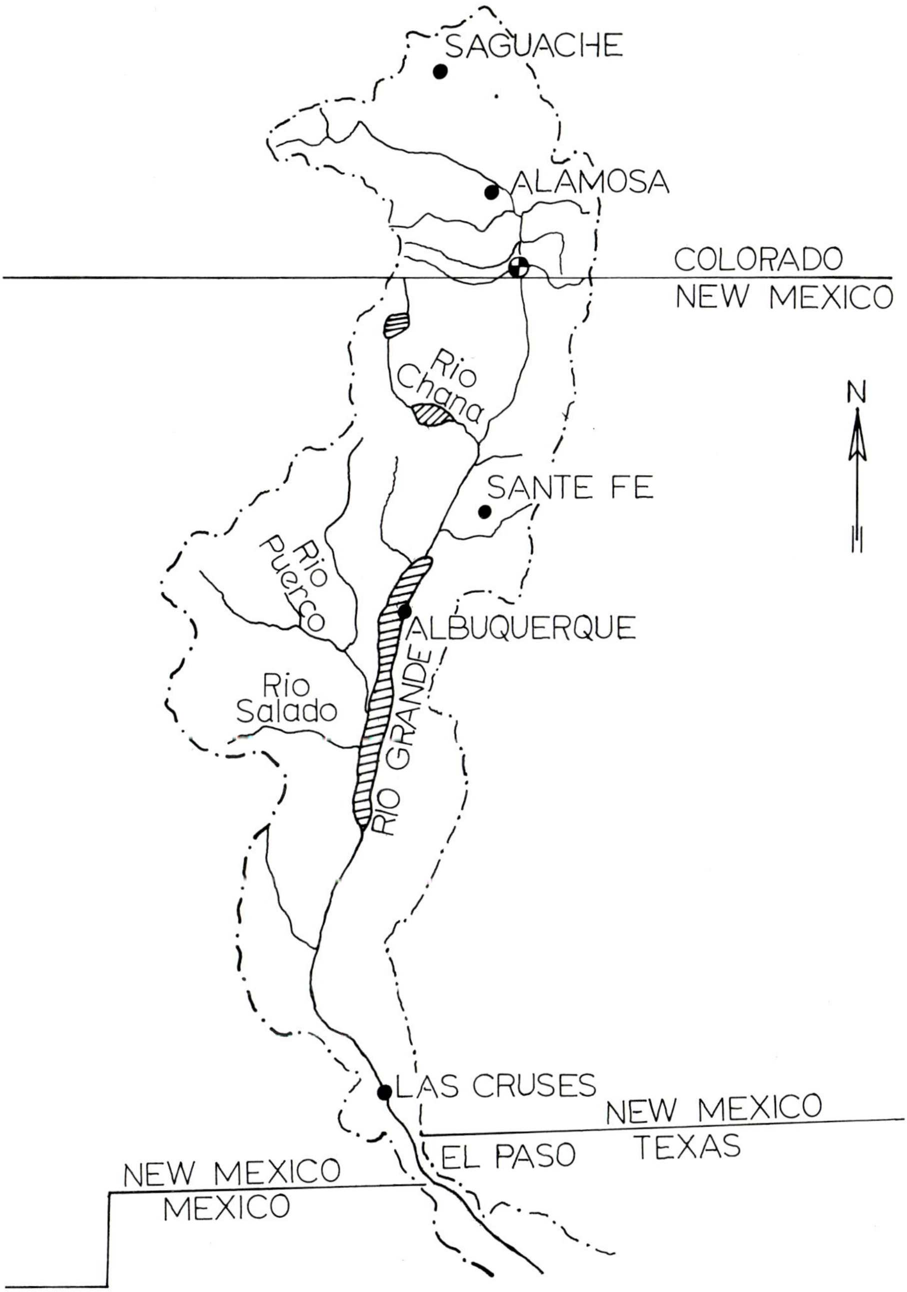

Figure 2 Upper Rio Grande River Basin.

Water management in the basin is complex because many of the water needs are incompatible. The current management of water resources is complicated by a multiplicity of governmental units on both sides of the border. On the Mexican side, the Federal government exercises the dominant role through a variety of agencies. On the US side, prime responsibility is placed in three different states, each having its separate, independent and different water laws. Regulations and laws of the Federal government and several claims by native American Indians further complicate matters. The Indians claim use of waters which are located on or along the boundaries of Indian reservations. All Indian reservations have water rights protected by the Federal government. Quantification of Indian and Federal water rights are lacking and need to developed in the upper Rio Grande basin (Lord et al., 1993).

A prolonged drought would impose severe hardships on water users in the upper Rio Grande and the basin as a whole, requiring the establishment of an appropriate public policy. In general, water resources planning ignores climate as a variable due to its assumed invariance. However, there is growing evidence that climate fluctuation, even climate change, may be a necessary premise for water resource plans that are projected 50 to 100 years into the future (US WRC, 1978).

Allocation changes designed to meet a diminishing water supply would be difficult to implement because of the complex web of international treaties, interstate compacts and local, state and federal laws (Waterstone, 1994). The high degree of complexity makes it difficult for a decision maker to comprehend the problem fully and eventually reach a reasonable decision. Increasing water demands will require significant conservation efforts (Thomson and Thompson, 1991); furthermore, transfers from water intensive uses such as agriculture will be required to meet industrial and municipal needs. Economic activities and population for the region are expected to increase, especially with the recent signing of the North American Free Trade Agreement (NAFTA) with the Republic of Mexico and Canada. These developments, which will place greater demands on water resources, may require changes in public policies presumably through negotiation. A multiple objective decision making framework for resolving these water resources problems, including uncertain or imprecise elements, is provided using the Fuzzy Composite Programming technique. To sum up, sources of conflict appear to be as follows:

(1) Sensitivity to climate fluctuation which would be acting on one of the most vulnerable regions in the USA
(2) The absence of agreements on ground water usage and water supply, especially since increased demands may lead to increased pumpage and a possible lowering of the water table
(3) Water quality problems because the arsenic content of surface water exceeds standards
(4) The non-quantification of Indian rights which as a necessary requirement will require an adjudication
(5) The development of the recent free-trade agreement with Mexico, which will probably increase the pressure on water resources. We believe that MCDM techniques could be used to provide a structure for the systematic modeling of the problem.

We have identified four objectives and eighteen sub-objectives of water resources management over the next 50 years, as illustrated in Figs. 3 to 7 (US Army Corps of Engineers, 1989) namely;

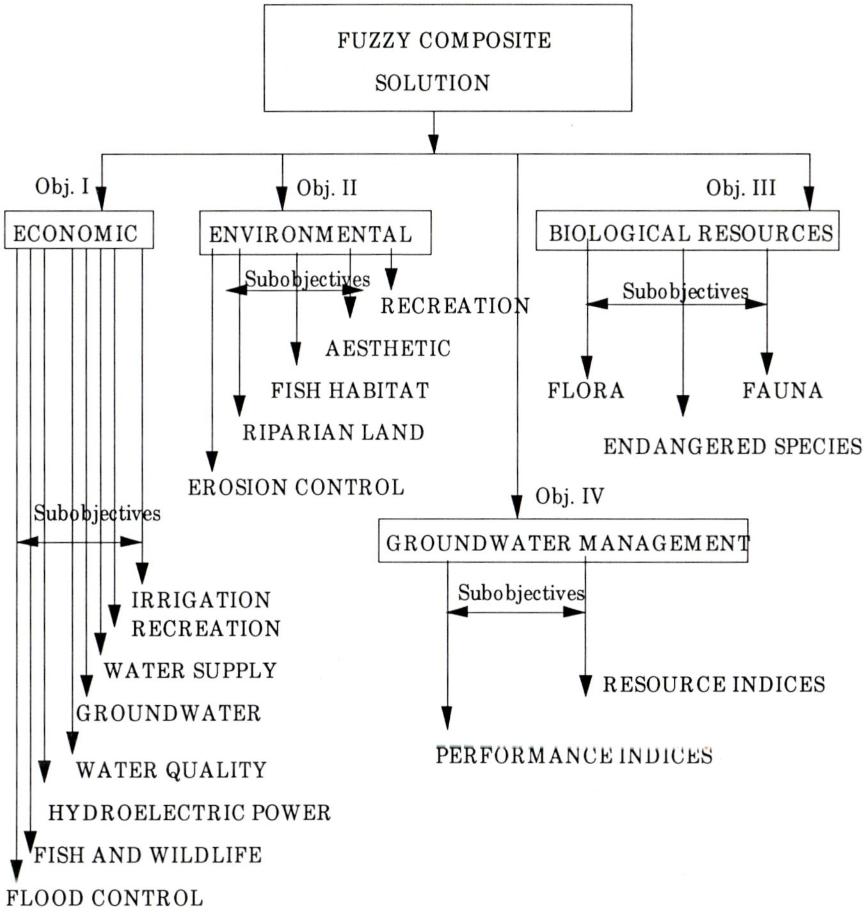

Figure 3 Composite Decision Tree.

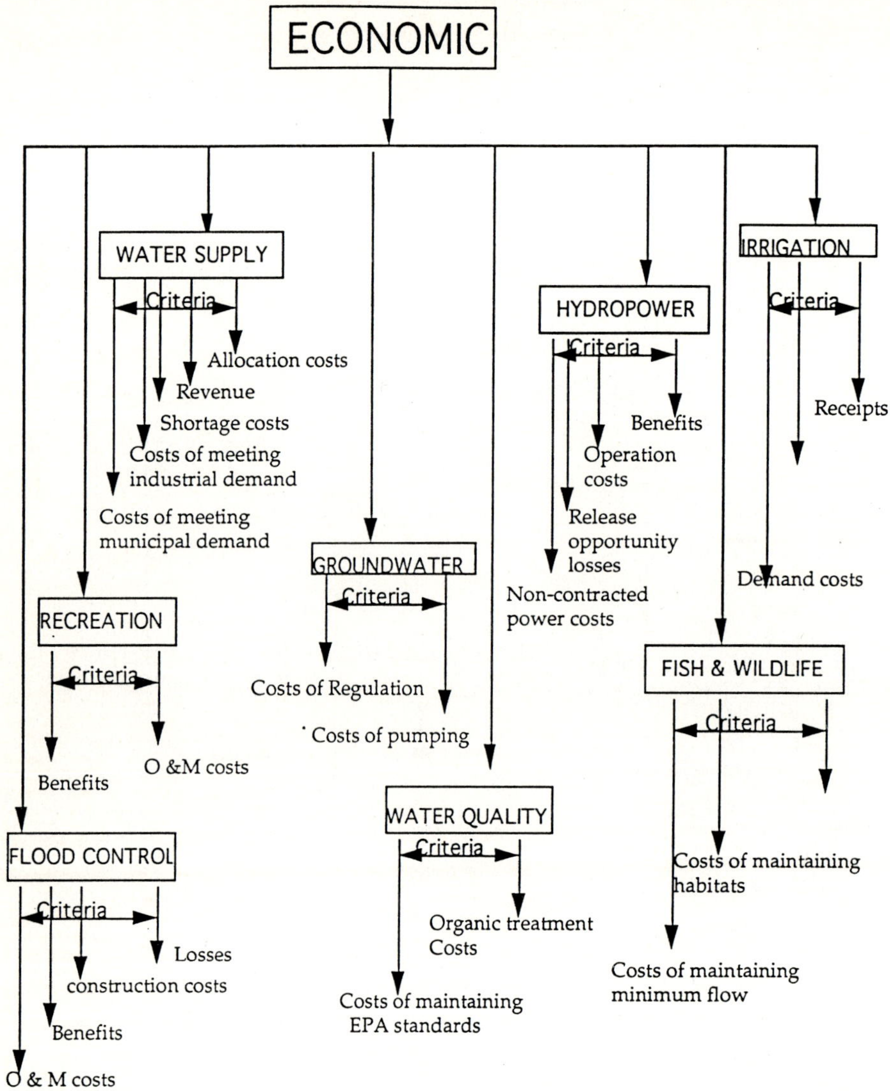

Figure 4 Economic Enhancement Factors.

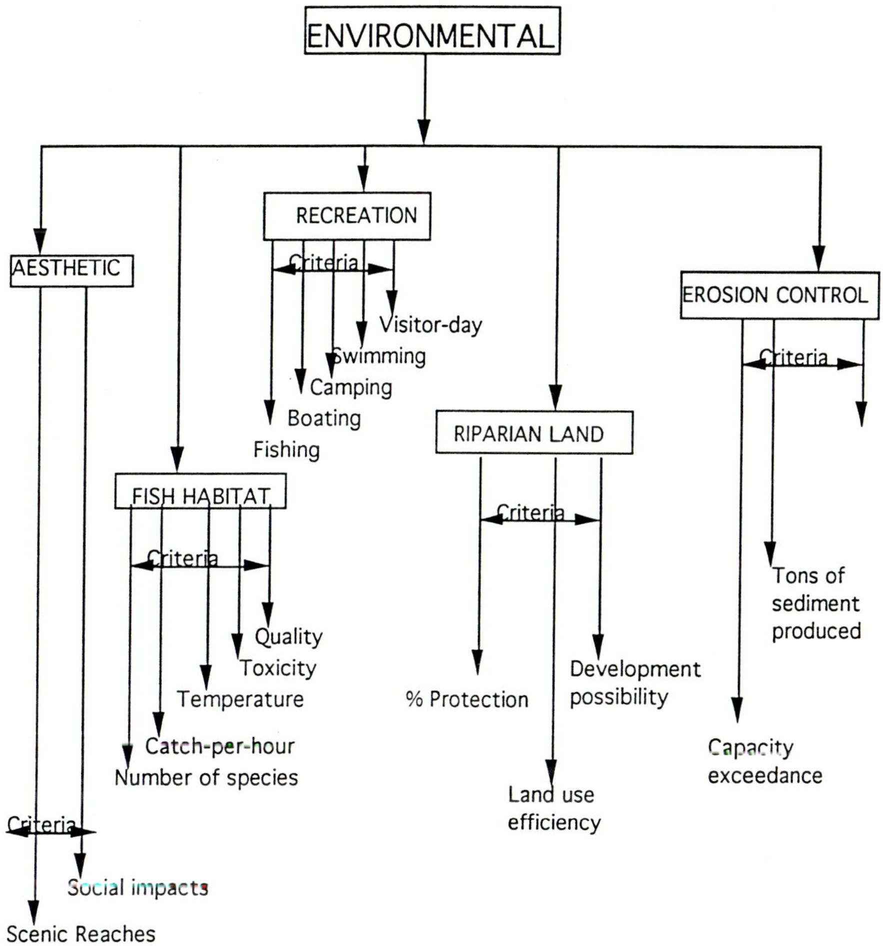

Figure 5 Environmental Protection Factors.

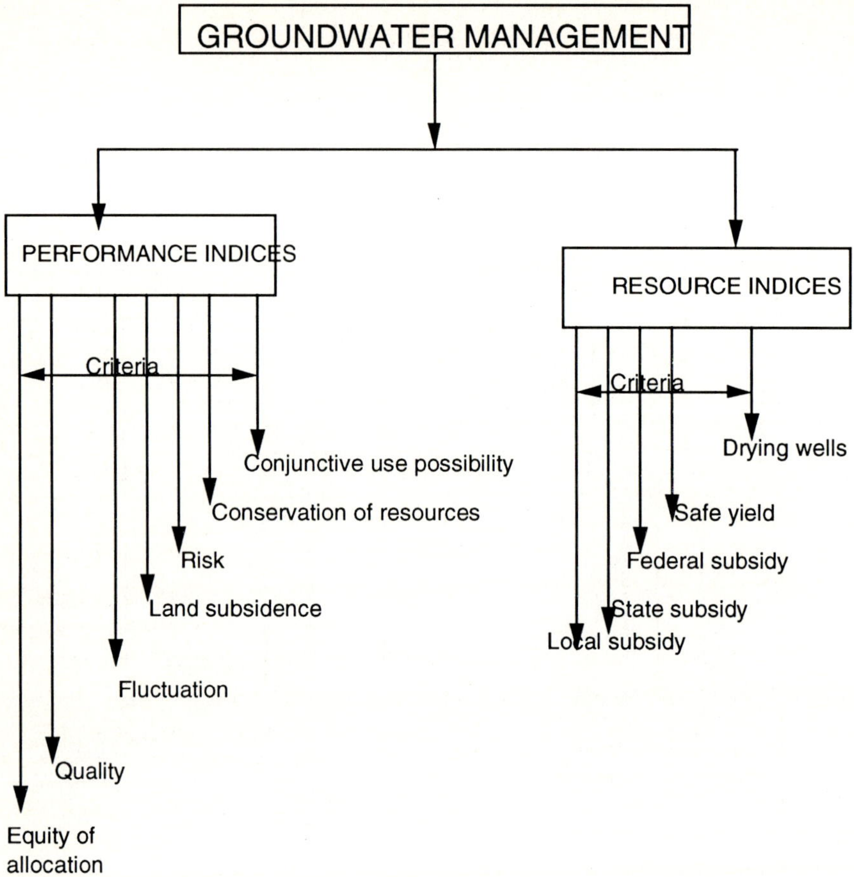

Figure 6 Groundwater Management Factors.

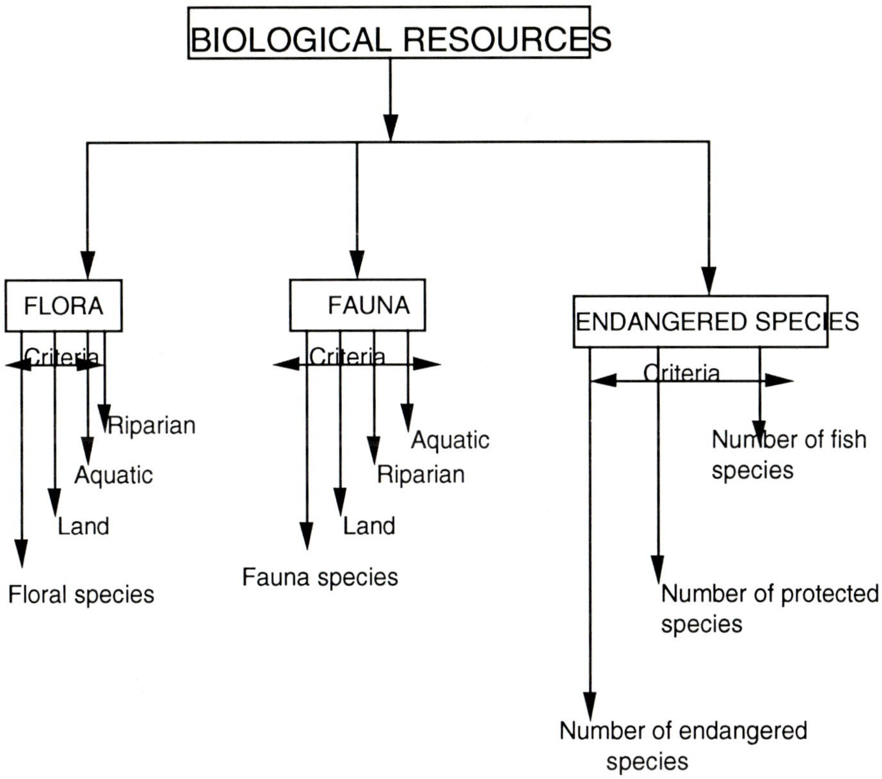

Figure 7 Biological Resources Factors.

(1) Enhance economic factors in the region (Fig. 4). Eight subobjectives including flood control, irrigation, hydropower, recreation, water supply, fish and wildlife, water quality and groundwater have been selected to represent this category. These subobjectives are all expressed in millions of dollars.

(2) Protect the environment (Fig. 5). Category includes five subobjectives, namely, aesthetics, fish habitat, recreation, riparian land and erosion control. These subobjectives are expressed in numerical form or else in linguistic terms such as poor to excellent, which are then transformed into numerical values using for example the concept of value functions.

(3) Promote sustainable groundwater management policies (Fig. 6) expressed in terms of performance and resource indices measured non-numerically as linguistic variables (poor to excellent).

(4) Conserve biological resources (Fig. 7) consisting of flora, fauna and endangered species all expressed in terms of numbers or quantities.

The list of criteria or basic indicators, their ideal and worst points as well as their weights or coefficients of importance have been defined. The objectives are first divided into subobjectives, and then these subobjectives are further divided into criteria. Therefore, we determined the importance or weights of the criteria in a three-step procedure. First the weight for each objective was assigned based on the preferences of the interest groups. Then the resulted weights were further divided to obtain the weights for the subobjectives, and similarly, the weights for the subobjective were further divided to obtain those for the individual criteria. The parameters include weights and balancing factors (explained later) for the objectives and subobjectives. We have assigned these parameters necessary for the fuzzy composite programming analysis on the basis of available information in the water resources management problems in the region (US Corps of Engineers, 1989; US Bureau of Reclamation, 1992). The basic indicators, as explained later, are taken as trapezoidal fuzzy numbers. The set of alternatives, to be ranked using the given criteria, represent preferences on the outcomes available to 15 different decision makers, which have been identified earlier. For example, we have assumed that Pueblo Indian Tribes prefer the plans which include pure actions A1 and A7 since these actions ensure water quality and control floods. Selected alternatives (US Bureau of Reclamation, 1992) consist of a mixture of pure actions, which include:

. Enhancing flood control,

. Constructing hydropower plants,

. Moving project storage to different location,

. Reducing sediment by pool stabilization,

. Controlling water quality in surface streams,

. Applying prior appropriation doctrine.

Considering the compatibility of the combination of pure actions, the alternatives have been further reduced, yielding a limited number of mixed actions. The total number of combinations of pure actions forming mixed alternatives is 2^{12} when we consider 2 levels for each alternative namely, do and do not do. Grouping actions by common goals yields a reduced set of five elementary plans, with a total number of $2^5 = 32$ mixed actions.

Notation for the plans is as follows:

P(i) is used for individual action selection, where i=0,1,2,....,5 .

P(i,j) is used for selecting pair of actions, where i=1,2,...,4.

P(1,2,3,4,5) for all the five actions using available reports (US Corps of Engineeers 1989, US Bureau of Reclamation 1992).

We have then constructed the payoff matrix representing the rating of alternatives by criteria (or basic indicators).

Proposed Solution

The fuzzy composite programming (FCP) algorithm (Bogardi et al., 1994; Duckstein et al., 1994) is used for this multiple objective problem context because some basic indicators are not known with certainty. The details and relation of this method to other MCDM techniques are given in the above references. FCP identifies solutions closest to the ideal solution, which is the solution point where all objectives are at their best possible values, or as far as possible from the worst point, which is also called the status quo point. That is, each component of the ideal point is obtained by maximizing each individual objective. If two or more objectives are in conflict, this solution is not feasible and is sometimes called "utopia". The measure of closeness used in this study is a most frequently used measured of closeness namely, the L_p norm given by the following expression for objective functions not expressed in commensurate terms (Zeleny, 1973; Duckstein and Opricovic, 1980):

$$S_i(x) = \frac{Z_i^* - Z_i(x)}{Z_i^* - Z_i^{**}} \tag{1}$$

where $Z_i^{**} = \text{Min } Z_i$; $i = 1,2,...,I \quad x \in X$

With the transformation of equation (1), the fuzzy compromise solution X_i^* is given as follows:

$$\underset{x \in X}{\text{Min}} \left\{ L_p = \sum_{i=1}^{I} \alpha_i S_i(x)^p \right\}^{1/p} = L_p(x_i^*) \tag{2}$$

where equation (2) represents the fuzzy compromise solution and $1 \leq p \leq \infty$. Values of balancing factor p are taken as 1,2 and ∞. When $p = \infty$:

$$L_\infty = \lim_{p \to \infty} L_p = \underset{i}{\text{Max}} \, S_i(x) \tag{3}$$

For $p = 1$ there is full compensation between weighted deviations since any change such that $\Delta(\alpha_i S_i(x)) = -\Delta(\alpha_j S_j(x))$ for $i \neq j$ leaves L_1 unchanged. For $p = 2$, each deviation is weighted in proportion to its magnitude. The larger the deviation, the larger the weight. We assume that the basic indicators such as demand costs, non-contracted power costs, catch-per-hour, temperature, percent protection, capacity exceedance, conjunctive use possibility and drying wells are subject to uncertainty or imprecision.

We are representing this imprecision by trapezoidal fuzzy numbers such as the one represented in Fig. 8 for percent protection, which may be interpreted as "fully in fuzzy set from .4 to .6, possibly belonging to the set from .1 to .7". Some of the basic indicators have been taken as fuzzy having selected membership functions. Results have been obtained using the fuzzy composite programming program.

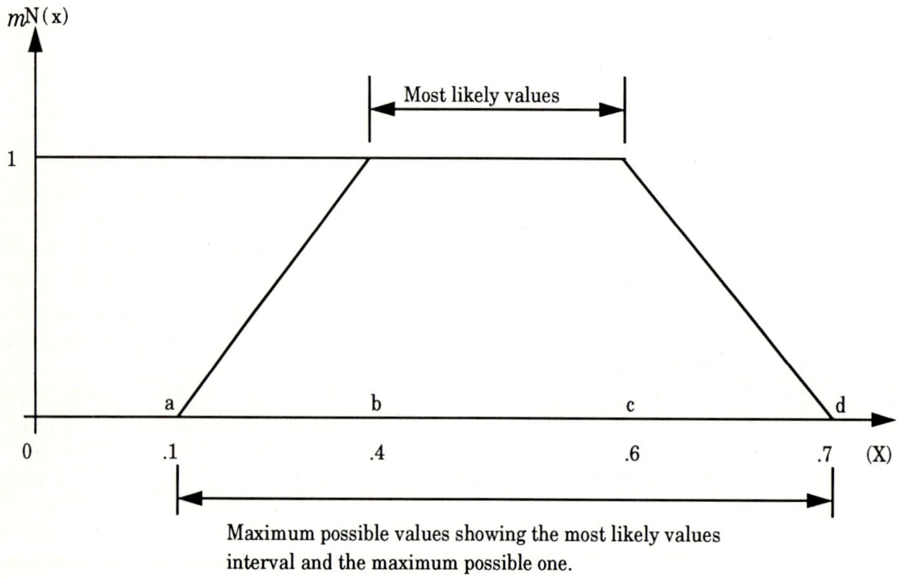

Figure 8 Membership function mN(x) of trapezoidal fuzzy number (a, b, c, d) representing the criterion C38: % protection of riparian land.

Interpretation of Results

The ranking of alternatives for subobjectives, also called second level objectives has been obtained. The ranking consists of the original objectives, also called first level objectives. The fuzzy composite solution technique ranks action Q_{22} first, which comprise the union of reduced actions 1,4, and 5. Action Q_{11} is ranked last. Action Q_{11} consists of adopting the combination of reduced actions 2 and 3. In other words the reduced action consisting of selecting the union of "increasing water supply and irrigation efficiency, reducing sediment by pool stabilization and establishing groundwater regulation" is the most preferred plan. On the other hand the union of action of "meeting water quality, protecting the public interest, developing hydropower and enhancing flood control" is the least preferred one. It should be noted that the last two actions Q_{31} and Q_{32} were not judged acceptable by the decision makers.

Discussion and Conclusions

This paper has only considered the MCDM viewpoint given by one particular fuzzy distance-based technique which is applied hierarchically. The main results found in this paper concern the setting of a general MCDM problem framework including 32 reduced actions, 66 criteria or basic indicators and 15 decision makers and an approach to produce ranking of the action plans that appear to be most desirable to the decision makers. Setting the problem in a multi-level framework seems to provide a good way to achieve this goal as well as highlight the different horizons of the various decisions. One may thus consider that the basic indicators are evaluated in the short range or tactical horizon, the subobjectives in the medium range, and the objectives in the long range or strategic decision horizon.

Distance-based techniques are usually well received by decision makers (Duckstein et al., 1982); yet, fuzzy concepts for one of the most used technique, compromise programming, has only been fully developed and extended to fuzzy composite programming quite recently. The methodology can also be applied to the evaluation and comparison of conflicting hydrologic regions on the basis of the same basic indicators or criteria as used herein.

Returning to the rankings that correspond respectively to the second level ranking of alternatives by separate objectives and top or overall ranking of alternatives using all 3 hierarchical compromise levels, it can be concluded that the selection of a unique value of p can lead to a ranking which would be difficult to interpret and defend. Note that a sensitivity analysis on the value of p should be performed as well as on the membership functions of the indicators, since they are selected on a subjective basis by the decision maker, therefore they should be considered uncertain. Once this is done, decisions that exhibit stability in case of hypothesized climate change scenarios can be identified.

Acknowledgments

This research has been supported in part by grants from the US Army Corps of Engineers, the National Science Foundation and the National Institute for Global Environmental Change.

References

Bardossy A and Duckstein L (1992) Analysis of a karstic aquifer management problem by fuzzy composite programming. Water resources Bulletin. 28(1): p. 63.

Bogardi JJ, Gupta AD, and Jiang HZ (1991) Search beam method - A promising way to define non-dominated solutions in multi-objective groundwater development. Water Resources Development, 7(4), 247-258.

Bogardi I, Bardossy A and Duckstein L (1994) To appear: Conflict analysis using multiple criterion decision making under uncertainty. In: Preceedings NATO ASI Transboundary Water Resources: Management and Environmental Issues Conference, Skopelos, Greece, May 15-25, 1994.

Duckstein L, Goicoechea A and Hansen DR (1982) Multiobjective decision analysis with engineering and business applications John Wiley & Sons, New York, p. 462.

Duckstein L, Bardossy A, Tecle A and Bogardi I (1994). Fuzzy composite programming applied to waste water management under changing physical condition. In: Fuzzy Optimization, Recent Advances.M. Delgado, J. Kacprzyk, J. L. Verdegay and M. A. (eds.). Physica-Verlag, Heidelberg, Germany.

Duckstein L and Opricovic S (1980) Multiobjective optimization in river basin development. Water Resources Research, 16:14-20.

Lord B, McGuire TR, and Wallace MR (1993) Indian water in the west. The University of Arizona Press. Tucson & London.

McQuillan D M (1984) Water quality concerns in the Albuquerque South Valley: Environmental Improvement Division, Report EID/GWH-84/2, 18 p.

Szidarovszky F, Gershon ME and Duckstein L (1986) Techniques for multiobjective decision making in aystems management. Elsevier, New York.

Tecle A and Duckstein L (1992) A procedure for selecting MCDM techniques for forest resources management. In: Multiple Criteria Decision Making and Support at the Interface of Industry, Business, and Finance, A. Goicoechea, L. Duckstein, and Zionts (editors), Springer Verlag, New York, New York.

Thomson B and Thompson A (1991) Rio Grande Laboratory/ Consortium Concepts. Proceedings of the Rio Grande Basin - Global Climate Change Scenarios, WRRI Rep. M24, pp. 33-36.

U.S. Army Corps of Engineers (1989) Reevaluation of the Operating Plan. Albuquerque Project Office. Albuquerque, New Mexico.

U.S. Bureau of Reclamation (1992) Annual Operating Plan. Albuquerque Projects Office. Albuquerque, New Mexico.

U.S. Water Resources Council (1978) The Nation's Water Resources 1975-2000. Second National Water Assessment. U.S. Government Printing Office, Washington, D.C., Vol. 1.

Waterstone M. (1994) Institutional Analysis of Water Resources. In: Engineering Risk in Natural Resources Management with Special References to Hydrosystems under changes of Physical or Climatic Environment. L. Duckstein and E. Parent (eds). Kluwer, Amsterdam. Series E: Applied Sciences Vol. 275, pp. 59-75.

Wilkens D W, W B Scott, and C A Kaehter (1980) Planning Report for the Southwest Alluvial Basin (East) Regional Aquifer System Analysis, Part of Colorado, New Mexico, and Texas, USDS Open-File Report. 80-564.

Zeleny M (1973) Compromise Programming. In: Multiple criteria Decision Making. J. L. Cochrane and M. Zeleny (Editors). Univ. of South Carolina Press, Columbia, pp.263-301

Chapter IV: Applications of the Engineering Approach

IV.1 The Upper Rhine along the German-French Border: Problems and Solutions in Water Resources and Ecology

Werner Buck
Institute for Hydrology and Water Resources Planning (IHW)
University of Karlsruhe, Kaiserstr. 12, D-76128, Karlsruhe
Germany

Geographical Situation and Water Resources Measures

On its course from the Alps to the North Sea, the Rhine crosses and/or follows the borders of Switzerland, Austria, Germany, France and The Netherlands. With a length of almost 1.300 km, the river has a total catchment area of approx. 185,000 km^2, which can be divided into six parts:

Alpine Rhine - from the main headwaters to Lake Constance,
High Rhine - from Lake Constance to Basle,
Upper Rhine - from Basle to Bingen,
Middle Rhine - from Bingen to Cologne (Köln),
Lower Rhine - from Cologne to Lobith/The Netherlands, and
Delta of the Rhine - from Lobith to the estuary.

In the Upper Rhine Valley (Fig. 1), the river has a slope of 0.8 m per km in the southern part and nearly 0.1 m per km in the northern part. The valley bottom is filled mainly with alluvial gravels originating from detritus which the Rhine transported down from the Alps during the last ice-age. The deposits of the quarternary period, together with the deeper lying deposits of the tertiary period, represent the fill of the Upper Rhine Rift, which is 300 km long, approx. 40 km wide and in general one hundred or so meters deep. Since the last ice-age, the Upper Rhine carved a valley from the deposits and shaped a natural flood plain with an erosion rim of up to 15 m height.

In the southern part of the valley, the Upper Rhine forms for a stretch of 180 km the border between Germany and France from Basle/Switzerland in the south and Lauterbourg/France - Karlsruhe/Germany, respectively, in the north. This is the area being investigated herein. The runoff regime of the Upper Rhine is essentially determined by the climate and altitude of the alpine catchment (mainly in Switzerland) and the low mountain ranges bordering the Upper Rhine Valley, namely the Vosges in France and the Black Forest in Germany. The minimum, mean and maximum discharges (1931-89), as well as the discharge ratios for a gauging station near Karlsruhe (with a catchment area of 50.200 km^2), indicate that the Upper Rhine has a rather balanced runoff regime:

minimum flow	mean flow	maximum flow
397 m^3/s	1310 m^3/s	4367 m^3/s
Nov. 1947:		Febr. 1970:
0.3	1	3.3

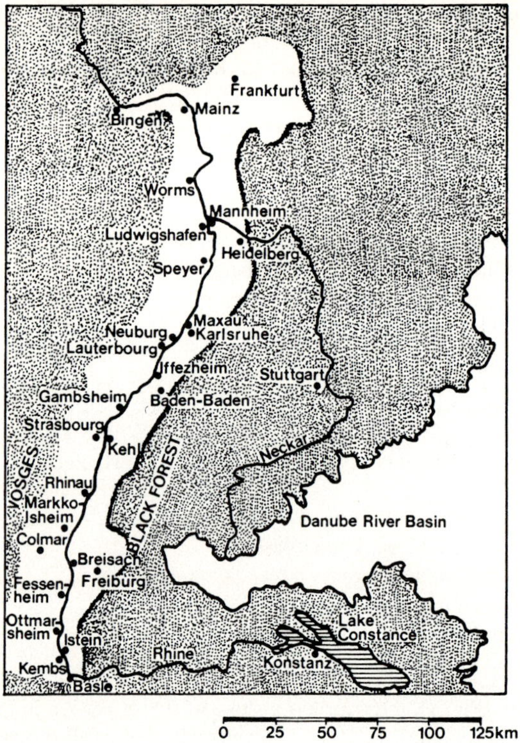

Figure 1 The Upper Rhine Valley.

Flood flows may occur in the early summer as well as in winter, due to snowmelt in the Alps in conjunction with heavy rain in the low mountain ranges caused by basin wide rainfall events of long duration. Due to erosion and sedimentation processes, the Upper Rhine originally offered a system of parallel braided flow channels in its southern part, whereas it meandered in the northern part. The erosive forces of the river are a result of its slope, of the bed load deficit downstream of Lake Constance and of the High Rhine with its rocky river-bed. The mean annual bed load transport capacity of the natural river in the upper stretch has been estimated to be approx. 500,000 m^3; this amount will definitely cause erosion of the river bed.

Since the beginning of the 19th century, a variety of intensive river regulation measures have been carried out as listed in Table 1 by date, location, type, main purposes and side effects; see also Fig. 2 (Buck and Schiffler 1988, CHR/KHR 1993). It can be seen that goals shifted from single-purpose technically- and economically-oriented goals to multiple purpose goals, incorporating resource management goals. Fig. 3 gives a sketch of this remarkable change of values, shift in priorities and relative importance of various issues. These preferences are determined by the overall social and political conditions. In the Upper Rhine region, an additional fact has been instrumental for a long time: after World War I, the French Republic was granted the right to make use of the water power potential of the Upper Rhine. As a matter of fact, the regulations of the Versailles Treaty are still valid, but they are alleviated by the official German-French friendship and the existence of the European Union. The energy production of the river barrages at Gambsheim and Iffezheim is divided into equal shares for the French and the German power supply companies.

After World War I, negative impacts on nature and agriculture were neglected because of other preferences and conditions. This resulted in heavy bed erosion of up to 8 m in the river bed of the Rhine parallel to the Grand Canal d' Alsace, with a corresponding decrease of the groundwater level in the Rhine valley, producing disastrous effects on the vegetation. For mitigation purposes in the Rhine bed, a conservation weir (Kulturwehr) upstream from Breisach and bottom sills have been constructed; such measures have also been taken in connection with the so called "loop" solution (Schlingenlösung) downstream from Breisach. The measures following the construction of the Iffezheim barrage are described in the next section.

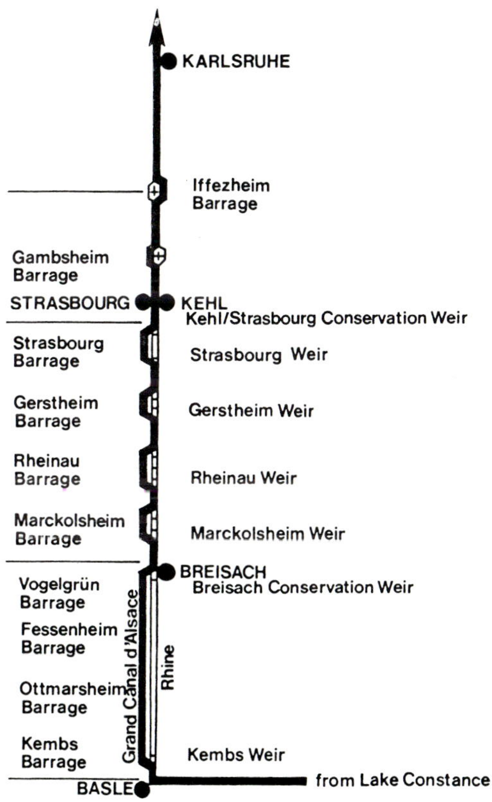

Figure 2 Location of human-made structures on the Upper Rhine.

Table 1 River regulation measures at the Upper Rhine and their effects

starting date	**location; measures**	**main purposes**	**main side effects**
1817	*downstream from Basle:* corrective measures according to the plans of J. G. Tulla: guide banks, training levees, dams; cut-offs Æ fixed straightened river course	flood protection and land improvement, fixed boundary line	drawdown of surface and groundwater levels caused by bed erosion
1906	*downstream from Basle:* low water regulation by M. Honsell: spurs, training walls	all-year navigation for large barges	continuation of bed erosion
1927	*between Basle and Breisach:* Rhine Lateral Canal (Grand Canal d'Alsace) with 4 barrages: Kembs (1932), Ottmarsheim (1952), Fessenheim (1956), Vogelgrun (1959)	hydroelectric power generation (navigation)	further drawdown of groundwater table due to low discharges in the river; downstream increase of flood peaks
1961	*between Breisach and Strasbourg:* "loop" solution with 4 short lateral canals and barrages: Marckolsheim, Rheinau, Gerstheim, Strasbourg (1967), bottom-sills and 2 conservation weirs (Breisach, Kehl)	hydroelectric power generation (navigation), preservation of river water levels	further increase of flood peaks due to loss of flood plains
1970	*between Strasbourg and Iffezheim:* 2 river barrages: Gambsheim, Iffezheim (1977)	prevention of bed erosion (navigation, water power)	further acceleration of flood waves, downstream bed erosion
1978	*downstream from Iffezheim barrage:* artificial bed-load supply	compensation of bed erosion: preservation of existing conditions	gravel pits in the Upper Rhine Valley
1982	*downstream from Basle:* flood retention measures: - operation of power plants - retention weirs in the river - retention in the former flood plain (Polder)	flood control, ecological restoration (Integrated Rhine Program: begin in 1988)	on other uses: such as forestry and agriculture

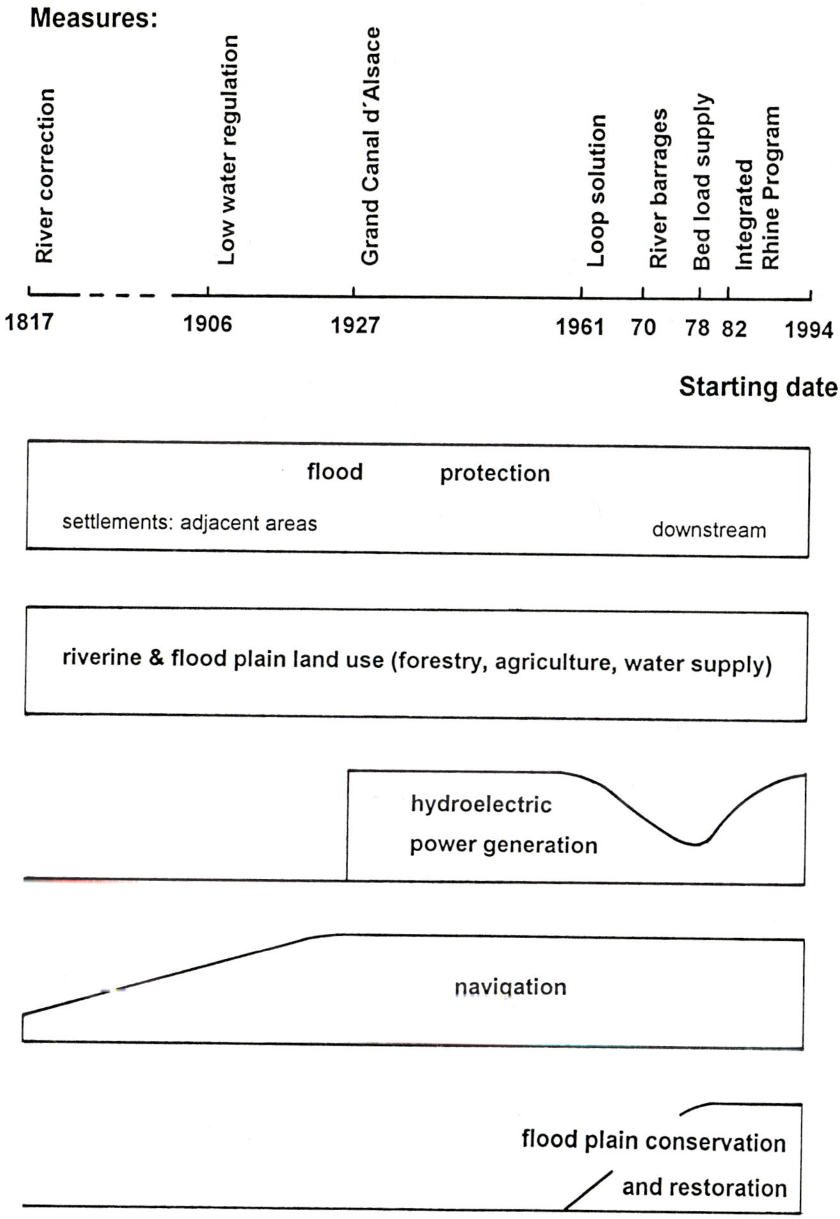

Figure 3 Function of the Upper Rhine for its users: shift in priorities of the main objectives from the 1820'ies until today.

Current Water Resources Characteristics and Activities

The main impacts of river training works in the Upper Rhine, up to the bringing into service of the Iffezheim barrage are

- bed erosion directly downstream from the last barrage, endangering the structures themselves, the adjacent areas in the Rhine valley (groundwater, flood plain forest, agriculture, water supply, ...) and shipping;
- a further increase of the Rhine floods both in magnitude (maximum flood peak) and speed of wave propagation. Peaks of these floods are now closer to those of the next downstream tributary of considerable size, the Neckar River, than they were under natural conditions.

Four possible variants of counter-measures against bed erosion have been investigated by the German Federal Waterways and Shipping Administration:

(1) protection of the river-bed by means of a layer of coarse gravel,
(2) construction of bottom-sills,
(3) building of additional barrages, and
(4) local action against the bed load deficiency by feeding artificial bed load (Geschiebezugabe) downstream of the Iffezheim barrage.

By means of "in situ" experiments and the use of physical simulation models, it was found that some of the tested protection measures were unfavorable; a layer of coarse gravel would not withstand the erosive forces induced by the ships moving upstream and downstream (variant (1)); submerged weirs would restrict river navigation unduly (see variant (2)).

Variants (3) and (4) have been investigated thoroughly in respect to technical aspects and further evaluated by cost-benefit analysis and environmental impact assessment. In the extended cost comparison, the following costs and benefits have been taken into account, omitting the benefits of preventing bed erosion:

- costs of construction and investment, operation and maintenance, reinvestment,
- navigation (changes in travel time and fuel consumption),
- hydro-electric power generation,
- flood protection, and
- road connection across the barrage(s).

The cost-benefit analysis revealed that solutions with further barrages were the most economical, showing the lowest cost-benefit difference. However, in comparing the results with other waterway and traffic projects in the Federal Republic of Germany, the Ministry of Traffic came to the conclusion that other projects are to be preferred to the investment cost-intensive barrage solutions and that the rather inexpensive artificial bed-load supply solution should be chosen. Besides, the non-structural solution resulted in a less negative statement in the final environmental impact report.

In 1982, Germany and France agreed upon the further regulation of the Upper Rhine. The decision was made to postpone the construction of an additional river barrage and to prevent bed erosion downstream from the Iffezheim barrage by adding artificial bed load. Every year, an average of 150.000 m^3 of a sand-gravel mixture is poured onto the river bed from hydro bottom-dump scows (Kuhl 1992).

Furthermore, both parties agreed to prepare and to carry out flood retention measures according to proposals drawn up by an International Flood Studies Commission made up of experts from Switzerland, Austria, France and Germany. The flood control measures consist of

- special operation rules for the French hydro-electric power plants in order to increase flood retention,
- construction of a few conservation and flood retention weirs in the river bed of the Rhine, and
- construction of quite a number of retention basins ("Polder") in the original flood plains.

The above mentioned agreement is additional to the 1969 and 1975 agreements and treaties on Rhine regulation. Since 1969, a Standing German/French Commission follows the preparation and realization of the measures. Members, who are the waterways managers, water resources administrators and representatives of the power supply companies in both countries, meet three times a year. The commission is supported by a Technical Committee and several Working Groups. The latter prepare agreements and instructions for the operation of the hydraulic works.

A new task is the planning of an additional passage for migratory fish across the Iffezheim barrage, especially for salmon. This measure is part of the 1987 international Rhine Action Program (RAP). Following the serious accidental spillage of toxic chemicals near Basle in November 1986, the countries bordering the Rhine have set the following goal to be achieved by the turn of the century: "... the ecosystem of the Rhine must become a suitable habitat to allow the return to this great European river of the higher species which were once present here and have since disappeared, such as salmon.". The other objectives of the RAP are: Rhine water must remain useful for drinking-water purposes, the sediments must be free of pollution by hazardous chemicals and the North Sea environment shall be protected (Dieperink 1993). The target "Salmon 2000" is the slogan for the "Ecological Master Plan for the Rhine". It has been elaborated by ICPR, the International Commission for the Protection of the Rhine (Schulte-Wülwer-Leidig 1993). The ecological master plan focusses on two points:

- the restoration of the main stream as the backbone of the complex Rhine ecosystem, with its main tributaries forming habitats for migratory fish (sea trout, salmon, allice shad and others);
- the protection, conservation and improvement of ecologically important stretches of the Rhine and the Rhine valley with a view to increase the species diversity of indigenous animals and plants.

Apart from the Standing German/French Commission, which focuses on river regulation measures, other international institutions exist that also deal with water resources management problems in the Upper Rhine area. Fig. 4 shows the region in question; Fig. 5 gives an overall view of the bi- and trilateral cooperation in the triangle of Switzerland - France - Germany. In the "Environment" working group, experts prepare and follow with technical and scientific advice joint projects, most of which are partly financed by programs of the European Union, like INTERREG and LIFE. Amongst other projects, in the coming years a map of groundwater contour lines of the Upper Rhine valley aquifer and of groundwater quality will be published, and transboundary geological sections will be drawn up. In a LIFE project, a comprehensive groundwater model of the Upper Rhine aquifer between Basle and Karlsruhe - Heidelberg will be established (various personal communications 1994).

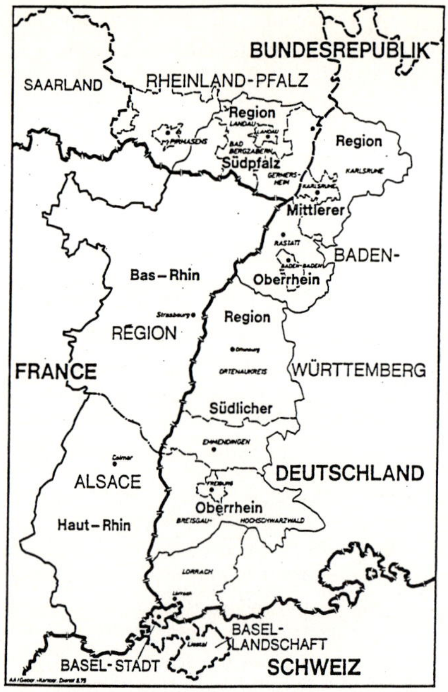

Figure 4 Region of the Trilateral Government Commission.

German-French-Swiss Intergovernment Commission

Trilateral Regional Committee: Basle, CH Freiburg, D Strasbourg, F	Upper Rhine Conference	Bilateral Regional Committee: Karlsruhe/Neustadt, D Strasbourg, F

Working groups:						
cultural relations	electronic media/ communi-cation	spatial planning	transport-ation	economy	environ-ment	education

Figure 5 Structure of international cooperation in the Upper Rhine Region.

Hydro-Ecological Planning and Projects

Presently, the main problems in the field of water resources in the Upper Rhine Valley directly adjacent to the river are the rather low degree of flood protection downstream of the last river barrage, Iffezheim, and the degradation of former flood plains. At least in the foreseeable future, the bottom erosion downstream from the Iffezheim barrage can be controlled by artificial bed-load feeding.

Combined flood control and ecological measures

Since the German-French agreement of 1982, flood control measures have been planned in France and in the Federal Republic of Germany in the States of Baden-Württemberg (east bank) and Rheinland-Pfalz (west bank). From a total of 212 million m^3 of retention volume, the State of Baden-Württemberg has to provide 127 million m^3, and the French Republic 55 million m^3.

Thanks to the already implemented measures, the mean recurrence interval of the design flood of the river dikes downstream of the the Iffezheim barrage is now about 100 years. This has to be raised to 200 years, representing the degree of flood protection when the natural flood plains were still in existence. Immediately after the completion of the last river barrage in 1977, the mean recurrence interval had been only 50 years.

While planning the flood retention measures proposed by the International Flood Studies Commission, it became evident that the filling of the former flood plains up to more than 2.5 m for days at a time would damage the remnants of the flood plain vegetation, namely trees and shrubs. Stagnant water is not favorable to natural ecological conditions either.

The State of Baden-Württemberg in 1988 thus declared the Integrated Rhine Program with the aim of achieving both the agreed upon degree of flood protection and conservation and restoration of the natural state of the flood plain. Consequently, the necessary flood retention volume will increase from 127 to 168 million m^3, resulting in a cost of approximately 700 instead of 330 million DM.

Pilot study with application of decision support methods

One of the possible areas for construction of a retention basin is located upstream of the conservation weir Breisach, parallel to the Rhine and the Grand Canal d'Alsace. Figure 6 shows one of the 5 alternatives which have been investigated, together with the no-action (status quo) alternative in a pilot study carried out by a working group of the German Associaton for Water Resources and Land Improvement (Deutscher Verband für Wasserwirtschaft und Kulturbau DVWK) (Pflügner 1991, Buck and Pflügner 1991). The intention of the study was to gain experience on the application of non-monetary evaluation methods which would also be useful tools for Environmental Impact Assessment (EIA). In Germany, EIA has been required legally for projects affecting river ecology and environment as late as 1990.

The five alternatives differ (1) in the location of the dike, delimiting the flood detention area from the hinterland and (2) in the operation of the flood polders.

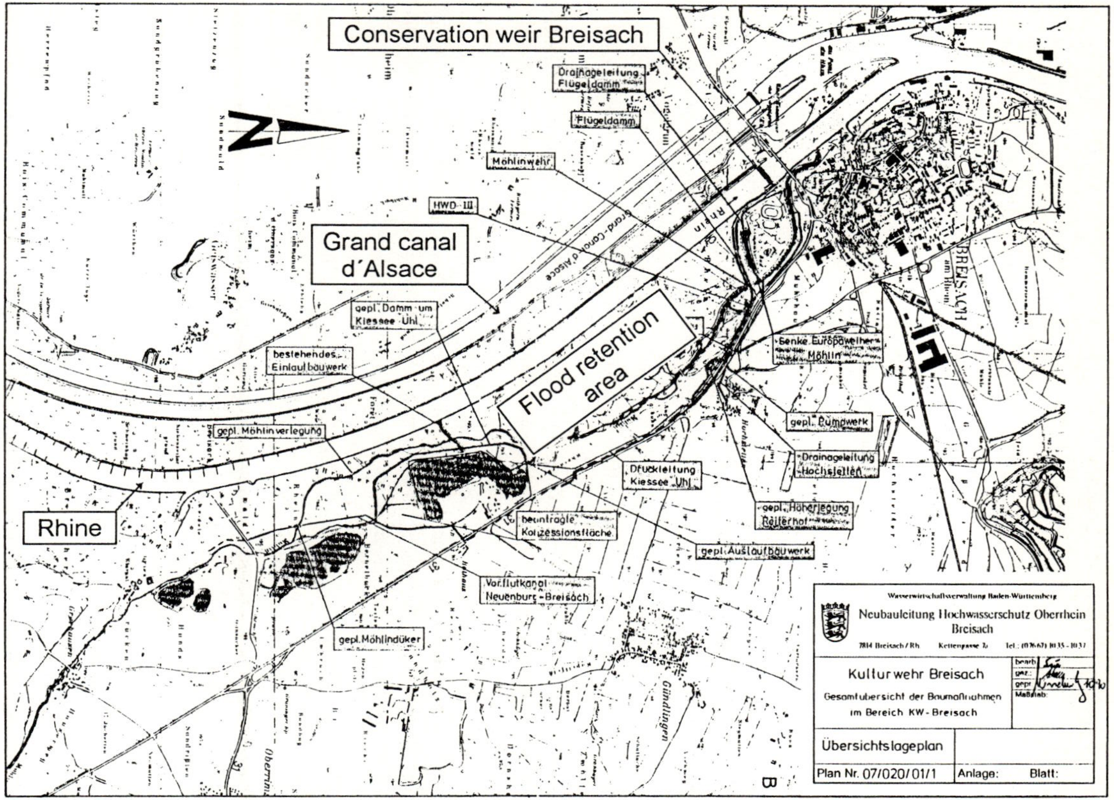

Figure 6 Location of the flood retention project along the Rhine at the conservation weir Breisach.

The areas can be filled through inlet structures either exclusively in case of a large flood or, for ecological reasons, in every instance when flooding would occur under natural conditions ("ecological flooding") The retention volume ranges from 3 to 8.4 million m^3. According to the German-French agreement, the retention volume at this site should amount to 10 million m^3. The five alternatives at 3 different locations A, B and C are denoted in the following tables with "A-without" (*without* ecological flooding), "B-without", "B-with" (*with* ecological flooding), "C-without", and "C-with".

The pilot study focused on the ecological effects in the flood "polder" area in relation to the estimated contribution to flood protection downstream of the last river barrage, which is roughly 100 km downstream. Costs and effects on forestry, agriculture, recreation and leisure have not been considered in this study.

Data and other information were provided primarily by the water resources administration and were taken from an environmental impact study carried out under the overall auspices of the WWF Institute for Floodplains Ecology in Rastatt (10 km downstream of Iffezheim).

The pilot study can be divided into the following steps:

- deriving a hierarchical goal system for the site specific ecological conditions and deduction of evaluation criteria;
- determination of the ecological effects of the alternatives for all criteria;
- development of transformation functions (value functions) to convert cardinal values for effects to normalized ratings in accordance to the sub-goals;
- weighting the relative importance of the criteria;
- aggregation of normalized ratings and weights for criteria by two multi-objective decision support methods.

The overall ecologically oriented goal is the strengthening and/or the restoration of site-specific natural conditions. This goal has been divided into the sub-goals "abiotic factors" and "biotic factors", which were assumed to be both equally important. The further subdivision of the abiotic part into 7 criteria is shown in Fig. 7. Value function 4 expresses that the optimum mean groundwater level below surface in the vegetation period is 0.5 m; the worst values are 1.2 m or more and zero. n our case, all criteria could be measured on cardinal scales.

The biotic part of the goal system is divided into on the one hand terrestrial and semi-terrestrial components and on the other hand, flora (plants) and fauna (animals). The best rating is assigned to natural conditions (under today's overall prevailing environmental conditions: water quality, climate, etc.).

Table 2 shows the normalized ratings and the weights which have been assigned by expert opinion. Equal weights for the ecological part and for flood control are justified by the policy statement in the Integrated Rhine Program.

criteria [units]	transformation function
① **area of potential flood plain** [ha]	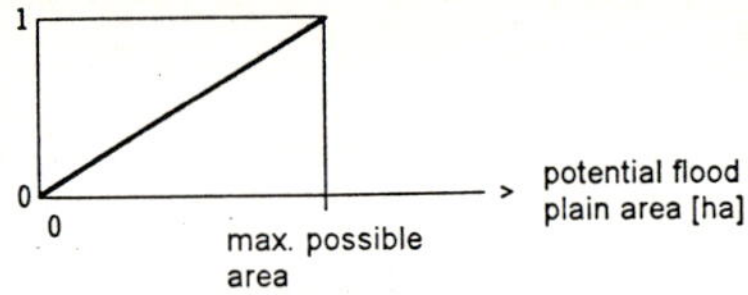
② **relief factor** = real area / horizontally projected area [-]	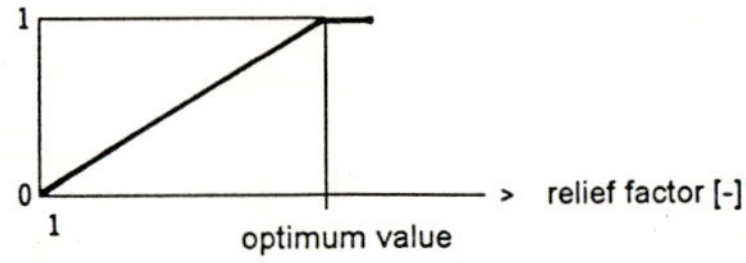
③ **soil factor** = area covered with flood plain type soil / project area [-]	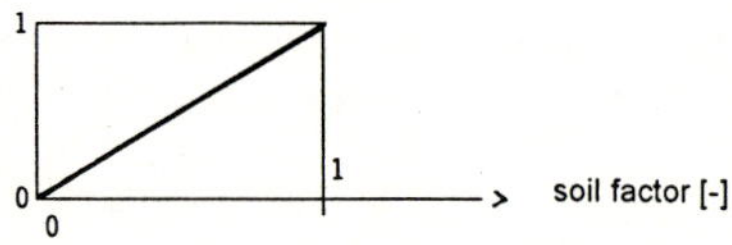
④ **mean distance of groundwater level to surface*** [m]	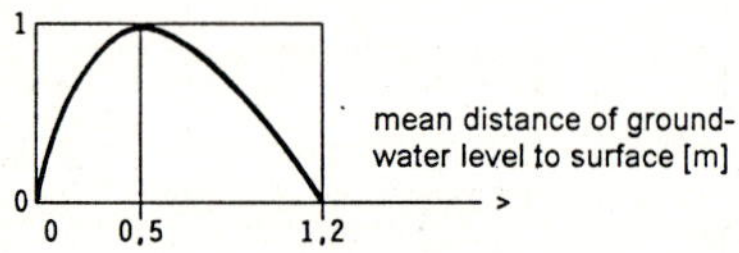
⑤ **amplitude of groundwater level fluctuations*** [m]	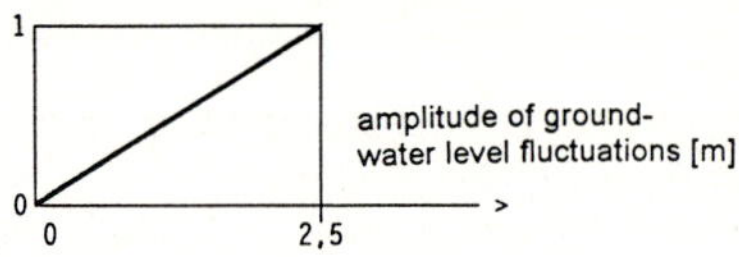
⑥ **duration of flooding*** [days per year]	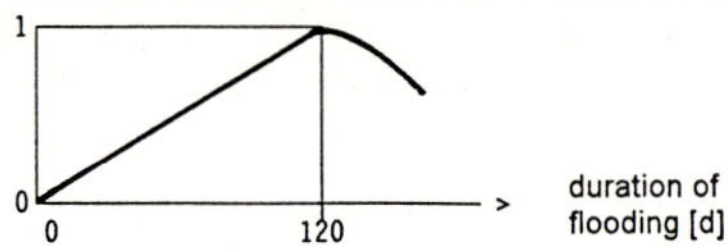
⑦ **exceedance time of flooding higher than 2.5 m*** [days per event]	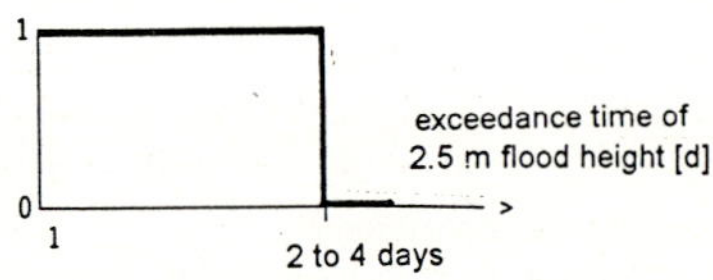

* in the vegetation period

Figure 7. Criteria and transformation functions for the sub-goal "Strengthening/restoration of site-specific abiotic factors".

Table 2. Normalized ratings and weights for criteria and alternatives.

criteria	weights	normalized ratings for alternatives					
		status quo	A-without	B-without	B-with	C-without	C-with
① floodplain area	0.0375	1.00	0.52	0.52	0.52	1.00	1.00
② relief	0.0375	0.80	0.73	0.73	0.75	0.80	0.80
③ soil	0.0250	0.60	0.54	0.54	0.60	0.55	0.65
④ groundwater level	0.0250	0.40	0.40	0.40	0.50	0.40	0.60
⑤ groundw. amplitude	0.0375	0.59	0.59	0.59	0.77	0.59	0.77
⑥ duration of flooding	0.0500	0.00	0.10	0.10	0.70	0.10	0.70
⑦ exceedance time	0.0375	1.00	1.00	1.00	1.00	1.00	1.00
❶ terrestric flora	0.0625	0.45	0.40	0.40	0.50	0.40	0.55
❷ terrestric fauna	0.0625	0.50	0.30	0.35	0.55	0.50	0.65
❸ aquatic flora	0.0625	0.45	0.35	0.40	0.45	0.45	0.50
❹ aquatic fauna	0.0625	0.45	0.35	0.40	0.45	0.45	0.50
⓪ flood control	0.5000	0.30	0.65	0.65	0.55	0.84	0.64

One of the multicriteria methods applied with a very simple structure is the additive weighting method (AWM). The normalized values for every criterion are multiplied by their weights and added up to an overall score. Although the procedure has been used extensively for practical applications in Germany, it has been criticized because of its frequent misuse and inappropriateness (among others: cardinalisation of ordinal scales, weighting procedures, unjustified compensation between criteria), but apparently these difficulties may be avoided (Pflügner 1991).

In the pilot study reported here, another multicriteria decision technique has been applied: Composite Programming (CP). It looks similar to the additive weighting method, but uses in addition so-called balancing factors (Bárdossy et al. 1985). These factors reflect the degree of possible balancing out of low score results on one criterion by excellent results on another criterion. For instance, a relief typical for flood plains in the "Polder" area is not really effective if the polder is not filled at all or only scarcely. Similarly, a high degree of flood protection is not sufficient by itself; it has to coincide with acceptable ecological conditions in the flood plain - in accordance with the two main goals of the Integrated Rhine Program. The applied basic balancing values are shown in Fig. 8. A value of 6 means a rather low degree of possible compensation. In the additive weighting method, compensation or substitution is not controlled, as all balancing factors are 1; in principle, every substitution is allowed. Another difference is that the weighting method is taken as an additive utility function, whereas CP uses the Lp-norm as a proxy function.

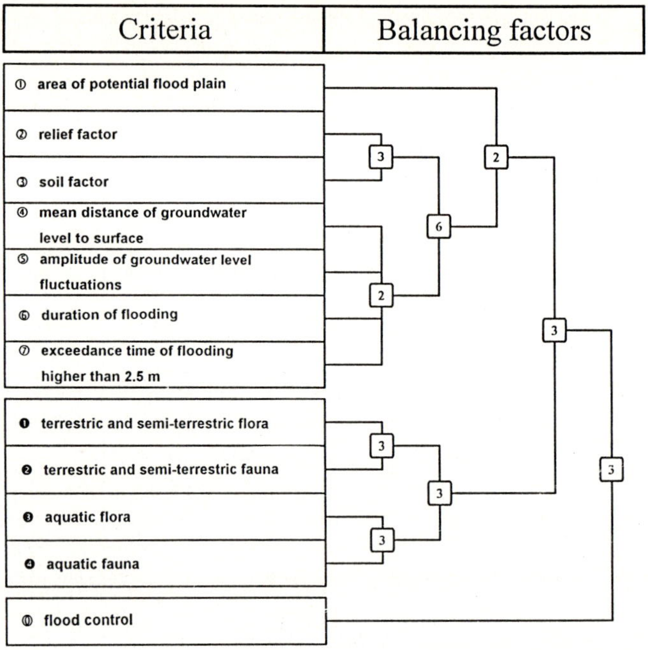

Figure 8 Balancing factors for Composite Programming.

The results of the two decision support methods are shown in Table 3. The results of a sensitivity analysis are also given; for this analysis, weights and balancing factors were varied. Column 7 shows the outcome for the AWM, column 5 the results for CP for the basic assumptions concerning weights and balancing factors. The best solutions are different for the two methods: AWM prefers alternative "C - without ecological flooding", whereas CP points to alternative "C - with ecological flooding". However, it may be noted that for AWM, alternative "C - with" has a rather similar score than alternative "C - without". The reason for the different results is the tendency of AWM to compensate less favorable outcomes for the ecological criteria by a higher score for flood protection (Table 2). CP takes into account the more balanced results in the soil/relief-related criteria, both in the water-related criteria (high balancing factor 6) and between the overall ecological score and the flood-protection score; these values are rather different for alternative "C - without" and to the contrary of similar order of magnitude in alternative "C - with".

Table 3 Results of Additive Weighting Method and Composite Programming with sensitivity analyses

Alternatives	weight given to goal "flood protection"							
	0.2	0.4	0.6	0.8	**0.5**	0.5	**0.5**	0.5
	balancing factors							
	basic values according to fig. 8: e.g. ecology – flood control = 3					ecology – flood control = 10	all factors = 1 (AWM)	all factors = 10
	overall score and *rank*							
	(1)	(2)	(3)	(4)	(5)	(6)	(7)	(8)
status quo	0.409 *6.*	0.378 *6.*	0.350 *6.*	0.324 *6.*	0.363 *6.*	0.341 *6.*	0.418 *6.*	0.239 *6.*
A-without	0.424 *5.*	0.464 *5.*	0.511 *5.*	0.569 *4.*	0.486 *5.*	0.430 *5.*	0.548 *5.*	0.319 *5.*
B-without	0.443 *4.*	0.481 *4.*	0.524 *4.*	0.578 *3.*	0.501 *4.*	0.450 *4.*	0.557 *4.*	0.325 *4.*
B-with	0.555 *2.*	0.553 *2.*	0.522 *3.*	0.551 *5.*	0.553 *3.*	0.553 *2.*	0.573 *3.*	0.522 *2.*
C-without	0.496 *3.*	0.540 *3.*	0.595 *2.*	0.672 *1.*	0.566 *2.*	0.494 *3.*	0.688 *1.*	0.330 *3.*
C-with	0.622 *1.*	0.627 *1.*	0.631 *1.*	0.635 2.	0.629 *1.*	0.628 *1.*	0.658 2.	0.581 *1.*

Changes in numerical values assigned to weights of the sub-goal flood control and to the balancing factors shows that the decision for alternative "C - with ecological flooding" is very stable. Even alternative "B - with" is preferred to variant "C - without" for some combination of factors and weights. Decision makers may use the information supplied by the planners according to their needs. This information ranges from the affected goals and predicted impacts to the results of Multiple Criterion Decision Making (MCDM) techniques and sensitivity analyses.

Conclusions

The southern part of the Upper Rhine forms at a stretch of 180 km, between Basle and Karlsruhe, the German-French border. Since the beginning of the 19th century, a variety of river regulation measures have been carried out, causing considerable side effects on downstream flood protection, flood plain ecology, etc. on both sides of the river. Conservation and restoration of flood plains has gained much more importance in the last decades because even the remnants of the natural flood plains are endangered.

Several bilateral and trilateral institutions (with members also from Switzerland) are active in exchanging information and treating common water resources and environmental problems. Cooperation has been well established for decades on the technical and regional political level and and is continually developing. Nevertheless, it must be realized that a lot of work still remains to be done, especially concerning environmental protection. Multiple objective planning methods may provide the required support for decision makers.

A study was presented to demonstrate the application of two decision support techniques. Conflicting objectives are flood control and flood plain restoration. Methods of simple compensation of positive and negative results , e.g., Additive Weighting Method (AWM) often find little acceptance in real life decision situations. On the other side, Composite Programming (CP) offers a suitable alternative. It is strongly recommended to increase the use of sound planning and decision support techniques.

Acknowledgements

The author would like to acknowledge the contribution of the reviewers. Valuable comments by my colleague Dr Charlotte Kämpf are much appreciated.

References

Bárdossy A, Bogardi I, Duckstein L (1985) Composite Programming as an extension of Compromise Programming. In: Serafini P (ed.) Mathematics of Multiple Objective Optimization: 375-408, Springer Wien.

Buck W, Schiffler GR (1988) River regulation measures on the Upper Rhine, their side effects and counter-measures. In: Volker A, Henry JC (eds) Side effects of water resources management. IAHS Publ. No. 172: 95-108, Wallingford.

Buck W, Pflügner W (1991) Nutzwertanalytische Bewertung auenökologischer Wirkungen - Pilotstudie für eine Hochwasserschutzmaßnahme. Wasserwirtschaft 81: 578-587.

CHR/KHR (1993) Der Rhein unter der Einwirkung des Menschen - Ausbau, Schiffahrt, wasserwirtschaft - Bericht Nr. I-11 Lelystad/Niederlande.

Dieperink C (1993) Salmon, sustainibility and transboundary river basin management, some reflections. Proc. Transboundary river basin management and sustainable development, IHP UNESCO Paris.

Kuhl D (1992) 14 years artificial grain feeding in the Rhine downstream from the barrage Iffezheim. In: Proc. 5th Int. Symp. on River Sedimentation Karlsruhe, Vol III: 1121-1129.

Personal communications (1994) by representatives and members of Integrated Rhine Program (IRP), Bilateral German-French Commission and Regional Board Middle Upper Rhine (RVMO), Karlsruhe.

Pflügner W (1991) Pilotstudie zur Anwendung nutzwertanalytischer Verfahren. Mitteilungen Heft 22, Deutscher Verband für Wasserwirtschaft und Kulturbau, Bonn.

Schulte-Wülwer-Leidig A (1993) International Commission for the Protection of the Rhine against Pollution. The integrated ecosystem approach for the Rhine. Proc. Transboundary river basin management and sustainable development, IHP UNESCO Paris.

IV.2 Conflict Resolution Between Two Riparian Countries: A Case Study About a Hydropower Conflict

H.P. Nachtnebel
Institute for Water Management, Hydrology and Hydraulic Engineering
Universität für Bodenkultur, A-1190 Vienna
Austria

Introduction

The Danube basin has an area of 817 000 square kilometres with eleven riparian countries. Small catchments from four additional countries are draining to the Danube basin, to. More than 40 large dams and barrages have been constructed on the main river and its tributaries utilizing the hydropower potential. While these facilities provide important opportunities for the utilization of the river's resources, they also are illustrative examples of the difficulties of balancing these important economic activities with environmentally sound and sustainable management. These aspects are addressed in a general form by GOLDSMITH and HILDYARD (1984), PEARCE and TURNER (1990), COMMON and PERRINGS (1992), TURNER (1993), van DAMM and WESSEL (1993).

Hydropower schemes are multipurpose schemes. Besides power generation they support navigation and often they reduce flood risk along the impounded section. At the same time adverse environmental impacts might be observed in the accompanying groundwater system, in both alluvial and karstic aquifers. Also, the viability of the riverine forest which constitute a large and ecologically important biological diversity, is affected. It can be easily concluded that any further hydropower development should be evaluated in a multiobjective framework considering economic, ecological and social objectives. Due to the degree of economic development the national preference structures are different. And this is one of the reasons that hydropower schemes which are located along the border may lead to conflicts among countries.

The objective of this paper is to analyze a hypothetical conflict among to countries. The impacts of alternative hydropower schemes are assessed and evaluated in a multicriterion framework including 33 criteria references. Because of different impact levels and also different preference structures in the involved countries the evaluation procedure results in different rankings of the alternatives.

A conflict resolution technique is proposed which ranks is each alternative with respect to its generalized distance from the preferred alternative in each country. The approach is supplemented by the evaluation of the integrated impacts irrespectively from national borders. The data base is from an Austrian case study (NACHTNEBEL, 1990) which is typical for the Danube.

Next, navigation and hydropower utilization in the Danube basin is discussed and subsequently the groundwater resources and flood plain forests, both sensitive to any change in the hydrological regime, are described.

Navigation on the Danube

The Danube is an international waterway and the Danube commission re-established in 1948 has the responsibility for providing guidelines for the navigation the Danube. Subsequently these guidelines are adopted by the riparian countries and are implemented at the national level.

The main traffic comes from countries in the Commonwealth of Independent States (CIS) and non-Danube countries (EQUIPE COUSTEAU, 1993). The destinations remain largely concentrated on the Romano-Ukrainian part of the river, and especially in the iron and steel centers near the delta. This overall concentration of the traffic flow in the downstream region of the Danube is supplemented by the large local traffic in crude minerals, particularly sand and gravel, often taken from river beds such as that near Komarno.

Table 1 1989 Country-to-country navigation estimated in ton-km.

Origin	Destination									Total
	AUT	BUL	SLOV	GER	HUN	ROM	CIS	ex-YUG	Others	
AUT	110 210	18 816	70 272				2 262 942	60 795	41 200	2 564 235
BUL	91 392	436	34 600	133 392	1 712		250 700	7 800		521 832
SLOV	166 608	115 700	117 480	11 753	10 398	10 398	1 448 800	180 600	194 000	2 245 339
GER	4 148	22 232		4 175	1 464	22 440	41 796	54 405		150 660
HUN	277 184		366	365 268	279 200		1 450 750	44 091		2 416 859
ROM	112 000		138 640	518 364	51 832	3 617 400	13 000	142 554	418 200	5 011 990
CIS	5 350 850	1 325 440	2 562 565	67 338	2 383 500		738 184	2 996 367	226 800	15 651 044
ex-YUG	246 075		161 700	486 018	30 033		1 084 336	1 370 883		3 379 045
Others	338 200		63 000			2 061 200	296 900			2 759 300
Total	6 696 667	1 482 624	3 080 151	1 656 580	2 747 741	5 711 438	7 587 408	4 857 495	880 200	34 700 304

Table 1 summarizes the traffic flow according to the main origins and destinations mixes cabotage and traffic between the different countries, based on Annual reports of the Danube Commission. Local navigation is important. For example, it accounts for 73 % of Slovak traffic with 5,874 million tons carried over a distance of less than 19 km. Although national cabotage volumes, mainly sand and gravel, are considerable, the Danube trade is structured by certain long-distance traffic flows. This can be concluded from the analysis of the traffic volumes expressed in ton-km.

Navigating conditions on the Danube are typical of those to be found on a river containing large free-current sections. Thresholds remain in unstable sectors because of the formation of sand banks, settlement of alluvium or the presence of dangerous rocky beds. Moreover, the regulated or channeled sections of the mainstream of the Main or the Rhine and of the Rhine-

Main-Danube canal do enable the same convoys to be used from the Black Sea to the North Sea.

Hydropower Potential and its Utilization

Large scale hydroelectricity (with a capacity of more than 10 MW) represents an important potential resource in the Danube countries, but its intensive exploitation entails very serious consequences for the environment. This is the case particularly when dams located in the Danube plain regions - or the plain regions of the largest tributaries - are concerned. The Gabcikovo dam, in Slovakia, is a perfect example (VODOHOSPODARSKE VYSTAVBA BRATISLAWA, 1990).

Table 2 Hydroelectric power in the Danube countries (TWh).

	AUS	CZECH	SLOV	HUN	ex-YUG	ROM	BUL	Total
Electricity consumption	43.5 (1990)	56.7 (1990)	29.3 (1990)	37 (1991)	70.7 (1988)	74 (1990)	48.6 (1990)	360
Electricity production	44 (1990)	65.2 (1990)	21.4 (1990)	27.4 (1991)	78.4 (1988)	64 (1990)	44.2 (1990)	345
Economically operable hydraulic potential	53.7	3.5	7.3	4.6	71.8	40	15.3	196
% consumption	> 100	6.1	24.9	12	> 100	54	31.4	79.4
Total hydroelectricity production (1991)	30.5	1.4	2.2	0.178	24	16.5	4.6	
% consumption	70	2.4	7..5	0.4	34	22.2	9.5	13.4
Economically operable SHPP potential *)	> 4	1.66	1.2	0.032	?	4	> 2.5	
% consumption	> 9.2	3	4	0.09	?	5.4	> 5.1	3.7
Small hydroelectricity production (1991)	2	0.476	0.25	0.015	?	0.4	0.62	
% consumption	4.5	0.8	0.8	0.04	?	0.5	1.3	

* SHPP: Small Hydropower Plant

Overall, economically operable hydraulic potential represents 196 TWh for a six Danube countries including Austria, ex-Czechoslovakia, Hungary, ex-Yugoslavia, Romania and Bulgaria which were recently investigated in some detail (EQUIPE COUSTEAU,1993). In Austria and the former Yugoslavia, hydroelectricity could theoretically cover 100 % of electricity consumption (Table 2) while in Czechoslovakia it could only take a 6 % share, the figures for the other countries totaling between 12 % and 54 %. Only 40.5 % of this total hydroelectric potential is currently being exploited (79.4 TWh). Several proposals developed at the national level exist to increase the utilized rate (IUCN, 1994; EPDRP, 1993, 1994)

Groundwater Resources

Major groundwater resources are located in the alluvial plains of the Danube and the tributaries. The spatial distribution of humid zones (EQUIPE COUSTEAU, 1993; IUCN, 1994) gives a good indication about the location of shallow groundwater bodies which together with deep layered aquifers supply a high percentage of the population (80-95 % varying from country to country) with drinking water. Additionally, vulnerable karstic water resources are of special importance for drinking water supply.

Due to agricultural practice, landfill sites, inefficient sewage systems and interaction with polluted surface water systems these water resources are endangered. Pollution of groundwater is a common problem and nitrate, chloride, sulfate, ammonia, phenol are often locally exceeding the respective country standards. The low recharge rate of groundwater bodies in the lower part of the basin delays remarkably the sanitation and recovery processes. Due to large impoundment works the interaction between surface and groundwater has been decreased and in the medium term qualitative impacts on the groundwater resources have been also observed.

Flood Plain Forests

The river and the aquatic and terrestrial habitats of the floodplain (Tab. 3) constitute one of the most complex ecological systems in the world. It is made up of a group of highly diverse, constantly interacting ecosystems: running waters of the river and its active arms, stagnant waters connected permanently or periodically to the river, temporary waters, terrestrial plant formations subject to variable flooding periods, and an underground aquatic ecosystem.

The diversity of European floodplain forests is a the result of hydrological conditions. The plant density is reminiscent of some of the structural features of dense tropical forests. Their dendrological richness (50 species of trees and shrubs) and specific diversity are remarkable in comparison with non-floodplain temperate forests (diversity index of 0.6 to 0.8 for the latter, and greater than 3 for floodplain forests). This explains a number of remarkable features such as continuous flowering from January to October. The number of species flowering at the same time is twice as high as in non-floodplain forests.

In 1967, 103 fish species were recorded in the Danube. This can be compared with 40 and 61 for the Rhone and the Rhine respectively. This taxonomic richness can be related to the number of endemic species and to the remaining high diversity of floodplain biotopes. However, the local structure and distribution of the fish communities reflect the impact of human modifications of the system, such as in the German-Austrian section. Furthermore, the Iron Gate plant has stopped the upstream movement of migratory species.

Besides the Delta area the most important flood plains are located between Vienna and Belgrad. The Austrian floodplain forests which is already under environmental protection covers approximately 10,000 ha and is suffering under desiccation. Several hydropower alternatives from which one is planned as a binational project have been elaborated for this river section. The Hungarian Szigetköz and Slovak Csallaköz comprise an exceptional meandering zone along the border between Slovakia and Hungary. Today, the active covers 6,000 ha on the Hungarian side and 23,000 ha on the Slovak side. However, by the entire zone will be affected severely the Gabcikovo dam and its canal to Bratislava. The Hungarian Gemenc region covers approximately 47,000 ha, 18,000 of which are classified as protected land. It is planned to restore former hydraulic conditions within a stepwise restoration program within this floodplain. The Kopacki Rit, which represents some 50,000 ha lying between the Drava and the Danube - now split between Croatia and Serbia - is the richest floodplain of the Danube.

Table 3 Area covered by floodplains in the Danube riverine corridor (IUCN, 1994).

Country	Area in hectares
Austria	27.500
Bulgaria	80.000
Croatia	350.000
Czech Republic	9.600
German (Bavaria)	45.662
Hungary	51.553
Romania	1,028.000
Slovakia	5.000
Ukraine	150.000
Total	1,747.315

Description of the Case Study

Several large hydropower schemes are going to be implemented in the next years in the Danube basin (EQUIPE COUSTEAU, 1993; EPDRB, 1993, 1994). Some of them are international projects, affecting two countries. Although the value of hydropower utilization is recognized there is an ongoing discussion about its adverse environmental impacts. In some national studies (HARY and NACHTNEBEL, 1989) the ecological and socio-economic changes are summarized (UNEP and UNESCO, 1990; STAROSOLSZKY; 1986; PETRASOVITS, 1986; IUCN, 1994). Due to recent political and economic transition in several Danube countries there is a different perception of risk and environmental damage. Also, there is a different attitude in emphasising economic development and environmental preservation. Or in other words there exists a different preference structure in the Danube countries which in term may lead to conflicts in project assessment.

Characteristics of the Case Study

A hypothetical case study is used to analyze the conflict raised by hydropower utilization between two countries. To provide a more reliable data base a data set referring to a real world example which has been investigated in detail is selected (NACHTNEBEL, 1990). The project area (ISD, 1989; NACHTNEBEL et al., 1989; KANIAK, 1986; DOKW, 1989) on the edge of the Danube between Vienna and Hainburg, covers about 10,000 ha. The river section is bordered by extended riverine forests and is already environmentally protected. It is planned to declare it as a national park to minimise adverse human impacts. The groundwater system along the river is already serving domestic water requirements and plans are already elaborated to extend its utilization. Further, a karstic groundwater system feeding a medicinal spring is located in the vicinity of the river. Hydrogeological investigations indicate that the yield of the spring is strongly dependent on the water table conditions of the regional groundwater system and on the water table of the Danube. With respect to navigation this river section constitutes one of bottle necks of the Austrian Danube. Frequent dredging works are necessary in order to maintain navigable conditions. Especially during low flow periods the recommendations of the DANUBE COMMISSION(1988), which promote a navigable depth of 2.50 m cannot be guaranteed.

It can be concluded that this example has many features in common with other hydropower projects along the Danube. In the hypothetical case study it is assumed that the national borderline follows the Danube and that country 1 and 2 are in dispute about development plans in the Danube section.

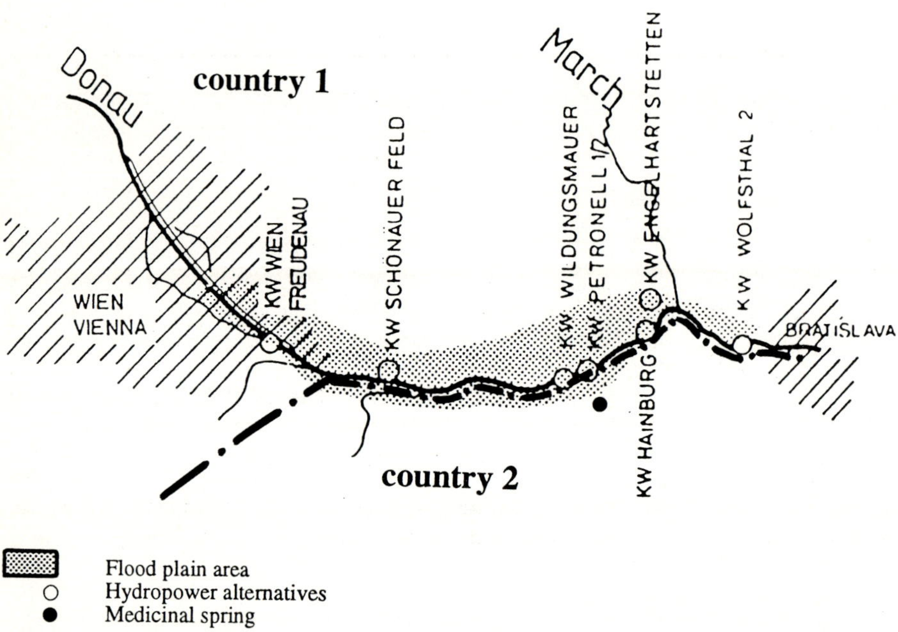

Figure 1 Case study area with location of hydropower schemes.

Development Alternatives

The hydropower alternatives are characterized (Table 4) by their location and their capacity. Additionally, every alternative is split into a pure economical alternative and a "softer alternative" including secondary measures for the compensation of hydrological impacts caused by the hydropower scheme. The status quo is considered by alternative A1 which itself is split up into two sub-alternatives. The first excludes any hydropower utilization, while (A2) additionally substitutes the nonutilised hydropower by energy import.

The runoff river schemes upstream trap the sediments of the Danube, especially the coarse material. As a consequence degradation downstream of the last power station is observed. The mean annual degradation of the river bed has been estimated to 1,5-2,5 cm/a and measures have to be taken to cope with this process. Several measures including artificial armoring of the river bed and addition of coarse grained bed material to decrease the degradation process were analyzed. Due to a high uncertainty in the efficiency of this measures none of them is considered in this paper.

Table 4 List of hydropower alternatives.

No of Alternative	**A3& A4**	**A5&A6**	**A7&A8**	**A9&A10**	**A11&A12**
Names of alternatives	Hainburg	Schönau Petronell Wolfsthal	Petronell Wolfsthal	Wildungs- mauer Wolfsthal	Engelharts tetten
Location (km)	1883	1906 1890 1873	 1890 1873	1892,5 1873	1883,0
No of power stations		3	2	2	1
Installed capacity (MW)	360	247	327	327	352
Annual power generation (GWh)	2075	1700	1990	1920	2035
Investment costs (Mrd. ATS)	11,4	24,9	15,9	15,6	12,2

In the long-term the degradation of the river bed will be also reflected in the decrease of the surface water level, in a corresponding lowering of the groundwater table and in a decreased inundation frequency. Therefore, a pure preservation of the free flowing section without any measures to protect the river bed, will also result in some ecological adverse impacts.

It assumed that the Danube constitutes between location 2 and location 6 the border between country 1 and country 2 (Fig. 1). Country 1 is strongly interested in hydropower utilization while country 2 emphasizes ecological objectives. Country 1 is willing to finance fully the hydropower project. To compensate for environmental impacts in country 2 a major percentage of power generation reaching up to thirty percent of power generated along the joint border section will be allocated to country 2

Identification of Goals for this Section of Danube

The main goals of water related development are in support of navigation, utilization of hydropower potential, preservation of water resources for domestic requirements and environmental preservation because of the unique characteristics of the flood plain area. Subsequently, these goals are specified and criteria are developed to quantify the goals.

Hydropower

Governmental statements in both countries underlined the importance of hydropower utilization but simultaneously in country 2 some environmental groups asked for planning steps to establish a national park in the flood plain area of Danube.

Goals related to power generation and energy management are included in energy reports which are regularly issued by the ministries of the countries and in international studies (EQUIPE COUSTEAU, 1993; KOVACS, 1986; IUCN, 1994). The principles of governmental energy and environmental policies include the following set of guidelines:

- reduction of primary energy consumption
- increased utilization of renewable resources, especially of hydropower
- minimization of environmental impacts related to power generation and consumption.

Navigation

The Danube section from Braila (170 km) to Kehlheim, FRG (2414,7 km) is classified as category IV according to the European waterways standards (FEKETE, 1990; DANUBE COMMISSION, 1988). This requires for unimpounded sections a minimal depth of 2,50 m and a width of 40-180 m for navigation. In impounded sections the minimal prescribed depth is 3.5 m. For the respective stretch of the Danube the recommendations of the Danube commission include a minimum depth of 2,50 and a width of 150 m. During low flow periods several fords with a depth of 2 m and less restrict economical navigation and frequent dredging works are required to maintain the waterway. All the Danube countries have adopted these recommendations and their governments declared it as an important goal to guarantee at least the minimal requirements for navigation throughout the year. Recent programs and recommendations (ECE, 1994) even propose a navigable depth of 3.2<m for international European water ways.

Drinking Water Supply

One of the goals of the regional water management is the protection of the extended alluvial aquifers bordering the Danube. This resource partly serves the regional drinking water supply including some villages in the vicinity of the Danube. In this context the emphasis is also on the protections of springs which are supplied from a karstic aquifer closely located to the Danube. It is worth to note that 80-95 % of domestic water requirements in the basin are covered by groundwater.

Environmental Preservation

In 1978 and 1979 major areas of the flood plain forests were legally protected. Because of the unique ecological characteristics of this area planning activities were initiated in the last years to delimit a natural preserve worthy of becoming a national park. Obviously, the preservation of the flood plain forests and of the riparian wetlands constitutes an important objective for this region.

Social Objective

The social objective refers to satisfy drinking water requirements, to increase the employment opportunities, and to increase the water related facilities for recreation. The increase of the employment opportunities is important for both countries, especially the creation of long-term jobs is attractive.

Definition of Criteria

Hydropower development as a critical environmental intervention will be considered in the context of sustainable development. Sustainable water resources development aims at identifying economically attractive, technically feasible, socially acceptable and ecologically sound water resources projects in a form that "meets the needs of the present without compromising the ability of future generations to meet their own needs" (WCED, 1987).

The question arises what kind of tools or methodologies are available to consider simultaneously these four key issues in the planning process. Although it has been evident that multiple criterion decision making, risk analysis and conflict resolution techniques are appropriate tools for this purpose (HARTMANN, 1986; HIGLER, 1986; LOUCKS, 1994; HAIMES, 1994; NACHTNEBEL et al., 1994) The specific elements for identifying sustainable water resources projects have been unknown especially for the engineering practice. Evidently, acceptable level of high reversibility, low risk and high equity are the conditions of sustainable water resources development.

Reversibility can be measured by the degree how an engineered natural resource system such as a contaminated groundwater system can be remedied to its original, unengineered state. It may take long time and considerable effort to clean up a contaminated aquifer; thus the reversibility level of the original groundwater development could have been quite low. Here, reversibility is expressed by the degree to which specific habitats are preserved. Flood plain forests require at least hundred to two hundred years to develop their typical plant composition and spatial pattern. Especially, the preservation of river morphology, of flood plain forests and the diversity of faunistic species is used as an indicator for reversibility.

TABLE 5 Goals, criteria and units

Goals	Subgoals	Criteria	Units
maximisation of	maximise power	annual power output	GWh
economical utilisation	generation	investment costs	Mrd
ATS			
of resources	minimise costs	maintenance costs	ordinal
Increase social welfare	increase of employment	created jobs during	man-
years			
	rate	construction	
	increase of recreational	recreational facilities	ordinal
	opportunities		
	improved navigation	duration of restricted navigation	
	days/year		
	protections of the	risk	ordinal
	medicinal spring		
preservation of	preservation of the	losses due to construction	ha
the specific ecosystem	flood plain forests		
in this region		area of initial vegetation	ha
		losses of inundated area	%
		area of flood plain forests	ha
		forest edges	km
		timber galleries	km
	preservation of typical	impact an water fowl	ordinal
	faunistic populations	impact on other populations	ordinal
		compatibility with national park	ordinal
		requirement	
	preservation of the morpho-		ratio of
impoundment to	km/km		
	metric variability	free flowing section	
	of riverbanks	length of remaining riverbanks	km
		length of water-bank line at low flows	
	km		
		length of water-bank line at mean	
		discharge	km
		shallows water zones at low flows	ha
		shallows water zones at mean discharge	ha
		gravel banks at low flows	ha
		gravel banks at mean discharge	ha
		connectivity between mean	number
		river and oxbows	
		rate of degradation of	ordinal
		the river bed	
	improvement of water	saprobic scale	ordinal
	quality	change in groundwater quality	ordinal
	preservation of the	length of impervious dams	km
	groundwater system	area with changes in the	
		mean groundwater table (>0.5m)	km^2
		area with changes in the	
		gw.dynamics (0.5-1.0m)	km^2
		area with changes in	
		the gw.dynamics (>1.0m)	km^2

Risk can be defined as the possible adverse consequences of uncertainties facing water resources development. Risk will occur because planning criteria such as economic benefits or just reversibility can be estimated for the planning horizon of water resources development only with more or less uncertainty. Various types of risks such as economic, social or ecological should be defined and combine these risk elements in order to select sustainable water resources alternatives under minimum risk. Risk is not considered explicitly in this paper but it is obvious that any major losses in typical habitats will increase the probability of irreversible changes in the riverine ecosystem. Many red list species already endangered by extinction are only found in the remaining flood plain forests. Therefore, the preservation of species can be seen as a risk reducing objective which might be achieved by natural protection of large areas of the river corridor.

Equity can be defined as the degree of fair distribution of benefits and losses among various parties influenced by water resources development or it can refer to perception of long term impacts. For instance, this refers to equity between present and future generations or equity among societal parties having quite different preference structures. Here in this paper, equity is seen as the outcome of a trade-off procedure between two different countries and simultaneously between ecological and economical interests within each country.

The list of criteria characterizing the economically, ecologically and socially related goals in the context of sustainable water resources development is given in Table 5. This list is one of the outcomes of an expert working group analyzing a hydropower conflict at a national level (ECOLOGICAL ADVISORY BOARD, 1985). Obviously, the number of criteria is quite large and some of the criteria may be interrelated. In EDER et al. (1993) a procedure was proposed to derive a subset including only highly discriminating criteria.

Preference Structure

The two countries share a joint resource but have different preferences. To support a comparison of the two evaluation procedures the set of criteria given in Table 4 is used for both countries. The 33 criteria C_i are grouped with respect to subgoals, such as preservation of aquatic habitats and preservation of riverine flood plain forests. The subgoals SO_k express targets of either economic or ecological objectives, named O_1 and O_2.

Given this information an impact assessment study can be carried out independently for each country. A certain number of points (or a sum of weights) is assigned to the various subgoals to consider the different number of criteria for the goals and to avoid any artificially introduced bias in the preferences. A balanced preference structure would be reflected by allocating the same number of points, or weights to the objectives, say economical development and ecological preservation.

The aggregated weights are given in Table 6 reflecting the different preference structure of the two involved countries, and the table is completed by an independent position like that one of a referee.

TABLE 6 Preferences of the countries expressed by the weights of the main objectives.

	Economy	Ecology
country 1	140	60
country 2	80	120
referee	100	100

Multicriterion Approach for Group Decision Making

Methodology

Numerous methods have been elaborated to solve a multi-criterion problem (KEENEY AND RAIFFA, 1976; GOICOECHEA et al., 1982; ZELENY, 1973, 1982, 1983; HAIMES et al., 1982; ROY, 1985; SZIDAROVSZKY et al., 1986; BOGARDI and NACHTNEBEL, 1994) but not to many address group decision making approaches (HARSANYI; 1977; KRYSZTOFOWICZ, 1979; KRYSZTOFOWICZ and DUCKSTEIN, 1979; ORTOLANO, 1976; WAGNER and ORTOLANO, 1975; BASAR and OLDSER, 1982; BRAMS and KILGOUR, 1988; FANG et al., 1986; 1989; 1993; FRASER and HIPEL 1990). Harsanyi divides decisions with more than one decision maker or groups into rational or ethic approaches which look for a compromise among parties involved, and game theoretical methods where pure individual benefits are pursued. The rational approaches (ARROW, 1951; KEENEY and RAIFFA, 1976; COHON, 1978; HAIMES, 1977, 1982, 1994; KRYSZTOFOWICZ, 1979) try to aggregate utilities and attempt to identify solutions which maximise the groups expected utilities. Game theory is a mathematical study of conflict resolution within a framework of normative rules (von NEUMANN and MORGENSTERN, 1947; HOWARD, 1971; FRASER and HIPEL, 1979; RADFORD, 1988, 1989; FANG et al., 1993).

In this paper a sequential evaluation procedure based on composite programming (BARDOSSY, 1992) is applied to rank the alternatives. First, each country assesses and evaluates the impacts with respect to its preference structure, elaborates a plan evaluation table and achieves a ranking of alternatives. If both countries would rank the same alternative as the best one then there would be no conflict among the countries. Otherwise, in a second step those alternatives which are as close as possible to both favorite alternatives can be seen as compromise solutions for the parties in conflict. Not necessarily, these identified alternatives are nondominated solutions. The multicriterion approach which is applied for each country tries to identify compromise solutions by a simultaneous trade-off at several levels. At the beginning like in many other multicriterion techniques the elements of the plan impact matrix $\{c_{ij}\}$, which express the output of alternative A_j with respect to criterion C_i, are scaled to obtain the plan evaluation matrix $\{x_{ij}\}$.

$$x_{ij} = (c_{i,best} - c_{ij})/(c_{i,best} - c_{i,worst}) \qquad (1)$$

The performance indicators x_{ij} are allocated to subgoals SO_k which reflect for instance the preservation of aquatic habitats, which are in turn allocated to economy and ecology related objectives O_1 and O_2. Compromise programming identifies solutions by a simultaneous trade-off at the level of performance indicators, subgoals, and main objectives expressed by a distance measure or Lp-norm. Considering alternative j the outputs are obtained for each level by

$$L_j(p_k) = SO_{jk} = \left(\sum_i w_j (x_{ij})^{p_k} \right)^{1/pk} \qquad i \in \{G_k\} \tag{2}$$

$$w_k = \sum_i w_i \quad \text{with} \quad \sum w_i = 1 \tag{3}$$

$$L_j(q_1) = O_{j1} = \left(\sum_k w_k (SO_{jk})^{q_1} \right)^{1/q_1} \qquad k \in \{H_m\} \tag{4}$$

$$w_1 = \sum_k w_k \tag{5}$$

$$L_j(r) = O_j = \left(\sum_{l=i}^{2} w_l (O_{jl})^r \right)^{1/r} \tag{6}$$

where the set $\{G_k\}$ defines the binary relationship between criterion i and subgoal SO_k and respectively the set $\{H_m\}$ describes the binary relationship between subgoals and the two main objectives. Here, $\{H_m\}$ consists of two sets, namely an economy related objective with the subgoals five and six, and an ecology related objective with the subgoals 1-4 as given in Tables 7-9. The ranking from a country's perspective is simply achieved by arranging the values of the L(r)-norm in ascending order. The preferred alternative is obtained by

$$\text{Min}\,\{L_j(r)\} = \text{Min}\,\{O_j\} = O^*. \tag{7}$$

The distance D_j of alternative j from O* is expressed by

$$D_j = \left(\sum w_l \left| O_{jl} - O_l^* \right|^r \right)^{1/r} \tag{8}$$

Considering two countries the final goal is to select an alternative which minimizes the generalized distance measure

$$L_j(s) = \sum_{l=1}^{2} \left({}_1w_l \left| {}_1O_{jl} - {}_1O_j^* \right|^s + {}_2w_l \left| {}_2O_{jl} - {}_2O_l^* \right|^s \right)^{1/s} \tag{9}$$

${}_1O_l^*$ components of the preferred alternative in country 1

${}_2O_l^*$ components of the preferred alternative in country 2

where the prefix index refers to the country. Weights are assumed to be equal for the countries. This approach is supplemented by an evaluation of the alternatives from an overall viewpoint irrespective from any national preferences. The project impacts are integrated (Table 9) and assuming equal weights for the two main objectives the methodology is repeated by applying equations 1-8.

The metric of the distance measure L(.) is defined by the exponent (.) which is here either p_k, q_l, r or s. Assuming a value of one would imply that losses in one performance indicator x_{ij} can be compensated by high performance levels in another indicator $x_{i'j}$. With increasing value of the exponent the lower performance levels define the distance L(.) and for (.) = ∞ the worst outcome defines the distance. In other words, the exchange among different performance indicators is decreasing with an increasing exponent (.).

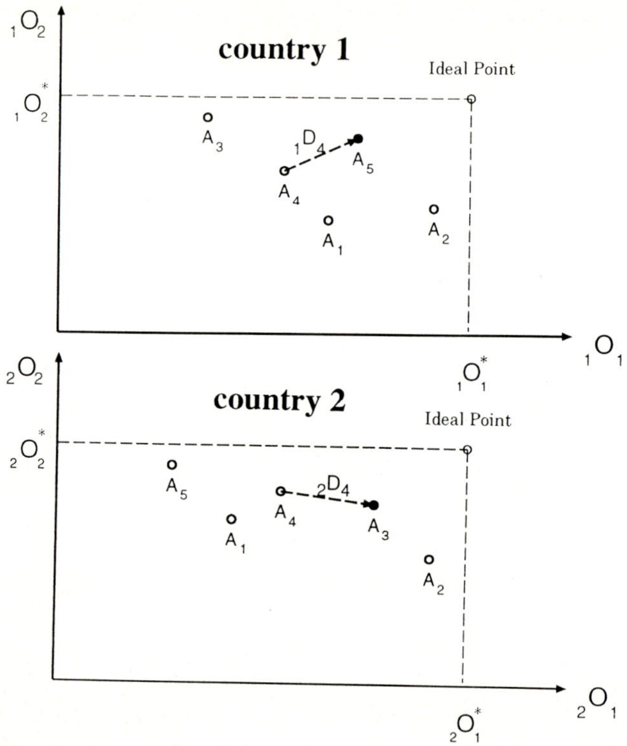

Figure 2 Schematic representation of the performance of the alternatives within a bi-objective framework for both countries.

Application

In tables 7 and 8 the plan impact matrices are given together with the set of weights for country 1 and 2, and Table 9 includes the data for the whole area irrespectively from any national borderline. The tables also include the best and worst figures for each criterion, $c_{i,best}$ and $c_{i,worst}$. The trade-off within the subgoals is performed in the same way for both countries following the proposed methodology. The weights for the main objectives are given in Table 6; the exponents are 2 for socially and economically related subgoals while a value of three is assumed for ecologically related subgoals. However, quite similar results are obtained by setting all exponents equal to 2.

The results for the two countries are given in Table 10 and 11. Country 1 prefers alternative 10 while country 2 which emphasizes ecological objectives prefers alternative 2. The tables include also the figures for O_1 and O_2 and also the respective figures in comparison to the preferred alternative. A graphical interpretation of these differences is given in Fig. 2. Analyzing the figures for O_1 and O_2 in Table 10 it can be concluded that in both countries the ecological impacts of alternatives 1 and 2 are low compared to the other projects. The economic performance of A1 and A2 is rather poor. From a pure economic viewpoint country 1 would prefer A7 while country 2 would select A6 or A5. The ordinal criteria are expressed in tables 7, 8 and 9 by very good (1.0), good (2.0), satisfactory (3.0), poor (4.0) and bad or unsatisfactory (5.0)

Criteria	Unit	A_1	A_2	A_3	A_4	A_5	A_6	A_7	A_8	A_9	A_{10}	A_{11}	A_{12}	Weight	Best	Worst	Subgoal
Investment costs	Mrd. ATS	0.0	14.6	11.4	12.0	24.9	26.1	15.9	16.7	15.6	16.4	12.2	12.8	25.0	0.0	26.1	6
Power generation	GWh	0.0	1362.0	1362.0	1362.0	1174.0	1174.0	1319.0	1319.0	1220.0	1220.0	1362.0	1362.0	25.0	2075.0	0.0	6
Operation costs	ordinal	2.0	2.0	3.0	3.0	5.0	5.0	5.0	5.0	5.0	5.0	3.0	3.0	11.0	1.0	5.0	6
Medicinal springs	ordinal	1.0	1.0	1.0	1.0	1.0	1.0	1.0	1.0	1.0	1.0	1.0	1.0	10.0	1.0	5.0	5
Navigation	days	111.0	111.0	116.0	116.0	11.0	11.0	11.0	11.0	11.0	11.0	116.0	116.0	17.0	11.0	116.0	5
Recreation	ordinal	1.0	1.0	4.0	4.0	3.0	3.0	4.0	4.0	4.0	4.0	4.0	4.0	17.0	1.0	5.0	5
Employment	manyears	300.0	300.0	15400.0	16200.0	33620.0	35240.0	21470.0	22550.0	21060.0	22140.0	16740.0	17280.0	20.0	39100.0	300.0	5
longterm jobs	number	15.0	15.0	70.0	70.0	210.0	210.0	140.0	140.0	140.0	140.0	70.0	70.0	15.0	210.0	15.0	5
Impervious dams	km	0.0	0.0	40.3	40.3	10.2	10.2	27.9	27.9	20.7	20.7	40.3	40.3	2.0	0.0	57.8	4
Change in gw.table	ha	50.5	50.5	54.5	5.0	53.5	22.9	51.9	18.1	47.7	16.0	54.5	5.0	2.0	1.3	69.4	4
Gw. dynamics < 1m	ha	39.4	39.4	3.1	3.1	10.8	7.7	9.4	7.0	9.4	7.0	3.1	3.1	2.0	51.8	0.8	4
Gw. dynamics > 1m	ha	27.7	27.7	3.0	3.2	6.2	6.6	6.7	5.7	6.7	5.7	3.2	3.2	3.0	35.8	0.8	4
Groundwater qual.	ordinal	2.0	2.0	5.0	4.0	3.0	2.0	4.0	3.0	3.0	2.0	5.0	4.0	3.0	1.0	5.0	4
Water fowl	ordinal	4.0	4.0	4.0	4.0	2.0	2.0	2.0	2.0	2.0	2.0	4.0	4.0	1.0	1.0	5.0	3
Faunistic populat.	ordinal	1.0	1.0	5.0	5.0	3.0	3.0	4.0	4.0	4.0	4.0	5.0	5.0	8.0	1.0	5.0	3
Compatibility n.park	ordinal	1.0	1.0	1.0	1.0	1.0	1.0	1.0	1.0	1.0	1.0	1.0	5.0	3.0	1.0	5.0	3
Free flowing part	%	100.0	100.0	26.0	26.0	18.0	18.0	100.0	100.0	19.0	19.0	26.0	26.0	4.0	100.0	10.0	2
River banks	km	63.0	63.0	15.1	15.1	23.9	23.9	23.3	23.3	26.3	26.3	15.1	15.1	2.0	96.0	7.9	2
Bankline (1)	km	74.7	74.7	54.9	54.9	53.5	53.5	54.3	54.3	55.8	55.8	54.9	54.9	2.0	113.8	28.0	2
Bankline (2)	km	65.6	65.6	54.9	54.9	53.5	53.5	54.3	54.3	55.8	55.8	54.9	54.9	2.0	100.0	28.0	2
Shallow water (1)	ha	108.6	108.6	37.9	37.9	30.8	30.8	26.1	26.1	31.9	31.9	37.9	37.9	2.0	165.5	13.7	2
Shallow water (2)	ha	109.0	109.0	34.4	34.4	30.8	30.8	26.1	26.1	31.9	31.9	34.4	34.4	2.0	166.2	13.7	2
Gravel layers (1)	ha	110.3	110.3	0.9	0.9	0.0	0.0	0.0	0.0	8.5	8.5	0.9	0.9	2.0	168.1	0.0	2
Gravel layers (2)	ha	4.5	4.5	0.0	0.0	0.0	0.0	0.0	0.0	1.4	1.4	0.0	0.0	2.0	6.8	0.0	2
Connectivity	number	28.0	28.0	7.0	7.0	17.0	17.0	13.0	13.0	15.0	15.0	8.0	8.0	2.0	42.0	4.0	2
Surface water qual.	ordinal	2.0	2.0	2.0	2.0	2.0	2.0	2.0	2.0	2.0	2.0	2.0	2.0	2.0	1.0	5.0	2
Degradation river	ordinal	5.0	5.0	2.0	2.0	3.0	3.0	2.0	2.0	2.0	2.0	2.0	2.0	3.0	1.0	5.0	2
Conctruction area	ha	0.0	0.0	740.0	740.0	550.0	550.0	495.0	495.0	315.0	315.0	496.0	496.0	2.0	0.0	740.0	1
Initial vegetation	ha	11.9	11.9	0.5	0.5	4.7	4.7	0.9	0.9	2.0	2.0	0.7	0.7	2.0	18.2	0.3	1
Floodplain forests	ha	78.9	78.9	12.0	12.0	36.1	36.1	21.3	21.3	25.5	25.5	15.7	15.7	1.0	120.2	6.3	1
Forest edges	km	22.2	22.2	1.8	1.8	10.4	10.4	7.9	7.9	12.4	12.4	2.2	2.2	1.0	33.8	0.9	1
Forest galleries	km	10.4	10.4	3.9	3.9	5.6	5.6	4.9	4.9	6.7	6.7	4.1	4.1	1.0	16.0	2.0	1
Inundation losses	%	0.0	0.0	86.4	86.4	25.0	25.0	62.0	62.0	52.0	52.0	87.8	87.8	4.0	0.0	87.8	1

TABLE 7 Plan impact matrix for country 1.

Criteria	Unit	A_1	A_2	A_3	A_4	A_5	A_6	A_7	A_8	A_9	A_{10}	A_{11}	A_{12}	Weight	Best	Worst	Subgoal
Investment costs	Mrd. ATS	0.0	7.7	0.0	0.0	0.0	0.0	0.0	0.0	0.0	0.0	0.0	0.0	15.0	0.0	26.1	6
Power generation	GWh	0.0	714.0	714.0	714.0	526.0	526.0	671.0	671.0	572.0	572.0	714.0	714.0	15.0	2075.0	0.0	6
Operation costs	ordinal	1.0	1.0	1.0	1.0	1.0	1.0	1.0	1.0	1.0	1.0	1.0	1.0	8.0	1.0	5.0	6
Medicinal springs	ordinal	2.0	2.0	3.0	3.0	1.0	1.0	1.0	1.0	1.0	1.0	3.0	3.0	8.0	1.0	5.0	5
Navigation	days	111.0	111.0	116.0	116.0	11.0	11.0	11.0	11.0	11.0	11.0	116.0	116.0	8.0	11.0	116.0	5
Recreation	ordinal	1.0	1.0	4.0	4.0	3.0	3.0	4.0	4.0	4.0	4.0	4.0	4.0	10.0	1.0	5.0	5
Employment	manyear	300.0	300.0	1700.0	1800.0	3700.0	3900.0	2400.0	2500.0	2300.0	2500.0	1800.0	1900.0	8.0	39150.0	300.0	5
longterm jobs	number	15.0	15.0	10.0	10.0	10.0	10.0	10.0	10.0	10.0	10.0	10.0	10.0	8.0	210.0	30.0	5
Impervious dams	km	0.0	0.0	17.5	17.5	5.3	5.3	14.6	14.6	10.8	10.8	17.5	17.5	3.0	0.0	57.8	4
Change in gw.table	ha	15.0	15.0	13.8	1.9	23.9	6.8	15.4	5.4	15.2	4.8	13.8	1.3	3.0	1.3	69.4	4
Gw. dynamics < 1m	ha	11.9	11.9	0.8	0.8	3.3	2.3	2.8	2.1	2.8	2.1	0.8	0.8	3.0	51.8	0.8	4
Gw. dynamics > 1m	ha	8.2	8.2	0.8	0.8	1.8	1.9	2.0	1.7	2.0	1.7	0.8	0.8	5.0	35.8	0.8	4
Groundwater qual.	ordinal	2.0	2.0	5.0	4.0	3.0	2.0	4.0	3.0	3.0	2.0	5.0	4.0	5.0	1.0	5.0	4
Water fowl	ordinal	4.0	4.0	4.0	5.0	2.0	2.0	2.0	2.0	2.0	2.0	4.0	4.0	2.0	1.0	5.0	3
Faunistic populat.	ordinal	1.0	1.0	5.0	5.0	3.0	3.0	4.0	4.0	4.0	4.0	5.0	5.0	14.0	1.0	5.0	3
Compatibility n.park	ordinal	1.0	1.0	5.0	2.0	3.0	3.0	4.0	4.0	4.0	4.0	5.0	5.0	10.0	1.0	5.0	3
Free flowing part	%	100.0	100.0	36.0	36.0	18.0	18.0	14.0	14.0	19.0	19.0	36.0	36.0	14.0	100.0	10.0	2
River banks	km	33.0	33.0	7.9	7.9	12.1	12.1	12.2	12.2	13.8	13.8	7.9	7.9	2.0	96.0	7.9	2
Bankline (1)	km	39.1	39.1	28.8	28.8	28.1	28.1	28.4	28.4	29.2	29.2	28.7	28.7	2.0	113.0	28.0	2
Bankline (2)	km	34.4	34.4	28.8	28.8	28.1	28.1	28.4	28.4	29.2	29.2	28.7	28.7	2.0	100.0	28.0	2
Shallow water (1)	ha	57.3	57.3	18.1	18.1	16.1	16.1	13.7	13.7	16.8	16.8	18.1	18.1	2.0	165.5	13.7	2
Shallow water (2)	ha	57.3	57.3	15.9	0.4	16.2	16.2	13.7	13.7	16.7	16.7	15.9	15.9	2.0	166.2	13.7	2
Gravel layers (1)	ha	57.8	57.8	0.4	0.0	0.0	0.0	0.0	0.0	4.4	4.4	0.4	0.4	2.0	168.1	0.0	2
Gravel layers (2)	ha	2.3	2.3	0.0	0.0	0.0	0.0	0.0	0.0	0.7	0.7	0.0	0.0	3.0	6.8	0.0	2
Connectivity	number	15.0	15.0	4.0	4.0	9.0	9.0	7.0	7.0	8.0	8.0	4.0	4.0	3.0	42.0	4.0	2
Surface water qual	ordinal	2.0	2.0	2.0	2.0	2.0	2.0	2.0	2.0	2.0	2.0	2.0	2.0	4.0	1.0	5.0	2
Degradation river	ordinal	5.0	5.0	2.0	2.0	3.0	3.0	2.0	2.0	2.0	2.0	2.0	2.0	5.0	1.0	5.0	2
Conctruction area	ha	0.0	0.0	0.0	0.0	110.0	110.0	160.0	160.0	150.0	150.0	0.0	0.0	5.0	0.0	740.0	1
Initial vegetation	ha	6.3	6.3	0.2	0.2	2.4	2.4	0.5	0.5	1.1	1.1	0.3	0.3	5.0	18.2	0.3	1
Floodplain forests	ha	41.3	41.3	6.3	6.3	18.9	18.9	11.2	11.2	13.4	13.4	8.2	8.2	4.0	120.2	6.3	1
Forest edges	km	11.6	11.6	0.9	0.9	5.4	5.4	4.1	4.1	6.5	6.5	1.1	1.1	4.0	33.8	0.9	1
Forest galleries	km	5.5	5.5	2.0	2.0	3.0	3.0	2.6	2.6	3.5	3.5	2.2	2.2	4.0	16.0	2.0	1
Inundation losses	%	0.0	0.0	86.0	86.0	25.0	25.0	62.0	62.0	52.0	52.0	88.0	88.0	12.0	0.0	88.0	1

TABLE 8 Plan impact matrix for country 2.

Criteria	Unit	A_1	A_2	A_3	A_4	A_5	A_6	A_7	A_8	A_9	A_{10}	A_{11}	A_{12}	Weight	Best	Worst	Subgoal
Investment costs	Mrd. ATS	0.1	22.3	11.4	12.0	24.9	26.1	15.9	16.7	15.6	16.4	12.2	12.8	20.0	0.1	26.1	6
Power generation	GWh	0.0	2075.0	2075.0	2075.0	1700.0	1700.0	1990.0	1990.0	1791.0	1791.0	2075.0	2075.0	20.0	2075.0	0.0	6
Operation costs	ordinal	2.0	2.0	3.0	3.0	5.0	5.0	5.0	5.0	5.0	5.0	3.0	3.0	6.0	1.0	5.0	6
Medicinal springs	ordinal	2.0	2.0	3.0	3.0	1.0	1.0	1.0	1.0	1.0	1.0	3.0	3.0	5.0	1.0	5.0	5
Navigation	days	111.0	111.0	116.0	116.0	11.0	11.0	11.0	11.0	11.0	11.0	116.0	116.0	12.0	11.0	116.0	5
Recreation	ordinal	1.0	1.0	4.0	4.0	3.0	3.0	4.0	4.0	4.0	4.0	4.0	4.0	12.0	1.0	5.0	5
Employment	manyears	500.0	500.0	17100.0	18000.0	37300.0	39100.0	23800.0	25100.0	23400.0	24600.0	18300.0	19200.0	15.0	39100.0	500.0	5
longterm jobs	number	30.0	30.0	70.0	70.0	210.0	210.0	140.0	140.0	140.0	140.0	70.0	70.0	10.0	210.0	30.0	5
Impervious dams	km	0.0	0.0	57.8	57.8	15.5	15.5	42.5	42.5	31.5	31.5	57.8	57.8	4.0	0.0	57.8	4
Change in gw.table	ha	65.5	65.5	68.2	6.3	69.4	29.7	67.3	23.4	61.8	20.8	68.2	6.3	4.0	6.3	69.4	4
Gw. dynamics < 1m	ha	51.8	51.8	4.1	4.1	14.3	10.1	12.4	9.2	12.4	9.2	4.1	4.1	4.0	51.8	4.1	4
Gw. dynamics > 1m	ha	35.8	35.8	4.1	4.1	8.0	8.5	8.7	7.4	8.7	7.4	4.1	4.1	5.0	35.8	4.1	4
Groundwater qual.	ordinal	2.0	2.0	5.0	4.0	3.0	2.0	4.0	3.0	3.0	2.0	5.0	4.0	2.0	1.0	5.0	4
Water fowl	ordinal	4.0	4.0	4.0	4.0	2.0	2.0	2.0	2.0	2.0	2.0	4.0	4.0	2.0	1.0	5.0	3
Faunistic populat.	ordinal	1.0	1.0	5.0	5.0	3.0	3.0	4.0	4.0	4.0	4.0	5.0	5.0	16.0	1.0	5.0	3
Compatibility n.park	ordinal	1.0	1.0	5.0	5.0	3.0	3.0	4.0	4.0	4.0	4.0	5.0	5.0	6.0	1.0	5.0	3
Free flowing part	%	100.0	100.0	20.0	20.0	10.0	10.0	10.0	10.0	10.0	10.0	20.0	20.0	8.0	100.0	10.0	2
River banks	km	96.0	96.0	23.0	23.0	36.5	36.5	35.5	35.5	40.1	40.1	23.0	23.0	3.0	96.0	23.0	2
Bankline (1)	km	113.8	113.8	83.7	83.7	81.6	81.6	82.7	82.7	85.0	85.0	83.7	83.7	1.0	113.8	81.6	2
Bankline (2)	km	100.0	100.0	83.7	83.7	81.6	81.6	82.7	82.7	85.0	85.0	83.7	83.7	1.0	100.0	81.6	2
Shallow water (1)	ha	165.5	165.5	57.8	57.8	47.0	47.0	39.8	39.8	48.6	48.6	57.8	57.8	3.0	165.5	39.8	2
Shallow water (2)	ha	166.2	166.2	52.5	52.5	47.0	47.0	39.8	39.8	48.6	48.6	52.5	52.5	3.0	166.2	39.8	2
Gravel layers (1)	ha	168.1	168.1	1.4	1.4	0.0	0.0	0.0	0.0	12.9	12.9	1.4	1.4	2.0	168.1	0.0	2
Gravel layers (2)	ha	6.8	6.8	0.0	0.0	0.0	0.0	0.0	0.0	2.1	2.1	0.0	0.0	1.0	6.8	0.0	2
Connectivity	number	42.0	42.0	11.0	11.0	25.0	25.0	20.0	20.0	23.0	23.0	12.0	12.0	4.0	42.0	11.0	2
Surface water qual.	ordinal	2.0	2.0	2.0	2.0	2.0	2.0	2.0	2.0	2.0	2.0	2.0	2.0	4.0	1.0	5.0	2
Degradation river	ordinal	5.0	5.0	2.0	2.0	3.0	3.0	2.0	2.0	2.0	2.0	2.0	2.0	7.0	1.0	5.0	2
Conctruction area	ha	0.0	0.0	740.0	740.0	660.0	660.0	655.0	655.0	465.0	465.0	496.0	496.0	3.0	0.0	740.0	1
Initial vegetation	ha	18.2	18.20	0.8	0.8	7.1	7.1	1.4	1.4	3.1	3.1	1.1	1.1	4.0	18.2	0.8	1
Floodplain forests	ha	120.2	120.2	18.3	18.3	55.0	55.0	32.5	32.5	38.9	38.9	24.00	24.0	1.0	120.2	18.3	1
Forest edges	km	33.8	33.8	2.9	2.9	1.6	1.6	12.1	12.1	19.0	19.0	3.4	3.4	1.0	33.8	2.9	1
Forest galleries	km	16.0	16.0	5.9	0.6	8.7	8.7	7.6	7.6	10.3	10.3	6.4	6.4	1.0	16.0	5.9	1
Inundation losses	%	0.0	0.0	86.4	86.4	25.0	25.0	62.0	62.0	52.0	52.0	87.8	87.8	10.0	0.0	87.8	1

TABLE 9 Plan impact matrix for the whole project area.

TABLE 10 Ranking of alternatives for country 1.

	Distance to Ideal Point			Distance to Preferred Alternative			
ALTERNATIVE	ECONOMY	ECOLOGY	TOTAL	ECONOMY	ECOLOGY	TOTAL	RANK
1	0.737	0.452	0.665	0.200	+0.260	0.220	8
2	0.665	0.452	0.609	0.129	+0.260	0.178	6
3	0.611	0.877	0.701	0.074	0.165	0.109	12
4	0.611	0.856	0.694	0.074	0.144	0.100	9
5	0.555	0.707	0.605	0.018	+0.005	0.015	5
6	0.568	0.696	0.609	0.031	+0.016	0.027	7
7	0.529	0.741	0.601	+0.008	0.029	0.017	4
8	0.532	0.728	0.598	+0.005	0.014	0.010	2
9	0.534	0.718	0.595	+0.003	0.006	0.004	3
10	**0.537**	**0.712**	**0.595**	**0.000**	**0.000**	**0.000**	**1**
11	0.610	0.865	0.696	0.073	0.153	0.104	10
12	0.611	0.867	0.698	0.074	0.155	0.105	11

TABLE 11 Ranking of alternatives for country 2.

	Distance to Ideal Point			Distance to Preferred Alternative			
ALTERNATIVE	ECONOMY	ECOLOGY	TOTAL	ECONOMY	ECOLOGY	TOTAL	RANK
1	0.708	0.578	0.633	0.067	0.000	0.042	2
2	**0.641**	**0.578**	**0.604**	**0.000**	**0.000**	**0.000**	**1**
3	0.702	0.905	0.830	0.061	0.327	0.257	11
4	0.701	0.867	0.805	0.060	0.289	0.227	9
5	0.585	0.752	0.690	+0.056	0.174	0.140	4
6	0.584	0.749	0.688	+0.057	0.171	0.138	3
7	0.608	0.827	0.747	+0.033	0.249	0.194	8
8	0.607	0.822	0.743	+0.034	0.244	0.190	7
9	0.618	0.789	0.726	+0.023	0.211	0.165	6
10	0.617	0.788	0.724	+0.024	0.210	0.163	5
11	0.701	0.906	0.830	0.060	0.328	0.257	12
12	0.701	0.896	0.824	0.060	0.318	0.250	10

Based on this data it can be concluded that alternatives 11, 12 and 3 are unattractive solutions for both countries, while 5, 10 and 6 exhibit in both good rankings. Alternatives 2 and 1 are only attractive for country 2 while they are classified on position 6 and 8 in country 1.

Conflict Resolution Among the Countries

Now the goal is to identify an alternative which is acceptable for both countries. Such an alternative should be as close as possible to both of the individually preferred alternatives, namely 10 and 2. The two distances have to be expressed according to the national preference structure

$$ {}_1L_j(s) = {}_1D_j = \left(\sum_{l=1}^{2} {}_1w_l \left| {}_1O_{jl} - {}_1O_l^* \right|^s \right)^{1/s} $$

$$ {}_2L_j(s) = {}_2D_j = \left(\sum_{l=1}^{2} {}_2w_l \left| {}_2O_{jl} - {}_2O_l^* \right|^s \right)^{1/s} $$

$$ L_j(s) = \left({}_1D_j^s + {}_2D_j^s \right)^{1/s} $$

The above given distance can be understood as a generalized distance because it expresses the L(.)-norm in a non-orthogonal system. The results are given for s=2 in Table 12.

TABLE 12 Ranking of alternatives for compromise among countries.

ALTERNATIVE j	COUNTRY1 ${}_1D_j$	COUNTRY2 ${}_2D_j$	DISTANCE $L_j(s=2)$	RANK
1	0.220	0.042	0.224	8
2	0.178	0.000	0.178	5
3	0.109	0.257	0.279	10
4	0.100	0.227	0.248	9
5	0.015	0.140	0.141	2
6	**0.027**	**0.138**	**0.140**	**1**
7	0.017	0.194	0.195	7
8	0.010	0.190	0.191	6
9	0.004	0.165	0.165	4
10	0.000	0.163	0.163	3
11	0.104	0.257	0.270	12
12	0.105	0.250	0.263	11

According to this analysis good and compromising solutions would be A6 and A5, followed by A10 and A9. Again A11 and A12, A3, A4 and A1 are not within the set of attractive solutions. A1 represents the status quo or the zero alternative and interestingly it fails due to economic inefficiency.

Referee's Viewpoint

Next, the individual country objectives will be discarded and a unified approach is applied. All the outcomes of alternatives (Table 9) are aggregated independently from any national border. and then a composite programming approach is applied. Equal weights are assigned to economical and ecological objectives to achieve a sound trade-off. The ranking results are given in Table 13. It can be seen that from an overall viewpoint alternatives A6 and A2 would be the preferred ones. Of interest are also alternatives A10 and A5, while A3, A4, A11 and A12 are unattractive solutions. It can be concluded that this last approach yields similar results like the previous approach. Even a sensitivity analysis within a wider range of weights yields the same subset of preferred alternatives.

TABLE 13 Ranking from a single decision makers viewpoint.

	Distance to Ideal Point			
ALTERNATIVE	ECONOMY	ECOLOGY	TOTAL	RANK
1	0.741	0.483	0.626	6
2	**0.704**	**0.483**	**0.604**	**1**
3	0.594	0.899	0.762	12
4	0.594	0.883	0.752	10
5	0.528	0.684	0.611	4
6	**0.545**	**0.658**	**0.604**	**1**
7	0.488	0.771	0.645	8
8	0.492	0.752	0.635	7
9	0.490	0.718	0.615	5
10	0.494	0.705	0.608	3
11	0.594	0.890	0.575	11
12	0.595	0.869	0.744	9

Summarising, it can be concluded that there are some alternatives which are "stable" in the sense that they are close to the individual rankings of the countries and to the integrated evaluation. It was not the objective to achieve a unique ranking but rather to select compromising solutions.

Summary and Conclusions

Hydropower development is of great economical value for some Danube countries and it has also some additional effects like flood protection and improved navigation. At the same time adverse environmental impacts might be observed. Thus, no solution exists which simultaneously satisfies all objectives, related to economic development and environmental preservation. A methodology was described which assists in conflict resolution among different interest groups; either among interests within a society or between two different states with different preference structure. The methodology is based on composite programming techniques related to compromise programming. The compromise solution is identified in two steps. First, a ranking is performed for each country and then an alternative is identified which is as close as possible to the favourite country alternatives. A generalised distance measure has been introduced to define the distance between individual solutions for each country and the compromise solution. Supplementary, an overall viewpoint is also considered where the project impacts are assessed without consideration of national borders.

To demonstrate the methodology a case study from the Austrian section of the Danube was selected. The advantage is seen in the fact that many analytical steps can be independently carried out in each country and that finally a rational approach is presented to satisfy both parties. Also, the conflicts within a country can be tackled in a similar way.

The main result of the application is seen in the selection of a reduced set of alternatives which satisfy to a large extent the expectations of both countries. Both the overall approach and the compromise approach between the countries yield similar results. This reduced set of alternatives would provide a basis for a more focused discussion and negotiation process. The disadvantage is in the somewhat arbitrary definition of a distance and in the possibility that dominated solutions are obtained.

Acknowledgements: This study was partially supported by the project No 5064 funded by the Jubiläumsfond of the Austrian National Bank.

References

Arrow, K.J. (1951) Social Choice and Individual Values. Cowles Commission Monograph 12, Wiley, New York.

Bardossy, A. and Duckstein, L. (1992) Analysis of a Karstic Aquifer Management Problem by Fuzzy Composite Programming. Water Res. Bulletin, Vol. 28,1.

Basar, T. and Oldser, G.J.(1982) Dynamic Noncooperative Game Theory. Academic Press, London.

Bogardi, J.J. and Nachtnebel, H.P. (1994) Multicriteria Decision Analysis in Water Resources Management, UNESCO Publ., Water Science Div. UNESCO, Paris.

Brams, S.J. and Kilgour, D.M.(1988) Game Theory and National Security. Basil Blackwell, New York.

Cohon, J.L. (1978) Multiobjective Programming and Planning. Academic Press, New York.

Common, M. and Perrings, C. (1992) Towards an Ecological and Economics and Sustainability. Ecological Economics 2, S. 57-76.

Danube Commission for Navigation (1988) Recommendations for the Danube Water Way. Budapest, Hungary.

Danube Commission for Navigation (.) Shipping Transport among Danube Countries. Annual Reports published by the Commission. Budapest, Hungary.

DOKW (1989) Donaukraftwerke AG. Hydropower Utilisation Downstream of Vienna. Final Report. Donaukraft, Vienna, Austria.

EPDRB (1993, 1994) Environmental Program for the Danube River Basin. Pre-investment Studies for River Basins in the Danube Catchment Submitted to PCU, c/o UNIDO, Vienna, Austria.

Ecological Advisory Board (Ökologiekommission der Bundesregierung) (1985) Schlußbericht Arbeitskreis Donaugestaltung, Arbeitskreis Nationalpark Donau-March-Thaya. Bundesministerium für Umwelt, Jugend und Familie, 1030 Wien, Österreich.

ECE (1994) Economic Commission for Europe. European Inland Water Ways. UN, N.Y. and Geneva.

Eder, G., Duckstein, L. and Nachtnebel, H.P. (1993) Worth of Criteria in Multicriterion Water Resources Planning using MCQA-III: An Austrian Case Study. Report. Dept. of Systems and Ind. Eng. Univ. of Tucson, Arizona.

Equipe Cousteau (1993) The Danube: for Whom and for What ? Report to the IBRD; Paris, France.

Fang, L., Hipel, K.W. and Kilgour, D.M. (1986) A Comprehensive Decision Support System for Two-player Conflicts. Large Scale Systems, Vol. 11, No. 1, pp. 19-29.

Fang, L., Hipel, K.W. and Kilgour, D.M. (1989) Conflict Models in Graph Form: Solution Concepts and Their Interrelationship. European Journal of Operational Research, Vol. 41, No. 1, pp. 86-100.

Fang, L., Hipel, K.W. and Kilgour, D.M. (1993) Interactive Decision Making. J. Wiley and Sons Inc. N.Y., USA.

Fekete, G. (1990) Die Donau im Rahmen der europäischen Schiffahrt. In: Schiffahrt und Strom, 36. Jg., Heft 4, S.6-10, Wien, Österreich.

Fraser, N.M. and Hipel, K.W. (1990) DecisionMaker: The Conflict Analysis Program. Copyright owned by N.M. Fraser and K.W. Hipel. Departments of Management Sciences and Systems Design Engineering, University of Waterloo, Waterloo, Ontario, Canada, N2L 3G1.

Goicoechea, A., Hansen, D.R. and Duckstein, L. (1982) Multiobjective Decision Analysis with Engineering and Business Applications, J. Wiley & Sons, N.Y.

Goldsmith, E. and Hildyard, N. (1984) The Social and Environmental Effects of Large Dams. A Sierra Club Book, San Francisco, USA.

Haimes, Y.Y. (1977) Hierarchical Analysis of Water Resources Systems. McGraw-Hill, New York.

Haimes, Y.Y. (1982) The Surrogate-Worth Trade-Off Method. Elsevier.

Haimes, Y.Y. (1994) Risk Management: A Requisite for Sustainable Development. In: Proc. International UNESCO Symposium 'Water Resources Planning in a Changing World', p. II/3-II-16, Karlsruhe, June 1994. IHP/OHP National Committee, Bundesanstalt für Gewässerkunde, Koblenz, Germany.

Hartman, L. (1986) Methodological Guidelines for Integrated Environmental Evaluation of Water Resources Development. In: The Impact of Large Water Projects, p. 467-484. UNEP-UNESCO 1990, Paris.

Harsany, J.C. (1977) Rational Behaviour and Bargaining Equilibrium in Games and Social Situations. Cambridge University Press.

Hary, N. and Nachtnebel, H.P. (1989) Ecosystemstudie Donaustau Altenwörth. Mitt. der Österr. Akademie der Wissenschaften, Univ.Verlag Wagner, Innsbruck, Austria.

Higler, L.W.G. (1986) Classification of Water Systems for the Environmental Evaluation of Water Resources Development. In: The Impact of Large Water Projects, p. 487-506. UNEP-UNESCO, 1990, Paris.

Howard, N. (1971) Paradoxes of Rationality. MIT Press, Cambridge, MA.

ISD (1989) Interdisziplinäre Studie Donau. Under Contract of ÖWAV, ARG, 1190 Vienna, Austria.

IUCN (1994) International Union for Conservation of Nature. Analysis and Synthesis of National Reviews for the Danube River Basin Environmental Program, IUCN Headquarters, Gland, Suisse.

Kaniak, J. (1986) River Engineering Measures and Potential Sites for Hydropower Utilization of the Danube Downstream of Vienna. Report to the Austrian Ministry of Environment, Vienna, Austria.

Keeney, R.L. and Raiffa, H. (1976) Decisions with Multiple Objectives: Preferences and Value Tradeoffs. Wiley, New York.

Kovacs, G. (1986) Decision Support System for Managing Large International Rivers. In: The Impact of Large Water Projects, p. 435-448. UNEP-UNESCO, 1990, Paris.

Krzysztofowicz, R. (1979) Group Utility Assessment Through a Nominal-Interacting Process, unpublished working paper, Dept.. of Civil Engineering, MIT, Cambridge, Mass.

Krzysztofowicz, R. and Duckstein, L. (1979) Preference Criterion for Flood Control Unter Uncertainty. Water Resources Reseearch, 15 (3), p. 513-520.

Loucks, D.P. (1994) Sustainability: What Does This Mean For Us And What Can We Do About It. In: Proc. International UNESCO Symposium 'Water Resources Planning in a Changing World', p. I/3-I/12. Karlsruhe, June 1994. IHP/OHP National Committee, Bundesanstalt für Gewässerkunde, Koblenz, Germany.

Nachtnebel, H.P.(1990) Comparison of Hydropower Utilization and Environmental Impacts Along the Danube Downstream of Vienna. Ingenieurs et architectes suisse, No,18, 412-415.

Nachtnebel, H.P. and Haider, S. (1989) Report About the Hydrological Regime in the Flood Plains Downstream of Vienna. In: Interdisziplinäre Studie Donau, p. 51-92. ISD. Vienna, Austria.

Nachtnebel, H.P., Eder, G. and Bogardi, I. (1994) Evaluation of Criteria in Hydropower Utilization in the Context of Sustainable Development. In: Proc. International UNESCO Symposium 'Water Resources Planning in a Changing World', p. IIV/13-IV/24. Karlsruhe, June 1994. IHP/OHP National Committee, Bundesanstalt für Gewässerkunde, Koblenz, Germany.

Ortolano, L. (1976) Water Plan Ranking and the Public Interest. J. of Water Resources Planning and Management Division, ASCE, April.

Pearce, D.W. and Turner, R.K. (1990) Economics of Natural Resources and the Environment. Harvester Wheatsheaf, Hempstead, UK.

Petrasovits, I. (1986) Evaluation of the Agricultural Environmental Impacts of a Danube Barrage System in Hungary. In: The Impact of Large Water Projects, p. 55-62. UNEP-UNESCO, 1990, Paris.

Radford, K.J. (1988) Strategic and Tactical Decisions, 2nd ed. Springer-Verlag, New York.

Radford, K.J. (1989) Individual and Small Group Decisions. Springer-Verlag, New York.

Roy, B. (1985) Méthodologie Multicritère d'Aide à la Décision. Economica, Paris.

Starosolszky, O. (1986) Impact of Peak Load Operation of Run-of-the-River Power Plants. In: The Impact of Large Water Projects, p. 103-109, UNEP and UNESCO, Paris, 1990.

Szidarovszky, F., M.E. Gershon and Duckstein, L. (1986) Techniques for Multiobjective Decision Making in Systems Management. Elsevier, Amsterdam.

Turner, R.K. (1993) Sustainable Environmental Economics and Management. Belhaven Press, NY., USA.

UNEP and UNESCO (1990) (Eds.) The Impact of Large Water Projects on the Environment. Paris.

Van Damm, J.C. and Wessel, J.(1993) (Eds.) Transboundary River Basin Management and Sustainable Development. Lustrum Symposium in Delft, May 1992. Proceedings of IHP-IV Project M-5-2, UNESCO, Paris.

Vodohospodarska Vystavba Bratislava (1990) Gabcikovo-Nagymaros Project. Vodohospodarska Vystavba Bratislava, Slovakia.

von Neumann, J. and Morgenstern, O. (1947) Theory of Games and Economic Behaviour, 2nd Ed., Princeton Univ. Press, Princeton.

Wagner, H.M. and Ortolano, L. (1975) Principles of Operation Research with Applications to Managerial Decisions, Prentice-Hall, Englewood Cliffs, N.J.

WCED (1987) World Commission on Environment and Development. Our Common Future. Oxford University Press. (1987)

Zeleny, F. (1983) Multiobjective Decision Making. Springer Vlg, Wien.

Zeleny, M. (1973) Compromise Programming, in J.L. Cochrane and M. Zeleny (eds.) Multiple Criteria Decision Making, Univ. of South Carolina Press, Columbia.

Zeleny, M. (1982) Multiple Criteria Decision Making. McGraw-Hill, New York, 563 pp.

IV.3 Operation and Management of the Lake Ontario - St. Lawrence River System: Present and Future Options

Anthony J. Eberhardt
Chief, Water Control Center, U. S. Army Corps of Engineers, Buffalo District,
1776 Niagara Street, Buffalo, NY 14207, U.S.A.

Jean-Claude Rassam
Chief, Water Resources Systems Division, Hydro Quebec,
75 Rene Levesque Boulevard West, Montreal. Quebec, H2Z 1A4, Canada

Lucien Duckstein
Professor, The University of Arizona, Department of Systems and Industrial Engineering,
Tucson, Arizona 85721, U.S.A.

Introduction

Lake Ontario has been regulated since 1960. The present decision support system (DSS) is called Regulation Plan 1958-D and is essentially a set of seasonally adjusted rule curves bound by seasonal constraints. An alternative DSS, called the Interest Satisfaction (IS) Mode, is a multi-objective optimization model that is currently being tested.

The purpose of this paper is to describe these two DSS for the weekly operation of the Lake Ontario- St. Lawrence River System in order to meet various regulation objectives, e.g. flood reduction, commercial navigation, power production, environmental concerns and recreational boating in an international transboundary context.

Since the regulation of the Lake Ontario-St. Lawrence River System falls under the jurisdiction of the International Joint Commission (IJC), any DSS that is used for that system should account for the political will in a quasi- real time situation. The issue raised by this paper is whether or not this could be achieved by a multi-objective optimization tool. Further possible improvements of the approach are also suggested.

Description of the Lake Ontario-St. Lawrence River System

Geographical Location

The Lake Ontario-St. Lawrence River System is a part of the Great Lakes-St. Lawrence River System. The St. Lawrence River extends from the outlet of Lake Ontario to Quebec City, at which point the river forms an estuary that ends in the Gulf of St. Lawrence as shown in Fig. 1. The system under consideration includes Lake Ontario and part of the St. Lawrence River down to its entrance to Lake St. Pierre at Sorel. A schematic representation of the system is shown in Fig. 2.

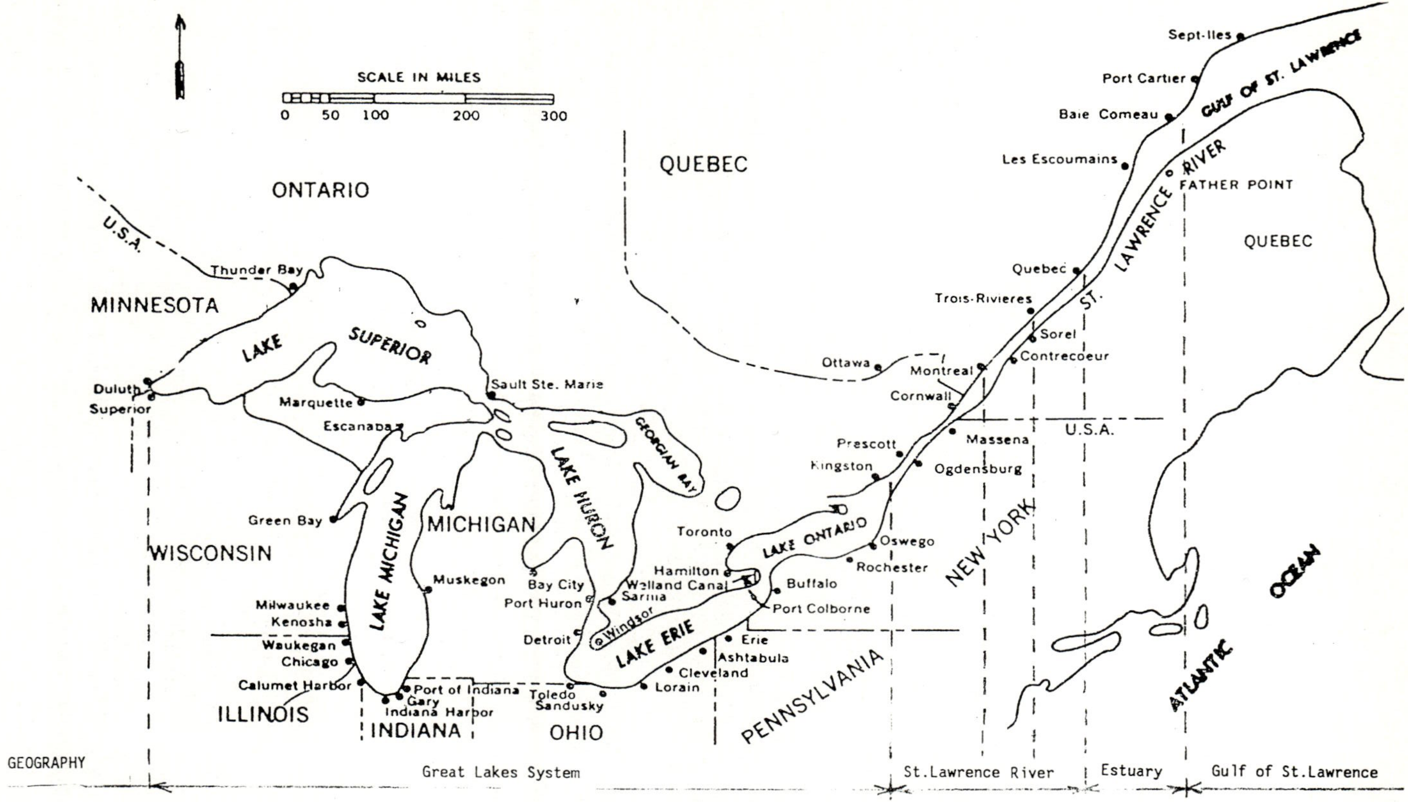

Figure 1 Location of the St. Lawrence River within the Great Lakes-St. Lawrence River System.

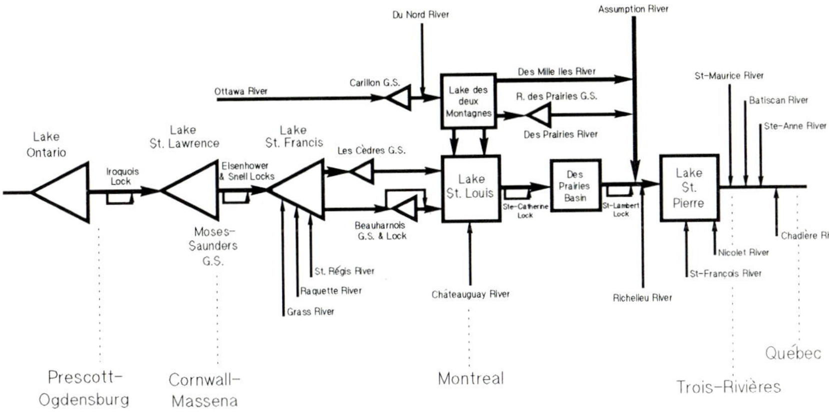

Figure 2 Schematic Configuration of the St. Lawrence River System.

Components of the System

The system is comprised of four power plants operated by Hydro Quebec, Ontario Hydro and the New York Power Authority (NYPA). Table 1 provides a brief description of these plants. The navigation sub-system is divided into two components: the Navigation Channel extends from the Gulf of St. Lawrence up the St. Lambert Locks in the Montreal area; the St. Lawrence Seaway starts from that point up to Duluth, Minnesota. A brief description of the navigation locks and canals is provided in Table 2. There are two spillways, one at Long Sault and another at Ile Juille, and five level control structures, which are located at Coteau, St. Timothée, Pointe-du-Buisson, Pointe-des-Cascades and Sorel. The system is also equipped with many ice control structures, i.e., the Iroquois Dam, the Laprairie Basin Structure, the Lake St. Pierre artificial islands and twenty ice booms.

Table 1 Characteristics of the Power Plants

Power Plant	Operator	Inst.Capacity MW
Robert Moses	NYPA	928
Robert Saunders	Ontario Hydro	876
Beauharnois	Hydro-Quebec	1657
Les Cèdres	Hydro-Quebec	162
TOTAL		3623

Table 2 Characteristics of the Navigation Locks and Canals.

Navigation Lock	Level Drop m	Navigation Cana	Length km
Iroquois	2.3	Wiley-Dondero	16
Snell	13.7	Beauharnois	21
Eisenhower	11.6	South Shore Sea	8
Beauharnois	24.5		
Ste. Catherine	9.5		
St. Lambert	5.6		
TOTAL	67.2		45

Objectives of the System

The system is mainly operated to facilitate navigation in the St. Lawrence River and to provide flood protection for riparians on Lake Ontario and Lake St. Louis. Lake Ontario is regulated within a range of 1.5 m, while its natural range is 2.1 m. The system regulates the Lake Ontario outflow to enhance power production. Lately, other interests have emerged, and system operation reflects those interests; i.e., recreational boating in the Thousands Islands Reach of the St. Lawrence River at the outlet of Lake Ontario, protection of wetlands and other habitats around Lake Ontario and Lake St. Louis and flood protection of agricultural lands around Lake St. Pierre.

Lake Ontario Regulation Plan 1958-D

Background

Lake Ontario has been regulated since April 1960. Plan 1958-D was put into operation in October 1963 and has remained the Lake Ontario regulation plan since that time. For information regarding Lake Ontario regulation from an historical perspective, the reader should refer to Eberhardt, 1990.

The plan consists of a supply indicator, two basic rule curves, seasonal adjustments and a number of maximum and minimum outflow limitations (International St. Lawrence River Board of Control, 1963). The regulated Lake Ontario outflow is obtained in the following manner. The basic regulated outflow is derived from a family of rule curves which show outflow as a function of end-of-period lake level and "adjusted supply indicator".

Fig. 3 shows the rule curves for February through July. The basic regulated outflow is adjusted by applying seasonal adjustments. The resultant seasonally adjusted outflow is compared to maximum and minimum outflow limitations which vary throughout the year. If the seasonally adjusted outflow is between minimum and maximum limitations for the period, it is adopted as the regulated outflow. If it is higher than the maximum or lower than the minimum limitation, the applicable outflow limitation is adopted as the regulated outflow.

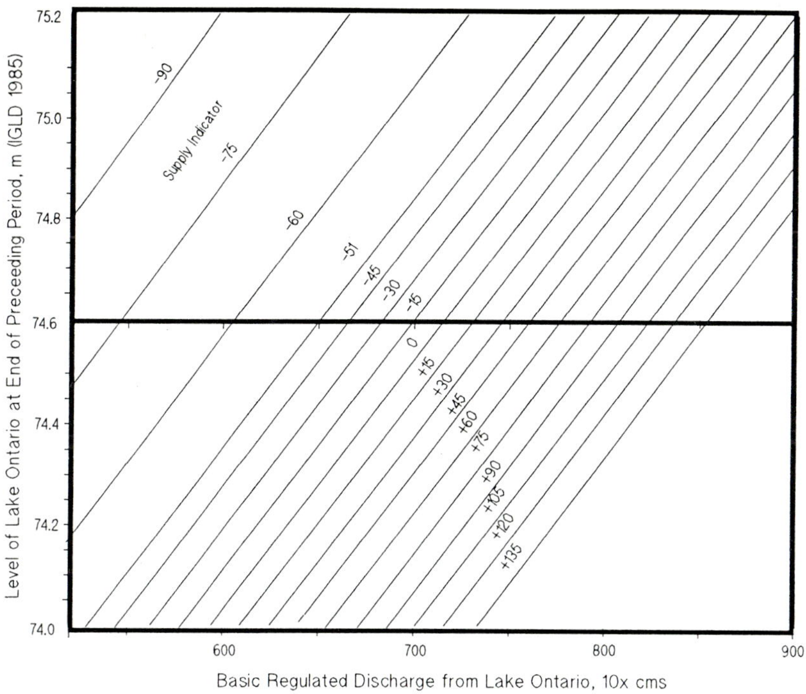

Figure 3 Basic Rule Curve for February through July for Lake Ontario Regulation Plan 1958-D.

The adoption of Plan 1958-D was based on its successful performance in meeting the requirements of ten established criteria as compared to preproject conditions. The criteria are described in Table 3.

An additional criterion, Criterion (k), was included, which must be invoked or revoked by the IJC to allow outflows to be specified in response to conditions outside the range upon which the plan was developed. Since regulation began, discretionary deviations under Criterion (k) occurred during the 1960s in response to low supplies providing assistance to power and navigation interests and during the 1970s and 80s in response to high supplies providing assistance to riparians. The most recent invocation of Criterion (k) by the IJC was between February and August 1993.

Protocol and Procedures for Operating the Plan

The weekly situation on Lake Ontario and the St. Lawrence River is discussed, and the regulated outflow is recommended by the Operations Advisory Group (OAG) after a review of data provided by the offices of the U.S. and Canadian Regulation Representatives (Reg. Reps.); the U.S. Army Corps of Engineers, Buffalo District and Environment Canada, Cornwall. The Reg. Reps. can accept or reject the OAG recommendation. If the Reg. Reps. cannot agree on a regulated outflow, the matter is elevated to the Board. If the Board cannot come to a full agreement on an outflow, Plan flows dictate unless otherwise directed by the IJC.

Table 3 Existing Criteria within Regulation Plan 1958-D.

Criterion	Description
(a)	"The regulated outflow from Lake Ontario from 1 April to 15 Dec such as not to reduce the minimum level of Montreal Harbor belo have occurred in the past as adjusted"
(b)	"The regulated winter outflows from Lake Ontario from 15 Decem be as large as feasible and shall be maintained so that the difficulti are minimized"
(c)	"The regulated outflow from Lake Ontario during the annual sprin Harbor and in the river downstream shall not be greater than woul assuming supplies of the past as adjusted"
(d)	"The regulated outflow from Lake Ontario during the annual flood Ottawa River shall not be greater than would have occurred assum as adjusted"
(e)	"Consistent with other requirements, the minimum regulated outfl Ontario shall be such as to secure the maximum dependable flow f
(f)	"Consistent with other requirements, the maximum regulated outfl shall be maintained as low as possible to reduce channel excavatio
(g)	"Consistent with other requirements, the levels of Lake Ontario sh the benefit of property owners on the shores of Lake Ontario in t which have been experienced"
(h)	"The regulated monthly mean level of Lake Ontario shall not exce feet (75.37 meters) with the supplies of the past as adjusted"
(i)	"Under regulation, the frequency of occurrences of monthly mean approximately 246.29 feet (75.07 meters) and higher on Lake Ont
(j)	"The regulated level of Lake Ontario on 1 April shall not be lower feet (74.15 meters). The regulated monthly mean level of the lake November shall be maintained at or above elevation 243.29 feet (7

Performance of Plan 1958-D

Since the initiation of Lake Ontario regulation, the Great Lakes have been subject to periods of low as well as high supply periods. Precipitation on the Great Lakes basin was generally below average during the early 1960s, most notablyin 1963 when precipitation was 17 percent below average. Between 1965 and 1988, above average precipitation prevailed. Only five of the twenty-four years recorded below average precipitation, and only in 1974 was precipitation more than slightly below average. Throughout the dry and wet periods, Lake Ontario regulation has performed as intended as shown in Fig. 4.

While record low levels were occurring on Lakes Michigan-Huron in 1964 and 1965, the levels on Lake Ontario were higher than would have occurred with preproject conditions. Similarly, levels were maintained below pre-project levels during high supply periods in the mid-1970s and late 1980s. Strict application of the procedural plan was not responsible for mitigating these extreme situations, but rather discretionary actions under Criterion (k).

The sequence of supplies which occurred in 1985 and 1986 was previously not recorded on the Great Lakes (during its short period of record in terms of geologic time). Record high levels were established on each of the upper Great Lakes. However, lower net basin supplies on Lake Ontario and extraordinarily high outflows resulted in no new record levels on the lake. During January 1987, outflows of 10190 m^3/s were released, which were possible due to the relatively mild winter and ice-free conditions on the St. Lawrence River.

The period November 1986 to July 1987 was the driest nine-month period on the Great Lakes during this century. The strategy to discharge the greater of plan or pre-project during the prevailing high supply period (December 1985 invocation of Criterion (k)), resulted in levels dropping rapidly during the sudden unexpected shift in supply situations. The levels which resulted during the summer of 1987 were below average; a condition which, although not intrinsic to the plan, had not been experienced during that time of year for decades.

The "low" levels resulted in a great deal of criticism regarding the plan's ability to respond to changing supply conditions, particularly from the recreational boating industry, which had developed extensively within the 1000 Islands area of the St. Lawrence during the high water years.

A wet summer in 1992 and record water supplies to Lake Ontario during early 1993 resulted in very high levels on the lake. The actions taken by the Board during the winter and spring of 1993 substantially reduced the levels on Lake Ontario during a period of extreme supplies. Although no precise figures are available, the damages prevented were also substantial underscoring the benefits to Lake Ontario riparians through levels regulation.

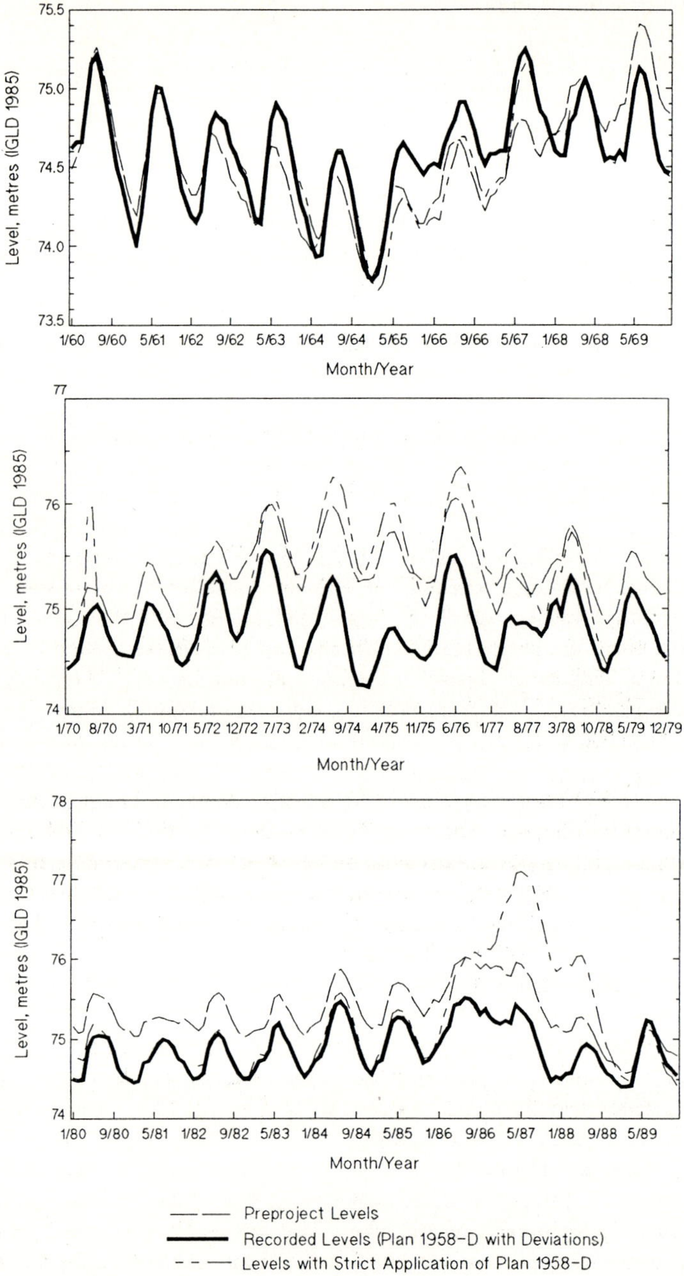

Figure 4 Lake Ontario Levels - Comparison of Plan 1958-D with and without Outflow Deviations.

Interest Satisfaction (IS) Model

Background

The Interest Satisfaction (IS) Model is a new approach for the regulation of Lake Ontario. The model is based on multi-objective, multi-criteria techniques and is an Expert Systems procedure which specifies an outflow that provides the maximum summation of satisfaction to all interests considered. It was developed by the Buffalo District Corps of Engineers and the Departments of Civil and Industrial Engineering of the State University of New York at Buffalo (Eberhardt, 1992 and 1994). The IS Model is being considered by the Board for implementation as an alternative to Plan 1958-D.

Review of Multi-Objective, Multi-Criteria Optimization Models

Multi-criterion decision making (MCDM) has been applied relatively frequently to water resources planning or system design, which has been essentially taken as a static problem (Cohon, 1978; Goicoechea et al., 1982; Tecle et al., 1988; Eder et al., 1994). In contrast, considerably fewer applications to operation of water systems, especially reservoir operation, can be found; these are stochastic dynamic problems fraught with both conceptual and numerical difficulties (Krzysztofowicz and Duckstein, 1979, 1980; Szidarovszky and Duckstein, 1986; Haimes and Li, 1989; Rassam, 1994).

In dynamic operation, a sequential decision model is constructed whenever possible along dynamic programming principles. Non-dominated trajectories--or policies--may be generated, but the choice of a metric to rank the policies is not an easy matter, even in the deterministic case. For example, Rakshit et al. (1989), following Szidarovszky and Duckstein (1986), propose to use an LP-norm to rank order trajectories, but there are several ways to choose an ideal trajectory; one may select the best decision at every stage, but then the corresponding policy is usually not feasible, even if decisions would be feasible at some stages considered in isolation. Another possibility is to reduce the policy to an index such as present value of the objective function vector.

The IS Model is a semi-dynamic deterministic multi-objective, multi-criteria optimization model. The objective function is optimized for each stage separately using a linear programming algorithm.

Description of the IS Model

The IS Model specifies quarter-monthly outflows based on the degree to which various interests are satisfied with a particular level or outflow; i.e., interest satisfaction relationships are used as constraints. Several aspects of the existing regulation plan, Plan 1958-D, are retained. Outflow limitations are retained or modified. Outflow and level restrictions specified by Plan 1958-D criteria are incorporated into the model through the IS relationships. The performance of the model compared to other plans is evaluated in terms of existing plan Criteria.

Database

The data used in the model are based on coordinated quarter-monthly values which are standard basis-of-comparison files used in most Great Lakes regulation model studies, which represent the net total supply (NTS) to Lake Ontario. The period of record is 1900-93. Historic quarter-monthly flows (1900-93) into Lake St. Louis from the Ottawa River are taken from previous update studies and evaluations.

Interest Satisfaction

The model uses ten IS relationships to determine the weekly outflow. The outflow specified provides the maximum satisfaction based on the addition of ten IS values. The relationships consider:

- Lake Ontario and Lake St. Louis Riparians
- Lake Ontario and Lake St. Louis Recreational Boating
- Commercial Navigation on Lakes Ontario, St. Lawrence and St. Louis
- Hydropower Generation at Moses-Saunders and by Hydro Quebec
- Montreal Harbour.

All of the IS relationships are functions of water elevation and/or outflow.

Constraints

The following constraints were used for the ten interests considered:

Lake Ontario & Lake St. Louis Riparians

The upper target level set by Plan 58D for Lake Ontario is 75.37 m, which is about 35 cm above the seasonal high level which occurs in June. The incidence of flood and erosion damage is greatly increased above this level. The IS relationship for riparians has 100% satisfaction occurring when levels are equal to the seasonal long-term (1900-90) average and declining to 0% when levels are 35 cm above or below average. Satisfaction declines to -100% when the deviation from average is more than 35 cm. Fig. 5 shows the value function for the Lake Ontario Riparian interest. On Lake St. Louis, 100% satisfaction occurs between 21.0 and 21.6 m, the extent of the range depending upon the time of year. Satisfaction declines linearly to -100% at 20.1 and at 22.5 m.

Lake Ontario & Lake St. Louis Recreational Boating

Lake Ontario levels ranging from 75.1 to 75.2 m between April 1st and October 31st provide 100% satisfaction. Satisfaction declines linearly to -100% when levels are at 74.15 or at 75.68 m. On Lake St. Louis, 100% satisfaction occurs between 21.4 and 21.6 m and declines linearly to 20.0 and 22.5 m.

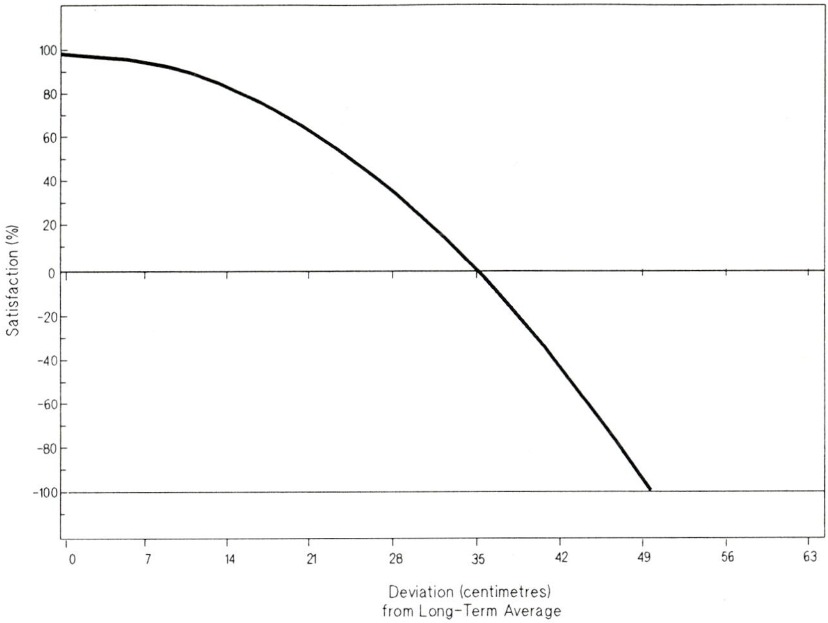

Figure 5 Lake Ontario Riparian Interest Satisfaction Relationship.

Navigation on Lake Ontario, Lake St. Lawrence and Lake St. Louis

Lake Ontario levels between 74.57 and 75.37 m between April 1st and December 31st provide 100% satisfaction. Satisfaction declines to -100% when levels are at 73.89 m and at 76.2 m. St. Lawrence River levels and outflows are also critical to navigation. Outflows greater than 9630 m^3/s result in velocities greater than 1.2 m/s, making transits hazardous in certain reaches. Levels between 72.78 and 74.0 m/s on the St. Lawrence River near the Moses-Saunders powerhouse in Lake St. Lawrence, roughly 160 km from the mouth of Lake Ontario, result in 100% satisfaction at this location. On Lake St. Louis, 100% satisfaction results from levels between 20.9 and 22.5 m.

Hydropower Generation at Moses-Saunders and by Hydro Quebec

At the Moses-Saunders powerhouse, outflows between 7930 and 8780 m^3/s provide 100% satisfaction. Satisfaction declines linearly below or above these outflows. For Hydro Quebec, outflows between 7012 and 7361 m^3/s result in 100% satisfaction, except during the ice formation periods when the optimum outflow is 6000 m^3/s. Satisfaction for this interest declines linearly outside of these values.

Montreal Harbour

Satisfaction is optimal when levels in the harbour are between 6.15 and 8.0 m, the exact range depending upon the time of the year. The value declines linearly above or below these levels.

Environmental

The factors considered are those impacted by levels (e.g., marshes). Levels which fluctuate periodically from low to high are considered most satisfactory. Although not considered in the IS Model formulation, model performance was assessed on the variation of levels in running ten-year periods compared to pre-project (PP, assuming no regulation) fluctuations; 100% satisfaction occurs if this range is equal to 2.26 m.

Outflow Limitations

The maximum outflow set by the model is 9340 m^3/s. Minimum outflows are as defined in Plan 1958-D for power generation varying monthly between 5320 and 5950 m^3/s. The maximum change in outflow from week to week is 850 m^3/s.

Procedures

The following describes the evaluation of the IS Model over a 94-year period of record. Using historic Lake Erie outflows and net basin supplies for Lake Ontario from 1900 through 1993, predicted net total supplies (NTS) with various exceedance probabilities were generated assuming a Log Pearson Type III distribution. Starting in January 1900, an NTS is assumed for the following quarter-month based on the trend of historic supplies for the previous twelve quarters.

Outflows (Q) are tried from minimum to maximum allowable in 10 m^3/s increments and corresponding elevations are generated using the standard storage equation,

Δ = (predicted NTS-Q)/SF; where SF = 29732 m^3/s-qtr-mo. per SF 1 metre change on Lake Ontario elev $_{i+1}$ = elev $_i + \Delta$; where i = qtr-mo. (1 to 4512)

The outflow providing the maximum value in the following Objective Function is chosen for the following quarter-month:

$$P = (F_{lorp}P_{lorp}) + (F_{lorb}P_{lorb}) + (F_{locn}P_{locn}) + (F_{lscn}P_{lscn}) + (F_{hp}P_{hp}) + (F_{llrp}P_{llrp}) + (F_{llrb}P_{llrb}) + (F_{hq}P_{hq}) + (F_{mh}P_{mh}) + (F_{llcn}P_{llcn})$$

where interests satisfaction (IS) values are defined by linear functions as described above with:

P_{lorp} = satisfaction for Lake Ontario riparians;

P_{lorb} = satisfaction for Lake Ontario recreational boaters;

P_{locn} = satisfaction for Lake Ontario commercial navigation;

P_{lscn} = satisfaction for commercial navigation in Lake St. Lawrence;

P_{hp} = satisfaction for hydropower at Moses-Saunders;

P_{llrp} = satisfaction for Lake St. Louis riparians;

P_{llrb} = satisfaction for Lake St. Louis recreational boaters;

P_{hq} = satisfaction for hydropower generation for Hydro Quebec;

P_{mh} = satisfaction for Montreal Harbour interests; and,

P_{llcn}= satisfaction for Lake St. Louis commercial navigation.

The IS values vary from -1 to 1.

The F-factors are applied to individual IS values in order to provide the best results in terms of existing plan criteria. With the chosen outflow, a new change in elevation is calculated based on the actual basis of comparison supply,

$$\Delta_{new} = (\text{Actual NTS - Q}) / \text{SF}$$

$$\text{elev}_{i+1} = \text{elev}_i + \Delta_{new}$$

The procedure is continued quarter-monthly through December 1993.

Performance of the IS Model

The performance of the IS Model was evaluated in terms of its ability to comply with existing 1958D criteria. Factors were applied to each of the IS relationships in order to best satisfy the criteria. Factors could be chosen to provide equal satisfaction for all interests; however, the existing criteria, which do not specifically consider recreational and environmental interests, would have to be revised. The IJC is presently considering such revisions.

The performance of the IS Model compared to Plan 1958D, 1958DV (Plan 1958-D with discretionary deviations) and PP (Preproject conditions) in terms of Lake Ontario water levels,is shown in Fig. 6. The F-Factors chosen were determined by trial comparing results to other Plans. Although the Factors provide very good results, optimization techniques such as Goal Programming would guarantee that the Factors were optimal. Figure 6 shows that the IS Model results in levels very near to those which occurred with Plan 1958-D with discretionary outflow deviations, which essentially formalizes such deviations. During low supply periods, such as occurred during the 1960s, the IS Model results in higher levels and the converse is true in periods of high supply as is demonstrated in the 1970s and 1980s. In 1976, the level with the IS Model is slightly higher than actual due to the limit on outflow of 9340 m^3/s. In actuality, deviations might still occur with IS and the actual outflow of 9910 m^3/s could have occurred lowering the Lake Ontario level. A higher than actual level also occurred in 1987 for this reason. However, with outflows specified by IS in 1987, levels in late summer do not decline below average; a benefit to recreational boating interests.

Improvements in the Operation of the Lake Ontario-St. Lawrence River System

Although it is presently in a test mode, the IS Model has shown a capability to match the same level of performance of a rule curve-based DS, such as Plan 1958-D. The advantage of the IS Model over Plan 1958-D lies in the ease with which its own parameters (e.g. F-Factors, IS relationships, etc.) could be modified to better reflect the shift of interests or the variation of the political will over time and space.

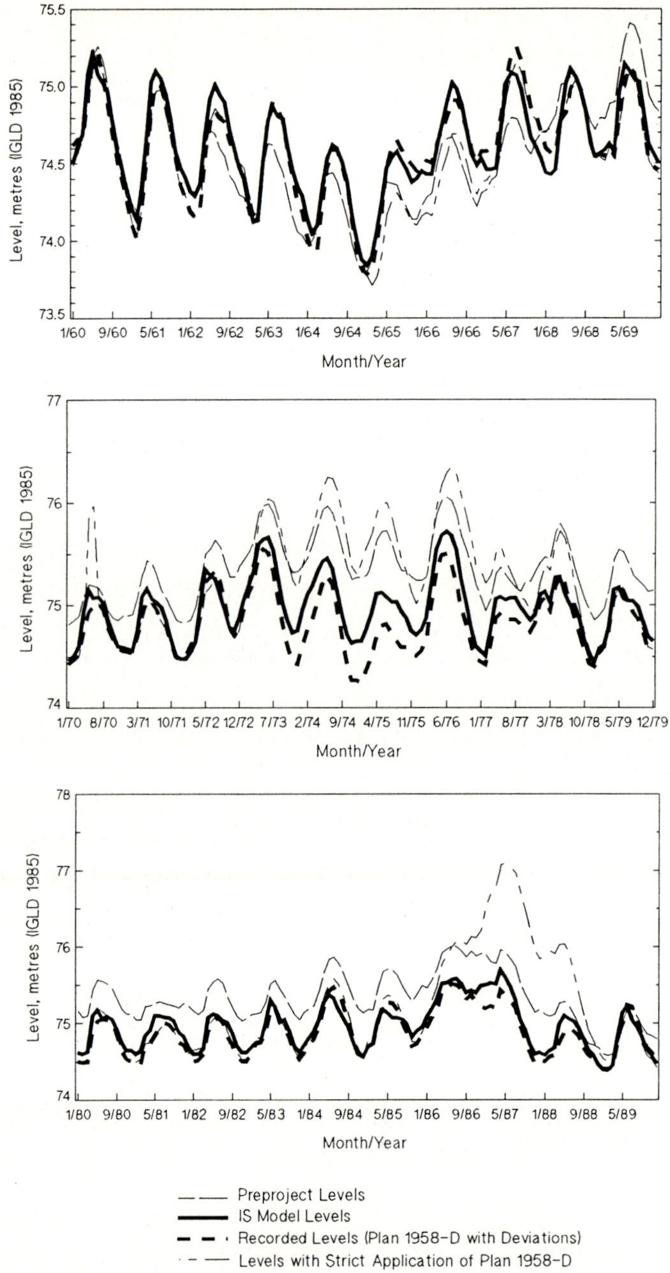

Figure 6 Lake Ontario Levels - Comparison of IS Model with Other Plans.

As a DSS, it should, however, be used with other tools to represent the hydrology and hydraulics of the Lake Ontario-St. Lawrence River System. The IS Model is an additive value-type, which implies full weighted compensation between criteria. Specifically, P stays the same value if $\Delta(F_iP_i) = -\Delta(F_jP_j)$, for any two criteria $i \neq j$. In order to allow for models with partial or no compensation, an LP-norm could be used. Furthermore, risk analysis should be included. The following is a list of possible improvements:

- Hydraulic river modeling for the St. Lawrence portion of the System.
- Multi-attribute utility theory for a single reservoir which allows the embedding of short- range operation into long-range decisions and includes risk (Krzysztofowicz and Duckstein, 1979 and 1980);
- Bayes decision theory, which provides a stopping rule for switching from an observation to a decision-making mode (Krzysztofowicz, 1994);
- Stochastic forecasting of net basin supplies (Croley and Hartman, 1986);
- Use of Distance-based, multi-objective techniques in particular, the multilevel or hierarchical version of compromise programming called composite programming (Bogardi et al., 1994, Bardossy and Duckstein, 1992), which aggregates successively basic indicators, sub- objectives and objectives and accounts for compensation;
- Fuzzy techniques, in particular, rule-based control and rule-based modeling (Bardossy and Duckstein, 1994), which make it possible to include both objective and subjective information into decision-making, while keeping the models simple or at least manageable and thus appear to be suitable for large scale management modeling.

Summary and Conclusion

This paper describes two Decision Support Systems (DSS) for the weekly operation of the multi-objective Lake Ontario- St. Lawrence River System that directly impacts interests in New York State in the United States of America and the Provinces of Ontario and Quebec in Canada. The first DSS is the Regulation Plan 1958-D, which is basically a set of seasonally adjusted rule curves bound by seasonal constraints. The use of the plan has attained its maturity in terms of flexibility and technical improvement. However, its performance is judged satisfactory only with quasi-continuous discretionary deviations as shown in Fig. 4. An alternative DSS, also discussed in this paper, is based on multi-objective, multi-criteria techniques and uses an Expert Systems procedure to specify the weekly outflow of Lake Ontario that maximizes, through LP, the sum of satisfaction of all interests considered, hence, its name: Interest Satisfaction (IS) Model. Although still in a test mode, the IS Model compares favorably to the outcome of Plan 1958-D with discretionary deviations, which essentially reflects the preference of the decision makers, as shown in Fig. 6. However, the advantage of the IS Model over Plan 1958-D lies in its potential for improvement so that the hydraulic-hydrologic characteristics of the system are better represented, and the complexity of the decision-making process is more accurately modeled.

In conclusion, there is a very strong argument for utilizing and further investigating multi-objective, multi-criteria techniques for the future regulation of Lake Ontario.

References

Bardossy, A. and L. Duckstein (1992), Analysis of a Karstic Aquifer Management Problem by Fuzzy Composite Program Water Resources Bulletin, 28(1):63-74

Bardossy, A. and L. Duckstein (1995), Fuzzy Rule-based Modeling in Geophysical, Economic, Biological and Engineering Systems. CRC Press, Boca Raton, FLA.

Bogardi I., A. Bardossy and L. Duckstein (1994), Conflict Analysis using Multiple Criterion Decision Making under Uncertainty. Proc.NATO ASI (this volume)

Cohon, J.-L. (1978), Multiobjective Programming and Planning, Academic Press, N.Y.

Croley, T. E. II and H. C. Hartman (1986), Near-Real-Time Forecasting of Large Lake Water Supplies, Great Lakes Environmental Research Laboratory, April.

Eberhardt, A. J. (1990), The Regulation of Lake Ontario Levels, Proc. of the Symp. on International and Transboundary Water Resources Issues, American Water Resources Association and Canadian Water Resources Association, Toronto

Eberhardt, A. J. (1992), Interest Satisfaction Model for Lake Ontario, Dissertation Submitted to the Faculty of the Graduate School of the State Univ. of New York

Eberhardt, A. J. (1994), Lake Ontario Regulation Utilizing an Expert Systems Approach Constrained by Interest Satisfaction Relationships, Proceedings ASCE Water Resources Planning and Management Conference, Denver, May.

Eder G., L. Duckstein and H.P. Nachtnebel (1994), Ordering Criteria by MCQA with Application to an Austrian Water Resources Planning Case. Presented at the EURO XIII Conference, July 19-22, United Kingdom.

Goicoechea A., D.R. Hansen and L. Duckstein (1982), Multiobjective Decision Analysis with Engineering and Business Applications, John Wiley and Sons, N.Y

Haimes, Y.Y. and E. Li (eds) (1989), Control Theory and Advanced Technology. Special Issue on Multi objective Discrete Dynamic Systems, **5** (4), pp. 403-442

Krzysztofowicz, R. and L. Duckstein (1979) (1980), Assessment Errors in Multiattribute Utility Functions. Organizational Behavior and Human Performance, 26, pp.326-348.

Krzysztofowicz, R. (1994), Strategic Decisions Under Nonstationary Conditions: A Stopping-Control Paradigm, In: Engineering Risk and Reliability in the Management of Natural Resources under Physical Change: with Special Emphasis on Climate Change, L. Duckstein and E. Parent (eds.), Klumer, Amsterdam

International St. Lawrence River Board of Control (1963), Regulation of Lake Ontario, Plan 1958-D, Report to the International Joint Commission, July.

Duckstein (1989), A Stagewise Compromise Programming Framework for Multi-Stage, Multiple Objective Decision Making Problems, In: Y.Y. Haimes and D. Li, Eds., Control Theory and Advanced Technology, Special Issue on Multiobjective Discrete Dynamic Systems, Vol. 5, No. 4, pp. 413-441, December.

Rassam, J.-C. (1994), Gestion du risque pour des systemes hydriques de grande envergure, Ph.D. Dissertation, Faculty of the Graduate School, Laval University, Quebec, 564 p. + 1 appendix, February.

Szidarovszky, F. and L. Duckstein (1986), Dynamic Multiobjective Optimization: A Framework with Application to Regional Water and Mining Management, European Journal of Operational Research, Vol. 24, pp. 305-317.

Tecle A., M. Fogel and L. Duckstein (1988), Multicriterion Selection of Wastewater Management Alternatives, ASCE J. Water Resources Planning and Management, Vol. 114, No. 4, pp.383-398, July.

IV.4 Evaluation of the Discrimination Capability of Criteria by MCQA and Application to an Austrian Water Resources Planning Case

G. Eder
IWHW - BOKU: Institute for Water Management, Hydrology and Hydrological Engineering
Universität für Bodenkultur, A-1190 Vienna, Austria

H.P. Nachtnebel
IWHW - BOKU: Institute for Water Management, Hydrology and Hydrological Engineering
Universität für Bodenkultur, A-1190 Vienna, Austria

L. Duckstein
Department of Systems and Industrial Engineering, University of Arizona
Tucson, Arizona 85721, USA

Introduction

In many multicriterion applications a broad set of criteria is proposed to evaluate a small set of discrete alternatives. Often, the criteria are correlated to each other and their application many lead to a bias evaluation. Also, an impact analysis is required for each criterion which will increase costs and manpower requirements. Therefore, it is essential to evaluate the criteria with respect to the discrimination power in order to obtain a smaller but efficient set of criteria.

The purpose of this paper is to develop a multicriterion concept to investigate the worth and the discriminating power of criteria and to illustrate it by means of a real Austrian case study. Two versions of the Multicriterion Q-Analysis technique (MCQA-I, MCQA-II) have been developed before, whereas MCQA-III is a new contribution. The Q-Analysis concept gives a description of a geometrical structure formed by a simplical and a vertical complex. The technique was proposed by Atkin (1974). Based on his work, Multicriterion Q-Analysis (MCQA-I) was improved and extended in Duckstein et al. (1982) on the basis of a concordance index only, so that the preference order of projects is computed according to the number of satisfied criteria. A second version of MCQA, called MCQA-II, developed in Hiessl et al. (1985), further reflects the decision maker's negative opinion about an alternative via a discordance index. In the present investigation the set of criteria is represented by the simplex set. The new computational procedure MCQA-III, introduced Eder (1993) assists in find a preference order among criteria, based on their discriminating power for ranking alternatives. Discordance, concordance, eccentricity and criteria evaluation concepts are combined in an Lp-norm. The objectives and the grouping of thirty-three criteria are first briefly presented. A comprehensive description of the set of alternatives, including ten different hydroelectric power station plans, a National Park project and a "no action" plan follows. In the second part, the transformation of evaluation data is shown. A brief description of MCQA-III is given and subsequently the methodology is applied to the Danube case study. The ranking of criteria obtained by MCQA-III is given and conclusions are finally given with emphasis on the new technique of criteria ranking.

Description of the Case Study

Nine hydroelectric power stations are located along the Austrian section of the Danube. The discussed study area downstream of Vienna is one of the few free flowing parts and is one of the most ecologically attractive sections of the Danube. In 1984 a proposal for an in-stream type of hydropower station located in this area was authorized by the Water Law Authority. Severe opposition by the public and environmental groups stopped the clearing works and then led to the cancellation of the water law concession by the Administrative Supreme Court. Since then, several possible projects for further management of the Danube downstream of Vienna as far as the Slovakian border have been decision has been taken. This board also determined a large set of criteria considered to be of high importance for evaluating any project.

In 1985 the Ministry of Environment established an Ecological Advisory Group to discuss the further development of the flood plains. This board, composed of engineers, regional planners and biologists concluded that the unique characteristics of this Danube sections, should be preserved and proposed a set of criteria to assess the impacts for any project.

Description of Criteria

In general, criteria reflect the performance of alternatives with respect to the objectives, and may be expressed on either an ordinal or a cardinal scale. Most of them can be measured and estimated by empirical data and models and are given as cardinal numbers. Some of these criteria especially ecological and sociological ones, can only be described verbally; a five point scale has been selected: "very good (V=1)", "good (G=2)", "satisfactory (S=3)", "poor (P=4)", "unsatisfactory or bad (B=5)". An ordinal scale can only show whether an alternative is better or worse than another one, but cannot express in a measurable fashion differences in the performance index.

Description of Alternatives

In 1986 the Republic of Austria published the summary report from the Ecological Advisory Board on further development of the Danube east of Greifenstein (Ökologiekommission, 1986) which, together with Kaniak (1980) and Nachtnebel et al. (1988, 1989), serve as a basis for the application of MCQA. Twelve alternatives have been defined. Five different plans in regard to the location and capacity of hydroelectric power stations have been considered as technical possibilities (A3, A5, A7, A9, A11). In addition to each of these plans, complementary measures such as ground water level control in the surrounding area are considered to reduce environmental impacts (A4, A6, A8, A10, A12). Preservation of the Danube as a free flowing river downstream of Vienna to the border of Slovakia is an alternative described by a "no action" plan (A1). The evaluation of this alternative with respect to economic means is not possible because it does not include any energy aspect, neither generation nor import. For this reason, a second alternative has been generated which includes energy import (A2) corresponding to the energy production of the hydroelectric power station in Hainburg. This possibility couples the free river concept with energy import. The costs for this import over a period of 25 years is counted as investment cost in alternative A2. A complete list of alternatives with a brief description by a comparison of the projects is given below.

Table 1 Goals, criteria, units and weights for hydropower impact assessment.

Goals	Subgoals	Criteria	Units	Number	Weights
economic interests	maximise energy production, minimise relative costs	investment costs	GWh	C1	20
		energy	Mrd.ATS	C2	20
		maintenance costs	ordinal	C3	6
ecological interests	preservation of flood plains and riparian woods	required area due to construction	ha	C4	3
		preservation of native vegetation	ha	C13	41
		preservation of flood plain forests	ha	C14	1
		preservation of Danube riparian woods	m	C15	1
		preservation of gallery forests	m	C16	1
		loss of flood plain forests	%	C17	10
	preservation of structural diversity of river morphology and surrounding area production of good water quality	proportion of reservoir length to undisturbed river length	km/km	C5	8
		length of undisturbed bank	km	C6	3
		length of bank-water line at regulated low flow	km	C7	1
		length of bank-water line at mean flow	km	C8	1
		preservation of riparian area at regulated low flow	ha	C9	3
		preservation of riparian area at mean flow			
		preservation of gravel bars at regulated low flow	ha	C10	3
		preservation of gravel bars at mean flow	ha	C11	2
		number of links remaining between tributary and the main river	ha	C12	1
		stream water quality	number	C18	4
		stream bed erosion	ordinal	C25	4
			ordinal	C28	7
	preservation of typical Danube fauna	changes in existing animal populations	ordinal	C20	16
		population number and species of water birds	ordinal	C19	2
		compatibility with National Park plans	ordinal	C27	6
	preservation of existing ground water flow and ground water quality	length of dams, which are reinforced down to ground piezometric level	km	C21	4
		area with permanent ground water level changes > 0.5 m	km^2	C22	4
		area with phreatic fluctuation of .5-1.0 m	km^2	C23	4
		area with phreatic fluctuation > 1.0 m	km^2	C24	5
		ground water quality	ordinal	C26	2
sociological interests	job creation and protection	new positions resulting from implementing the project	man-years	C32	15
		permanent management jobs	number	C33	10
	interference with thermal springs	influence on therapeutic springs of Bad Deutsch Altenburg	ordinal	C29	5
	creation and preservation of recreation	recreational opportunity	ordinal	C31	12
	Danube as a waterway	obstruction to shipping caused by flood, low flow and ice	days	C30	12

Alternative names refer to the townships near the weir site (Fig. 1). Some projects consist of up to three different hydropower schemes; therefore, their names are combinations of more than one township name. A detailed description of alternatives is shown in Appendix (Tab. A.1).

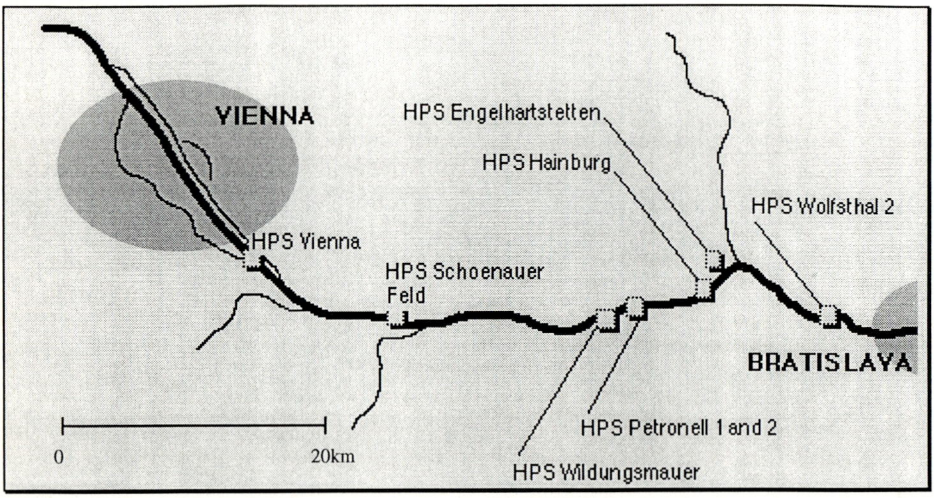

Figure 1 Map of the case study region.

Transformation of Evaluation Data

The basis of a discrete multicriterion decision-making concept is the Evaluation Matrix (Appendix 1, Table A.1). It describes each alternative in terms of criterion values defined on an ordinal, cardinal or verbal scale and expressed in the corresponding unit. The Evaluation Matrix is transformed into the Payoff Matrix in order to quantify the non-cardinal evaluations. Units of criterion values have been given in the previous description of criteria. Verbal classifications in the Evaluation Matrix are now expressed using a scale from one to five.

The Payoff Matrix B with elements b(n,m) describing criterion n from the viewpoint of alternative m is transformed into

$$B = \left[b(n,m); m = 1,\ldots,M; n = 1,\ldots,N \right]$$

n criterion index
m alternative index
M number of alternatives
N number of criteria

The Preference Matrix A with elements a (n,m) by using the range between maximum and minimum value of each criterion.

Table 2 Characteristics of alternative plans

(A1):	Free river without energy import (no action) - establishment of a National Park area - no utilization of hydropower - no protection against bed erosion
(A2):	Free river with energy import - in addition to alternative 1: political dependency on foreign countries because of energy import
(A3):	Hainburg without complementary measures - highest energy exploitation - high ecological disadvantages
(A4):	Hainburg with complementary measures
(A5):	Schoenauer Feld-Petronell 1-Wolfsthal 2 without complementary measures - less ecological disadvantages than alternative 3
(A6):	Schoenauer Feld-Petronell 1-Wolfsthal 2 with complementary measures
(A7):	Petronell 2-Wolfsthal 2 without complementary measures - second highest energy exploitation - longer reinforced dams as A5 and A6
(A8):	Petronell 2-Wolfsthal 2 with complementary measures
(A9):	Wildungsmauer-Wolfsthal 2 without complementary measures - The Wildungsmauer location destroys fewer ecological sites than project Petronell 2 - nearly the same energy exploitation as alternative 5 but with smaller investment costs
(A10):	Wildungsmauer-Wolfsthal 2 with complementary measures
(A11):	Engelhartstetten without complementary measures - dry construction method - same energy exploitation as alternative 3 - less deforestation than in project A3 and A4
(A12):	Engelhartstetten with complementary measures

$$a(n,m) = \frac{b(n,m).\, b_{min}(n)}{b_{max}(n) - b_{min}(n)} \tag{1}$$

with:
a(n,m) = element of Preference Matrix
b(n,m) = element of Payoff Matrix
$b_{min}(n)$ = minimum value of criterion n (worst)
$b_{max}(n)$= maximum value of criterion n (worst)

To apply an MCQA technique the Preference Matrix has to be transformed into an Incidence Matrix by determining, different threshold levels or slicing levels a(k ;k =1,...,K). Elements a(n,m) of the Preference Matrix, which are greater than or equal to the threshold a(k), define a binary relation R between two finite sets in the Incidence Matrix, which are the set of alternatives A = {A(m); m = 1, ... M} and the set of criteria C = { c(n), n =1 ,...,N }, namely C1 to C33.

The kth Incidence Matrix (k = 1 ,...,K) is then defined as:

$$1(k) = \begin{cases} i(n,m) \;\; k = 1 \text{ if } a(n,m) \geq \alpha(k) \\ i(n,m) \;\; k = 0 \;\; \text{otherwise} \end{cases} \tag{2}$$

$$1(k) = \left[i(n,m)_k ; m = 1,\ldots,M; n = 1,\ldots,N\right] \tag{3}$$

M Number of alternatives
N number of criteria
where i(n,m)k is the description of criterion n from the viewpoint of alternative m.

The Q-Analysis procedure is performed at each slicing level a(k):=1,..., K,. In order to find a final ranking, the various preference orders of alternatives at different threshold levels have to be synthesised.

Methodology of Multi-criterion Analysis With a Criterion Ranking Concept (MCQA-III)

Only Incidence Matrices, slicing levels and weight sets are necessary to apply the MCQA technique. Once these have been determined, no further information about the system is needed. MCQA-I, MCQA-II and MCQA-III indices are based on the results of Q-Analysis (Atkin, 1974; Casti et al., 1979). Q-Analysis is the description of a system seen as geometrical structure called complex G(s). The complex is formed by simplices with dimension q which show the relation between two sets, in this case, as discussed above, the set of alternatives P (A1 to A12) and the set of criteria C (C1 to C33) at level a(k). In alternative ranking procedures simplices refer to the set of alternatives A and vertices to the set of criteria C. It is done in incremental fashion for analyzing the set of criteria. The complex G(m,k) at slicing level a(k) of alternative A(m) and criterion C(n) can be expressed as follows:

$$G(n,k) = \sum_{m=1}^{M} i(n,m) \tag{4}$$

This simply means that in the Incidence matrix I(k) the performance of an indicator is expressed by 1 as long as it satisfies a certain threshold level x(k). It can be easily concluded that the stepwise increase of x(k) should result in a stepwise reduction Alternatives A_m satisfying the threshold for criterion C_n. also, highly discrimination criteria should not single out alternatives simultaneously because this would indicate that these criteria are somehow correlated.

To formalize the procedure

G(n,k) = simplex of criterion n at slicing level a(k)
i(n,m) = binary relation R between two finite sets (alternatives and criteria) in the Incidence Matrix
k = index of slicing level a(k):=1,..., K,.

Eccentricity:

This eccentricity measure has been proposed in Chin et al. (1991)

$$\text{ecc } 2(n,k) = \frac{\sum_i \frac{q_i}{\sigma_i}}{\frac{q_{max}(q_{max}+1)}{2}} \qquad (5)$$

q_i = every q-level (= dimension) at which (criterion) n appears
s_i = number of elements in equivalence class at q-level i
q_{max} = maximum q-level in analysis

Usually, at low threshold levels a(k), only one equivalence class of alternatives exists. The eccentricity index ecc 2 determines which criteria are grouped more strongly in the case where there are no criteria in singleton equivalence classes. Q-values of one mean that there exists one equivalence class at a certain q- dimension only. Q is the number of distinct equivalence classes. An equivalence class is formed by q- connected subsets of a simplical complex.

PCI - Project Comparison Index:

PCI involves comparison of a given alternative with all the others in an alternative rating concept. It reflects a relative ranking of alternatives based on their q - connectivity. PCI(n) measures for how long a criterion is in an own equivalence class by changing threshold levels a(k) A measure of the relationships between simplices with respect to shared vertices is called q - connectivity, as explained in Atkin (1974).

$$PCI(n) = \sum_{k=1}^{K} \alpha(k)[q_{max}(n,k) - q^*(n,k)] \qquad (6)$$

PCl(n) = Project Comparison Index used to evaluate criterion n
$q_{max}(n,k)$ = highest q- level of criterion n at threshold level a(k)
$q^*(n,k)$ = q - level at which criterion n is for the first time in the same equivalence class as any other criterion, at threshold level a(k)

The Normalized Project Comparison Index (PCIN) is respectively.

$$PCIN(n) = \frac{PCI(n)}{PCI_{max}} \qquad (7)$$

PDI - Project Discordance Index:

The PDI reflects the decision maker's negative opinion in relation to a project in an alternative ranking procedure. The PDl(n) in a criterion rating concept is computed on the complementary Incidence Matrix I_c using the same algorithm as in PCI, in which the original Incidence Matrix I is replaced by I_c written as:

$$I_c = [i_c(n,m); n = 1 \ldots N, m = 1 \ldots M] \tag{8}$$

Now the relation R is defined as "no satisfaction" of criterion n from the viewpoint of alternative m

$$i_c(n,m) = \begin{cases} 1 & \text{if } i(n,m) = 0 \\ 0 & \text{if } i(n,m) = 1 \end{cases} \tag{9}$$

which in turn yields the PDI:

$$\mathrm{PDI}(n) = \sum_{k=1}^{K} \alpha(k)\left[q_{c,\max}(n,k) - q_c{}^*(n,k)\right] \tag{10}$$

with:
$q_{c,max}(n,k)$ = highest q - level of criterion n at threshold level a(k) in the complementary Incidence Matrix
$q_c{}^*(n,k)$ = q - level at which criterion nis for the first time in the same equivalence class as any other criterion, at threshold level a(k) in the complementary Incidence Matrix

The Normalized Project Discordance Index (PDIN) is written as:

$$\mathrm{PDIN}(n) = \frac{\mathrm{PDI}(n)}{\mathrm{PDI}_{\max}} \tag{11}$$

CEI - Criterion Evaluation Index:

CEI is a new concept referring to the changes in the binary relations between simplex and vertex from one slicing level to another. It is a measure of the discriminating power of each criterion. The CEI, which does not involve any comparison between criteria is defined as:

$$\mathrm{CEI}(n) = \sum_{m=1}^{M} \left[i(k,m)_K = i(n,m)_1\right] \tag{11}$$

with:
$i(n, m)_K$ = binary relation between set C and set P at slicing level k
k = slicing level k = 1 . . . K here k = K
n = criterion index
m = alternative index

This means that CEI shows how sensitive the binary relations of criterion n are to a change of threshold levels. The normalized Criterion Evaluation Index (CEIN) is obtained by:

$$\mathrm{CEIN}(n) = \frac{\mathrm{CEI}[n]}{\mathrm{CEI}_{\max}} \tag{12}$$

CDI - Criterion Discrimination Index:

CDI, which is another new concept, selects those criteria which appear to be of no use for ranking projects. Such criteria either have no relevance to any of the alternatives or satisfy all alternatives at every slicing level. The CDI is taken as a large enough number, say 10, to make sure that this criterion will be ranked poorly:

$$CDI(n) = \begin{cases} 10 \; \textit{is simplex } c(n) \textit{at the lowest slicing level} \\ \textit{has the} \dim \textit{ension } M-1 \textit{ or } 0 \\ \\ 0 \; \textit{otherwise} \end{cases} \tag{13}$$

where:
M = number of alternatives
c(n) = simplical complex representing criterion n

ECCI - Eccentricity Index:

For computing this index (ECCI) the eccentricity measure proposed in Chin et al.(1991) is used: A high eccentricity value means that criterion n is very different to others in evaluating alternatives.

$$\mathrm{ECCI(n)} = \sum_{k=1}^{K} \alpha(k).\mathrm{ecc}\ 2(n,k) \tag{14}$$

The normalised Eccentricity Index (ECCIN) is:

$$\mathrm{ECCIN(n)} = \frac{\mathrm{ECCI(n)}}{\mathrm{ECCI}_{max}} \tag{15}$$

CRI - Criterion Rating Index:

CRI to be minimised embeds the concepts of concordance and discordance, ranks criteria from the viewpoint of their discriminating power and assigns a high value (low ranking) to apparently useless criteria.

$$\mathrm{CRI(n)} = \left(|1-\mathrm{PCIN(n)}|^p + |1-\mathrm{PDIN(n)}|^p + |1-\mathrm{CEIN(n)}|^p + |\mathrm{CDI(n)}|^p + |1-\mathrm{ECCIN(n)}|^p \right)^{1/p} \tag{16}$$

MCQA Procedure Applied to the Case Study

Application of MCQA-techniques to evaluate criteria

Based on the Criterion Ranking Index MCQA can also be applied in order to rank or evaluate criteria. The aim of this procedure is to find a reduced set of criteria which is still comparable for ranking of the alternatives. The normalization process is determined by the definition of MCQA, which states that the Payoff Matrix has to be normalized by the ranges in the simplical complex. This concept has been used to rank alternatives, but its application as a means to investigate the structure of the set of criteria is doubtful. The values in the Payoff Matrix are scale dependent and usually show very large differences. Therefore, most of the normalized values in the Preference Matrix may be either very large or very small. This means that a large set of slicing levels is required in order to obtain a satisfactory number of relations between the set of criteria and the set of projects at small slicing levels, but such a large set introduces noise into the procedure.

Calculation procedure

In a criterion rating concept (MCQA-III) the Payoff Matrix is transformed into the Preference Matrix by the range in the vertex; in other words by normalization based on the set of criteria. This means that the values of each simplex formed by projects are between 0 and 1 but the maximum does not necessarily occur at 1 nor the minimum at 0. Nevertheless, when considering ranking criteria it is a much better procedure than normalizing on the basis of alternative ratings. A satisfactory ranking can already be found with a small set of slicing levels.

The procedure is applied using three slicing levels, namely, a(1)=0.25, a(2)=0.50 and a(3)=0.75. At each slicing level the Incidence Matrix is computed from the values in the Preference Matrix. The Shared Face Matrix reflects the q-connectivity between simplices with respect to shared vertices (Atkin, 1974, Eder, 1993). One of the thirty-three criteria (C25) satisfies all alternatives at the lowest a-level. Therefore, the Q-values are equal to one for the whole procedure. The detailed computational procedure can be found in Eder (1993). Computing MCQA-III provides a satisfactory ranking of criteria taken as simplices by means of CRI1 (p - value 1).

Table 4 Ranking of criteria. Criteria Rating Index (CRI) and weights of criteria

Rank	Criteria	CRI 1	Weight
1	C1	2.0663	20
2	C22	2.4218	4
3	C32	2.9333	15
4	C26	3.1069	2
5	C15	3.2219	1
6	C17, C21	3.2276	10,4
7	C18	3.2358	4
8	C2	3.3755	20
9	C28	3.4185	7
10	C3	3.4196	6
11	C33	3.4714	10
12	C4	3.4776	3
13	C16	3.4992	1
14	C12, C13, C14, C31	3.7463	1, 4, 1, 12
15	C29	3.8603	5
16	C30, C19	3.9245	12,2
17	C20, C27	3.9776	16,6
18	C5, C6, C7, C8, C9, C10, C11, C23, C24	3.9984	8, 3, 1, 1, 3, 3, 2, 4, 5
19	C25	12.000	4

Discussion

A normalization of the set of alternatives discriminates against criteria which are measured on large scale ranges or estimated and given on a subjective scale. The fact that criteria are evaluated with large values does not mean that they are better criteria for ranking alternatives than criteria defined on a large scale range but having small values. Therefore, a normalization of the set of alternatives is not a good way to calculate the Preference Matrix. Another approach has to be found to prepare the input data for the MCQA procedure using the Evaluation Matrix. Normalizing the data on the set of criteria as done in the MCQA procedure for ranking alternatives is one possibility. Another way may be to apply a two step data normalization. First, the Payoff Matrix is normalized on the range of criteria. Next, a normalization is performed on the set of alternatives. This two step data transformation has not been applied in this study. It might have the consequence of sensitivity loss in the input data set. The ranking of criteria should only lead to a better view as to the structural characteristics of the set of criteria. It should give information about the usefulness of criteria in order to rank alternatives.

Two approaches may be used for describing the structural characteristics of the set of criteria. In the first approach criteria rankings based the results of MCQA-I and MCQA-II are compared with the information implemented in the Incidence Matrices, eccentricity measures and Q-vectors at each q-dimension and slicing level. This takes a lot of time and is a difficult technique, because a good knowledge of Q-Analysis and MCQA is required; in fact, this has motivated the development of Criterion Rating Index CRI which gives an order of preference on criterion discriminating property. This second approach, which is much simpler than the first one and yields the same results, is recommended. In conjunction with the new measure CRI, the first approach may be performed for the sake of comparison only. In any case, MCQA provides indices for examining the usefulness of criteria. The new measure CRI describes the discriminating power of criteria. From the viewpoint of ranking alternatives highly discriminating criteria are needed, especially when the set of alternatives is large. The set of criteria should not be too large so as to avoid noise. In every ranking, criterion C25 "Water quality of the stream" has received the highest CRI value. Looking at the Incidence Matrix corresponding to the lowest threshold level (Eder, 1993), we see that criterion C25 satisfies all alternatives. Therefore, water quality is a useless criterion when establishing a ranking among projects in a MCQA procedure based on four slicing levels. Criteria C5, C6, C7, C8, C9, C10, C11, C23 and C24 determine the preference level between hydroelectric power stations and the National Park projects. None of them is used for ranking different power stations or determining the best choice between alternative A1 or A2. The binary relations between the set of criteria and the set of projects vary for criteria C5, C6, C7 C8, C9, C10, C11, C13, C14, C23, C24, C29, C30 and C31 in exactly the same way up to a slicing level of a(k)= 0.50. A large subset of these criteria, namely C5, C6, C7, C8, C9, C10, C11, C23 and C24 change dimension in a similar way up to the highest threshold level a(k)= 0.75. All these criteria have high CRI values because their ability to discriminate between projects is low. Criterion C1 "Investment costs" is in fact a good criterion in terms of available choices. Therefore it also ranks first in the CRI's (Table 4). The results show that the ten best discriminating criteria are C1, C2, C15, C17, C18, C21, C22, C26, C28 and C32.

Conclusions

The new MCQA-III concept applied to rank criteria is a good tool in order to find out whether or not a criterion is useful in an alternative ranking process. Criteria could be exchanged for new ones. Highly discriminating criteria could be retained and refined. MCQA-III procedure has to be based on a Preference Matrix which has been normalized on the range of criteria. Normalization's on the set of alternatives discriminate against values given on a large scale range as compared to those measured on a small scale. Criterion C25 "Water quality" could be deleted in a three threshold level MCQA procedure. It satisfies every alternative at all slicing levels. Criteria C5, C6, C7, C8, C9, C10, C11, C23 and C24 should be analyzed at greater depth to reach a higher sensitivity in evaluating projects. Comparing C5, C6, C7, C8, C9, C10, C11 to each other and C23 to C24, we see that they are very similar criteria. As long as their binary structure at any slicing level does not differ, each group may be replaced by only one criterion. The same is true for the groups of criteria <C17, C21>, <C12, C13, C14, C31>, <C19, C30>, and <C20, C27>. Generally, many criteria contain the same information. Therefore, a satisfactory ranking of alternatives could be found with fewer criteria. Criteria should be developed which allow both a ranking of hydropower stations and of no power station projects. More criteria which are specified in greater detail should be developed.

In summary, MCQA-I and MCQA-II techniques, which have already been used in several case studies of smaller scale, usually are well developed for ranking alternatives. The application of MCQA to find the structure of the criterion set is a new idea. MCQA-III, including Criterion Rating Index CRI, has been developed in order to gain more information on criteria. The case study demonstrates that MCQA is also a useful tool for evaluating the set of criteria in the context of MCDM.

Acknowledgments

This research has been supported in part by a grant from the Austrian Chamber of Commerce. The Department of Systems and Industrial Engineering, University of Arizona, has provided the technical facilities. The Universität für Bodenkultur, Vienna, Austria, Department of Water Resource Management, Hydrology and Hydraulic Engineering, provided the problem formulation and data base for the case study. Additional support was obtained from project No 5064 funded by the Jubiläumsfond of the Austrian National Bank.

References

Atkin, R.H. (1974). Mathematical Structure in Human Affairs, Heinemann,London.

Atkin R.H. and Casti, J. (1977). Polyhedral Dynamics and the Geometry of Systems, RR-77-6, International Institut for Applied Systems Analysis, Laxenburg, Austria.

Casti, J., Kempf, J., Duckstein, L., Flogel, M., (1979). Lake ecosystems: A Polyhedral Dynamics Representation, Ecological Modelling, Vol. 7: 223-237.

Chin, C., Duckstein, L., Wymore, M.L., (1991). Factory Automation Project Selection Using Multicriterion Q-Analysis, Applied Mathematics and Computation, Vol. 46 (2): 107 - 128.

Duckstein, L., Bobee, B., Ashkar, F., (1991). A Multiobjective Criteria Decision Modelling Approach to Selection of Estimation Techniques for Fitting Extreme Floods, Stochastic Hydraulics and Hydrology, Vol. 5: 227- 238.

Duckstein, L and Opricovic, S. (1980). Multiobjective Optimisation in River Basin Development, Water Resour. Res. 16(1): 14-20.

Eder, G. (1993). Multicriterion Q-Analysis of Danube Water Resource Projects, Department of Water Resource Management, Hydrology and Hydrological Engineering, Universität für Bodenkultur, Nußdorfer Lände 11, 1190 Vienna, Austria.

Gershon, M., Duckstein, L., Plate, E.J., (1982). Multiobjective River Basin Planning With Qualitative Criteria, Water Resour. Res., 18: 193-202.

Hiessl, H. et al, (1985). Multiobjective Q-Analysis with Concordance and Discordance Concepts, Applied Mathematics and Computation, Vol. 17: 107-122.

Kaniak, J. (1986).Moegliche Varianten für die Gestaltung und Energiewirtschaftliche Nutzung von Greifenstein. Gutachten im Auftrag des BMfHGI, Wien.

Nachtnebel, H.P. (1988). Wasserwirtschaftliche Planung bei mehrfacher Zielsetzung, Wiener Mitteilung, Bd. 78, Department of Water Resource Management, Hydrology and Hydrological Engineering, Universität für Bodenkultur, Nußdorfer Lände 11, 1190 Vienna, Austria.

Nachtnebel, H.P., Aigner, R., Ortmann, M. (1989). Anwendung von Mehrzielentscheidungsverfahren am Fallbeispiel Donauraum östlich von Wien. Der öffentliche Sektor. (ed. Matzner, H.) Institut für Finanzwissenschaft und Infrastruktur der TU Wien, 15.Jhg., Heft 2, S.116-132. 1040 Wien, Austria.

Ökologiekommission der Bundesregierung, (1986). Arbeitskreis Donaugestalltung (Kaniak), Arbeitskreis Nationalpark Donau - March - Thaya Auen (Loetsch), Protokolle, Arbeitspapiere, Schlussbericht, Referate, Vienna, Austria.

Pfaff, R.T. and Duckstein, L. (1981). Ranking of alternative plans of the Santa Cruz River Basin by Q-Analysis, Proceedings, Joint AZ Sect., AWRA, and Hydrology Sect., AZ-Nevada Acad. of Sciences, May 1.-2., Tucson, AZ.

Tecle, A., Duckstein, L. Fogel, M.M., (April 1987). Multicriterion Decision Making in waste Water Management: Application and Comparison of Techniques, Systems and Industrial Engineering Department, University of Arizona, SIE report # 87-019, Tucson, AZ.

APPENDIX:

Table A.1: Evaluation Matrix

Criterion		A1	A2	A3	A4	A5	A6	A7	A8	A9	A10	A11	A12
C 1	Mrd ATS	0.1	22.3	11.4	12.0	24.9	26.1	15.9	16.7	15.6	16.4	12.2	12.8
C 2	GWh	0.0	2075.0	2075.0	2075.0	1700.0	1700.0	1990.0	1990.0	1791.0	1791.0	2075.0	2075.0
C 3	ordinal	G	G	S	S	B	B	B	B	B	B	S	S
C 4	ha	0.0	0.0	740.0	740.0	660.0	660.0	655.0	655.0	465.0	465.0	496.0	496.0
C 5	%	1.0	1.0	0.3	0.3	0.2	0.2	0.1	0.1	0.2	0.2	0.3	0.3
C 6	km	96.0	96.0	23.0	23.0	36.5	36.5	35.5	25.5	40.1	40.1	23.0	23.0
C 7	km	113.8	113.8	83.7	83.7	81.6	81.6	82.7	82.7	85.0	85.0	83.7	83.7
C 8	km	100.0	100.0	83.7	83.7	81.6	81.6	82.7	82.7	85.0	85.0	83.7	83.7
C 9	ha	165.5	165.5	57.8	57.8	47.0	47.0	39.8	39.8	48.6	48.6	57.8	57.8
C10	ha	166.2	166.2	52.5	52.5	47.0	47.0	39.8	39.8	48.6	48.6	52.5	52.5
C11	ha	168.1	168.1	1.4	1.4	0.0	0.0	0.0	0.0	12.9	12.9	1.4	1.4
C12	ha	6.8	6.8	0.0	0.0	0.0	0.0	0.0	0.0	2.1	2.1	0.0	0.0
C13	ha	18.2	18.2	0.8	0.8	7.1	7.1	1.4	1.4	3.1	3.1	1.1	1.1
C14	ha	120.2	120.2	18.3	18.3	55.0	55.0	32.5	32.5	38.9	38.9	24.0	24.0
C15	km	33.8	33.8	2.850	2.850	15.850	15.850	12.050	12.050	18.950	18.950	3.800	3.800
C16	km	15.95	15.950	5.900	5.900	8.650	8.650	7.550	7.550	10.250	10.250	6.380	6.380
C17	%	0.0	0.0	86.4	86.4	25.0	25.0	62.0	62.0	52.0	52.0	87.8	87.8
C18	number	42.0	42.0	11.0	11.0	25.0	25.0	20.0	20.0	23.0	23.0	12.0	12.0
C19	ordinal	P	P	P	P	G	G	G	G	G	G	P	P
C20	ordinal	V	V	B	B	S	S	P	P	P	P	B	B
C21	km	0.0	0.0	57.8	57.8	15.5	15.5	42.5	42.5	31.5	31.5	57.8	57.8
C22	km	65.5	65.5	86.2	6.2	69.4	29.7	67.3	23.5	61.8	20.8	68.2	6.2
C23	km	51.8	51.8	4.1	4.1	14.2	10.1	12.4	9.1	12.4	9.1	4.1	4.1
C24	km	35.8	35.8	4.1	4.1	8.0	8.5	8.7	7.4	8.7	7.4	4.1	4.1
C25	ordinal	G	G	G	G	G	G	G	G	G	G	G	G
C26	ordinal	G	G	B	P	S	G	P	S	S	G	B	P
C27	ordinal	V	V	V	B	B	S	S	P	P	P	P	B
C28	ordinal	B	B	G	G	S	S	G	G	G	G	G	G
C29	ordinal	G	G	S	S	V	V	V	V	V	V	S	S
C30	day	111.0	111.0	116.0	116.0	11.0	11.0	11.0	11.0	11.0	11.0	116.0	116.0
C31	ordinal	V	V	P	P	S	S	P	P	P	P	P	P
C32	year	0.5	0.5	17.1	18.0	37.5	39.2	23.9	25.0	23.4	24.6	18.3	19.2
C33	number	30.0	30.0	70.0	70.0	210.0	210.0	140.0	140.0	140.0	140.0	70.0	70.0

ordinal criteria are expressed by

V	very good
G	good
S	satisfactory
P	poor
B	bad or unsatisfactory

PART 3

CURRENT PRACTICES AND OPERATIONAL EXAMPLES

INTRODUCTION TO PART 3

This part deals with more general water management issues in international river basins than Part 2, and, as a result, does not exhibit the same precise modeling of conflicts. It consists of two chapters: Chapter V provides examples on major international rivers and seas, namely the Danube and the Elbe, the Black Sea and the Aral Sea. Chapter VI considers the case of transborder regional rivers, such as the Balkan Rivers (Greece, Bulgaria and Turkey) and the Ob river in Siberia.

Transboundary water management along the Danube and its drainage basin is discussed in five papers. The Danube is indeed an international river, since nine countries have access to the river and six additional countries share its drainage basin for a total of fifteen countries involved. Nachtnebel (V.1) describes the Danube River Basin Environmental Program as an example of how all the countries sharing the basin have decided to work together toward an environmental renewal of the deteriorating river environment. The following papers show that it is a long way from problem recognition to actual solution. The good news is that there are existing bilateral agreements between several neighboring Danube countries, but mostly on releasing joint observations only. Three papers deal with the results of bilateral monitoring programs over the Danube. The observation data which are taken (at an operational level) can then be used to design both policy and implementation level decisions.

Literathy (V.2) summarizes pollution monitoring activities along the whole Danube and states that as a result of this national and international effort, the movement of nutrients and pollutants is fairly well known in the Danube basin. The role of bottom sediment has been identified as a major source of pollution through the release of absorbed contaminants. The most polluted sections and the main pollutants have been identified. Also, the role of hydraulic engineering structures as sites of pollutant "time bombs" has been clarified.

Bernardova, Mrazek and Fiala (V.3) report the pollution monitoring effort over the Czech portion of three rivers of different character. The Elbe River conveys water from the industrialized western portion of the Czech Republic through Germany to the North Sea and has also been examined in its lower part by Schumann and Simon (II.1). The Morava River drains more than one fourth of the countries' water into the Danube. Finally, the Odra River--the smallest of these three rivers--enters Poland and drains into the Baltic Sea.

Liska and Laszlo (V.4) describe in detail how an efficient bilateral monitoring program works over the Slovakian-Hungarian border section of the Danube. They point out that the cross-sectional distribution of the various pollutants should be monitored in order to have a correct picture of pollutant transport. It is hoped that if such an effective bilateral monitoring program can be implemented between two countries, transboundary water management issues, which are often in conflict between the same countries, can be addressed just as effectively (at the policy level).

Varduca and Mara (V.5) consider the Danube from the aspect of the heavily polluted final recipients, the Danube Delta and the Black Sea. As a consequence of the nutrient input from the upstream countries, the aquatic ecosystem of the Delta has been disturbed. Different control actions are considered and a simple procedure is presented to select the most attractive control actions. Here again the introduction of some of the multicriterion techniques presented in Part 2 may occur after certain time.

Mee (V.6) describes the current catastrophic environmental situation of the Black Sea that includes three major issues: eutrophication, induced alterations in the marine ecosystem and chemical and microbiological pollution. Although the Black Sea is deep, it is virtually isolated except for the narrow Bosphorus channel. One of the main causes of its ecological disruption is the Danube at the relatively shallow northwestern continental shelf, where the depth is less than 200 m. In addition to characterizing the existing situation and the main contributing factors, the paper highlights the socio-economic consequences to the riparian countries. The six riparian countries signed a convention (policy level) in 1992 for the protection of the Black Sea. As a consequence, the Black Sea Commission was established (implementation level) to develop an action plan for the implementation of the convention. Evidently, even if financial resources are available to accomplish the action plan, the realization of the action plan cannot lead to the restoration of Black Sea ecology without efficient transboundary water management, for instance through the Danube River Basin Environmental Program between the Danube River nations. This is an example when one institutional level system cannot design an effective polity without agreeing with another system at the same level.

The Aral Sea basin can be singled out as the most relevant example in the world of non-sustainable development. Causes of the crisis, present situation and possible remediation are described in the paper by da Cunha (V.7). Between 1960 and 1992, the surface of the lake decreased by about half and its volume decreased to less than 25%, while the salinity reached the level of oceans. The main causes of these drastic changes include non-sustainable irrigation development over the catchment area and possible climate change. This latter may be accelerated by a positive feedback due to the reduced surface area of the lake. Originally a one-country problem--that of the former Soviet Union--the solution of the Aral Sea problem now requires international cooperation among the five new Central Asian Republics sharing the catchment area. The paper describes the organizational structure of this cooperative effort (at the policy level) and summarizes the finding of a recent NATO Advanced Research Workshop on the identification of the research needs for the study of the Aral Sea basin problems.

Issues on water management in transborder regional river basins are presented in Chapter VI. Here, the term "border" is used to denote either international political borders or boundaries between regions in the same country.

Yevjevich (VI.1) has worked as a young engineer on water resources projects in the Former Yugoslav Republic of Macedonia (FYROM), where the major part of the Axios/Vardar river basin is located, the rest belonging to Greece. He advocates that cooperation between the two countries is the best solution for transborder water management and also to resolve current disagreements, such as the use by FYROM of the name Macedonia, which is the same as that of the Greek province of Macedonia.

In the Balkan peninsula a great challenge is about sharing the waters of the river Mesta/Nestos between Bulgaria and Greece. Three papers deal with water quantity and quality in this case. The first by Tzovaridis, Moutafis and Cavadias (VI.2) presents hydrological data of flowrates near the Bulgarian-Greek border, indicating that from 1970, there is a systematic decline in the quantity of water entering the Greek territory. This is due to water diversion by the Bulgarian side for irrigation and water supply purposes. Negotiations between Greece and Bulgaria are under way in order to resolve the problem, which is important for Greece as far as the production of hydroelectricity from the Nestos River is concerned.

The other two papers by Mihailov and Arsov (VI.3) and Argyropoulos, Ganoulis and Papachristou (VI.4) refer to water quality issues in the Nestos/Mesta River. The good news in the two papers is that both in the Bulgarian and Greek parts of the river the water quality is still relatively good. However, this positive aspect should be considered only temporary and only due to the drastic reduction of industrial activities on the Bulgarian side, due to the transformation of the Bulgarian economy to a free market. Bilateral efforts for monitoring, planning and water management should be continued and reinforced in order to prevent any future deterioration of the current situation.

The River Evros/Meric/Maritza flows from the Bulgarian mountains down to the Aegean sea and serves as a boundary between Greece and Turkey. As such, there is a tripartite concern about managing the water quantity and quality of the river between Greece, Bulgaria and Turkey. Samsunlu, Maktav and Kapdaslii (VI.5) report water quality data showing that the river may be classified as water quality class II. They also propose exchange of information and cooperation in the areas of remote sensing and data sharing in order to protect the river and especially its delta region, which is very vulnerable to pollution.

Karavitis (VI.6) discusses the very important problem of droughts, when inter-regional water transfer is in question. Such is the case of the water supply in the Athens Metropolitan Area. Mainly due to an extensive drought, the water supply problem was exacerbated in 1990. Structural and non-structural measures are proposed to alleviate the consequences which may be expected to occur more frequently in view of a possible global climate change.

Gubaidullin and Khamitov (VI.7) briefly present the enormous environmental problems the Siberian rivers face. Due to climatic conditions and pollution hazards mainly from the oil industry, environmental disasters are very likely to occur. Concentration of industrial activities in the Bashkortostan Republic, Russia, is another example put forth in this paper, where the lack of technological innovation and maintenance may have disastrous consequences to the water quality of regional rivers.

Chapter V: Water Management in Major International River Basins and Enclosed Seas

V.1 From Recognition towards Implementation: The Danube River Basin Environmental Program

H.P. Nachtnebel
Institute for Water Management, Hydrology and Hydraulic Engineering
Universität für Bodenkultur
A, 1190 Vienna, Austria

Introduction

The purpose of this paper is to present the Danube River Basin Environmental Programme as an example of cooperation among nations sharing a common watershed. The word, "example" does not intend to mean that this endeavor is necessarily the best approach to follow but as a possible vehicle how environmental degradation in an international river basin can be addressed from the recognition of the environmental issues towards the implementation of the necessary measures in close cooperation of the riparian countries. We will put emphasis on the process of Programme development; therefore hydrological, ecological or economic conditions of the basin will be only briefly described.

The Danube river basin (Kresser, 1973) (see Fig. 1) is the heartland of South Central and South Eastern Europe. The river flows over a distance of 2857 km and drains an area of 817,000 square kilometers including all of Hungary and Romania, most of Austria and of the former Yugoslavia, nearly half of the Czech and Slovak Federal Republics, a third of Bulgaria, significant areas of Germany, smaller areas of the Ukraine and Moldavia.

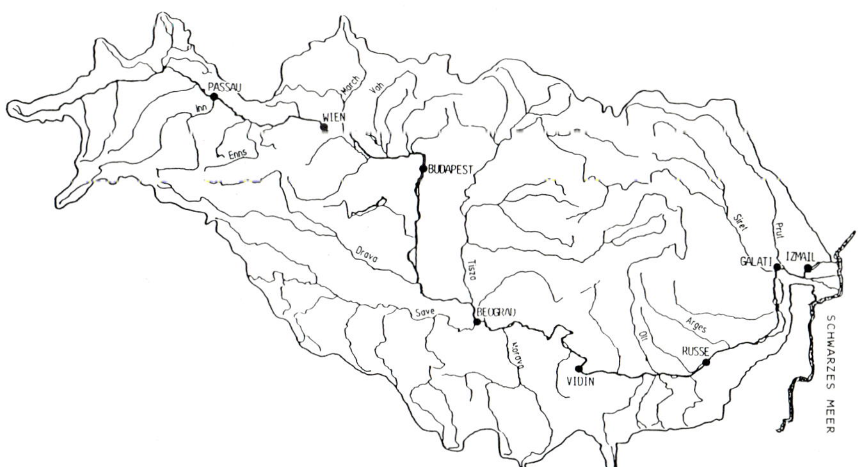

Figure. 1 Danube River Basin.

About half of the average total volume of water of the Danube (216 km³/year) comes from Austria and Yugoslavia, and the entire flow of the Danube discharges into the Black Sea through an extensive delta.

According to the geomorphological structure and the longitudinal profile (Fig. 2) of the Danube the basin can be conveniently divided into four regions:

- the stretch from the springs of the Danube down to Bratislava
- the section from Bratislava through Slovakia, Hungary and the former Yugoslavia to the Iron gate
- the third section from downstream the Iron gate to the Delta
- the Delta itself.

Table 1 Characteristics of the three main parts of the Danube catchment.

	gauging station	stream location kilometer	catchment area km^2	mean discharge m^3/s	runoff mm
Upper Basin	Bratislava	1869	131 338	2 020	485
Middle Basin	Orsova	955	576 232	5 699	312
Lower Basin	Ceatal Ismail	72	807 000	6 550	256

The delta is a huge region covering an area of 591200 ha. It has the shape of a triangle with sides of 70 km. Almost two thirds of the delta area are permanently submerged due to the low absolute altitude (200000 ha below the mean level of the Black Sea and 150000 ha only slightly above). The three main branches are divided close to the estuary into numerous smaller branches creating their own deltas.

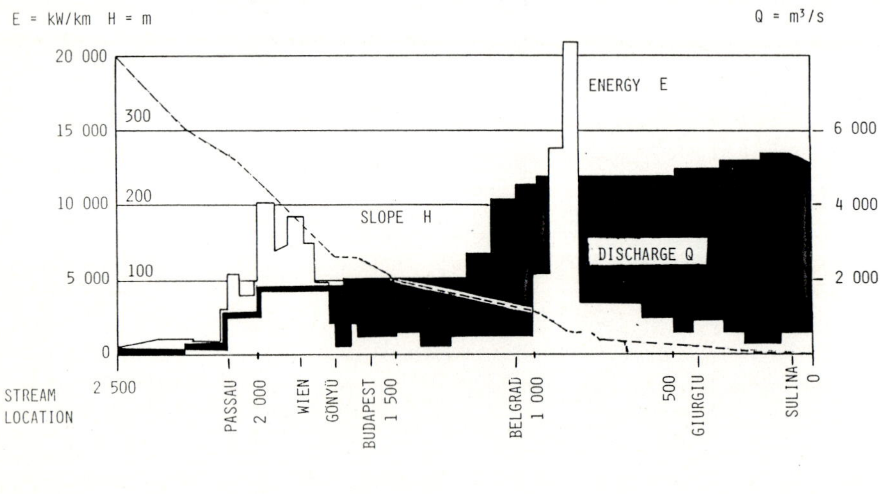

Figure 2 Longitudinal Profile of the Danube.

The delta and the banks of the Black Sea were designated by UNESCO as a "Biosphere Reserve". A part of the delta (53%) has been declared a "wet zone of international significance" in 1991 and designated as a "World Heritage" site under the Ramsar Convention. The seasonality of the discharge and the longitudinal development of the discharge is given in Fig. 3 and 4, respectively.

Land use in this large basin is highly diversified, including a wide range of agricultural practices, forestry, mining, natural areas, settlements and industries. Throughout its length the Danube river provides a vital resource for drainage, communications and transport, power generation, fishing, recreation and tourism, as well as an ecosystem with irreplaceable environmental values. More than 40 dams and large barrages plus several hundred smaller ones have been constructed on the main river and its tributaries. While these facilities provide important opportunities for the utilization of the river's resources, they also are illustrative examples of the difficulties of balancing these important economic activities with environmentally sound and sustainable management.

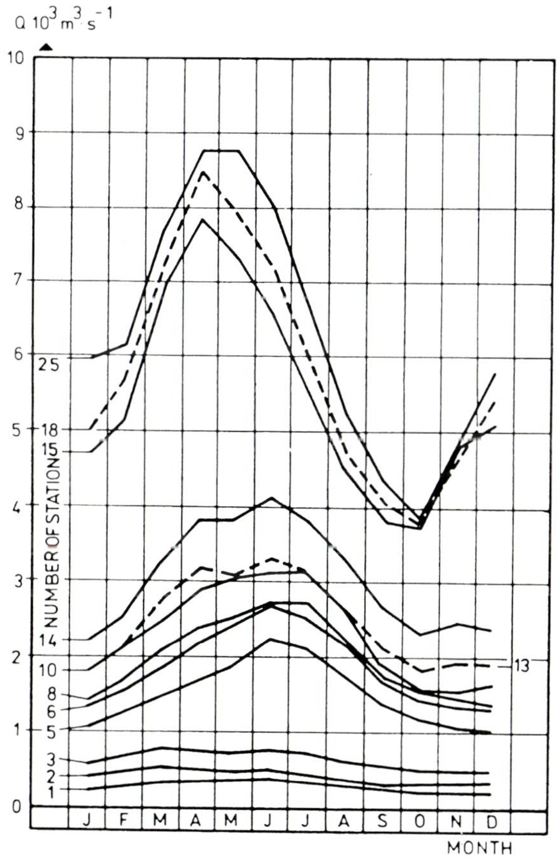

Figure 3 Seasonality of Discharges.

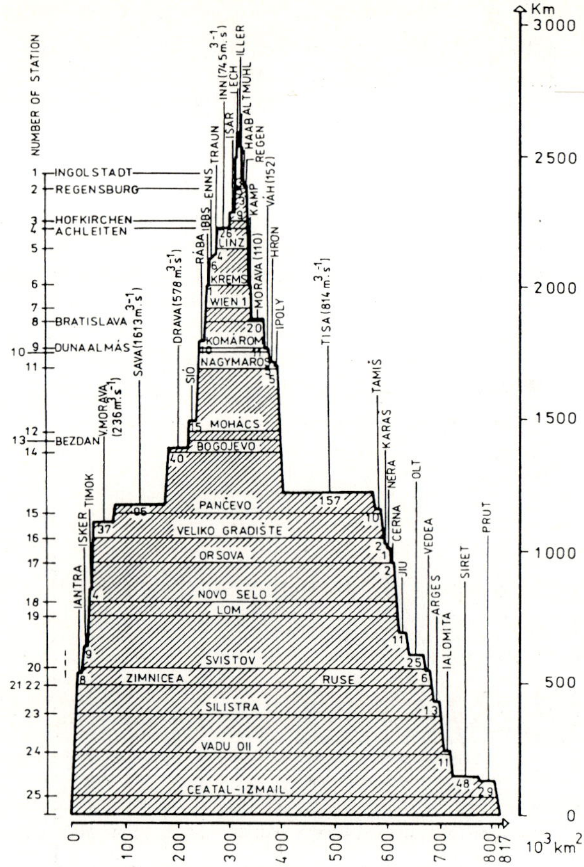

Figure 4 Longitudinal Development of the Discharge.

Surface Water Quality

In general, the water quality on the main stream is better than in the tributaries where especially during low flow periods, major quality problems are faced. Due to different quality standards in the countries no basin wide classification of the Danube waters is available and therefore a few parameters, including BOD, nutrients and hydrocarbons are described . Fig. 5 includes the BOD concentration of some major tributaries and the respective value in the Danube at the confluence. Fig. 5 exhibits also the buffering and self purification capacity of the main stream which is reflected by a decrease of BOD in its longitudinal profile although some heavily loaded tributaries after the main river. Dissolved oxygen concentrations ranged in the period 88-90 from 10 mg/l in the upstream section to about 8 mg/l at the mouth of the Danube.

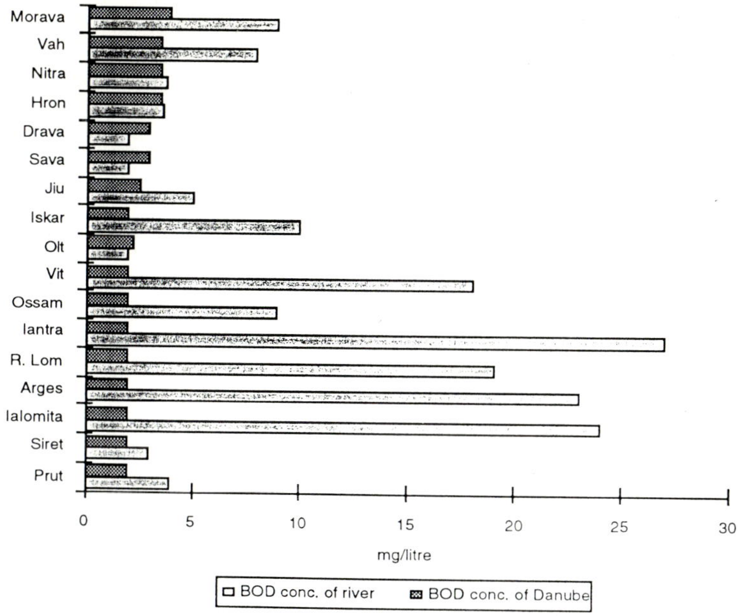

Figure 5 BOD Concentrations in the Major Tributaries and in the Main River.

Nutrient concentrations (nitrate and phosphate) increase along the river course and are at least doubled in the lower sections of the Danube (Weber, 1993). The average annual loads are given in a longitudinal profile in Fig. 6.

Data about microbiological pollutants are available but not published in a standardized format. Mineral oil concentrations show a slight tendency to increase downstream Fig. 7 while some other chemicals (Equipe Cousteau, 1993) indicate an increase downstream (lindane, DDT and some heavy metals). Elevated oil concentrations are observed in three reaches of the main stream; in Germany, in the middle section between Bratislava and Budapest, and in the lower reach from the industrial areas of Bulgaria and Romania.

Bacteria and Viruses

The Danube is heavily polluted with faecal coliforms, entero viruses, and fecal streptococci. The scarce data base indicates no significant trend of deterioration of river water towards the mouth of the Danube. The worst conditions are found downstream of larger cities lacking adequate waste water treatment. This holds also for the water quality of the tributaries.

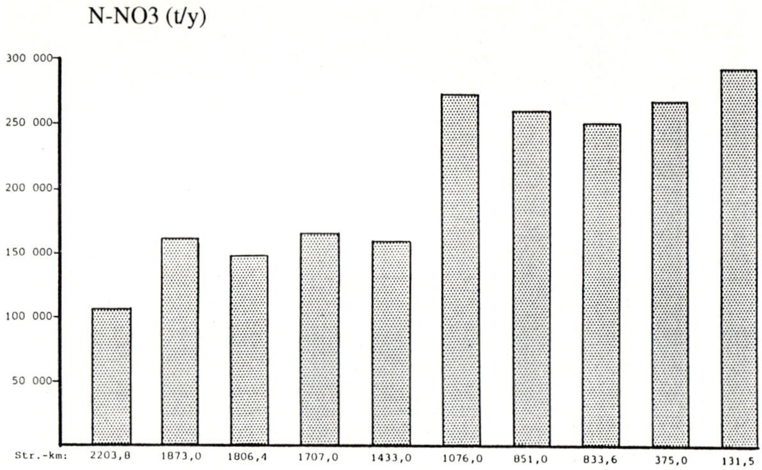

Figure 6 Nutrient Loads in the Danube.

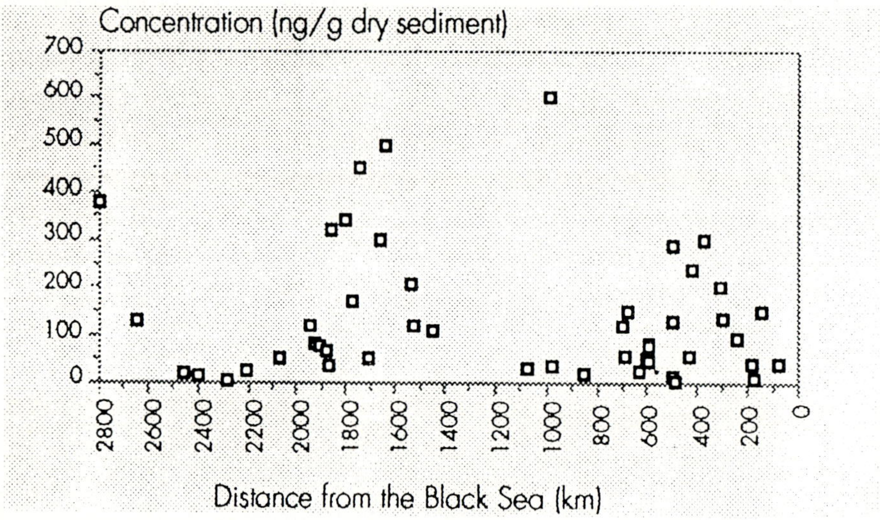

Figure 7 Mineral Oil Concentration.

Groundwater quality

Major groundwater resources are located in the alluvial plains of the Danube and the tributaries. The spatial distribution of humid zones indicated in Fig. 8 gives a good indication about the location of shallow groundwater bodies which together with deep layered aquifers supply a high percentage of the population (80-95% varying from country to country) with drinking water. Additionally, vulnerable karstic water resources are of special importance for drinking water supply. Due to agricultural practice, landfill sites, inefficient sewage systems and interaction with polluted surface water systems these water resources are endangered. Pollution of groundwater is a common problem and nitrate, chloride, sulfate, ammonia, phenol are often locally exceeding the respective country standards. The low recharge rate of groundwater bodies in the lower part of the basin delays remarkably the sanitation and recovery processes.

Biological Resources

In the Danube basin there is a broad variety of landscapes ranging from high alpine subcatchments to large alluvial flood plains and further downstream to estuaries in the Delta. Accordingly, an outstanding rich biodiversity was found and is still found in the basin. About 10 to 15 000 species including more than 100 fish species and about 180 bird species are still found in the basin. Only, within the river corridor about 2 000 plant species are reported. There is a distinct gradient in species richness along the river course reaching its maximum in the Danube Delta. Although many sites are under natural protection and several of them are approved by the RAMSAR convention many species are endangered or are already threatened to be extinct.

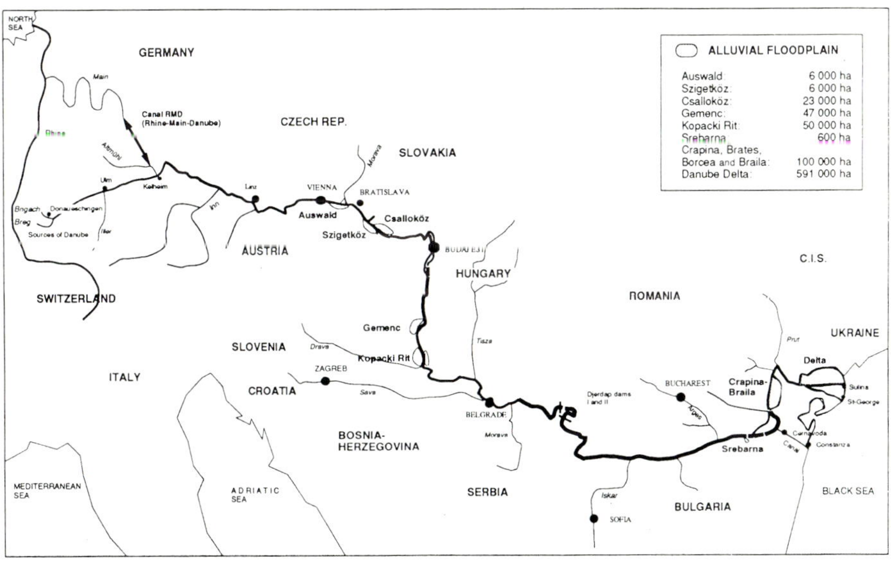

Figure 8 Humid Zones in the Basin.

About 40 fish species fall into this category and with respect to some bird species dramatic data are reported. 6 to 8 pairs of White-tailed Sea Eagle and 25-40 pairs of Dalmatian Pelicans are observed in the Delta. Similar figures hold for other sections of the Danube. Austria reports none breeding Sea Eagle while in Hungary 15 pairs are observed (IUCN, 1992). About 60 fish species are reported from Austria from which many are endangered or exposed to be extinct soon.

Recognition of Environmental Degradation

The environmental quality of the Danube River Basin is under great pressure from a diverse range of human activities. Urban populations are generating pollution from largely inadequate wastewater treatment and solid waste disposal facilities. A serious problem exists in relation to the disposal of industrial wastes into the air, water and on land within the basin. Of special concern are those wastes generated by agro-processing, chemical, pulp and paper, metal processing, mining and textile industries. In addition, modernization and intensification of agricultural practices and livestock production are major sources of non-point pollution of surface and groundwater.

Air

Inadequate attention to environmental impacts from industry and energy production has resulted in significant water, air and soil pollution at local, regional and transboundary levels. Airborne particulate matter from stationary and mobile sources may be a significant source of pollution of soil and water.

Water

Surface water quality is mainly endangered by point pollution from insufficiently treated domestic waste water, industrial effluent and by diffuse sources from agriculture. At the same time the water requirements in several Danube countries are mainly covered from surface water resources. About 30 % of water use in Bulgaria is provided by the Danube River and its tributaries, and the respective figure for Hungary is 90 %. It is also noteworthy that about 95 % of the surface water in Hungary originates outside the country. These figures include industrial requirements and domestic water supply. In Hungary, about 90 % of drinking water is derived from groundwater, of which about 50 % is from bank-filtered sources. Obviously, to ensure the quality of bank filtered water a broader spatial scale in water management has to be envisaged.

The concentration of pollutants in the main stem of the Danube (Weber, 1993; Buijs, 1992; Kreitner and Weber, 1989) is in general similar to the improved levels currently found in the Rhine, principally because the flow of the Danube is about three times greater than the Rhine. There are important exceptions to this which include oil, the concentration of which is much higher in the Danube than in the Rhine, and some heavy metals which create significant public health and ecological concerns. The situation on tributaries is much more severe because in many cases the concentration of pollutants greatly exceeds acceptable standards. This reduces the availability of water and results in very high treatment costs for municipal and industrial water supplies, causes significant ecological damage and presents a risk to public health.

Local action to control and manage waste discharges and land use is important, particularly on tributaries with limited discharges. However, the quality and sustainability of environmental conditions in most areas of the basin depends also on the effectiveness of practices adopted and implemented upstream by other local and national authorities. The critical interdependence of upstream and downstream neighbors for managing environmental quality can be seen at all levels of the basin, and in the important linkage between the Danube river, its delta, and the environmental quality of the Black Sea. Cooperative action at the regional as well as local levels must therefore be one of the guiding principles of sustainable management of the Danube river basin.

Soils

Agriculture and soil conditions. Agricultural soils are presently heavily loaded by nutrients, metals and pesticides. Cadmium, lead, copper and zinc threaten the quality of top soil through atmospheric deposition and the use of phosphate fertilizers that often contain excessive amounts of cadmium. The limited available data generally show that the agricultural quality of top soil in the Basin is decreasing due to erosion and pollution. Eroded soil carried in the rivers adsorbs metals and phosphorous and is trapped in reservoirs.

Ecological Sensitive Areas

Unique Wetland Areas. Wetlands are among the most threatened ecosystems in the Danube region. Of all habitats, wetlands best reflect the health of the environment, as almost all natural and artificial substances pass through or finish up in them. For this reason their conservation should be given high priority, not only for their unique environmental value, but also for their important and valuable function in maintaining a healthy riverine environment.

Floodplains. Many wetlands, especially floodplains, are very fertile but in their natural state are too damp for cultivation of standard crops. In the Danube region this fertility has been exploited by drainage. Most of the Danube valley has been drained or poldered for agriculture despite widespread recognition that these policies are agriculturally short-sighted. Much of drainage works in the Danube River Basin has been ill-planned and poorly-executed.

Riverine Forests. The region's riverine forests have been largely felled for farmland. The few that remain are of outstanding importance and careful management is vital. Meadows and fishponds seem the most severely affected. Wet forests, river floodplains and deltas are also particularly badly affected, presumably because the first are subject to management and the latter two are ideally fertile for agricultural uses.

There are still nearly 10,000 sq. km of protected riverine sites including forests, oxbow lakes, marshes, and reed belts. The river continuum concept which is concerned with the biology of running waters implies that the self purification and filtering capacity of the main stem of the river and the general condition of riverine ecosystems is strongly affected by any interruption of the longitudinal and also the transversal connectivity.

In the Danube river system there are more than 500 dams and reservoirs with capacity over 5 million m^3, and many more smaller seasonal reservoirs. Most of the length of the main stem of the Danube River and the major tributaries are confined by flood control dikes. Natural alluvial flood plain areas have declined from about 26,000 sq. km to about 6000 sq. km. over the last fifty years. There are also significant reaches with river training works and river diversion structures. These structures trap nutrients and sediment in the reservoirs, cut the ecosystems including habitat both longitudinally and transversely, cause changes in downstream flow and sediment transport regimes, and decrease the efficiency of natural purification processes.

Black Sea

Surface run-off from agriculture together with domestic and industrial effluent is drained via the river network to the Delta area and subsequently contributes about 50 % of the total nutrient load into the Black Sea. These discharges have led to massive and damaging eutrophication of the northwest shelf of the Black Sea and parts of the Delta. About 44 % of nitrogen and 35 % of phosphorous is generated from agricultural land use while municipal sewage contributes 27 and 37 % respectively. Any action to reduce nutrient input into the Black sea must therefore reply on a basin wide approach. A similar spatial scale has to be envisaged when measures are taken to increase sediment input into the Delta region to stop the progressing erosional processes along the Black Sea shore line.

As there is little exchange of water between the Black Sea and the Mediterranean the former accumulates the input of non-degradable contaminants from the various rivers that flow into it, as well as nutrient loads. The Black Sea, like the Danube itself, is a regionally important resource, and since it must absorb all of the pollution load of the Danube which contributes about 50 % of the nutrient load and several other major river basins, its protection can only be assured through regional cooperation.

The immediate consequence of increased nutrient supply by the Danube, was the intensification of phytoplankton blooms, particularly on the shallow northwestern continental shelf where nutrients are readily recycled. Such blooms may have initially been beneficial to the marine food chain but the longer-term effects were devastating.

Until recently, the Black Sea was unprotected by any common policy or legal regime. Efforts to negotiate a legal Convention for the Rehabilitation, Protection and Preservation of the Black Sea began in 1985 and was finally signed in April 1993 (Dogterom, 1993). The countries also requested the UNDP to assist them to design an Action Plan for implementation of the Convention and the Global Environment Facility (GEF) for a three year program for Environmental Management and Protection of the Black Sea with technical and financial inputs essential for fostering sustainable development in the region.

The Environmental Program

Background

The riparian countries of the Danube River Basin have been participating in environmental management activities on a regional, national and local level for several decades (Buijs, 1991; Westing, 1989). In accordance with the recommendations of the Bucharest Declaration of

1985[1] the first steps to create an international water quality monitoring network for the Danube have already been taken. Building on this collaboration, the riparian states decided in February 1951 to elaborate a convention on protection and management of the river and an ecological agreement for the entire basin.

In the context of various programmes of assistance for Central and Eastern European Countries (see Luzerne Declaration, 1993), the idea of launching a series of immediate operational activities to support these legislative initiatives in the form of a Regional Environment Programme was put forward and supported on several occasions. Of particular note was the endorsement given for starting such a Regional Programme in the conclusions of the conference of European Environment Ministers ("Environment for Europe Conference") held at Dobris Castle near Prague in June 1991. In response to this demand, a Programme Coordination Meeting was held in Sofia in September 1991 with the sponsorship of the cooperating international organizations. Important agreements were achieved at the Sofia meeting on the scope and direction of a regional environment programme.

In 1986 the Danube countries signed the Bucharest declaration which had the objective to initiate programs to improve the water quality of the Danube and to establish a network for standardized water quality measurements. Within the first implementation (1988-1990) phase 11 gauging sites were installed between Germany and the Delta to monitor water quality. This first important step provided the basis for a unified evaluation of the main stream. Subsequently, this network is planned to be extended.

Recognizing that new initiatives were needed to address the growing regional water management and related environmental problems in the Danube River basin, the riparian countries together with interested members of the international community met in Sofia in September 1991. The countries agreed to develop and implement the Environmental Programme for the Danube River Basin (hereafter referred to as the Programme) of priority actions and studies in preparation for the eventual agreement on a new convention In addition, the countries agreed to form a Task Force to oversee this programme.

The international community agreed to assist the countries to develop a three year action plan that would constitute the first phase of international cooperation. The Commission for the European Communities (CEC) agreed to provide support and coordination for the Task Force in its role as G-24[2] Coordinator.

In parallel two conventions related to environmental problems in the basin were prepared from which one the draft convention and Cooperation for the Protection and Sustainable Use of the Danube River, referred to herein as the draft convention (1993), will be signed soon.

The Preamble of the draft Convention identifies *the main problem* as the occurrence and threat of adverse effects on the environment, economies, and well-being of the Danubian States caused by changes in the condition of watercourses within the Danube River basin. Among the main adverse conditions noted in the draft Convention are pollution of water resources

---

[1] Declaration concerning the cooperation of Danubian countries regarding water management of the Danube, in particular for the protection of the Danube watercourse against pollution, signed in Bucharest, Rumania on December 13, 1985.

intended for drinking water supplies, and the degradation of riverine ecosystems and water quality by eutrophication (the excess growth of aquatic plants, particularly algae, stimulated by an excess of nutrients - riverine ecosystem degradation is brought about by the cycle of growth, death and decay of these plants):

To deal with these problems, the draft Convention establishes the intention of the contracting parties to intensify water management cooperation, to achieve lasting improvement and protection of the Danube River and of the waters within its catchment area, particularly in a transboundary context; to achieve sustainable water management, and to contribute to the protection of the marine environment of the Black Sea.

Article 2 of the draft Convention establishes three specific goals to be achieved by the actions and arrangements agreed in the draft Convention by the contracting parties.

* *sustainable and equitable water management*, including conservation, improvement, and rational use of surface and groundwater;

* *control of hazards* originating from accidental spills of hazardous substances, and natural phenomena such as floods and ice;

* *reduce the pollution load to the black Sea*

Due to the political and economical changes in central and Eastern Europe (CEE) the resources available for environmental improvement will be severely constrained over the next 5-10 years. An environmental action programme for CEE was developed and signed in Luzern 1993 (Luzerne declaration, 1993) to provide a framework for setting national environmental priorities and to enhance cooperation between East and West.

These documents, the Bucharest Declaration, the Environmental Action Programme, and the Draft Convention are the most recent and important agreements among the Danube countries.

Goals and Principles

To achieve the long term objective of the sustainable use and development of the Danube basin natural resources the main goal of the Regional Environmental Programme is to establish an operational basis for strategic and integrated management of the Danube river basin environment while focusing initially on priority environmental issues. The strategy is based on several key principles:

- The approach should protect and enhance environmental values, and contribute to economic welfare and safeguard public health;

- The approach should be integrated. To achieve long term benefit, the Danube River Basin must be seen as an ecological system in which the inter-relationships between air, water, soil and biological resources within and outside the basin must be considered. A similar interdependence is seen in the complex pattern of economic activities linked together by the river's course.

- The approach should be participatory. The approach must reflect riparian countries' own priorities and each phase of the programme should emphasize consensus building, information sharing, and joint decision making among the riparian countries including local authorities.

- The approach should promote a mix of actions in the public and private sectors that increases efficiency, reduces the cost of environmental management and protection including capital investment, maintenance and operation, and generally enhances environmental quality and the effectiveness of environmental management.

- The approach should be coordinated and among the international funding sources will be essential because of their interdependence and the large number of projects foreseen throughout the basin. The limited available resources must be allocated where they can have the greatest benefit.

Phases of the Regional Environmental Programme

The objectives of the Regional Environmental Programme will be achieved in two main phases:

* Phase I is the Three Year Action Plan which is outlined below; and

* Phase II which would serve to implement the recommendations of the Strategic Action Plan and pre-investment studies prepared under the first phase and support further institutional strengthening and human resources development.

The three parts of the Phase I three-year action programme are parallel or concurrent. Part A focuses on short-term, high priority investments, while Part B develops a strategic action plan and phased investment programme and Part C puts in place the essential policy reforms needed to effectively implement the strategic plan.

Part A: Short Term Actions

Immediate actions will be taken under the programme to address environmental problems of acute concern and obvious high priority including areas of known health hazard or areas where environmental values and critical resources such as water supplies are threatened irreversibly. Initially two activities have been formulated for this part of the programme:

(a) <u>Stage I of Pre-Investment Activities - High Priority Actions.</u> The focus of the Danube Investment Programme will be on the highest priority areas of acute environmental concern. These may include municipal and industrial waste water, solid waste, and surface and ground water pollution from point and non-point sources. Areas and sources of pollution will be screened to identify high priority areas for project preparation during Stage II and to select a small number of pilot demonstration projects.

Pilot projects will include not only civil works but the industrial and policy reforms essential for effective project financing, implementation and operation. This limited number of pilot investment projects would be considered for funding by participating multilateral and bilateral financing institutions. In view of the time required for project preparation and appraisal, and the uncertainty over priorities, these early investment activities are likely to consist of the completion, rehabilitation, and modernization of existing facilities provided that they are not obsolete. These pilot project would also be closely monitored and reported on to a Task Force in order to ensure the wide dissemination of lessons learned.

(b) Accident Emergency Warning System. As noted earlier, the Danube is the focus of intense economic activity as well as the location of many important settlements. An effective accident warning, alarm, and response system could contribute greatly to minimizing the serious environmental harm caused by industrial accidents such as oil spills and discharges of toxic and hazardous wastes, radioactivity and risks related to hydrotechnical works.

The program will support the formulation of such a system working through a special expert sub-group under the Task force. Communications as well as preparedness are likely to play critical roles in such a system, and the programme will undertake an assessment of current capabilities and facilities.

Part B: Development of a Strategic Action Plan

The Regional Environmental Programme is intended to be a comprehensive multi-year programme. To achieve this objective a Strategic Action Plan for environmental protection and natural resources management will be prepared during Phase I. The Strategic Action Plan would provide a comprehensive and integrated guide with which the riparian countries will be able to jointly coordinate the activities of local and national authorities, international financial institutions, and regional and non-governmental organizations towards a set of common, strategic goals.

The Plan would provide an assessment of the nature, extent and causes of environmental degradation in the Danube River Basin and identify priority areas for intervention in policy and legislation, institutional strengthening, human resources development and investment activities. It would assess the linkages between various activities, systems, and issues and their impacts on ecology, human health and the economy of the region. The Plan would provide a phased multi-year programme of environmental actions to be undertaken and will include preliminary cost estimates for the implementation of these actions.

Part C: Institutional and Technical Building Blocks

This Programme element will identify and implement priority actions to edge of the environmental situation and to strengthen the capacity of authorities, organizations and individuals involved in environmental management in the Danube Basin.

(a) Inventories. The inventories involve the systematic collection of information and its compilation in a form that facilitates study and analysis. They are an essential and fundamental step in strategic planning. It is anticipated that comprehensive inventories will have to be built up in phases. In this initial phase, the inventories will focus on areas of priority environmental concern. Three comprehensive inventories have been identified for implementation in the Three-year Action Plan. These include an inventory of existing data, infrastructure, and sources of pollution, to be conducted in two phases; the accumulation of pollutants in soils, groundwater and sediments; and biological resources:

(b) Analytical tools for planning and management. Effective management requires the efficient collection and storage and retrieval of reliable information on the state of the environment and appropriate standards to assist in the formulation of management programmes. The interdependence of the riparian countries in the basin requires that these data and standards be comparable. The Programme will support the demonstration of new technologies such as geographical information systems (GIS) and strengthen capacities in monitoring and data management, leading to the development of predictive analytical tools, such as models, which are required for effective planning and managing of the environmental resources in the basin.

(c) Strengthening, Networks, Institutions, Human Resources and Nongovernmental Organizations. The Programme would support a broad series of activities to strengthen institutional and human resource capacities within the region. These activities would be developed on the basis of needs assessments which would be supported by consultations with key parties. The activities under the Programme will place special emphasis on the development and expansion of networks in a variety of areas to link key groups horizontally. Local nongovernmental organizations in the region will receive support under this element which would be interactively designed to address their diverse needs.

The Programme will support these activities in four areas of concentration:

* Development of Networks;
* Strengthening Institutions;
* Human Resources Development;
* Enhancing Nongovernmental Organizations; and
* Workshops and seminars focusing on key issues such as privatization and financing, exchange of experience, developing strategies, and technology assessment.

(d) Applied Research. As an integral part of supporting the Programme, a structured series of applied research activities will be conducted. The research findings will be used to help define the legal, policy and management framework required to improve the environmental situation of the Danube river, Danube delta and areas of the Black Sea influenced by the Danube. These research activities will make an important contribution in support of the development and eventual implementation of international legal frameworks for the Danube Basin and the Black Sea. The programme will be coordinated with the Black Sea environmental programme.

Programme Phase II: Environmental Investment

Phase II of the Programme would support a large scale, multi-year investment programme to address a wide range of environmental issues in the Danube River Basin. It would focus on supporting activities to implement the recommendations of the Strategic Action Plan. Phase II would provide funding for policy and legislative measures, large-scale activities in institutional strengthening and human resources development, and a series of investment projects on both a regional and country specific level. Financial support for these activities would be obtained from a wide range of domestic and international sources. It is anticipated that international funding for Phase II would be provided by a variety of multilateral, bilateral, private sector and nongovernmental organizations. Priority investment activities identified in Phase I may be initiated by participating governments and international funding organizations in advance of the formal beginning of Phase II.

Strategic Action Plan (SAP)

Background Material for SAP

The Strategic Action Plan will be the key element for the Programme. The Strategic Action Plan will be based on a series of regional national and local studies, fact finding inventories and assessments, and analysis of options and integrated strategies to deal with priority problems. The studies outlined below constitute an initial programme intended to provide a comprehensive picture of environmental problems in the Danube River Basin, and to develop an initial phased investment programme.

(a) National Reviews. The riparian countries prepared a national review of environmental issues and proposed measures and actions within their portion of the Danube River Basin. The national reviews include an initial inventory of pollution sources, an assessment of environmental issues and problem areas, review of available data and studies, and assessment of existing infrastructure.

The national reviews consider also the need for policy, legal, regulatory and institutional interventions as well as infrastructure investments. They provide a comprehensive and integrated evaluation of environmental issues. The national reviews examine issues such as surface and groundwater quality, and wetland and habitat conservation issues. The national reviews are vital to the process of developing the SAP because they represent the initial statement concerning issues and priorities by each riparian country which can be used to guide other activities in the programme.

(b) Stage II Pre-investment Activities for Priority Areas - Project Preparation and Related Studies. A phased investment programme that responds to immediate needs for actions in areas of high priority and acute environmental sensitivity and provides a longer term programme of investment for Phase II will be developed. The scope of project preparation activities will include, in addition to the appropriate technology, project financing for construction and maintenance, and the overall institutional and policy framework that would enable sustainable project operation. In particular, innovative private sector and public/private sector joint ventures will be considered to overcome the present severe constraints on project financing, and new institutional arrangements proposed which overcome past problems of O&M effectiveness and financing.

(c) Regional Studies. There are a range of other important problems that have a distinctly regional or river basin character, for example, non-point sources of pollution, air pollution, and in some cases groundwater management. The SAP should provide an integrated framework to address these distinctly regional issues as well as local problems. The basis for this integrated framework is being developed through a series of regional and possibly sub-basin studies. The regional studies include systematic studies, surveys, and assessments of existing conditions, causes of problems, linkages among economic and social activities and environmental and natural resource management objectives and issues, and evaluate the range of viable management options and strategies.

Drafting group

The key milestone of the first phase of this international programme is the completion and adoption of an *initial* SAP. The SAP is intended to provide a framework for implementation of the various provisions that will be incorporated in the new Convention on Cooperation for the Protection and Sustainable Use of the Danube River which has now been agreed upon and will be signed in the near future.
To prepare the draft SAP report, the Task Force established a Drafting Group consisting of representatives from four riparian countries: Romania, Bulgaria, Hungary, and Austria, the PCU, the World Bank, and the United Nations Development Programme (UNDP).

Results of the first phase of the consultation process will be used by the Drafting Group to prepare the draft SAP report which will be presented to the Task Force in late May. This will be followed by detailed national reviews of the draft and a second round of consultation on its contents before seeking formal acceptance of a final revised version of the SAP by the Governments of the riparian countries later in the year.

Consultation Process

The consultation process includes at least two opportunities for representatives of various interest groups to provide feedback to the Drafting Group on the contents of the SAP report

* first, a series of consultation meetings were held in each country in February and March, 1994. The one day meeting provided an overview of the draft SAP report as well as an opportunity for representatives from relevant ministries and municipalities, research institutions and laboratories, private sector enterprises, environmental NGOs, and environmental journalists to discuss and comment on the objectives, principles, problem areas, and priorities being proposed for inclusion in the SAP report. Results of these consultation meetings are used by the Drafting Group to revise and continue drafting the SAP report.

* second, after the draft SAP report is completed in May, a second phase of the consultative process will include follow-up meetings with the same groups in June.

Relationship of the SAP to the Draft Convention

The SAP is intended to address problems of water quality management, and the related problems of protecting and restoring environmentally sensitive and valuable areas in the riverine environment. The SAP provides a framework for implementation of the provisions of the draft Convention. Thus, the two should share common goals.

The SAP will include a specific programme of objectives, priorities, and actions to be implemented through the mechanisms established in the draft Convention which stress joint action by the riparian countries to address common and shared or transboundary problems.

The SAP and the draft Convention are *regional* in character, providing a framework of agreements, strategies, and actions that enable the riparian countries to cooperate to achieve common goals and solve transboundary water management problems

Specific Objectives of the SAP

The broad and comprehensive goals of the draft Convention have been translated into specific operational SAP objectives. These SAP objectives specify the key water management problems to be addressed and identify the related and specific conditions to be monitored and measured.

The Goal of Sustainable and Equitable Water Management in the Danube River Basin

OBJECTIVE A. - Maintain and improve surface and groundwater quality and its availability for drinking water.

OBJECTIVE B. - Ensure adequate water resources of appropriate quality to meet other uses including irrigation, fisheries and recreation.

OBJECTIVE C. - Protect and restore the condition of Danube riverine ecosystems, including wetlands, habitat, and forest plains.

The Goal of Restoring the Unique Ecosystem of the Danube Delta and the Marine Environment of the Black Sea

OBJECTIVE D. - Reduce the negative impact of activities on the Danube River basin on the Danube Delta and the Black Sea.

The Goal of Controlling Hazards Originating from Accidental Spills

OBJECTIVE E. - Develop and implement regional systems and measurement to prevent, monitor, and improve the response to accidental spills.

The Goal of Enhanced Regional Cooperation Involves three Interrelated Sub-objectives Incorporated Together in Objective F

OBJECTIVE F - To introduce a regional approach to Danube water management problems; to provide a framework within which the basin countries can harmonize national objectives, priorities and actions; and to establish a regional framework for action that is agreed to by all countries and the international community.

Formulation of Strategies

Principles and Strategies.

The formulation of strategies and the selection of actions and measures for the SAP would be guided in part by certain principles and lessons developed from recent experience in central and eastern Europe, including a focus on:

* Regional and integrated approaches

* Control of pollution at the source

* Seeking win-win solutions

* Effects of economic, social and industrial restructuring on the environment in the countries in transition

* Environmental liability for clean up and restoration

* Phased approach to long term changes and problems - immediate actions should be those with the largest benefits at least cost

* Mobilization of local resources and initiative

* The polluter-pays and precautionary principles

Criteria for Formulating Strategies

The general criteria for the selection of highest priority actions would include the following considerations:

* actions which have greatest impact on the priority objectives taking into account that benefits may accrue from more than one objectives

* actions which are cost-effective and represent a least-cost approach

* viability (institutional, economical and financial): actions and projects that are in an advanced phase of restructuring, with regard to their management structures, procedures and financial independence, will have a higher priority than projects that survive on the basis of state interventions

* suitability for a phased approach, projects that can immediately contribute to the solution of problems through a step-by-step approach will have higher priority than projects that need multi-year programmes before impact on the objectives can be observed

* availability of needed project preparation studies, projects for which feasibility studies are finished or in an advanced state (assuming they meet other criteria and represent feasible and affordable schemes) will have a higher priority than projects for which studies have yet to start

* actions and projects with pilot or demonstration value; projects with important pilot or demonstration value will have a higher priority

* risks; projects with a high *risk in respect of its* management, the introduction of new technology operational *complications*, or ability to operate and maintain on a sound financial footing will have a low priority

* urgency of measures in regard to irreversibility of damages or risk to human health

* actions which further international cooperation and primarily address transboundary issues as opposed to those that require only national response or have localized impact

* actions which have a favorable effect on other environmental resources including air, soil and forests

The Scope of Possible Actions and Measures

Outlined below are possible generic measures and actions grouped into six broad categories. The SAP will include specific actions from these possibilities to address the priority objectives and problems discussed in the previous section.

Policy, legal and regulatory measures
Within this group, actions will be identified which will support the development and implementation of international and national environmental policies, laws and regulations. These include - development of the legal and administrative infrastructure, introduction of harmonized ambient and emission water quality objective or standards, planning procedures and economic measures, such as the application of the polluter pays principle, realistic user charges, fines and fees.

Institutional strengthening and capacity building
This group of actions will include - enhancement of environmental management capacity at the national and local level, expansion of environmental planning and management measures, regional exchange of information, improvement of enforcement programmes, improvement of monitoring, data collection, analysis and quality control.

Investment in pollution control
Investment needs and specific projects will be identified with emphasis on major "hot-spots" of regional and sub-regional importance, on rehabilitation, modernization; and improved efficiency and operation of existing facilities, on clean-up and restoration of hazardous sites, and on phased introduction of new schemes and capacity. Project preparation, financing, and related institutional issues and constraints will be analyzed and steps identified to overcome present constraints.

Management programmes for the control of water resources and for environmentally valuable areas including wetlands and forested floodplains
Management practices for the quantitative and qualitative control of groundwater and surface water resources need to be optimized and integrated. This concerns management practices with transboundary as well as with purely national impact. Actions will be included in the Plan directed at optimization of the operation of hydraulic structures with transboundary impact. Actions to support national programmes to improve or develop management of wetlands, forest plains and floodplains will be included. Regional aspects of groundwater management will be considered.

Pulblic awareness and participation
Implementation of the Plan will benefit greatly from public support particularly for schemes to be funded locally and to apply the polluter pays principle effectively. In order to obtain broad public support, international and national programmes to increase public awareness will be

integrated into the Plan. Mechanisms to include public participation in decision-making have to be developed.

Applied Research
Applied research to improve policy-making" to reduce management uncertainty or operational weaknesses will be identified. These programmes will include underpinning strategic research as necessary.

Toward Implementing the Programme

A Task Force, comprised of representatives of the riparians, international organizations, nongovernmental organizations and other relevant parties, was established to oversee Programme development and implementation.

In order to maximize the results and effectiveness of the Programme, the riparian countries and the international donor community should make a strong commitment to ensuring optimum coordination between the activities of the Programme and other relevant national programmes. This should include ensuring the close linkage and synergy between, for instance, the strengthening of national monitoring and data management systems supported through the European Community's PHARE national Environmental Sector Programmes as well as other activities supported by international financing institutions.

The estimated total cost of Phase I ("Three Year Action Plan") 57,6 million US Dollars of which 49,2 million US dollars is required from international sources. Present commitments to fund the proposed budget total 40,8 million US Dollars.

The participation of representatives of cooperating riparian countries is critical to the success of the Programme. Governments are appointing a Country Programme Coordinator and an institution to serve as the national Focal Point for its participation in the Programme. The Country Programme Coordinator is responsible for the day to day operations of the Programme within the country. A major role of the Coordinator is to coordinate the activities of participating ministries, local governments, academic and research institutions, nongovernmental organizations and the public. The participation of local Governments in the Programme is especially important given their increased responsibilities for environmental management as the result of administrative decentralization which is occurring in many of the participating countries.

It would be anticipated that national and local governments would support the activities of the Programme by: (a) preparing national reviews on a timely basis; (b) providing financial and other support to Programme activities; (c) providing full access to information required for the preparation and implementation of Programme activities; and (d) would authorize, subject to adequate prior notification and formal clearance, site visits by technical experts to support preparation and implementation of the Programme.

The Programme specifically includes the involvement of nongovernmental organizations which have or are developing active programmes in the Danube River Basin. On the basis of their current activities, the following organizations have indicated their strong interest: (a) Foundation Cousteau, (b) Regional Environmental Centre for Central and Eastern Europe

("Budapest Centre"), (c) World Conservation Union (IUCN), and (d) World Wide Fund for Nature (WWF). The Programme will include support for institutional strengthening and selected activities of local nongovernmental organizations.

Financing

Financial Resources

Financial resources for investment in new infrastructure, and for rehabilitation and modernization of existing infrastructure, to preserve and protect the river basin environment are limited, even though the likely number of funding sources is quite large. Hence the riparian countries and the international financial institutions will face important choices in deciding how to allocate these limited resources where they can produce greatest overall regional benefit. As well as balancing local needs and priorities with bringing the maximum environmental benefit to the region, the chosen strategy should also fully reflect the results of the ongoing economic and industrial restructuring in the central and lower riparian countries. While restructuring may result in reduced pollution loads from heavy industry, energy and mining, significant increases in pollution loads from agriculture, agro-industry and urban solid waste may occur unless preventive action is taken.

Restructured economies with modern production technology, appropriate prices for inputs, and less intensive energy and raw material use are likely to emerge. This may reduce pollution from ongoing production but will of course not affect pollution which has accumulated from past activities. It is essential that scarce resources be directed to those pollution problems which are likely to remain priorities even after economic restructuring. SAP financing needs fall into three categories:

* first, funds for technical activities including studies; demonstration and pilot projects, training and institutional strengthening; data collection and monitoring, programme design and implementation; and project identification, preparation, and feasibility studies.

* second, funds for capital investment in facilities to control pollution and improve water management,

* third, funds for project implementation including training and institutional strengthening, operation, and maintenance

* Domestic Resources - Local financial resources should be used wherever possible to finance SAP activities rather than foreign borrowing in order to avoid pressure on the country's balance of payments. Given the severe strain on public sector expenditures during the current stage of economic restructuring, emphasis should be placed on meeting expenditure requirements from revenues generated at the project level (e.g. municipal water services) before seeking national or international sources of funding. These revenue sources would be based in part on the enforcement of the Polluter Pays Principle which is incorporated in the general environmental laws in each of the riparian countries.

The principal local sources of funds available to support activities under the Programme include: user charges, pollution fees and fines, budgetary allocations and non-budgetary incentives, domestic loans and, potentially, local private sector investment. Domestic loans may not be a major factor in the near term because local financial mechanisms such as capital markets and banks have not been developed to support environmental improvements and services in the municipal and industrial sectors. Hence the use of so-called environmental or ecological funds as a source of technical assistance and investment financing has been necessary. Mobilization of local private investment is constrained by historical barriers to private ownership, a limited local banking and financial sector, and inexperience of potential investors with the types of activities included in the Programme.

* External Resources - Due to the present constraints on the availability of local financing and the need for foreign exchange to fund elements of most priority activities, the use of external financial resources will be very important in funding the Programme in the short and medium-term. Given the competing demands for limited external financial resources, particularly during the short-term, it is critical that their use be well planned, properly coordinated, and that available funds be allocated to address priority needs from both a national and regional perspective.

Applied Research

While the SAP is oriented towards taking near-term actions on the basis of available information, the long-term cost-effectiveness of the SAP will depend on overcoming the weakness of current data and knowledge concerning the key processes that affect the Danube environment. It is anticipated that the SAP will be revised periodically to reflect the current economic and social conditions in the basin, progress and achievements under the current Plan, and new data and knowledge. While it is possible to identify a programme of well targeted applied research, financing of such research to improve environment and water management must be considered as a long term commitment. While private sources of environmental research funding are increasing, long-term budgetary support is needed and is probably most efficiently organized and funded from budgetary resources at the national level. The ecological funds which commonly receive discharge fees and pollution fines are also a possibly important source of applied research funding. There are also strong incentives for regional and local authorities to contribute funds to applied research programmes to ensure the strengthening of local technical resources and to ensure that specific and possibly unique local problems are addressed.

Summary and Conclusions

The Danube river basin includes to a larger extent the territories of twelve countries and collects additionally the runoff from small catchments located in four other countries. So, the Danube, although neither the longest nor the largest river in Europe, is the most international river of Europe.

In the last decades several bilateral agreements were signed between neighboring riparian countries and a few international agreements were signed too or are nearly finalized. The Environmental Programme for the Danube river basin together with the Strategic Action Plan have the objective to improve the environmental state, especially water quality, in the basin. The programme is designed to assist the draft convention which is already agreed among the riparian countries. The main environmental problems refer to surface water quality, riverine ecosystems, and the nutrient load into the Black Sea.

The Strategic Action Plan is being developed by a drafting group composed of experts from the World Bank, the Programme Coordination Unit in Brussels and four participating Danube countries. An intensive consultation process ensures that the viewpoints and objectives of the riparian countries are properly considered. The joint goals for environmental management are seen in sustainable and equitable water management, the preservation of unique habitats and wetlands with emphasis on the Danube Delta, the control of hazardous and toxic spills, and of enhanced regional cooperation. To achieve these goals a Task Force supervising the activities of the drafting group and the Environmental Programme has been established. A detailed action plan, including a list of hot spots, is under preparation to improve water quality.

The criteria for formulating strategies and for establishing a priority list for implementation measures are defined. Also nonstructural measures such as institutional strengthening and capacity building will be emphasized. These measures are seen as being important for countries which establish new legislation and administrative structures. The funding of the whole programme is planned to be covered by the Danube nations themselves with support form international sources for selected projects.

References

Black Sea Convention (1992): Convention of the Protection of the Black Sea against Pollution, Bucharest.

Buijs, P.H.L. (1991): Environmental Management Programme for the Danube River Basin. Background Document for the Technical Expert Meeting, September 1991, Sofia.

Buijs, P.H.L., Uzunov, K., Tzankov, K. (1992): Water quality of the Danube river along the Bulgarian-Romanian stretch (Juni 1991). ICWS-Report 92/01, Amsterdam.

Commission du Danube (1992): Annuaire Hydrologique 1990. Budapest.

Danube Declaration (1985): Declaration about the cooperation of the Danube Countries concerning water mangement of the Danube and the protection against pollution. Bukarest, Romania.

Dogterom, J. (1991): The Future of the Danube and the Black Sea. Unpublished Paper.

Draft Convention (1993): Draft Convention on the Cooperation for the Protection and Sustainable Use of the Danube River, Vienna, Austria.

Equipe Cousteau (1993): The Danube for Whom and for What ?. Final Report to the European Bank for Reconstruction and Development Agreement. Èquipe Cousteau, Paris, France.

IUCN (1992): The Wetlands of Central and Eastern Europe. IUCN report, Gland, Suisse.

Kreitner, P., Weber, E. (1989): Die chemisch-physikalische Wasserbeschaffenheit der DONAU. Wasser und Abwasser, Bd. 33, Wien.

Kresser, W. (1973): Die Donau und ihre Hydrologie. Wasser- und energiewirtschaft, Revue mensuelle suisse, Bd. 65 Sonderheft Donau; Schweizerischer Wasserwirtschaftsverband Baden

Luzerne Declaration (1993): Environmental Action Programme for Central and Eastern Europe. Document submitted to the Ministerial Conference. Luzern, Switzerland.

Weber, E. (1993): Die Wasserbeschaffenheit der Donau von Passau bis zur Mündung. Ergebnisse der Donauforschung, Bd. 2, IAD, Wien.

Westing, A.H. (1989): Environmental Security for the Danube Basin. Environmental Conservation 16, S. 323-329, Geneva.

V.2 Pollution Monitoring Problems of Large International Rivers: the Danube

Peter Literathy
Institute for Water Pollution Control, Water Resources Research Centre (VITUKI Plc.)
H-1095 Budapest, Kvassay J. u. 1
Hungary

Introduction

The Danube river (2,857 km) and its catchment area (817,000 km) provide water resources for a population of about 70 millions. Disintegration of Yugoslavia and dissolution of the Soviet Union resulted in an increase in the number of countries in the Danube basin. Recently, there are nine countries having access to the Danube river and six additional countries in the Danube catchment.

In addition to the domestic wastes, activities of the population produce industrial (mining and productions), transportation and agricultural wastes, which pollute the surface and ground waters in the area through wastewater discharges, surface run-off, atmospheric fallout and leaching from solid waste disposal sites.

Usually conflict exists between the neighboring countries due to the differences in the interest of the upstream and downstream countries concerning pollution. This conflict could be solved by internationally accepted water quality guidelines and pollution control standards established by an expert group in which all riparian countries are represented, and the interest of each country is taken into account. Enforcement of the standards should be based on reliable monitoring data.

Inorganic and organic pollutants in aquatic systems are liable to precipitate or adsorb onto particulates and accumulate in the suspended and bottom sediments. Therefore, monitoring of these pollutants should be conducted by sampling and analyzing the bottom sediment in addition to the water column containing the pollutants in both dissolved and solid phases. The problems and difficulties of the bottom sediment monitoring has been discussed in several publications (Literathy, 1975; Literathy et al., 1987 and 1994).

The Danube, one of the largest river basin in Europe, is the recipient of large amount of nutrients and toxic pollutants, delivered from point and non-point pollution sources, including municipal and industrial discharges, and agricultural areas (Benedek and Laszlo, 1980; Hock and Laszlo, 1990). Pollution transport, particularly nutrients and petroleum compounds, from the Danube catchment to the Black Sea responsible for the eutrophication in the Northwestern sea areas (Mee, 1992). In addition to the chronic pollution inputs from point and non-point sources, the water quality along the river is influenced by accidental pollution, man-made structures and natural processes. The quality of the Danube influences most of the water uses in the catchment, particularly the drinking water supplies which abstract water from bank-filtration wells or from the river itself (Laszlo et al., 1990).

Anthropogenic inputs have modified and continuously modify the natural geochemical composition of the Danube sediment. Among the different pollutants nutrients, heavy metals and organic micropollutants are of major concerns which are introduced into the Danube practically in each riparian country. Despite international efforts during the past thirty years establishment of concerted pollution monitoring along the Danube is still a major goal in the future.

Facts and Problems

The Danube river ecosystem is exposed to pollution caused by

(a) nutrients, such as inorganic, nitrogen and phosphorous compounds, as well as easily biodegradable organics and bacteria from major cities and agricultural activities, and

(b) toxic pollutants, such as heavy metals, organics, including petroleum products and pesticides from industrial, agricultural and transportation origin.

Depending on the concentration levels of these pollutants, and the consequences of the environmental processes altering the properties and partitioning of the pollutants, resulted in several problems in the Danube basin as follows:

(1) Limitation in the water uses, including
- restriction in recreation, particularly body contact recreation, due to bacterial contamination,
- tainting the meat of the fish, which cause damage in the fishery,
- affecting drinking water supplies using either surface, or bank filtered water by contaminating with micropollutants, etc., and
- affecting industrial water uses differently depending on the type of industry and water use, such as cooling water systems could be damaged by biofauling.

(2) Disturbance of the dynamic equilibrium by eutrophication occurred
- in reservoirs before dams,
- in the Danube Delta, and
- in the Black Sea (N.B. The highly eutrophic situation in the northwestern marine area is proved to be caused by nutrients transported from the Danube basin).

(3) Shortcomings of the pollution monitoring programmes and the water quality characterization are related to the non-existence of accepted water quality guidelines and differences in the sampling, analytical and data interpretation methods in the riparian countries.

Findings of Previous Pollution Monitoring Programmes

The national water quality monitoring programmes provided a vast amount of data on water quality, but several results were not comparable due to differences in sampling and analytical techniques, and lacking quality control. Monitoring of the toxic pollutants was random concerning the particular compounds, sample matrix, i.e., water, sediment or biota, sampling sites and position, and frequency.

In Hungary, the first comprehensive national survey on pollutants in bottom sediment was carried out in 1978-1979. The measured values of heavy metal concentrations indicated that the Danube sediments were moderately polluted, the concentration range slightly exceeded the limit values for soils, as it is given later in Table 1.

In 1983, the mercury, cadmium, lead, arsenic, chromium, copper, zinc, iron, manganese, oil and polyaromatic hydrocarbon content of the bottom sediments were determined in an area of the River Danube where bank-filtration wells existed (Laszlo, 1984). The order of magnitude of the lowest measured concentrations was that of the background values. The mean contents did not exceed the allowed heavy metal concentrations in soils, however, the highest measured concentrations exceeded the limits of all components except copper. In 1986-1987 sediment pollution of different particle size fractions was studied along in the Hungarian Danube. The average concentrations of contaminants in the sediment fractions are listed in Table 1. For all analyzed components the highest mean contamination levels were found in the smallest particle-size fraction, i.e., clay and silt.

Table 1 Average concentration of contaminants in Danube sediment fractions (mg/kg).

Contaminants	Grain-	size	fraction
	d< 90 m	90 m >d< 250 m	250 m >d< 1000 m
Mercury	1.0	0.15	0.16
Cadmium	2.78	1.33	0.94
Lead	105	42.7	51.1
Chromium	50.9	15.1	15.4
Arsenic	9.6	2.1	3.2
Copper	228	34.3	30.5
Nickel	58	25.9	21.1
Iron	20600	8630	7830
Manganese	782	291	232
Zinc	381	77.0	85.9
Total organics	49700	20200	27400
Oil + grease	1050	428	477

Bucharest Declaration monitoring

The Bucharest Declaration was signed in 1985 and included the implementation of a water quality monitoring programme. Since 1988, water samples are being collected monthly at the border sections along the river, altogether at 11 sites, and analyzed for selected variables including heavy metals and a few organics.

Special attention is paid to the oil pollution because petroleum products are the most frequent pollutants along the Danube. Results of the oil pollution monitoring are summarized in Fig. 1 for the water together with the results obtained on the bottom sediment during the Cousteau Danube survey. This figure clearly demonstrate that monitoring water column alone will not be sufficient to identify pollutants, particularly those which are likely to associate with the particulate matter.

Lessons to be learned from the Bucharest Declaration monitoring include the need for: (1) extension of the sampling programme for monitoring heavy metals and selected organic micropollutants in bottom sediment and biota with less frequency than the water, (2) revision of the sampling sites, positions, (3) revision of the sampling and analytical methods, (4) incorporation of biomonitoring, at least for macrozoobenthon, and (5) enforcing training programmes and quality control measure; continuing the interlaboratory comparison (intercalibration) exercises in an extended manner.

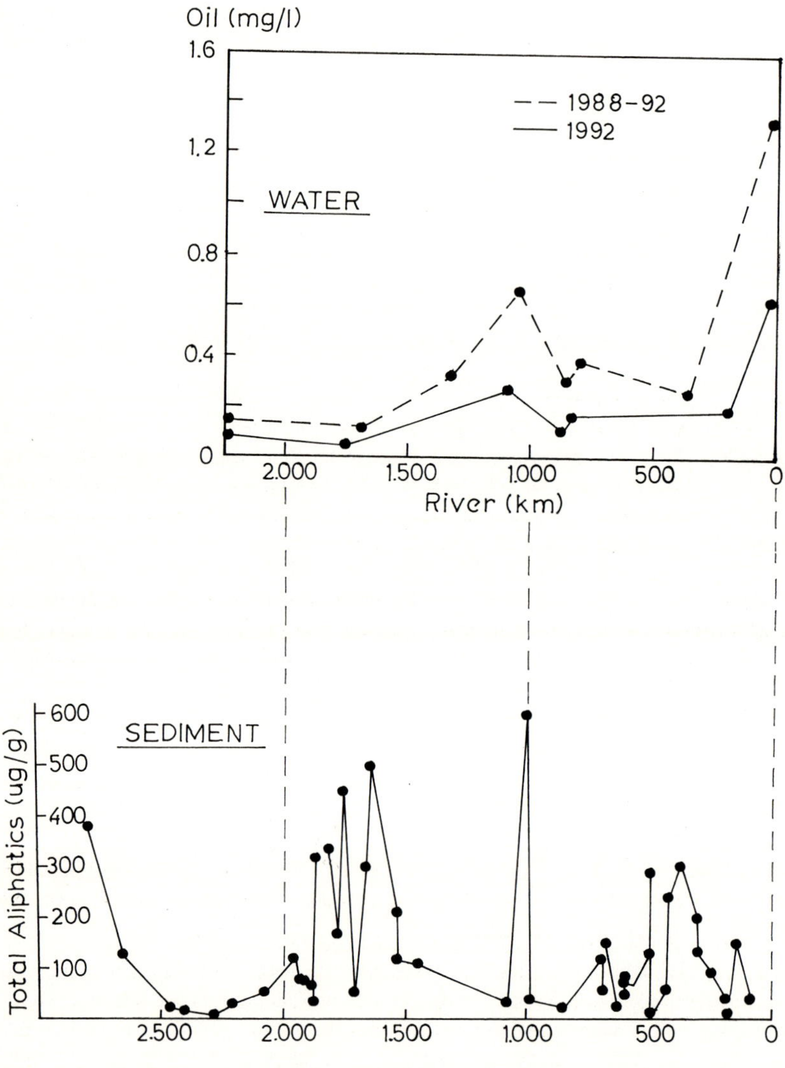

Figure 1 Variation in oil concentration in water and bottom sediment (Cousteau survey).

Cousteau Equipe's Danube survey

The latest comprehensive sediment pollution survey was carried out along the Danube from the source of the river to the Danube-Delta by a Cousteau Equipe project in 1991-1992 (Cousteau Equipe, 1993). Fifty sampling sites were selected which involved country border sections, hot spot areas associated with cities, big industrial complexes, dams or nuclear power plants and confluences in main tributaries of the Danube. Samples were collected by the Cousteau team, including one hydrobiologist from VITUKI, Hungary, and the samples were analyzed in the laboratories of the IAEA-MEL in Monaco and the VITUKI in Budapest.

The analyzed contaminants included:

(1) heavy metals,

(2) petroleum hydrocarbons, PAHs,

(3) chlorinated hydrocarbons and organophosphorus pesticides,

(4) PCBs, and

(5) fecal sterols.

Fig. 2 shows the concentration levels of selected pollutants in the sediment along the Danube.

Pollutant levels were generally low in Germany and Austria except fresh oil input characterized with the n-alkanes in Germany. The Slovakian-Hungarian Danube reach (1900-1450 river km) was found to be very polluted with fresh and weathered oil (phytane) as well, and the highest concentration of mercury was also observed here, downstream of the Vah river. The Irongate reservoir at 990 river km is considered as sink for most of the pollutants, however, high concentration of DDT and mercury was found along the Rumanian-Bulgarian Danube reach.

Very similar concentration patterns were observed for most of the pollutants including chlorinated hydrocarbons and fecal sterols. Contamination levels in the mussel *Anadonta anatina*, most abundant specimen along the Danube, correlated with those measured in the sediment in most cases. Using the sediment guidelines in Table 4., Fig. 3 shows the pollutant "hot-spots" along the Danube.

Survey in the Rackeve-Soroksar Danube-branch (RSDb)

The RSDb, a 57.6 km long branch of the Danube is located downstream of Budapest, controlled by sluices at both ends. The flow is maintained at 15 to 35 m^3/sec and the flow velocity is highly reduced compared to the Danube main stream.

The decrease in the suspended solid concentrations along the RSDb, such as 39.4, 21.3 and 12.6 mg/l at the 57.4, 38.0 and 0.5 river km, respectively, demonstrates significant sedimentation which occurs in the upper part of the RSDb. The major source of the bottom sediment in the RSDb is the suspended solids of the Danube.

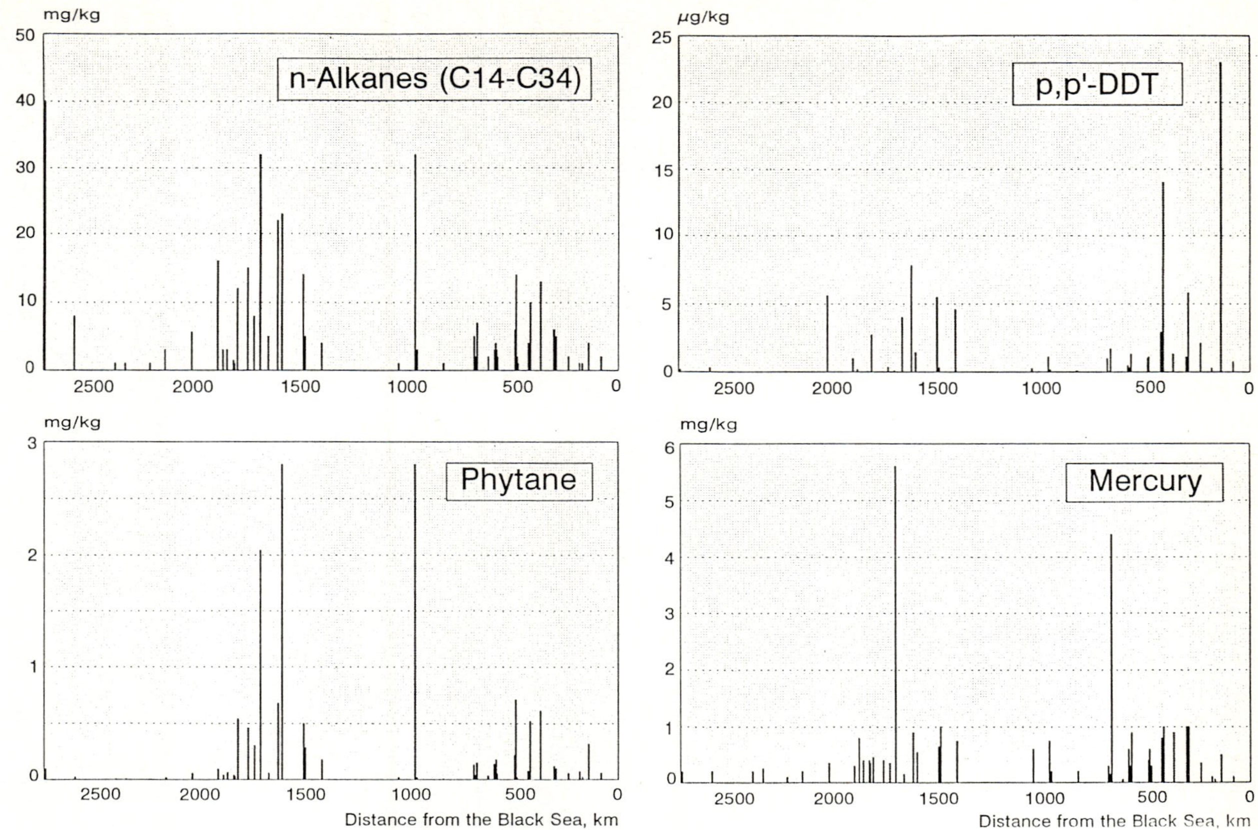

Figure 2 Petroleum hydrocarbons (n-alkanes and phytane), mercury and DDT levels in the Danube sediment (from Cousteau Equipe, 1993).

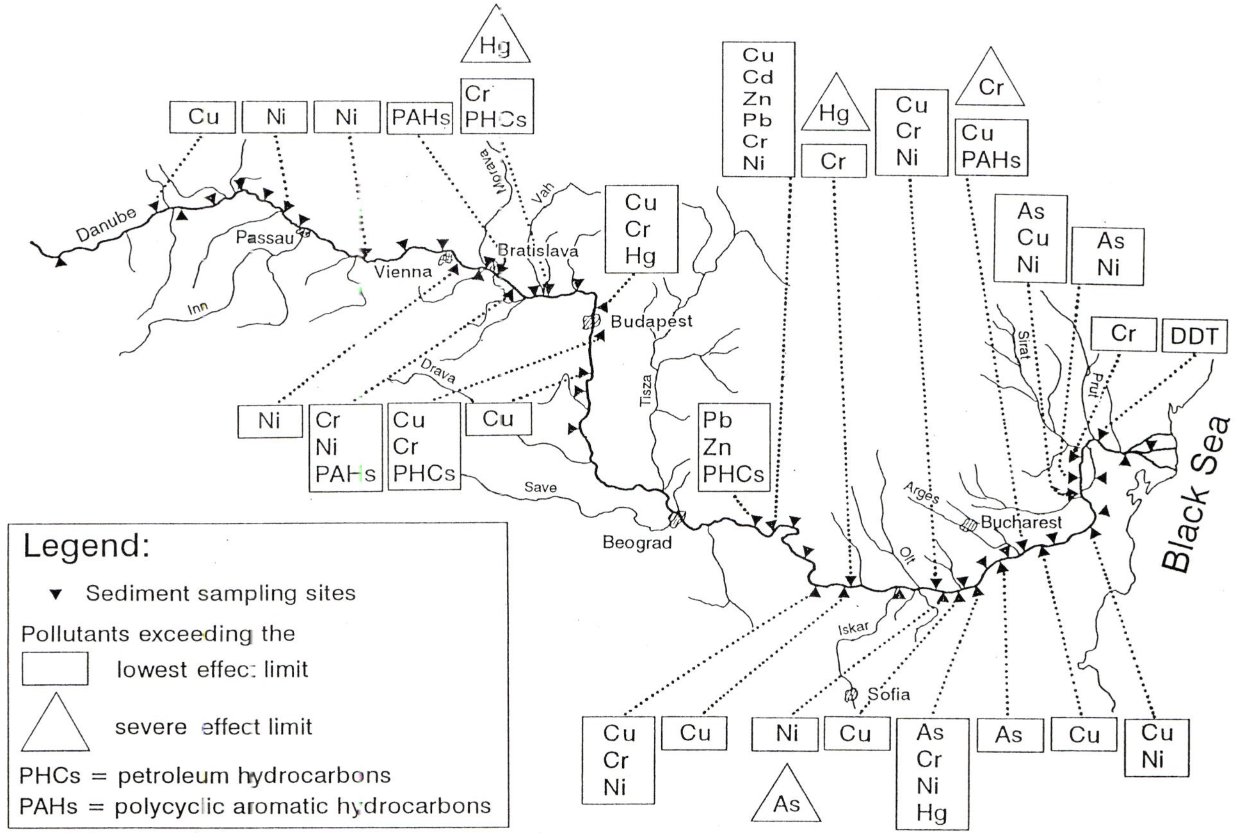

Figure 3 Sampling sites and sediment-bound pollutant "hot-spots" along the River Danube (from Cousteau Equipe, 1993).

Because suspended solids transport majority of the different micropollutants, therefore sediment cores taken from the RSDb can characterize historical pollution trends in the Danube. Table 2 shows some of the sediment characteristics in the upper reach of the RSDb.

Table 2 Concentration of contaminants in the sediment core taken at the 56.0 river km.

Sediment	CONTAMINANTS						
Layer	TOC	Cs-137	Hg	Cd	Pb	Cu	Cr
cm	%	Bq/kg ± 2	mg/kg	mg/kg	mg/kg	mg/kg	mg/kg
0- 5	0.6	38 ± 11	0.98	1.10	64	41.9	59
5-10	0.9	< 19	1.00	1.49	92	61.3	71
10-15	1.1	98 ± 9	1.12	1.29	86	55.0	57
15-20	1.0	252 ± 12	1.60	2.09	126	73.5	98
20-25	0.7	56 ± 9	1.04	2.20	112	70.8	119
25-30	1.1	32 ± 8	1.61	3.70	130	105	148
30-35	0.8	58 ± 12	1.54	3.30	125	89.9	122
35-40	1.1	75 ± 24	1.39	3.50	152	90.8	140
40-45	1.1	125 ± 14	1.60	2.90	140	87.5	131
45-50	0.8	105 ± 13	1.47	2.50	125	80.2	115
50-55	1.4	51 ± 10	1.35	2.59	128	80.5	126
55-60	0.9	< 21	1.23	2.19	124	83.3	237
60-65	0.6	< 27	0.78	1.80	104	68.2	109

The metal concentrations are similar to the finest fraction in Table 1, showing characteristics of the suspended solids. This is supported by the fact that the core samples contained more than 80 percent clay and silt.

It is interesting to note that the maximum measured Cs-137 level in the 15-20 cm layer could reflect the Chernobyl-disaster in 1986.

Survey on the macrozoobenthon community along the Danube

Kick samples (Furse et al. 1984) of the Danubian macrozoobenthon community were collected during the sediment and mussel collecting mission of the Cousteau Equipe in 1991 November-December in order to evaluate the practical environmental situation using the community structure data. These organisms consists of taxa belonging to a wide range of taxonomic groups representing different sensitivities to pollutants. The natural composition of the benthic community is therefore strongly influenced by the actual pollution of the water column and the sediment. Of course, physical conditions are also important such as the river flow, the temperature, the quality of substrate etc. For this reason, sampling sites must be chosen carefully in order to select comparable habitats and avoid those stresses to the community which are not related to pollution.

In order to use benthic survey for water quality assessment the modified Biological Monitoring Working Party (BMWP) Score system was applied where the different families present in the given sample are classified according to their relative sensitivities to pollutants and assigned a score from ten (most sensitive) to one (least sensitive). Summarized scores and calculated average scores per taxon were used to describe the biological water quality, as it is illustrated in Table 3, from the samples collected between Linz and Mohács Danube reach where the water flood conditions were suitable for benthos sampling.

TABLE 3 Biological water quality of the Danube based on macrozoobenthon data collected in 1991.

River Km	Number of scored families	Total score per Taxon	Average Score (ASPT)	Quality Class[1]
2010	9	39	4.3	III.
1950	9	34	4	III.
1940	17	72	4.2	III.
1920	3	9	3	V.
1900	10	51	5.1	II.
1880	4	14	3.5	V.
1857	3	9	3	V.
1807	14	59	4.2	III.
1765	5	15	3	IV.
1763	2	3	1.5	V.
1707	8	38	4.8	III.
1659	13	67	5.2	II.
1638	6	24	4	IV.
1533	12	48	4	III.
1526	12	49	4.1	III.
1446	11	43	3.9	III.

[1] Quality Classes: I = Excellent II = Good III = Moderate IV = Poor V = Very poor

In addition to the pilot scale application of the modified BMWP Scoring System the survey has also many new data of the zoogeographical distribution of the benthic macrofauna. The taxon list includes "non-indigenous" species (*Potamopyrgus antipodarum, Sinanodonta woodiana*) and new taxa that were unknown from the investigated sections before (i.e. *Theodoxus fluviatilis*).

Water Quality Management

Objectives

The main objectives of the water quality management for the Danube basin are as follows:

* To ensure balanced aquatic ecosystem by

- maintaining the dynamic equilibrium between production and decomposition of organic matter, thus avoiding eutrophication and development of anaerobic conditions,
- maintaining biological species diversity.

* To satisfy water quality requirements of water uses, like
  - fishery,
  - recreation,
  - irrigation, and
  - industrial (cooling water, etc.) uses,

which requirements could be satisfied with a generally good quality water without any treatments, however,

  - drinking water supplies, even from bankfiltered water, and
  - some industrial water uses

would require different levels of treatment, at least disinfection, before the safe use.

Characteristics of the aquatic ecosystems

It is important to emphasize that different polluting compounds behave differently in the aquatic environment, where they can be distributed between the abiotic and biotic compartments, or between the dissolved and solid phases within the abiotic compartment, as it is demonstrated in Fig. 4.

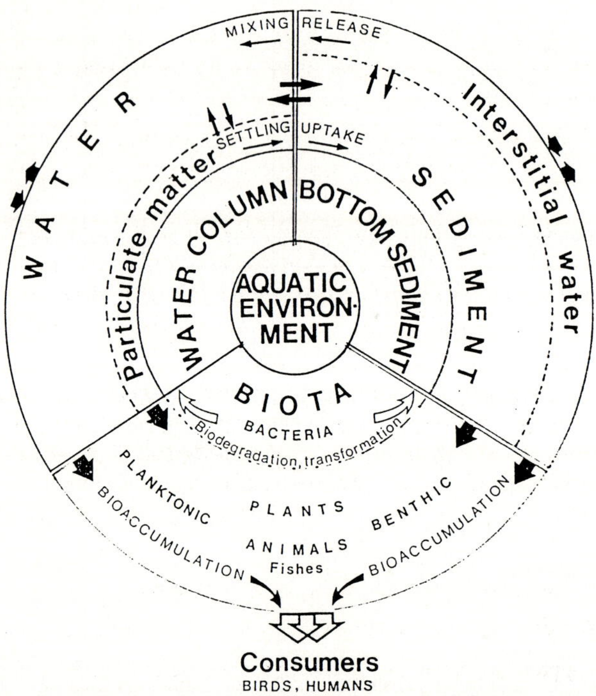

Figure 4 Compartments and their interactions in the aquatic ecosystem.

Depending on their abundance in the different matrices, the sample collection should be extended to those matrices where the pollutant concentration levels are expected to be significant. In the case of the toxic bioresistent, persistent compounds, which are also likely to result in bioaccumulation, sediment is considered as an important compartment where these pollutants reside.

Characterization of the quality/pollution of the aquatic ecosystem requires water quality criteria, standards, which could be outlined in the water quality guidelines developed for the given river system, and this is one of the future objectives of the regional Danube programme for the Danube river basin.

Assessment of the pollution in the sediment requires sediment quality objectives. Table 4. summarizes limit values for selected heavy metals and organics used in different countries or river basin, which have been considered during previous measurements in Danube sediments and could be considered when the sediment quality guidelines will be developed for the Danube.

Approaches

The effectiveness of the water quality management depends on the availability of reliable information on:

(1) the pollutants discharged,

(2) environmental weathering (degradation, transformation, partitioning) of the pollutants in the affected aquatic ecosystem, and

(3) transportation, migration and pathways of the pollutants within the aquatic ecosystem, and between the surface water and the groundwater (bankfiltered water) aquifer.

The harmonization of the water quality monitoring of the Danube river was delayed in the past due to the different political systems in the riparian countries. The water quality monitoring network operated in these countries mainly in an isolated manner, exceptions were limited to some border sections monitored by the neighboring countries in the framework of bilateral agreements.

Data reach - information poor, was the general result of such monitoring programmes. This fact called for the harmonization of the water quality monitoring on international level, which is one of the main objective of the Regional Environmental Programme for the Danube Basin, coordinated by the Programme coordination Unit at the European Community.

Concerning methodologies used in the water quality monitoring the following areas should be considered:

(a) selection of pollution characteristics, parameters to be monitored,

(b) selection of matrices, e.g., water, sediment and/or biota, to be sampled,

(c) selection of the appropriate sampling, sample treatment and analytical methods,

(d) quality control measures.

Table 4 Quality objectives, guidelines for contaminants in sediment.

Contaminants	Hungarian Soil Guidelines	Rhine River Suspended Sediment	Dutch Target Values	Sediment Quality Guidelines in Canada (Persaud et al., 1992)		
				Lowest Effect Level	Severe Effect Level	
	(µg/g)	(µg/g)	(µg/g)	(µg/g)	(µg/g)	(µg/g TOC)
Arsenic	15	-	29	6	33	-
Cadmium	5	1	0.8	0.6	10	-
Chromium	100	100	100	26	110	-
Copper	100	50	36	16	110	-
Iron (%)	-	-	-	2	4	-
Lead	100	100	85	31	250	-
Manganese	-	-	-	460	1100	-
Mercury	2	0.5	0.3	0.2	2	-
Nickel	100	50	35	16	75	-
Zinc	300	50	140	120	820	-
TOC (%)	-	-	-	1	10	-
TKN	-	-	-	55	4800	-
TP	-	-	-	600	2000	-
Aldrin	-	-	-	0.002	-	8
BHC	-	-	-	0.003	-	12
Chlordane	-	-	-	0.007	-	6
Lindane	-	-	0.05	-	-	-
DDT (total)	-	-	-	0.007	-	12
Dieldrin	-	-	0.5	0.002	-	91
Endrin	-	-	-	0.003	-	130
HCB	-	-	-	0.02	-	24
Mirex	-	-	-	0.007	-	130
Atrazine	-	-	0.05	-	-	-
Organophosphs.	-	-	0.02-0.07	-	-	-
Chlorophenols	-	-	1.0-2.5	-	-	-
PCBs (total)	-	-	-	0.07	-	530
PAHs (total)	-	-	-	2		11000
PAH individual	-	-	15-50	-	-	-

Elements of Pollution Monitoring

The reliable data sets on the water quality/pollution of the river could be obtained from the implementation of a properly designed ambient monitoring programme. The main components of such programme are discussed in the following.

Parameters to be monitored

Selection of the pollutants as target compounds for the pollution monitoring should be done on the bases of (1) pollutant inventories, (2) results of preliminary surveys to identify potential polluting compounds, and (3) by taking into account the water quality guidelines, criteria of the aquatic life and intended water uses, or priority pollutants as it is listed in Table 5. It is also important to consider that the physical state and the chemical structure of the pollutants could be altered in the environment (Fig. 5) which affects their behavior, and therefore, the pollutants can be categorized as follows: (a) primary polluting compounds are those emitted and discharged into the environment where they could be (1) degraded into harmless end products, such as water, carbon dioxide, etc., (2) transformed into resistant compounds, or (3) persistent during the physical, chemical and biological weathering processes; (b) secondary polluting compounds are those breakdown, conversion products, which are produced during the environmental weathering of the primary polluting compounds; they: (1) are usually polar, water-soluble compounds which can easily migrate through bank-filtration, (2) might be more toxic than their parents, (3) might be at least temporarily resistant to further degradation, (4) can affect the fate of other pollutants, and (c) tertiary polluting compounds are those primary and secondary pollutants which are accumulated in the ecosystem together with naturally occurring compounds, usually buried in the bottom sediment in inactive, non-bioavailable form but become active as a result of the presence of primary and secondary polluting compounds or extreme change in the environmental conditions. Tertiary polluting compounds in the sediment in reservoirs can behave as pollutant "time-bombs".

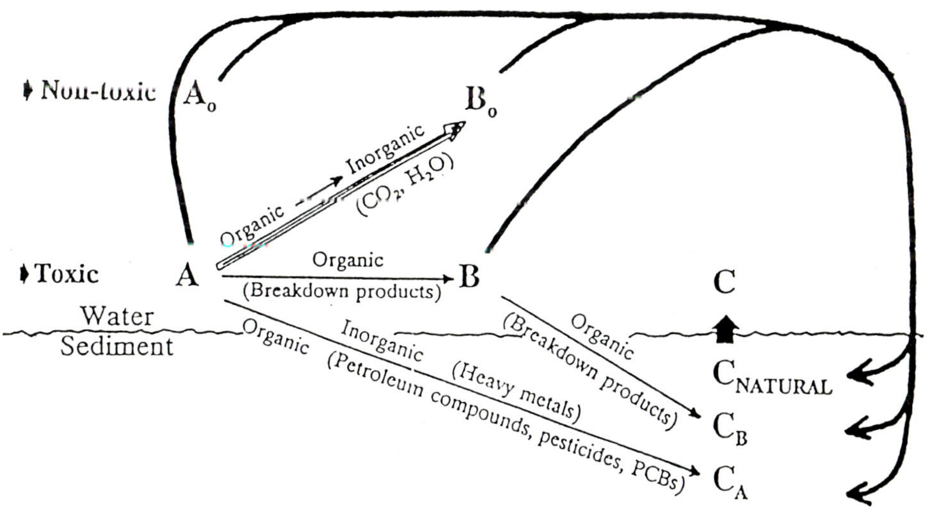

Figure 5 The behavior, degradation, transformation and pathways of the different pollutants in the abiotic aquatic environment.

Table 5 List of hazardous substances (Federal Register, Vol. 52, No. 74, 17 April 1987)

Priority Group 1	
CAS No.	Substance name
50328	Benzo(a)pyrene
53703	Dibenzo(a),h)anthracene
56553	Benzo(a)anthracene
57125	Cyanide
60571	Dieldrin/aldrin
67663	Chloroform
71432	Benzene
75014	Vinyl chloride
75092	Methylene chloride
76448	Heptachlor/heptachlor epoxide
79016	Trichloroethene
86306	N-nitrosodiphenylamine
106467	1,4-Dichlorobenzene
117817	Bis(2-ethylhexyl)phthalate
127184	Tetrachloroethene
205992	Benzo(b)fluoranthene
218019	Chrysene
1745016	P-Dioxin
7439921	Lead
7440020	Nickel
7440382	Arsenic
7440417	Beryllium
7440439	Cadmium
7440473	Chromium
11196825	PCB-1260, 54, 48, 42, 32, 21, 1016

Priority Group 3	
CAS N°	Substance name
71556	1,1,1-Trichloroethane
74873	Chloromethane
75218	Oxirane
75252	Bromoform
75343	1,1-Dichloroethane
84742	Di-N-butyl phthalate
88062	2,4,6-Trichlorophenol
91203	Naphthalene
98953	Nitrobenzene
100414	Ethylbenzene
107028	Acrolein
107131	Acrylonitrile
108907	Chlorobenzene
118741	Hexachlorobenzene
122667	1,2-Diphenylhydrazine
124481	Chlorodibromomethane
156606	1,2-Trans-dichloroethene
193395	Indeno(1,2,3-cd)pyrene
606202	2,6,Dinitrotoluene
1330207	Total xylenes
7221934	Endrin aldehyde/endrin
7440224	Silver
7440508	Copper
7664417	Ammonia
8001352	Toxaphene

Priority Group 2	
CAS No.	Substance name
56235	Carbon tetrachloride
57749	Chlordane
62759	N-nitrosodimethylamine
72559	4,4'-DDE, DDT, DDD
75003	Chloroethane
75274	Bromodichloromethane
75354	1,1-Dichloroethene
78591	Isophorone
78875	1,2-Dichloropropane
79005	1, 1,2-Trichloroethane
79435	1,1, 2,2-Tetrachloroethane
87865	Pentachlorophenol
91941	3,3'-Dichlorobenzidine
92875	Benzidine
107062	1,2-Dichloroethane
108883	Toluene
108952	Phenol
111444	Bis(2-chloroethyl)ether
121142	2,4-Dinitrotoluene
319846	BHC-alpha, gamma, beta, delta
542881	Bis(chloromethyl)ether
621647	N-nitrosodi-n-propylamine
7439976	Mercury
7440666	Zinc
7782492	Selenium

Priority Group 4	
CAS No.	Substance name
51285	2,4-Dinitrophenol
59507	P-Chloro-m-cresol
62533	Aniline
65850	Benzoic acid
67721	Hexachloroethane
74839	Bromomethane
75150	Carbondisulfide
75694	Fluorotrichloromethane
75718	Dichlorodifluoromethane
78933	2-Butanone
84662	Diethyl phthalate
85018	Phenanthrene
87683	Hexachlorobutadiene
95487	Phenol,2-methyl
95501	1,2-Dichlorobenzene
105679	2,4-Dimethylphenol
108101	2-Pentanone, 4-Methyl
120821	1,2,4-Trichlorobenzene
120832	2,4-Dichlorophenol
123911	1,4-Dioxane
131113	Dimethyl phthalate
206440	Fluoranthene
534521	4,6-Dinitro-2-methylphenol
541731	1,3-Dichlorobenzene
7440280	Thallium

Parameters usually determined during water quality/pollution monitoring programmes, and which will be considered as parameters in the water quality objectives, guidelines to be developed for the Danube basin, include:

*** Conventional water quality parameters**
such as temperature, pH, conductivity, alkalinity, acidity, TDS (total dissolved solids), SS (suspended solids), dissolved oxygen, Na, K, Ca, Mg, Cl, SO_4, HCO_3, BOD_5, COD, nutrients: NH_4, NO_2, NO_3, organic N, PO_4 and total-P.

*** Heavy metals**
such as Hg, Cd, Pb, Cu, Cr, Ni, Zn, As, Se, etc.

*** Organic micropollutants**
such as petroleum compounds (oil), PAHs, phenols, detergents, pesticides (chlorinated hydrocarbons, organophosphorus compounds, carbonates, triazines, etc.), PCBs, phthalates, fecal sterols, etc.).

*** Radioactivity indicators**
such as total beta and total alpha, Sr-90, Cs-137, etc.

*** Microbiological indicators**
such as *Fecal Coli*, *E. Coli*, pathogens (*Salmonella*, etc.), viruses.

*** Biological indicators**
such as algae, zooplankton, benthic organisms (mussels), fishes.

*** Water quantity parameters**
such as discharge, flow velocity.

In addition to the conventional water quality characteristics, the pollution monitoring programmes for international rivers should include all those heavy metals and trace organic compounds, as target compounds, which are proved to be or likely to be characteristic pollutants along the river or its particular reach.

In the case of the Danube nutrients, both nitrogen and phosphorous, petroleum compounds, some of the chlorinated hydrocarbons, different other pesticides, etc., are on the list of parameters to be considered. It is expected that breakdown products (usually polar compounds), as they are described as secondary pollutants, will be included in the list of monitoring parameters in the future as soon as validated methodologies will be developed.

Matrices to be sampled

Most of the national and international monitoring programmes aims to measure the quality and pollution in the water column, and majority of the results revealed that the concentration levels of heavy metals and some of the selected organic micropollutants, e.g., oils, lindane, are relative low and do not reflect the real quality of the aquatic ecosystem. Although the dissolved forms of pollutants in the water are directly (a) involved in biological processes (N.B. bioavailable forms), and (b) affecting most of the water uses, particularly the drinking water supplies, the major part of the toxic pollutants, e.g., heavy metals, hydrophobic organic micropollutants, are associated with the particulate matter, which plays important role in the pollution assessment. As a result of biological and chemical reactions forced by natural processes, toxic substances and nutrients could be mobilized, released from the sediment resulting in eutrophication, or detrimental effect on the aquatic life and during water uses.

Therefore, it is important to include all compartments of the aquatic environment as it is shown in Fig. 4., and the sampling should include matrices as follows:

* **Water column**
  - abiotic: dissolved materials,
    suspended materials, and
  - biotic: bacteria,
    phyto- and zoo-plankton,
    fishes.

* **Bottom sediment**
  - abiotic: grain-size distribution,
    specific grain-size fraction for pollutant analysis, and
  - biotic: macrozoobenthos,
    periphytone,
    mussels ("mussel-watch").

Uptake and release of inorganic and organic micropollutants in the bottom sediment of surface waters were subject of several studies (Literathy, 1975; Literathy and Laszlo, 1977; Literathy et al., 1987). It was concluded during the 1983 survey (Laszlo, 1984), that the suspended solid is enriched in micropollutants compared to the bottom sediment (without fractionation) as it is demonstrated by the following metal concentration ratios:

	Hg	Cd	Pb	Cu	Zn	Fe	Mn
Suspended sediment / Bottom sediment	1.9	19.6	4.3	9.0	4.7	0.75	5.7

Due to its larger surface area fine sediment is more liable to adsorb hydrophobic organic compounds, whereas the coarse sediment causes mainly dilution effect.

Comparability of the analysis of the sediment associated pollutants in the suspended sediment and in the bottom sediment can be achieved by analyzing the same grain-size fraction in both type of sediment. Fig. 6. shows the probability of the presence of the different particle sizes in the suspended and bottom sediments. The suspended solids are separated from the "dissolved" fraction in the water column by filtration through a 0.45 m pore size filter allowing the colloids to pass through into the solution. Usually due to dependence of the sedimentation rates on the particulate size it is not likely that suspended solids will contain particles larger than 200 m but definitely will include the clay particles, which on the other hand are not likely to settle to the bottom.

There is an international agreement to analyze the less than 63 m bottom sediment fraction for pollution monitoring purposes. Despite of the selection of this fraction for pollution assessment, suspended sediment may show higher concentration of the pollutants in the sediment due to the presence of the clay fraction which may contain even higher concentration than the silt. Therefore it might be topics of future discussions to limit pollutant monitoring in the silt fraction (between 4 and 63 m grain-size fraction) in both suspended and bottom sediment.

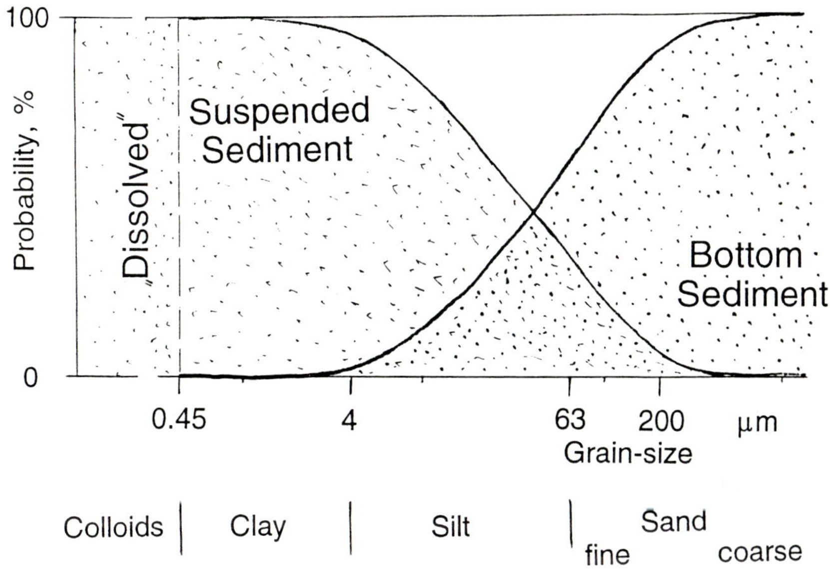

Figure 6 Probability of the abundance of difference grain-size particles in the suspended and bottom sediment.

Methodologies to be used

The selection of sampling and analytical procedures in the ambient water quality monitoring could follow two approaches depending on the aim of the monitoring programme, such as (a) regular monitoring for establishing pollution levels and trends by manual sampling, or (b) early warning monitoring by automatic field measurements and sampling. In the case of manual sampling the collection of samples is flexible and may involve a large number of sites and positions without extreme cost increase, on the other hand, establishment of automatic water quality monitoring stations is very costly, changing site/position is difficult and not feasible, and there are also limitations in the detection, analytical techniques. Because of these reasons, very careful cost-benefit analysis is needed prior to the decision on the construction of the automatic monitoring station.

As far as the analytical methods are concerned, there are internationally accepted procedures for most of the conventional water quality parameters, heavy metals, radioactive characteristics, microbiological and hydrobiological indicators, and some of the organic micropollutants, e.g., phenols, some pesticides. In most of the cases it is also important to agree on the sample pretreatment methods, particularly when we want to differentiate between dissolved and particulate matters, i.e., filtration is needed, or speciation of heavy metals in the sediment, i.e., fractionation by leaching test, is required.

Conflicts may be developed among laboratories (actors) representing the riparian countries in the international pollution monitoring programmes when the monitored parameter is a complex mixture of polluting substances, such as in the case of oil pollution, and its concentration is determined on the bases of the characteristics of a specific group of the complex mixture. One actor give preference to method A over Method B usually for the following reasons:

(a) expectation of favorable, or reliable analytical result biased by the measured signal proportional to the characteristic of a preferred group of compounds in a mixture of the polluting compounds, e.g., BOD versus COD characterizing the mixture of easily biodegradable and both biodegradable and bioresistent organic compounds, respectively, or the analysis of oil pollution with infrared versus fluorescence spectrometry, measuring the aliphatic or the aromatic petroleum hydrocarbons in the petroleum, respectively,

(b) availability of the instrumentation, e.g., one laboratory has instrument for method A but not for method B, and

(c) agreement on the used of common arbitrary calibration standard material.

Such conflict must be solved among riparian country laboratories by using agreed analytical protocol, having (providing) equivalent instrumentation, and training personnel. There are alternative approaches to solve such problems which includes:

(1) technical discussions which may conclude ranking of alternatives and selection of the most likely acceptable alternative,

(2) selection of the scientifically best procedure which may require significant financial input providing not available instrumentation to some participating laboratories, and

(3) intermediate solutions compromising with the shortcoming of available instrumentation and upgrading the instrumentation step-by-step as money will be available.

Following the first approach, a major step was taken in the Danube basin and the Black Sea when a sampling and analytical workshop was organized for characterizing oil pollution (11-15 April 1994, VITUKI, Budapest, Hungary, sponsored by the PHARE and the Black Sea programmes). Analysts from 11 Danube basin countries and five Black Sea countries attended the workshop which provided a platform to discuss related analytical problems. The workshop's output is being used to formulate agreed analytical method(s) for oil pollution monitoring and assessment in the Danube basin and in the Black Sea area.

Quality assurance/control

To assure the validity, reliability and comparability of the pollution monitoring data, quality control measures should be enforced during sampling, analysis and data processing. The design of the monitoring programme and the data interpretation and dissemination should be done in an internationally coordinated way.

Obviously the validated data set easiest to obtain from a single laboratory, or team, which collect samples at all sampling locations and then carry out analyses in the same laboratory. This approach, however, is not feasible in long-term and particularly due to the national interests of the riparian countries in an international river basin. Therefore, reliability and comparability of the monitoring results should be ensured by using agreed sampling and analytical methods and enforcing intralaboratory and interlaboratory (intercalibration) quality control measures as part of the international monitoring.

The major elements of the quality assurance programme should include:

- sampling and analytical protocols,
- uniform, or equivalent instrumentation,
- skilled personnel (training),
- intralaboratory quality control including regular analysis of (a) calibration standards,
  (b) Standard Reference Materials (e.g., NBS),
  (c) spiked samples.
- interlaboratory quality control, performance testing with analysis of check-samples in intercalibration exercises (on-going in the Bucharest Declaration monitoring since 1993 for the Danube basin).
- quality control in data processing, including expression of the analytical results and rounding, and interpretation.

Selection of the sampling sites

Sampling sites, positions should be selected to allow the characterization of the quality of the water body longitudinally and in cross-section by taking samples at different points of the cross-section from different depths. This is particularly important at those sites where the waste discharges, or tributaries are not fully mixed with the river.

Sedimentation and resuspension in the river controlled by the hydrodynamic characteristics of the river reaches, therefore deposition of the polluting compounds in the sediment may occur far downstream of the discharge. This must be taken into account during sediment-bound pollution monitoring and data interpretation. One of the major problem is to locate representative sampling sites. This always requires a preliminary survey to identify the sites where the bottom sediment contains significant clay-silt fraction. Such survey has been completed along the Hungarian Danube reach by collecting sediment from 415 transect, at each river km, on both sides of the river.

The results are shown in Fig. 7. Bottom sediment must contain at least 10 percent fine fraction and it can be concluded that the higher the fine fraction in the sample the more representative the sediment-bound pollutant characterization. This minimum requirement is satisfied at 36 sites on the right-bank and 30 sites on the left-bank of out of the 415 transects along the Hungarian Danube reach.

Pollution monitoring sampling sites should be established at

* border sections,
* upstream and downstream of major cities, tributaries,

* downstream of major industrial discharges, "hot-spots",
* water intakes, and
* ecologically vulnerable areas.

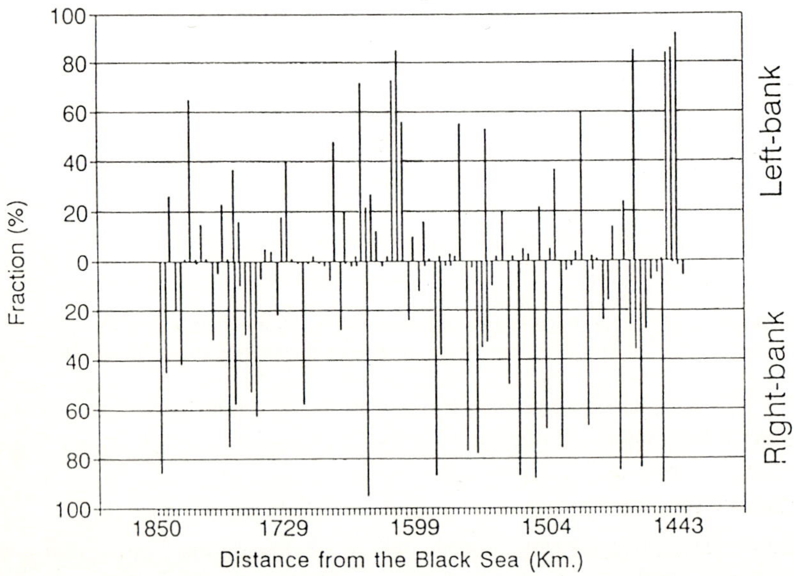

Figure 7 Abundance of the less than 63 m (clay-silt) grain-size fraction in the bottom sediment along the Hungarian Danube reach.

Sampling positions, horizontally and vertically, should be selected depending on the inhomogeneity of the cross-section caused by the incomplete mixing of waste discharges and tributaries. For such reason, samples are regularly collected from three positions, i.e., near the two banks and the main stream, in the border section monitoring between Slovakia and Hungary.

Sampling Frequency

Depending on the objectives of the pollution monitoring programme samples are collected in the case of

(a) manual sampling
(1) in the water column with daily, weekly, fortnightly, or monthly frequency, whereas (2) in the bottom sediment quarterly, or twice a year collection could be sufficient for biomonitoring, or pollutants monitoring, respectively;

(b) automatic sampling it could be continuous, or hourly, depending on in-situ sensors, on-line instruments, or samplers collecting discrete samples.

Conclusions

On the basis of the information available on the present status of the quality and pollution of the water resources in the Danube basin, priorities should be given to the

(a) protection of the aquatic ecosystems such as the

- Danube river and its tributaries,
- reservoirs before the dams,
- wetlands and national reserves,
- Danube Delta, and the
- Black Sea by maintaining dynamic equilibrium between productivity and decomposition, and biodiversity (species diversity);

(b) sustainability of the water utilization including

- drinking water supplies,
- fishery (same as protection of the aquatic ecosystem),
- recreation,
- irrigation, and
- industrial water supplies.

Water quality objectives, guidelines should be developed and achieved in the river basin by reduction of inorganic nutrient inputs, biodegradable organic loads, discharge of toxic heavy metals and organic micropollutants, as well as microbial contamination. Achievement of these objectives will require information as follows:

(a) pollutants inventory for nutrients and toxic pollutants from point-, and non-point (diffuse) sources, tributaries;

(b) ambient monitoring of the water and sediment by

- regular sampling for pollution levels and trends, and
- automatic stations for early warning;

(c) water quality characterization, environmental and public health risk assessment by taking into account

- water quality objectives, standards,
- nutrient, pollutant loads, and
- the fate and effects of pollutants in the aquatic environment;

(d) predictions by mathematical models for mixing, transport and fate processes; and

(e) quality assurance and quality control.

As far as the water quality/pollution monitoring elements are concerned, conclusions could be summarized as follows:

Inorganic and organic pollutants (nutrients and toxic pollutants) are liable to accumulate in suspended and bottom sediments. The major transport of these pollutants takes place in the suspended load. Sedimentation and resuspension are in dynamic equilibrium along the river influenced mainly by physical processes. Sediment-bound nutrients and pollutants undergo biological and chemical degradation and transformation, as well as bioaccumulation. After selection of the target compounds, characteristic for the studied river system, concerted efforts are needed to implement the pollution monitoring programme in the water column and in the sediment.

International and national water quality monitoring results revealed the nutrient and pollutant transport along the Danube. Surveys on the Danube sediment concluded that the Danube section between Vienna and Budapest is one of the most polluted one in addition to the Irongate reservoir as well as the downstream Danube reach down to the Danube Delta. The main toxic polluting compounds are heavy metals, e.g., mercury, chromium, arsenic, petroleum hydrocarbons, including PAHs, and chlorinated hydrocarbon pesticides. Reservoirs are major traps of sediment-bound pollutants and where pollutant "time-bombs" may be developed. Mobilization of the sediment-bound pollutants will result in biological uptake by the benthic organisms, as well as in a possible increase in the dissolved fraction in the water column. The macrozoobenthon monitoring provides valuable evidence for chronic biological stress associated with contaminated sites.

Pollutant monitoring in the water column requires the selection of representative sampling positions in the selected cross-sections due to the likely inhomogeneity, in the bottom sediment, however, the representative sampling sites should be establish where the sediment contains at least 10 percent clay and silt which is usually enriched in pollutants.

Harmonization between chemical measurements and biological observations leads to an effective Danube pollution monitoring system. This should include:

(1) selection of representative sampling sites,

(2) appropriate treatment of water, biota and sediment samples before analysis, e.g., grain-size fractionation,

(3) selection of inorganic and organic target contaminants, and biological species for monitoring,

(4) agreement on the analytical procedures and quality control measures, including regular intercalibration exercises, and

(5) appropriate data interpretation and dissemination. Achieving this objective could eventually lead to an effective pollution monitoring system for the Danube catchment, or for similar international river basin.

Acknowledgments

Thanks go to the Cousteau Equipe team and to the IAEA's Marine Environmental Laboratory in Monaco, particularly to Laurence D. Mee, for the fruitful collaboration during the implementation of the Danube sediment-biota collection and analytical programme, and to Ferenc Laszlo and Bela Csanyi for their contribution to the Danube studies.

References

Benedek, P. and F. Laszlo. 1980. A large international river: the Danube. *Progress in Water Technology*, **13**:61-76.

Bogardi, J. 1978. *Sediment transport in alluvial streams*. Akademiai Kiado, Budapest.

Equipe Cousteau. 1993. Concentration of chemical pollutants in sediments and mussels along the Danube. In: *The Danube... for whom and for what?* Equipe Cousteau's Final Report, Paris, 104-126.

Furse M.T., J.F. Wright, P.D. Armitage and D. Moss. (1981) An appraisal of pond-net samples for biological monitoring of aquatic macroinvertebrates. *Water Research*, **15**:679-689.

Hock, B. and F. Laszlo. 1990. Contribution to the water quality evaluations of the Danube. In: *Proceeding of the International Conference on Water Pollution Control in the Basin of the River Danube*, Novi Sad, Yugoslavia, 17-24.

Laszlo, F. 1984. Potential release of pollutants in riverbank filtration systems along the River Danube, Hungary. In: *Proceedings of the 3rd International Symposium on Interactions Between Sediments and Water*, CEP Consultants Ltd., Edinburgh, 264-267.

Laszlo, F., Zs. Homonnay and M. Zimonyi. 1990. Impacts of river training on the quality of bank-filtered waters. *Water Science and Technology*, **22**:167-172.

Literathy, P. 1975. Study of river pollution caused by micropollutants. *Water Research*, **9**:1001-1003.

Literathy, P. and F. Laszlo. 1977. Uptake and release of heavy metals in the bottom silt of recipients. In: *Interaction Between Sediments and Fresh Water*, H.L. Golterman, [Ed.], W.B. Junk Publishers, The Hague, 403-409.

Literathy, P., F. Laszlo and B. Csanyi. 1994. Approaches for sediment associated pollutant monitoring in the river Danube. *Water Science and Technology*, (in press)

Literathy, P., L. Nasser Ali, M.A. Zarba and M.A. Ali. 1987. The role and problems of monitoring bottom sediment for pollution assessment in the coastal marine environment. *Water Science and Technology*, **19**:781-792.

Mee, L.D. 1992. The Black Sea in crisis: a need for concerted international action. *Ambio*, **21**:278-286

Persaud, D., R. Jaagumagi and A. Hayton. 1992. *Guidelines for the Protection and Management of Aquatic Sediment quality in Ontario*. Log 92-2309-067, PIBS 1962, Queen's Printer for Ontario, Canada.

V.3 Water Quality Monitoring, Modeling and Protection Projects in the Czech Republic

Ilja Bernardová, Kvtoslav Mrázekx*, Michal Fiala
Water Research Institute Prague, Brno branch
Devaská 12, 657 57 Brno
*since 1992 IKM Meluzinova 21, 615 00 Brno
Czech Republic

Introduction

The position of the Czech Republic in the center of Europe predetermines this country for intensive cooperation with its neighboring countries, not only in politics and economics, but also in environmental protection. From a hydrological point of view, the Czech Republic (CR) is part of three large European river watersheds. The Elbe River conveys waters from the largest western portion of the country through Germany towards the North Sea. The Morava River, forming downstream the border between the Slovak Republic and CR and Austria, respectively, drains more than one fourth of the country's water into the Danube River. The smallest part of the country, in northeastern CR, belongs to the watershed of the Odra River, which flows through Poland towards the Baltic Sea. The main characteristics concerning these river basins are given in Table 1.

Table 1 The main characteristics for the basic river basins in CR.

Area in CR	Inhabitants	Long-term flows	River Basin	Territory		
	thousand km^2	Mio			Pop/km^2	
Elbe River	48.5	6.3	130	308	221	63
Morava River	20.7	2.7	130	108	74	18.8
Odra River	5.2	52	250	43.3	25.7	6.02

During the last thirty years, Czech streams have been significantly loaded with various pollutants caused by advanced industrial production using old technologies, dense settlement, increasing agricultural pollution, atmospheric deposition and unsuitable landfills and by insufficient wastewater treatment. The most polluted parts of rivers appear downstream from large towns and industrial enterprises that do not have their own wastewater treatment plants or that treat their wastewater insufficiently. Increasing concentrations of discharged micropollutants pose a risk to the further use of water and threaten the disturbance and deterioration of aquatic systems.

The opportunity of reversing the present negative trend was initiated by political and economical changes at the end of 1989. The change in water protection strategy is based on effective international cooperation started at the beginning of 1990, when negotiations began leading to the signing of the Agreement on the International Commission for the Protection of the Elbe River between the Czech and Slovak Federal Republics, Germany and the European Communities.

Water Quality Monitoring

The beginning of regular water quality control in a fixed network of sampling points was initiated in the period of significant water quality changes at the beginning of the 1960´s. After the experimental period of 1959-1961, nation-wide monitoring was initiated in 1963. Monitoring was focused on the significant streams that form about one fourth of all watercourses in CR and that serve as recipients for a major portion of discharged water and as drinking water sources. The aim of the above mentioned monitoring was to produce a basic overview of water quality and to gather data for long-term evaluation of water quality changes. The data also serve for policy making at central as well as regional levels. The former number of 180 sampling points regularly monitored at one month intervals has been gradually extended up to the present number of 289 points. Since organic pollution in CR is still considered to be the dominant pollution, the monitoring was focused on basic oxygen indicators, physio-chemical, biological and radioactivity indicators. The first systematic heavy metals screening was not introduced until the late 1980´s. Since the beginning of the 1990´s, heavy metals together with a limited number of chlorinated organic compounds, oil products and surfactants have been included in the list of monitored determinants. As the number of sampling points and parameters is rather high and requires a lot of money, a certain reduction of this sampling net is under discussion. The present state of organic pollution of Czech streams is shown in Fig. 1.

Besides systematic national monitoring, there is also a special transboundary rivers monitoring, which is focused on water quality control according to individual bilateral agreements with neighboring countries (Poland, Germany, Austria). The monitoring network covers more than 20 sites where the basic physico-chemical and biological indicators and a limited number of heavy metals are monitored.

Principles and Development of Saprobiological Monitoring in CR and Slovak Republic

The significance of water quality monitoring by means of biological methods has grown in importance with increasing loading of water and water ecosystems in the last decades. It has been demonstrated by the acceptance of biological monitoring in water quality evaluation in EC countries (de Pauw et al., 1992). The major objective at the beginning of the biological monitoring around 1975 was to complete chemical monitoring and water quality evaluation in streams with more detailed hydrobiological information.

The macroozoobenthic community was chosen as a basis of the biomonitoring in accordance with with other biomonitoring systems of flowing surface water used in other countries (see, for example, Sweeting,1992).

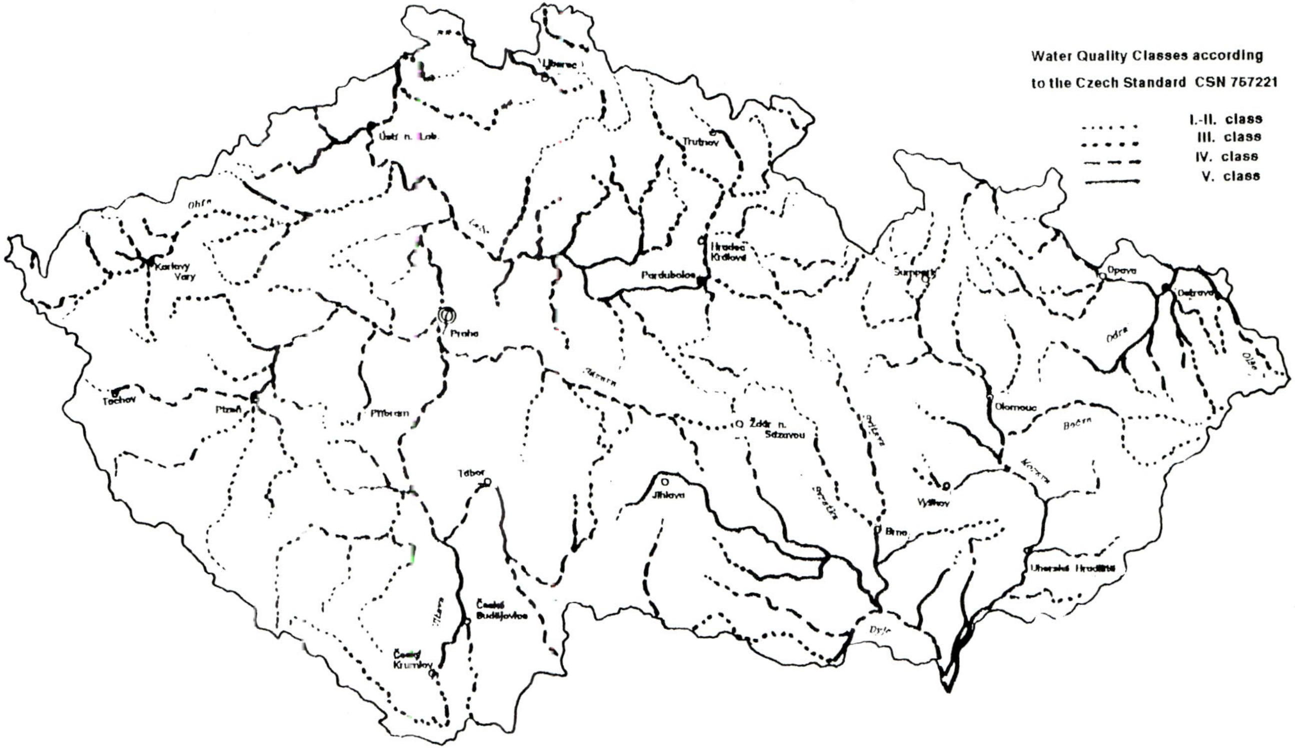

Figure 1 Water quality in streams in the Czech Republic: organic pollution during 1991-92.

This system is supported with bacterial periphyton data only in heavily polluted stretches with minimal appearance of benthic invertebrates. The evaluation is based on a precise determination to the species level (Sláde ek et al., 1981). Besides many advantages in the use of macrozoobenthos for water quality evaluation, significantly higher differentiation ability in comparison with the evaluation based on the periphyton has been demonstrated.

The basis of the mentioned biomonitoring system rests on a selected network of sampling points situated upstream and downstream from the pollution sources or nodes of pollution, supplemented by another point downstream in order to verify the self-purification process. The results of the aforesaid monitoring enables the detailed information on the pollution of more than four hundred streams evaluated in about 1600 sampling sites covering the stream length of app. 8500 km.

Water Quality Modeling

Effective regional water management requires not only the documentation of the present state but also use of a forecast model. In the biological model, the obtained saprobic index values represent one of the basic data types entered into a deterministic model of pollution changes in river longitudinal profiles. Due to a limited amount of measured data and to high costs on quantification of classical methods, the models of this type are also used in other countries. This model can be classified according to the Beck terminology (Beck, 1985) as a speculative simulative model. A typical feature of such models are special limitations forming a kind of a corridor with obstacles for a model to surmount.

The essential importance for the model function is given by the correlation between saprobic index and BOD/COD, or some other water quality indicators, as documented in a number of studies (Mrázek, 1984; Zelinka-Kubíek, 1985). In order to derive corresponding correlative relationship right in the modeled system, samples for the determination of the above-mentioned indicators were taken, and a statistically high significant correlation between saprobic index and BOD/COD values was established. The obtained correlation coefficients were between 0.7-0.9. The correlation of these indicators has been evaluated in Slovakia, where results from more than 1200 sites were used in the assessment (Kuníková,1991).

Water quality modeling is further based on theoretical assumptions of spatial changes in the relationship between water quality and flow rates. The general regularities verified in the course of the modeling process were applied when determining limiting factors. The obtained biomodel represents an effective tool of water quality regime evaluation. It enables the documentation of spatial changes of organic pollution by means of the saprobic index, water quality classes and probability of non-exceedance of BOD5, COD-Mn and NH4 and besides that simulation of the influence of changes in the amount of discharged pollutants and flow rates against model parameter changes.

Another approach is environmental reconstruction-quantification of environmental variables by weighted averaging regression (species-environment) and the use of these relationships in weighted averaging calibration to reconstruct values of environmental variables (like dissolved oxygen, nutrient loads etc.) for sites not covered by chemical monitoring. We use WACALIB v. 3.3 (Line, Break, and Birks, unpublished) for total nitrogen (N-NO3 + N-NO2 + N-NH4) load estimates (Fig. 2, concentrations are expressed as 90-percentile) in the longitudinal profile of the Morava River.

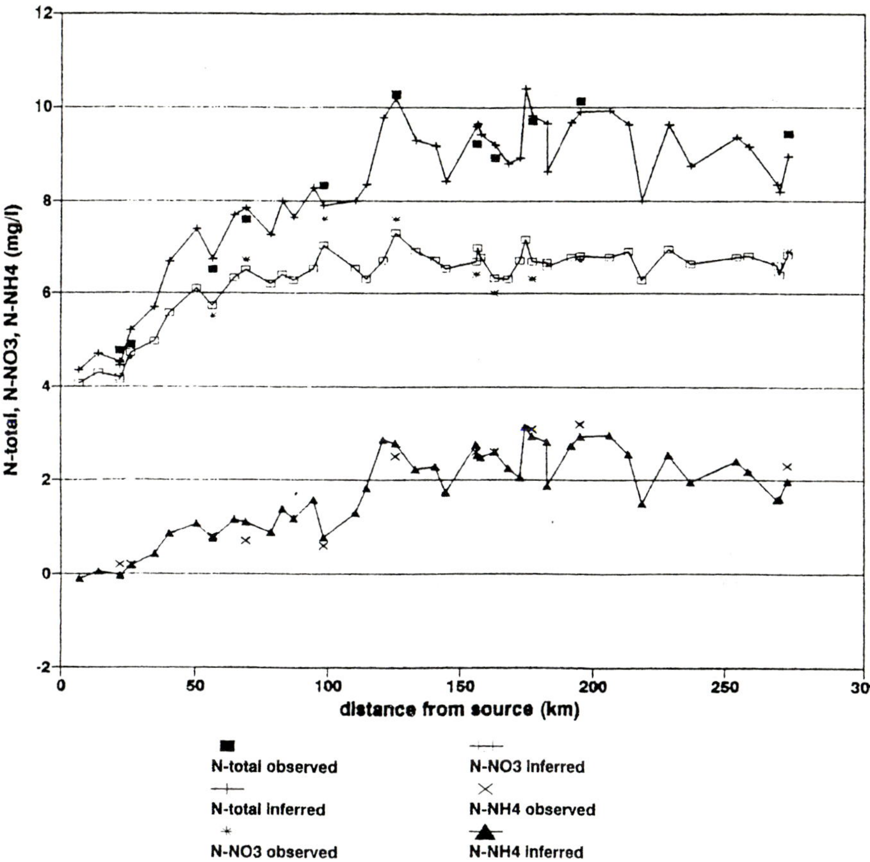

Figure 2 Nitrogen loading of the Morava river. Benthic macroinvertebrates values.

The idea of weighted averaging calibration is ecologically clear and simple: a species whose nutrient optimum is close to the nutrient concentration of the site will be most frequent at that site. A reasonable estimate of the species optimum is thus the average nutrient concentration of the sites where the taxon is present, weighted by its abundance (Birks et al.,1990).

Water Quality Protection Strategy in CR

The idea of weighted averaging calibration is ecologically clear and simple: a species whose nutrient optimum is close to the nutrient concentration of the site will be most frequent at that site. A reasonable estimate of the species optimum is thus the average nutrient concentration of the sites where the taxon is present, weighted by its abundance (Birks et al.,1990).

The evaluation of monitoring results and modeling of water quality represents one of the preparatory stages in the introduction of the water quality protection concept. In general, it is possible to use two principles, imission and emission, in order to fulfill goals in water quality protection. The method used in CR until 1992 was based solely on the imission principle. Relatively strict given limits for stream water quality were difficult to meet with respect to the low level of wastewater treatment and therefore, the situation was solved by the issuing of exceptions. There were hundreds of these exceptions, including the most important pollution sources in all regions.

The legislation was improved in 1992 when Governmental Regulation No.171/92 was introduced. According to this legislation, the emission values are prescribed. Besides obligatory emission standards, emission limits are also considered. Water Authorities are also allowed, with respect to local needs for protection of the stream, to use the imission limits stated in the Governmental Regulation as target values. Emission limits are given for two time horizons - to 2004 and from 2005 (the latter are more strict). Hopefully, this synthesis of two approaches will improve the situation for meeting the goals of water quality management in the country.

National Large Watershed Protection Projects in CR and Their Linkage to the International Water Protection Agreement

The new state policy of water quantity and quality protection in the river basins is supported by the results of three basic water protection projects. The most important one is the Elbe River Project, which was the first to be initiated and at present is the most important project, covering about 65 % of the Czech Republic. The broader objective of this national project is to propose such measures as ensuring requirements of state policy in the protection of the environment and improvement of water quality and simultaneously providing baselines for the fulfillment of international agreements.

The main activities are aimed at obtaining qualified data for creating significant water quality improvement measures for the Elbe River and for obtaining better knowledge of water quality, which serves the state water policy makers, international cooperation and related programs. A permanent network of 21 sampling sites was defined for the observation where the monitored indicators have been extended with additional parameters: AOX, TOC, DOC, PAH, volatile organic compounds, chlorinated phenols, pesticides and heavy metals. Three-year monitoring (1991-1993) enabled the evaluation of pollution within each indicator. Conclusions of the 1993 final report stressed the continuing heavy pollution of the Elbe River downstream from Pardubice, especially in indicators of organic pollution , nutrients and some specific organic pollutants ,including complex determinant AOX.

The decrease in several specific pollutants concentrations in 1992 and 1993 as compared to 1991 in the last sampling site of the Czech part of the Elbe River demonstrate the results stated in Table 2. The meeting of goals for the Elbe River water protection, i. e., the use of this water for drinking water purposes after infiltration and for irrigation and recovery of natural river ecosystem, is in respect to the aforesaid results problematic at present (Kalinová, 1993).

Table 2 Comparison of the specific organic substance content in the Elbe River upstream from the German border (Kalinová, 1993).

	1991	1992	1993	Substance in µg/l		
	n = 12	n = 12	n = 9			
Labe-H ensko	max	med	max	med	max	med
1.2-dichlorethane	47.3	10.7	11.4	6.6	1.8	0.8
Trichlorethylene	1.3	0.5	0.5	0.3	0.6	0.3
1.2-dichlorethane	3.9	1.1	5.6	0.5	2.9	0.3
Trichlorethylene	1.3.	0.1	0.1	0.1	0.1	0.1
Tetrachlorethylene	5.98	0.44	1.25	0.12	0.29	0.03
Chlorbenzene	1.69	0.2	0.21	0.06	0.03	0.03
1.2-dichlorbenzene	0.4	0.2	0.3	0.1	0.2	0.1
1.3+1.4 dichlorbenzene	9.6	1.7	5.8	0.8	5.7	0.1
Benzene	0.190	0.016	0.090	0.013	0.027	0.006
Toluene	0.04	0.01	0.04	0.01	0.03	0.01
Hexachlorbenzene	0.200	0.018	0.060	0.010	0.090	0.005
Lindan	0.450	0.013	0.031	0.011	0.062	0.020
PCB	0.067	0.005	0.005	0.005	0.012	0.005
Fluoranthene	0.3	0.1	0.1	0.1	0.2	0.1
Benzo(a)pyrene	0.9	0.95	1.8	0.3	0.17	0.16
Naphtalene	0.4	0.4	0.4	0.4	0.12	0.05
Atrazin	0.2	0.1	0.1	0.1	0.1	0.1
Simazin	43.9	0.2	0.6	0.1	0.17	0.05
Chlortoluens						
Dichlorfenols						

The Elbe River has not only a basic importance in the elaboration of a water quality protection concept in CR, but also simultaneously has become the first river for which an international cooperation has been introduced. The conditions of cooperation in the Elbe River protection are anchored in the Agreement on the International Commission for the Protection of the Elbe River signed in October, 1990. Signatories to this Agreement made obligations for joint efforts to improve current possibilities in Elbe River water use, to achieve renewal of a natural Elbe River ecosystem and to substantially reduce pollution loads of the North Sea. Concrete goals and measures in water quality protection during the period 1991-1995 were set by the First Action Plan approved in December, 1991. The completion of a monitoring station network in 1993 as part of a complex international water quality control system provides an emergency warning system, determination of monitored indicator values and the evaluation of pollutants transport.

The project dealing with the second largest watershed in CR, i. e., that of the Morava River, which is a part of the Danube River Basin, linked up this project to the Elbe Project after approximately a one year delay. The main goal of the Morava River Project is suggestion of measures leading to a substantial improvement of water quality in the watershed while maintaining rational water use and its natural function in a landscape ecosystem. The significantly negative influence of municipal sources of pollution, large-areal agricultural activities and related processing industries on water quality in streams are characteristic for this watershed. The selection of indicators was led by an effort to provide a wider water pollution survey with respect to specific organic pollutants and heavy metals. The micropollutants were also monitored in sediments, biomass and atmospheric depositions. The results of water quality evaluation during the period 1992-1993 conclusively show that ,despite the slight improvement as a result of decreased production of pollutants from the sugar industry, the Morava River is still heavily loaded with organic pollutants, nutrients, bacterial pollution and mercury from specific pollutant sources. Especially disturbing is a long-term significant increase in nitrite and phosphate pollution and a repeating occurrence of high concentrations of mercury along nearly the whole stream of the Morava River. Within the ecological tasks of the project, the state of ichtyocenosis was evaluated in connection with anthropogenic activities. The evaluation of heavy metals and PCB contents in fish muscular tissues (Leuciscus cefalus) shows that the mercury concentration exceeds acceptable values (0.1 mg/kg of wet weight).

After participating in several environmental management activities in an effort to address the growing regional water management and related environmental problems, the Danube riparian countries, together with interested members of the international communities, agreed during the meeting in Sofia in September, 1991 to develop and implement a program of priority actions and studies in preparation for agreement of a new convention. The first phase of this Environmental Program for the Danube River Basin focuses on the three-year action plan which includes fact finding, basin-wide studies, institutional strengthening and pre-investment studies. The Czech Republic entered this Program in 1993 and participates in the work of the Task Force, technical sub-groups (monitoring, laboratory analysis and information management, accident, emergency warning system and applied research program), in the fact finding and data support for pre-investment studies and in the preparation of Strategic Action Plan.

The third project--the Odra Project--addresses the water protection in the smallest watershed studied in CR. This watershed is characterized by low flows and high demands on water downstream, due to the high density of heavy industrial activity. The comparison of some data with nation-wide average values is interesting: population density is twice as great in this area, production is three times as great and natural flows only one third those in the republic as a whole. Besides the main goal of the Odra Project to create a new concept of environmental protection with respect to water quality protection and recovery of natural ecosystems in the Odra River Basin , the project will also form a baseline for both national policy makers and the fulfillment of Czech duties based on the international agreements. The monitoring is based on an effort to increase knowledge on the specific organic pollutants and heavy metals. First results of the water quality evaluation in the Odra Project show that the unsuitable state of water quality in streams of this watershed has been stabilized during the past three years. The most heavily polluted sections of the streams in CR are the downstream sites on the Odra and the Ole Rivers. The major influence on water pollution in the Odra River Basin is considered to be insufficiently treated municipal sewage and wastewaters from the paper-mill complex in Paskov.

Pollution of the Baltic Sea leads to the increasing phenomena of the disturbance of ecological balance. The pollution from a highly industrially developed Odra River watershed has contributed significantly to this situation recently. The Agreement of the International Commission for Protection of the Odra River, which is supposed to be signed in 1994, should contribute extensively to the improvement. The cooperation of signatories (Germany, Poland and EU) according to the Agreement proposal should be directed at a permanent decrease of pollution of the Odra River and the Baltic Sea with harmful substances, recreating the natural ecosystem and enabling the use of water in the Odra River basin for drinking water and agricultural purposes.

References

Beck, J . M. (1985). Water quality management: A review of the development and application of mathematical models. Springer Verlag Berlin, 29-33.

Bernardová, I. (1989): Application of multicriterial decisive analysis in setting up a list of planned waste water treatment plants in order of urgency (in Czech). Vodní hospodá ství, B 39,1,21-24.

Bernardová, I. (1994): Project Morava. General assesment and water quality models (in Czech). VUV Brno.

Bernardová, I., Mrázek, K., kollová, M. (1994): Development of the pollution of the River Morava evaluated from zoobenthic communities. Water science and Technology - in print.

Birks, H. J . B., Line, J . M., Juggins, S., Stevenson, A. C. et ter Braak, C. J. F.(1990): Diatoms and pH reconstruction. Philosophical Transactions of the Royal Society of London Series B 327,263-278.

Commission of the European Communities (1991): Project proposal on directive concerning the ecological quality of Community surface waters. HT/cg,5/3/91. Brussels.

Joerg, R. (1992): Emmission standards and River Water quality - combination of emission quality objectives with immission standards. River Water Quality. Ecological assesment and control. CEC Brussels Luxenbourg, 125-147.

Kalinová, M. (1993): Project Labe. Summary report on water quality in streams (in Czech). VUV Praha.

Loucks, D. P., Stedinger, J. R., Haith, D. A.(1981). Water resource systems planning and analysis. Prentice-Hall, Inc., Englewood Cliffs, New Jersey.

Kuníková, E.(1991): Influence of the pollution sources on water quality in streams expressed by saprobic index (in Slovac). VUVH Bratislava.

Line, J. M., ter Braak, C. J. F. et Birks, H. J. B., unpublished: WACALIB v. 3.3, University of Cambridge Computer Laboratory, Pembroke Street, Cambridge CB2 3QC, U.K., Agricultural Mathematics Group, Staringgebouw, P.O.Box 100, 6700 AC Wageningen, The Netherlands, and Botanical Institute, University of Bergen, N-5007 Bergen,Norway

Mrázek, K.(1984): Biological model of the rivers in CSR (in Czech). Vodní hospodá ství, B 34, 3, 57-61.

Neuwirth, A.(1993): Project Odra - II. (in Czech). VUV Ostrava de Pauw, N., Ghetti, P. F.,Manzini, P., Spaggiari, R. (1992): Biological assesment methods for running water. River Water Quality. Ecological assesment and control. CEC Brussels Luxenbourg, 217-248.

Sládeek,V. et al. (1981): Biological analysis of surface waters. Comment to CSN 830532 - part 6: Determination of saprobic index (in Czech). Ústav pro normalizaci a m ení, Praha.

Sweeting, R. A. et al. (1992): 1990 Biological assesment of rivers in the UK. River Water Quality Ecological assesment and control. CEC Brussels - Luxenbourg, 319-326.

Wright, J . F., Armitage, P. D., Furse, M. T. et Moss: (1989): Prediction of invertebrate communities using stream measurements. Regulated Rivers Research and Management 4, 147-155.

Zelinka, M., Kubíek, F. (1985): The principles of applied hydrobiology (in Czech). Státní pedagogické nakladatelství. Praha.

V.4 Organic and Inorganic Micropollutants in the Danube River along the Slovak-Hungarian Border

Igor Liska
Water Research Institute
Nabrezie Svobodu 5
812 49 Bratislava
Slovakia

Ferenc László
Institute for Water Pollution Control,
Water Resources Research Centre (VITUKI Plc.)
H-1095 Budapest, Kvassay J. u. 1
Hungary

Introduction

The Danube with a total length of 2 857 km and a longterm daily mean discharge of about 6300 $m^3.s^{-1}$ is the second biggest river in Europe. In terms of the length it is listed as the 21st biggest river in the world and in terms of drainage area it ranks as 25th with a drainage area of 817 000 km^2. The water regime of the Danube is uniform up to the mouth of the river Inn. At the upstream end of the Danube, the greatest flow is present during the May - June period (Alpine regime) as a result of the snow melting in the Alps, while the lowest flow is registered towards the end of the autumn and during the winter months (October - February).

The Danube enters Slovak territory at the km 1880.2. The Slovak - Hungarian part of the Danube with its length of 140.8 km has three major left bank tributaries: Vah, Hron and Ipel and four major right bank tributaries: Mosoni-Danube, Concó, Általér and Kenyérmez -creek. Another left bank tributary - Morava - has its mouth several kilometers before the start of the Slovak-Hungarian part, however, it influences the water quality in this part of the Danube. As already mentioned above, this part of the Danube has a mountain-like character with typical run-off. Summer flow maxima are 31.7% of the annual run-off, the winter maxima are 18.2% of the annual run-off. Water bearing is relatively constant within the annual cycle as well as in the long-term average.

The importance of the Danube river water quality in its transboundary portion between Hungary and Slovakia is stressed by the fact that the part of the Danube between Vienna and Budapest belongs to the most critical as to the pollution load. Common Slovak-Hungarian monitoring of the water quality has been performed for more than 30 years. Within this activity the working group on transboundary river water quality has been established which managed common monitoring. In 1976 the document on principles of the cooperation in transboundary river water quality monitoring has been accepted. According to this document the sampling frequency was increased from 4 to 12 per annum. This document has been revised several times and in 1988 the number of monitored parameters was adjusted to comply with the Bucharest Declaration. In 1991, nine additional parameters were involved.

The Danube is the dominant source of groundwaters in the surrounding area, therefore, river water quality has an essential impact on the quality of groundwaters influenced by the Gabcikovo water dam. This is the reason why a supplementary monitoring of the Danube river water quality with respect to this water dam is performed as a special project. Monitored area is not strictly limited to the Gabcikovo dam surrounding area but includes a long part of the Danube and its tributaries.

Heavy Metals

Water Pollution

Monitoring of heavy metals at the Slovak-Hungarian border section Szob started in the middle of the 1970s. Efforts were focused on revealing the mercury pollution because this heavy metal caused the greatest concern. Concentration distribution of mercury showed high variability in time and space. Fig. 1 illustrates the cross-sectional mercury concentration distribution at Szob at different flow discharges (Literathy, 1982). High concentrations and inhomogeneous cross sectional mercury distribution were characteristic mainly at low flow rates.

The mean concentrations of mercury along the Rajka- Budapest Danube reach indicated significant pollution in the period 1977-1979 (Table 1). The highest concentrations exceeded the limit value for drinking water (1μg/l) in most sampling sections.

Another toxic heavy metal, cadmium was also present at high concentrations in the Danube at the end of the 1970s although the mean concentrations did not surpass the WHO guideline for drinking water. The extent of pollution caused by lead and zinc was considerably lower.

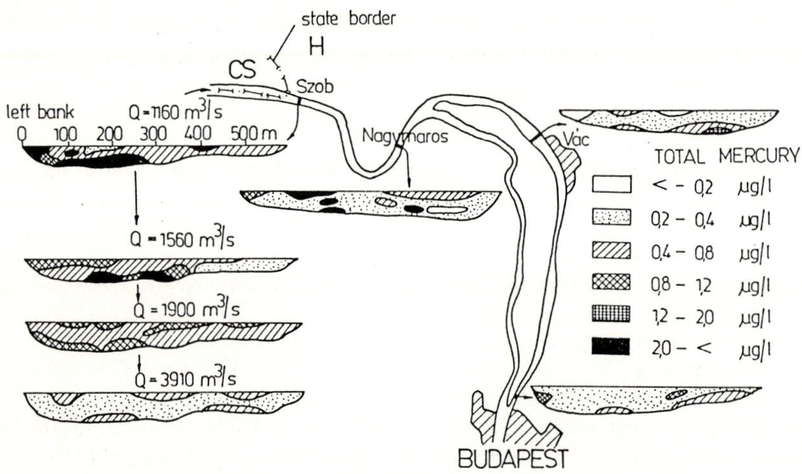

Figure 1 Cross-sectional distribution of mercury in the Danube river upstream of Budapest.

Table 1 Mean concentrations of heavy metals along the Slovak-Hungarian Danube section in the period 1977-1979

Sampling site	Mean concentration (µg/l)			
	Mercury	Cadmium	Lead	Zinc
Rajka 1848km	0.38	0.68	7.2	70
Medve 1806km	0.44	1.7	<7	120
Komarno 1768km	0.32	1.1	7.1	80
Dunaalmas 1752km	0.39	0.80	7.1	90
Szob 1708km	0.98	0.90	7.4	90

Extensive study of the heavy metal pollution was continued only several years later (SHCTWM, 1990-94). The analysis of the heavy metal pollution of the main channel of the Danube was supplemented by the survey of the main tributaries and side arms as well. It was revealed that the arsenic concentration in the tributary Hron was remarkably higher than in the Danube or in any other studied tributary. Presumably the arsenic content of the river Hron is of natural geochemical origin due to weathering of arsenic containing rocks.

The extended Slovak-Hungarian monitoring of heavy metals involved the analysis of the dissolved phase too during the 1989-1990 period. The detailed investigations showed (Fig. 2) that the dominant part of lead was present in the suspended sediment. In a less extent it was characteristic for the chromium and copper too. The dissolved cadmium fraction often represented the larger part of the total cadmium content. In the case of mercury the ratio of dissolved and total content varied in a wide range. Another conclusion of the detailed study was that heavy metal pollution of the Slovak-Hungarian Danube section decreased in the 1980s.

Sediment pollution

Inorganic and organic pollutants in aquatic systems are liable to precipitate or adsorb onto particulates and accumulate in the suspended and bottom sediments. Therefore, monitoring of these pollutants should be conducted by sampling and analyzing the bottom sediment in addition the water column containing the pollutants in both dissolved and solid phases.

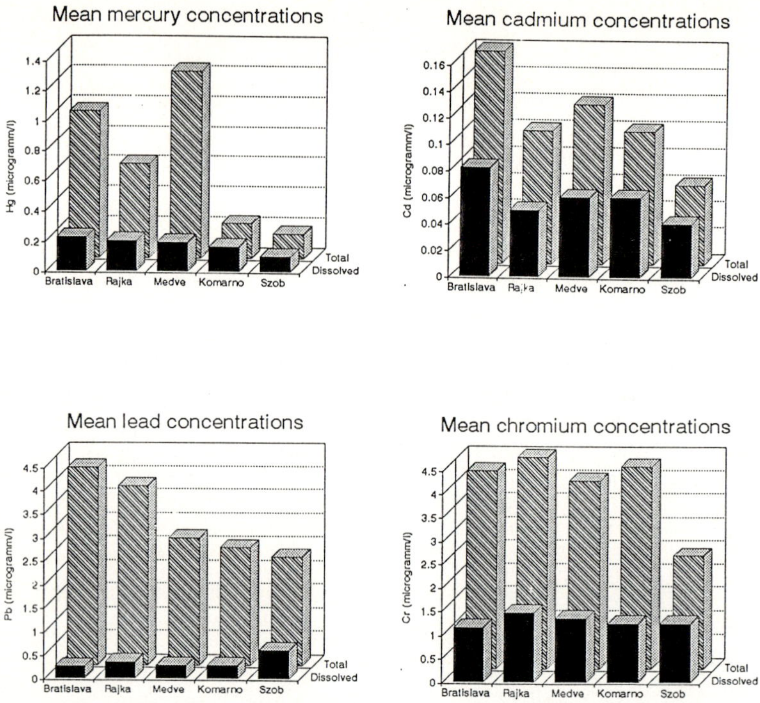

Figure 2 Dissolved and total heavy metal concentrations in the Danube in 1989-1990

In 1987 four series of bottom sediment samples were taken at 25 points from the bottom sediment of river Danube at locations between Rajka and Budapest. The sampling sites included also the Danube arms of Szigetköz. The purpose of the analysis was to determine accumulated heavy metals in the bottom sediments. Samples were taken from the upper 5 cm layer of the bottom material and the analysis was confined to the finer than 1 mm fraction of the particle size distribution. Concentration values were expressed as per unit dry material weight of the sediment.

In Fig. 3 the average, minimum and maximum values of the measured data are shown along with the allowable heavy metal content of soil. On the basis of the average concentration values calculated for the different metals it may be stated that toxic metal concentrations (above the limit values) were the highest along the Danube reach of Tát-Göd, while the lowest values were measured in the upper Danube reach.

These results are in agreement with those of earlier studies on bank-well drinking water resources along the Hungarian Danube reach namely: the lowest toxic metal concentrations are found in the upstream part of the Hungarian reach, while the highest ones fall into the middle of the reach.

In 1986-1987 sediment pollution of different particle size fractions was studied in the Danube section downstream of Komarno. The average concentrations of contaminants in the sediment fractions are listed in Table 2 (Hock and László, 1989).

For all analyzed components the highest mean contamination levels were found in the lowest particle-size fraction.Most of the metals exceeded the lowest effect value (Literathy, 1995) which indicates the anthropogenic input above the natural background.

Survey of sediment pollution in the Danube was extended to the tributaries in 1989. Arsenic, zinc, mercury, cadmium, chromium, manganese, nickel, lead, copper, iron, oil and polycyclic aromatic hydrocarbons were analyzed in the samples. Fig. 4. shows those sampling sites where the pollution concentration exceeded the limit value of the indicated component. Mainly the tributaries were polluted, but along the Danube downstream of Komarno the Danube sediment was also influenced by the pollution impact of the tributaries.

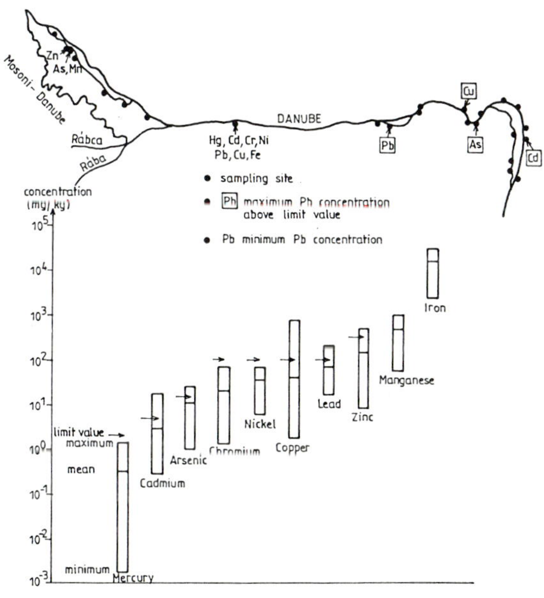

Figure 3 Contamination of the bottom sediment along the Danube.

Since 1992 the Slovak-Hungarian joint monitoring involves the analysis of suspended solid composition too in the Danube sections at Bratislava, Komarno and Szob. Tables 3 - 6 inform about the composition of bottom sediment and suspended solids. The characteristic pollutant concentrations are usually higher in the suspended sediments than in the bottom sediment.

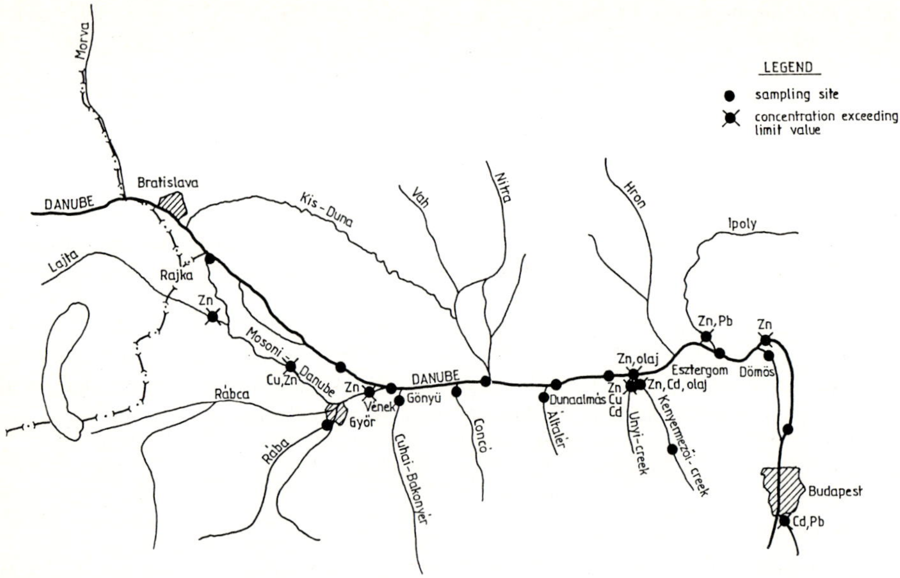

Figure 4 Polluted sediment sampling sites

Organic pollutants

Organic pollutants in the Danube river water are characterized by means of various types of parameters. The general information on the total organic load in the surface water is provided by several sum parameters such as TOC, BOD_5 and COD (Mn). The group parameters (phenols, oil pollution, anion-active detergents) inform about the presence of relevant groups of organic pollution originating mainly from industry, transport activities and public sewerage. Both of these two kinds of parameters, based mostly on simple spectrophotometric methods, have been monitored in the Danube for many years. Since 1990, also several most relevant specific organic pollutants have been included (volatile organic halogens, chlorinated pesticides, triazine herbicides, polychlorinated biphenyls, polycyclic aromatic hydrocarbons).

Moreover, screening of organic pollution using gas chromatography-mass spectrometry has been initiated. The mean annual values of selected sum, group as well as specific parameters of organic pollution in different sampling profiles in the year 1993 are shown in Figures 5-10. The mean values of sum and group parameters were calculated from 25 measurements (sampling biweekly). The mean values of specific pollutants are calculated from four values (sampling once per year season).

Table 2 Average concentration of contaminants in Danube sediment fractions (mg/kg).

Contaminants	Grain-size fraction		
	d < 90 µm	90 µm >d< 250 µm	250 µm >d<1000 µm
Mercury	1.0	0.15	0.16
Cadmium	2.78	1.33	0.94
Lead	1105	42.7	51.1
Chromium	50.9	15.1	15.4
Arsenic	9.6	2.1	3.2
Copper	228	34.3	30.5
Nickel	58	25.9	21.1
Iron	20600	8630	7830
Manganese	782	291	232
Zinc	381	77.0	85.9
Total Org. matter	49700	20200	27400
Oil + grease	1050	428	477

Table 3 Pollution of suspended sediment in the Danube between Bratislava-Szob in 1992.

Sampling site	Bratislava		Komárom						
Pollutant	Concentration of pollutant (mg.kg^{-1})								
	min.	mean	max.	min.	mean	max.	min.	mean	max.
Lindane	0,003	0,12	0,53	0,001	0,056	0,21	0,001	0,41	0,19
Heptachlor	-	0,0045	-	-	0,004	-	0,002	0,007	0,02
DDT	-	0,0028	-	-	0,0025	-	0,003	0,007	0,01
PCB	0,005	0,007	0,01	0,006	0,008	0,010	0,006	0,013	0,025
Benzo(a)pyrene	0,002	0,028	0,077	0,002	0,029	0,06	0,002	0,066	0,116
Fluoranthene	-	0,054	-	0,02	0,12	0,23	0,037	0,12	0,25
Iron	14500	16400	18400	11600	132000	15700	17000	19500	21300
Manganese	620	900	1100	480	750	930	480	1950	3000
Zinc	180	602	1700	90	980	3200	450	1000	1700
Copper	45	62	80	35	54	80	42	51	65
Chromium	19	24	28	13	22	27	22	32	43
Lead	38	42	47	28	79	156	48	66	97
Cadmium	0,5	2,2	5,8	0,25	1,1	2,2	1,1	1,9	2,4
Mercury	0,2	0,31	0,4	0,2	0,27	0,4	0,3	0,37	0,45
Nickel	19	22	25	14	17	20	19	22	26
Arsenic	8	12	18	10	17	30	13	18	25

As it can be seen from Figure 5 the concentration level of sum parameters was at low mg/l level. An increase of the organic pollution expressed by TOC, BOD_5 and COD(Mn) downstream the Danube can be distinguished. Apparently, the largest increase was found for BOD_5. A similar trend was observed for oil pollution which exceeded the 65 µg/l in Budapest.Other group parameters (phenols, detergents) did not have significant differences in their concentration in the monitored river section.

The results of measurements of specific organic pollutants are influenced by the limits of detection (LOD) for particular parameters. That is why the data in some Figures, containing concentration values below LOD, may seem incomplete. It was felt, that missing data would better match the description of the current status of pollution than rather problematic indication of concentration at the LOD level.

Table 4 Pollution of bottom sediment in the Danube at Bratislava in 1992.

Sampling site	Bratislava right bank		
Pollutant concentration	min.	mean	max.
Lindane ($mg.kg^{-1}$)	0,001	0,004	0,007
Heptachlor ($mg.kg^{-1}$)	0,002	0,003	0,004
DDT ($mg.kg^{-1}$)	-	-	-
PCB ($mg.kg^{-1}$)	0,02	0,046	0,097
Benzo(a)pyrene ($mg.kg^{-1}$)	0,009	0,056	0,075
Fluoranthene ($mg.kg^{-1}$)	0,15	0,19	0,25
Gross -activity ($Bq.kg^{-1}$)	580	683	8550
Cesium-137 ($Bq.kg^{-1}$)	7	120	301
Strontium-90 ($Bq.kg^{-1}$)	-	3	-
Iron ($mg.kg^{-1}$)	13900	17000	19700
Manganese ($mg.kg^{-1}$)	298	452	663
Zinc ($mg.kg^{-1}$)	87	123	166
Copper ($mg.kg^{-1}$)	13	23	34
Chromium ($mg.kg^{-1}$)	21	29	38
Lead ($mg.kg^{-1}$)	19	27	35
Cadmium ($mg.kg^{-1}$)	0,06	0,15	0,27
Mercury ($mg.kg^{-1}$)	0,10	0,16	0,20
Nickel ($mg.kg^{-1}$)	13	22	35
Arsenic ($mg.kg^{-1}$)	2,3	4,5	6,5

Mean concentrations of volatile chlorinated hydrocarbons in 1993 showed no relevant contamination by these compounds except of chloroform. The mean values of chloroform concentration in the river section between Bratislava and Medvedov exceeded 1 μg/l. More warning are maximum concentrations of chloroform in this river section which indicate serious seasonal pollution by chloroform downstream of Bratislava (Figure 7). The mean concentrations of polychlorinated biphenyls (expressed as Delor-103 and Delor-106) and Lindane were at negligible level below 10 ng/l except of the mean concentration of Lindane in Budapest exhibiting severe pollution of the river by this pesticide (Fig.8). The frequent use of Atrazine especially in corn fields in the Danubian

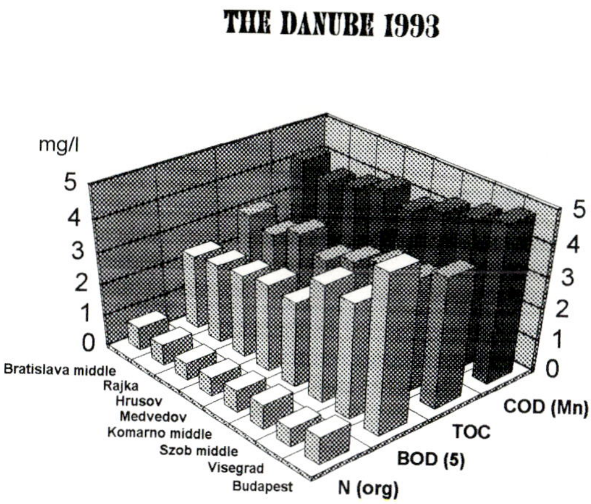

Figure 5 The mean values of sum parameters of organic pollution in the Danube between Bratislava and Budapest in 1993.

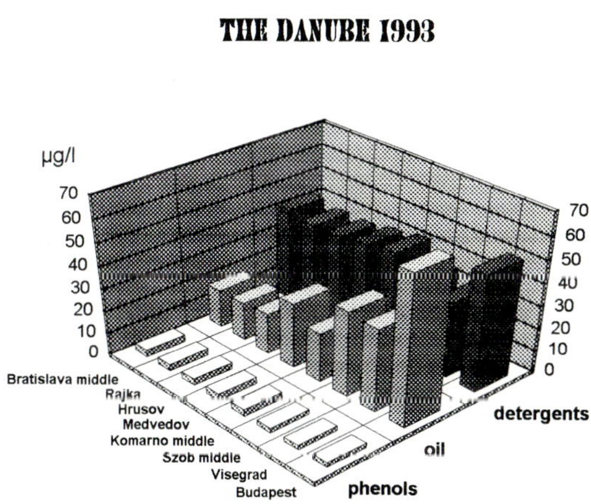

Figure 6 The mean values of group parameters of organic pollution in the Danube between Bratislava and Budapest in 1993.

Table 5 Pollution of bottom sediment in the Danube at Komárom in 1992.

Sampling site	Komárom					
	left bank			right bank		
Pollutant concentration	min.	mean	max.	min.	mean	max.
Lindane ($mg.kg^{-1}$)	0,0005	0,05	0,154	0,0005	0,04	0,083
Heptachlor ($mg.kg^{-1}$)	0,002	0,002	0,0025	0,001	0,003	0,0045
DDT ($mg.kg^{-1}$)	-	-	-	-	-	-
PCB ($mg.kg^{-1}$)	0,009	0,019	0,075	0,005	0,020	0,081
Benzo(a)pyrene ($mg.kg^{-1}$)	0,002	0,062	0,250	0,004	0,023	0,081
Fluoranthene ($mg.kg^{-1}$)	0,007	0,19	0,55	0,015	0,02	0,03
Gross -activity ($Bq.kg^{-1}$)	380	415	450	270	617	830
Cesium-137 ($Bq.kg^{-1}$)	4,3	5,4	6,3	3	16	41
Strontium-90 ($Bq.kg^{-1}$)	4	5	6	1,2	13,4	33
Iron ($mg.kg^{-1}$)	11700	16300	20900	4400	11800	19300
Manganese ($mg.kg^{-1}$)	453	510	568	230	356	475
Zinc ($mg.kg^{-1}$)	78	180	283	31	77	123
Copper ($mg.kg^{-1}$)	17	33	49	4	17	29
Chromium ($mg.kg^{-1}$)	26	235	445	15	26	36
Lead ($mg.kg^{-1}$)	20	30	40	9	22	36
Cadmium ($mg.kg^{-1}$)	0,13	0,43	0,73	0,06	0,13	0,17
Mercury ($mg.kg^{-1}$)	0,18	0,39	0,61	0,17	0,25	0,33
Nickel ($mg.kg^{-1}$)	15	22	29	4	14	24
Arsenic ($mg.kg^{-1}$)	4,3	12,6	21	2,5	4,5	6,5

Lowland influences the occurence of this herbicide in Danube river water. Mean concentrations ranged from 40 to 90 ng/l but the pollution had seasonal character with peak values above 100 ng/l (Fig.10). Improving of this situation can be expected after stopping of its use in Austria and implementing similar regulations in Slovakia. Pollution of the Danube with PAH was not found to be severe, the increased levels were detected in Budapest only.

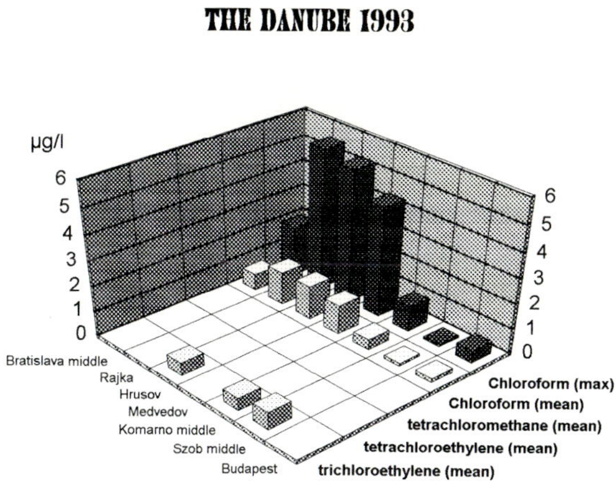

LOD = 0.5 µg/l (chloroform 0.1 µg/l)

Figure 7 The concentrations of chlorinated hydrocarbons in the Danube between Bratislava and Budapest in 1993.

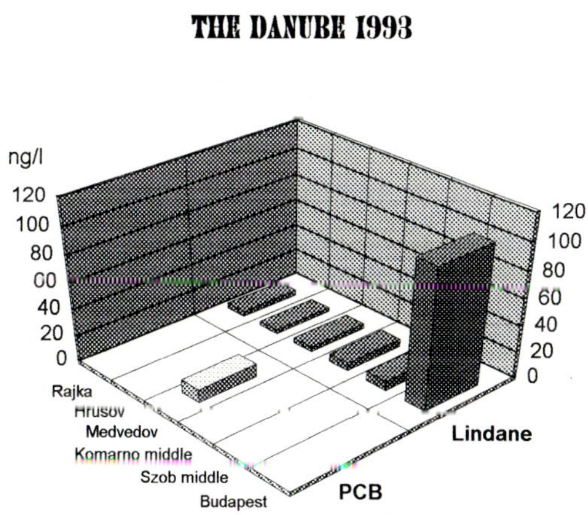

LOD (PCB) = 10 ng/l, LOD (Lindane) = 2 ng/l

Figure 8 The concentrations of polychlorinated biphenyles and Lindane in the Danube between Bratislava and Budapest in 1993

Table 6 Pollution of bottom sediment in the Danube at Szob in 1992

Sampling site	Szob					
	left bank			right bank		
Pollutant concentration	min.	mean	max.	min.	mean	max.
Lindane ($mg.kg^{-1}$)	0,0004	0,005	0,006	0,002	0,012	0,044
Heptachlor ($mg.kg^{-1}$)	0,002	0,004	0,006	0,002	0,004	0,007
DDT ($mg.kg^{-1}$)	-	-	-	-	-	-
PCB ($mg.kg^{-1}$)	0,005	0,017	0,032	0,0005	0,021	0,062
Benzo(a)pyrene ($mg.kg^{-1}$)	0,003	0,016	0,026	0,007	0,019	0,035
Fluoranthene ($mg.kg^{-1}$)	0,009	0,043	0,077	0,003	0,027	0,062
Gross -activity ($Bq.kg^{-1}$)	-	810	650	650	700	750
Cesium-137 ($Bq.kg^{-1}$)	7,1	11,6	16	12	25,7	39,5
Strontium-90 ($Bq.kg^{-1}$)	-	2,5	-	1,8	6,2	17
Iron ($mg.kg^{-1}$)	-	7500	-	8200	14700	21700
Manganese ($mg.kg^{-1}$)	-	295	-	187	377	545
Zinc ($mg.kg^{-1}$)	-	79	-	59	100	132
Copper ($mg.kg^{-1}$)	-	8	-	10	21	32
Chromium ($mg.kg^{-1}$)	-	11	-	14	24	35
Lead ($mg.kg^{-1}$)	-	16	-	16	24	32
Cadmium ($mg.kg^{-1}$)	-	0,23	-	0,08	0,23	0,32
Mercury ($mg.kg^{-1}$)	-	0,16	-	0,08	0,16	0,25
Nickel ($mg.kg^{-1}$)	-	6	-	8,5	17	27
Arsenic ($mg.kg^{-1}$)	-	3,5	-	3,8	6,1	8,7

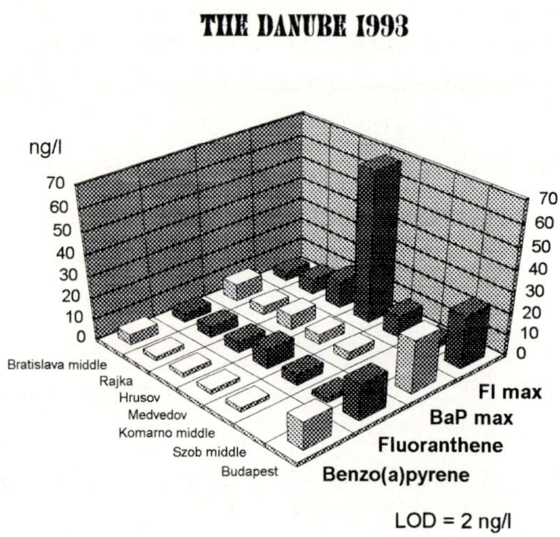

Figure 9 The concentrations of PAH's in the Danube between Bratislava and Budapest in 1993

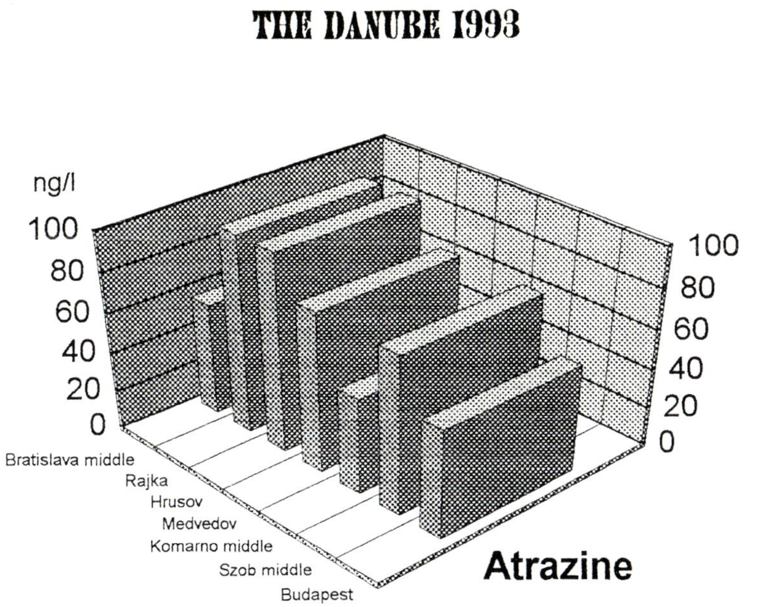

Figure 10 The concentrations of atrazine in the Danube between Bratislava and Budapest in 1993

The cross-sectional distribution of organic pollutants in the stream is demonstrated on the Szob sampling profile (Fig. 11-13). A slight increase of the concentrations of sum parameters in the left bank can originate from the fact that left bank tributaries in the Slovak-Hungarian part of the Danube (Vah, Hron, Ipel) drain relatively larger area than right bank tributaries and, therefore, their pollution load is more significant.

A similar explanation is applicable to cross-sectional profile of oil pollution and detergents. The distribution of non-polar persistent PAH's has a different characteristic with minimum concentrations in the stream line and higher concentrations (especially of fluoranthene) in the bank areas. The study of the cross-sectional distribution of selected organic pollution revealed that the differences in the mean values were not remarkable even though a certain pattern could be distinguished.

Much larger variations were found between particular measurements within one sampling (see maxima of fluoranthene concentration). Hence, the interpretation of data from one sampling campaign should be performed very carefully and, if possible, a complex set of data should be used for water quality evaluation.

Within transboundary river water quality monitoring also the quality of the major tributaries was evaluated. The sampling points were located in the mouth of tributaries. Since the data set for organic pollution of tributaries for 1993 is not complete, data from 1992 are presented (Fig. 14-16). Generally, concentrations of specific organic pollutants are higher than those in the Danube as a consequence of the lower water flow. The major load of PAH's entering the Danube in the monitored transboundary section is coming from the left bank tributaries. Pollution of tributaries by PCB's and Lindane in 1992 was low. Higher concentrations of atrazine in most of the monitored tributaries are the consequence of agricultural activities in the area drained by these rivers.

In addition to determination of selected groups of pollutants GC-MS screening was performed to obtain an overview on the character of organic contamination in the Danube. Considering the results from Table 7 it is obvious that the major part of identified compounds belong to oil pollution and its degradation products (aliphatic and aromatic hydrocarbons and their oxidation products - alcohols, acids).

The presence of ubiquitous phthalates is not surprising as they are widely used in the industry and frequently found in all kinds of water environment. Specific industrial contaminants of benzothiazole type are most probably originating from chemical industry in Bratislava as by-products in production of rubber accelerators. It must be pointed out that GC-MS analyses showed occurrence of much more chemical compounds, however, their identification was not possible.

The pollution of sediments varied in space and time having most of the highest values in the profile Komarom left bank. This was most probably the consequence of the contribution to the pollution from the Vah tributary. The concentrations of chlorinated pesticides found at several points exceeded the Guidelines for assessing of contaminants in sediments (Persaud et al., 1992). Contamination by PAH's was found higher in the upper part of the boundary reach of the Danube.

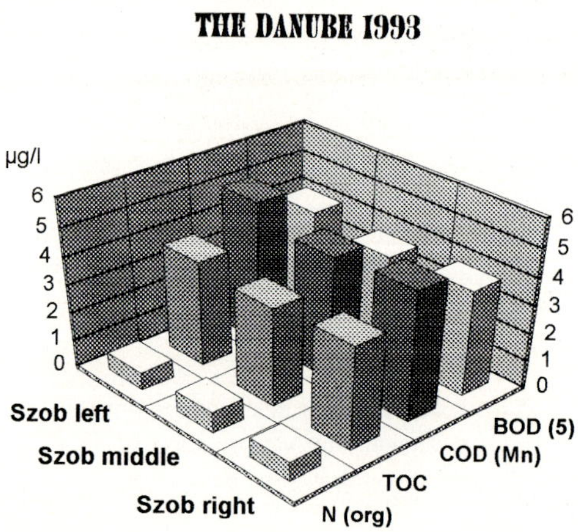

Figure 11 The concentrations of sum parameters in the Szob sampling profile of the Danube in 1993

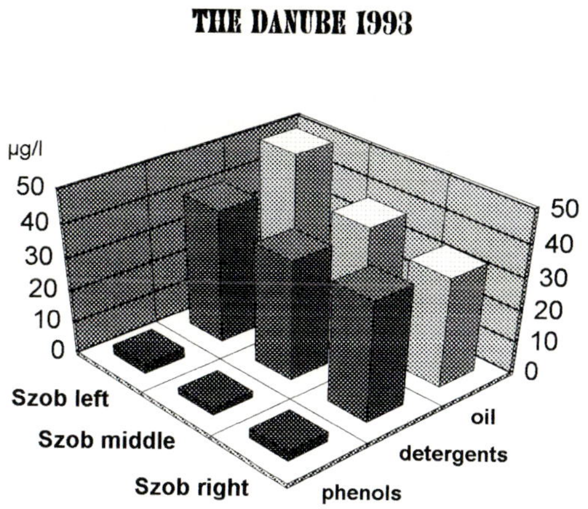

Figure 12 The concentrations of group parameters in the Szob sampling profile of the Danube in 1993.

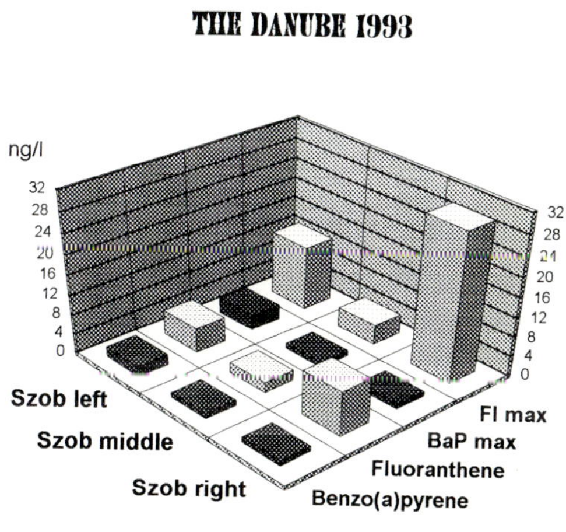

Figure 13 The concentrations of PAH's in the Szob sampling profile of the Danube in 1993.

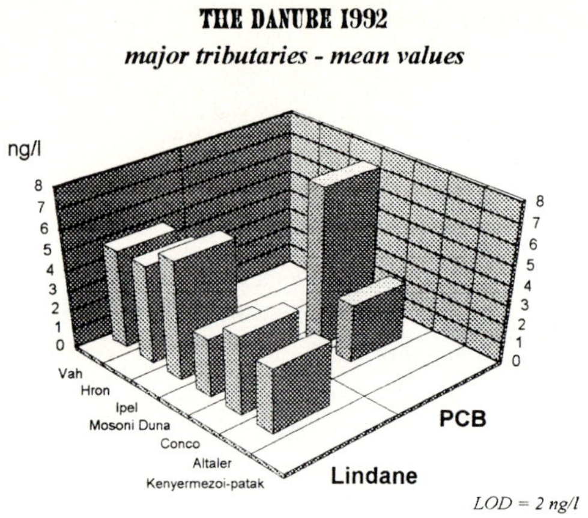

Figure 14 The concentrations of PCB's and Lindane in the major tributaries of the Danube in 1992.

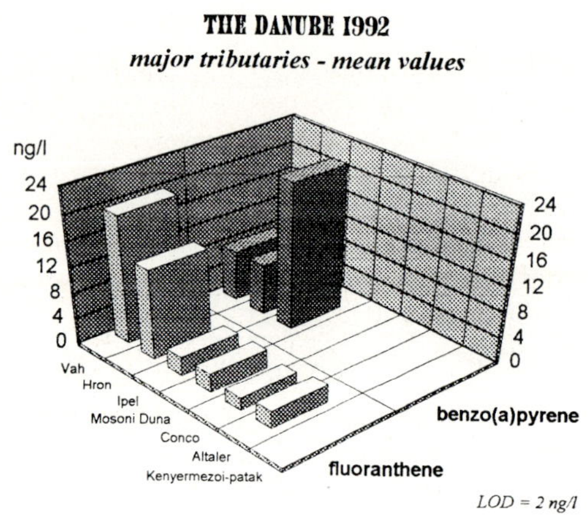

Figure 15 The concentrations of PAH's in the major tributaries of the Danube in 1992

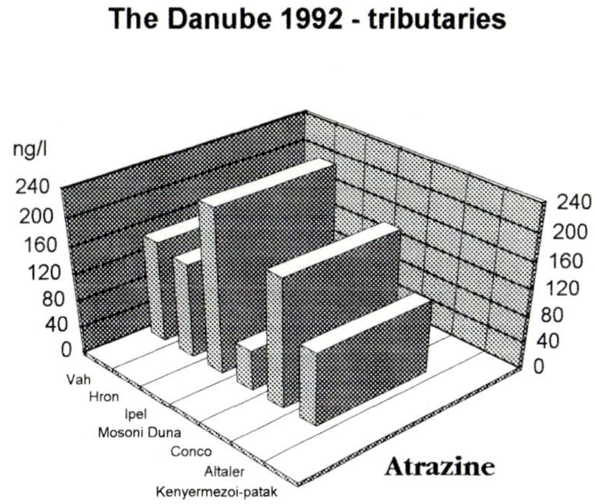

Figure 16 The concentrations of atrazine in the major tributaries of the Danube in 1992.

Radiological measurements

The first co-ordinated Slovakian and Hungarian radioecological measurements in the Danube were performed in the period 1977-1982 within a research programme supported by the International Atomic Energy Agency (IAEA, 1983). The study included the analysis of water, bottom sediment, suspended sediment algae and fish. The implementing institutes carried out gross alpha, gross beta, gamma spectroscopy analyses and measured some specific radionuclides, e.g. tritium, strontium-90, cesium-137 and radium-226 in different sample types. Results have shown that in the studied period radioactivity of the Danube was originated mainly from naturally occurring radionuclides such as ^{40}K, ^{226}Ra and other members of the uranium and thorium decay series. Of the artificial radionuclides only ^{3}H, ^{90}Sr and ^{137}Cs could be found in detectable concentrations. Some other fission products (^{95}Zr, ^{95}Nb, ^{131}I, $^{141,144}Ce$) could be identified in algae samples from time to time without simultaneous detection of these nuclides in water samples. It is known, however, that algae are good indicator organisms with high radionuclide accumulating capability.

The majority of artificial radionuclides found in the Danube were long lived fission products of fallout origin. That was the reason why no significant changes could be found in their activity concentrations downstream of the confluence of the major tributaries. No regular tendency of increases or decreases could be observed in activity concentrations of particular radionuclides in the period of 1977-1982, except for tritium. The monitoring programme revealed a 3-4-fold decrease in tritium concentration of Danube water during the six years period. A similar decrease was also been found in tritium content of the rain water. However, the activity of tritium per unit volume was about 50 per cent higher in Danube water than in rain water indicating that a considerable proportion of tritium in the Danube came from the glaciers rather than from the atmosphere directly. When comparing activity concentrations of radionuclides in water, sediment and fish obtained during a longer period (some years) of observations, concentration factors could be defined for ^{90}Sr and ^{137}Cs as follows: $1x10^2$ and $3x10^3$ in sediment/water relation, and 30-40 and $1.5\text{-}2.0x10^2$ in fish/water relation respectively.

After the implementation of the co-ordinated IAEA project the joint Slovak-Hungarian radiological monitoring of the Danube was continued again only in the year of 1988 when the "Bucharest Declaration" monitoring programme of the riparian countries started. The "Bucharest Declaration" monitoring programme included only the analysis of gross beta activity in water in the first years of the project, so a significant data collection of radiological parameters started with the implementation of the extended Slovak-Hungarian bilateral Danube monitoring project. The bilateral programme includes the analysis of radioactive parameters in water and sediment. Gross beta activity , tritium, strontium-90 and cesium-137 have been analyzed.

In 1986 the Chernobyl nuclear power plant accident influenced remarkably the water quality of the Danube. Immediately after the accident extended monitoring of the specific radioactive pollutants started in Slovakia and in Hungary. Some days after the accident the short lived ^{131}I was the dominant radionuclide in the Danube, but it disappeared within some weeks owing to its short half-life time. The tritium and ^{90}Sr content of the Danube didn`t change significantly, on the other hand $^{103,106}Ru$, $^{134,137}Cs$ which escaped from the reactor, appeared in the Danube water, especially in the sediment where the activity of these nuclides increased by order of magnitude. Even in the 1990s the cesium activity is higher in the Danube sediment than before the Chernobyl accident.

References

Hock, B., László, F. (1989): Contribution to the water quality evaluation of the Danube. In: Miloradov, M. (Ed.) Int. Conf. on Water Pollution Control in the Basin of the River Danube. 17-24

IAEA (1983): International studies on the radioecology of the Danube River. International Atomic Energy Agency, Vienna

Literáthy, P. (Ed.) (1982): Felszíni vizek nehézfém szennyezései. M szaki Könyvkiadó, Budapest

Literathy, P. (1995): Pollution monitoring of large international rivers: the Danube. Proceedings of NATO ASI Transboundary Water Resources Management

Persaud, D., A. Jaagumagi and A. Hayton (1992): Guidelines for the Protection and Management of Aquatic Sediment Quality in Ontario. Log 92-2309-067, PIBS 1962, Queen`s Printer for Ontario, Canada

SHCTWM (Slovak-Hungarian Committee of Transboundary Water Management) (1990, 1992, 1993, 1994): Evaluation of the joint water quality measurements. Official reports of the SHCTWM subcommittee for water pollution control

V.5 Transboundary Exchanges and Control of Nutrients on the Danube River and its Main Tributaries between Romania, Hungary, Bulgaria, Ukraine and Moldavia

Aurel Varduca and Septimius Mara
Water Quality Monitoring and Management Department
Environmental Research and Engineering Institute, Bucharest
Romania

Introduction

Transboundary pollution control involves at least the following minimum requirements:

(i) reliable water quality monitoring data, and

(ii) a joint transnational strategy based on the national scenarios.

In respect to this, the paper is related to:

(a) general transnational trend for the Danube River pollution generally and fertilizers problems particularly, and

(b) Romania's specific aspects.

The general approach model used in our research is based on a frame which includes the following segments: priority pollutants - target group - environmental themes - relevant functions. A general balance for N, P and C (BOD_5) was computed starting from the emission and immission. The specific weight of each hydrographical basin drainage area has been established as well for "hot spot" areas.

Assessment of the Danube River General Quality Trend: Evolution of Nutrients

Nevertheless, in applying one scenario, one should take into account not only the country's resources but also the general frame of input/output of pollutants. The following classes of parameters have been investigated: (i) dissolved oxygen - organic load regime, (ii) nutrients, (iii) buffering capacity, (iv) total dissolved salts, (v) toxic and micropollutants.

For the trend evolution assessment of all the parameters mentioned above , the ratio "R" between the value recorded in the last reference year (1992) and the value recorded in the initial year of Danube water quality monitoring activity for each control station was used. Some of the data, where it was possible, could be considered as a “starting point" (1966).

Transnational Monitoring Network

The actual transnational monitoring network of the Bucharest Declaration is based on 11 control stations (see Fig. 1).

Figure 1 Catchment area of Danube river: Map of sampling sites.

The Danube River water quality characterization was made by means of the annual mean parameter values, established at each control section based on monthly mean values. As a general overview, there is obviously a significant depreciation of Danube River quality at the entrance into the Romanian territory and in the Danube Delta.

(a) a decrease of % DO saturation with 25%-30% comparing with 1966 when recorded 80%-110% saturation was recorded;

(b) a fivefold increase of BOD_5 load during 1966-1988;

(c) a threefold increase in the COD-Mn load (1966-1989); and

(d) a double load of nitrogen considering the same period of time.

It should be mentioned that these figures are referring to station 6 and as a consequence, the same order of magnitude depreciation appears downstream also.

Dissolved Oxygen - Organic Loads

The following remarks concerning time-spatially trend could be made. There is obviously a decrease of dissolved oxygen saturation during the time and space. The ratio "R" between the last 1992 and initial year is in the range 0.78-1.04. The minimum values are in the Romanian territory. As an overview, R is higher for the upstream countries (better conditions) and lower for Romania. In the case of BOD_5 there are two parameters which should be taken into consideration:

(i) concentrations, and

(ii) loads.

Based on the selfpurification effect, there is no significant difference between reference year/station BOD_5 concentration (1988-2.6 mg/l) and the 1992 situation before discharge into the Black Sea (3.0 mg/l). The general trend for the upstream countries is a decrease in BOD_5 load, which is, on the other hand, correlated with DO % of increase. For the Romanian territory, the period 1985-1988 gave the maximum pollution load in BOD_5. Despite that, 1990 shows the minimum load; there is obviously an increase in 1991, the situation becoming steady state but with about 50 % less than in the previous period.

For the COD-Mn, roughly, there is a general tendency to decrease the concentration after the maximum values recorded in 1973. A better understanding of the organic (global approximately) trend could be obtained using the COD-Cr parameters. There are significant differences between "total" organic concentration in the Danube Delta /Black Sea discharge and reference year/station (1.3-1.4 higher). These "organic compounds" are not biodegradable, not possible to be "filtrated" by the Danube Delta and are caused by "import" from the upstream countries.

The general trend is a decrease in the COD-Cr for the upstream countries and for Romania too. The general trend in the upstream countries is a slight decrease for the COD-Cr load. A more accentuated decrease is recorded for the Romanian territory where the 1990 load represents about 60-65% from the 1988 correspondent values. What is interesting to remark is that for the Romanian territory, "R" presents a rough constant value 0.6. This observation confirms the hypothesis that the Danube in Romanian territory transfers the "imported" heavy biodegradable organic to the Danube Delta and after that into the Black Sea without significant local contributions.

Nutrients

a) Nitrogen

A general trend for standstill or decreased nitrogen load could be remarked in the case of upstream countries (see Fig. 2). For Romania, the time nitrogen load dynamics show a maximum in the 1988-1991 time frame and then a slight decrease.

A 2.43 ratio between the last station load and the reference year/station has been recorded for total nitrogen. The "R" curve shows an asymmetric shape with a maximum on the Romanian territory (7-8 areas) but with values over 1.0 (1.5) in comparison to some of the upstream countries (0.6-0.8). That means a specific local contribution with respect to the agriculture and other activities.

b) Phosphorus

A different situation could be recorded in the case of phosphorus load (see Fig. 3). The general trend for the upstream countries (except control station 5) is a decrease or steady state of phosphorous load. The "R" distribution shows a maximum in the Romanian territory (control station 7) and a asymmetric shape and decrease in space. However, the 7-9 areas show for nitrogen the local anthropogenic contributions. The report between the last/reference/year station is however more that 1.0 (2.0). The general trend-exception, control station 5, is a time decrease of phosphorus load, but with a slight space increase. In the case of Romania, there is a significant drop of P load after 1989 (1/3...1/4) which is followed by a tendency to a stationary value.

Buffering Capacity

As a general remark, alkalinity does not present significant time or space variation. However, there is a general trend of slight decrease for the 1973-1985 period of time and thereafter, a steady--state condition occurs. Similar behavior is proper to the pH value which is situated in the range 7.8-8.2. A slight tendency for the alkaline range could be seen. It is interesting to notice that in the years 1966-1973 for some areas of Romania, the pH of the Danube River was situated in the range 6.5-7.0.

Salinity

There is a general increase in the total dissolved salts from the first station to the entrance in the Danube Delta.

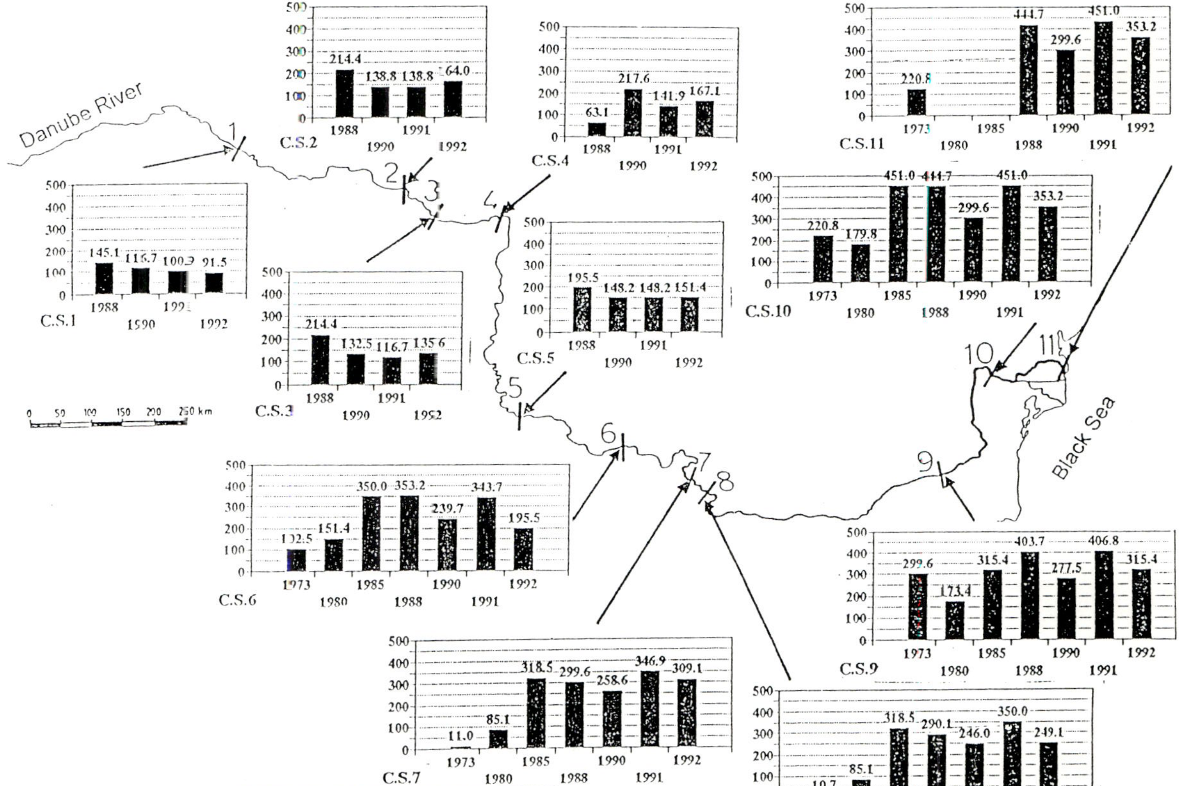

Central section	1	2	3	4	5	6	7	8	9	10	11
Last	91.5	164.0	135.6	167.1	151.4	195.5	309.1	249.	315.4	353.2	353.2
Initial	145.1	214.4	214.4	63.1	195.5	102.5	11.0	10.7	299.6	220.8	220.8
R.	0.63	0.76	0.63	2.64	0.77	1.907	28.1	23.23	1.05	1.59	1.59

Figure 2 Time evolution of fertilizers loads (annual average values: N kt/year)

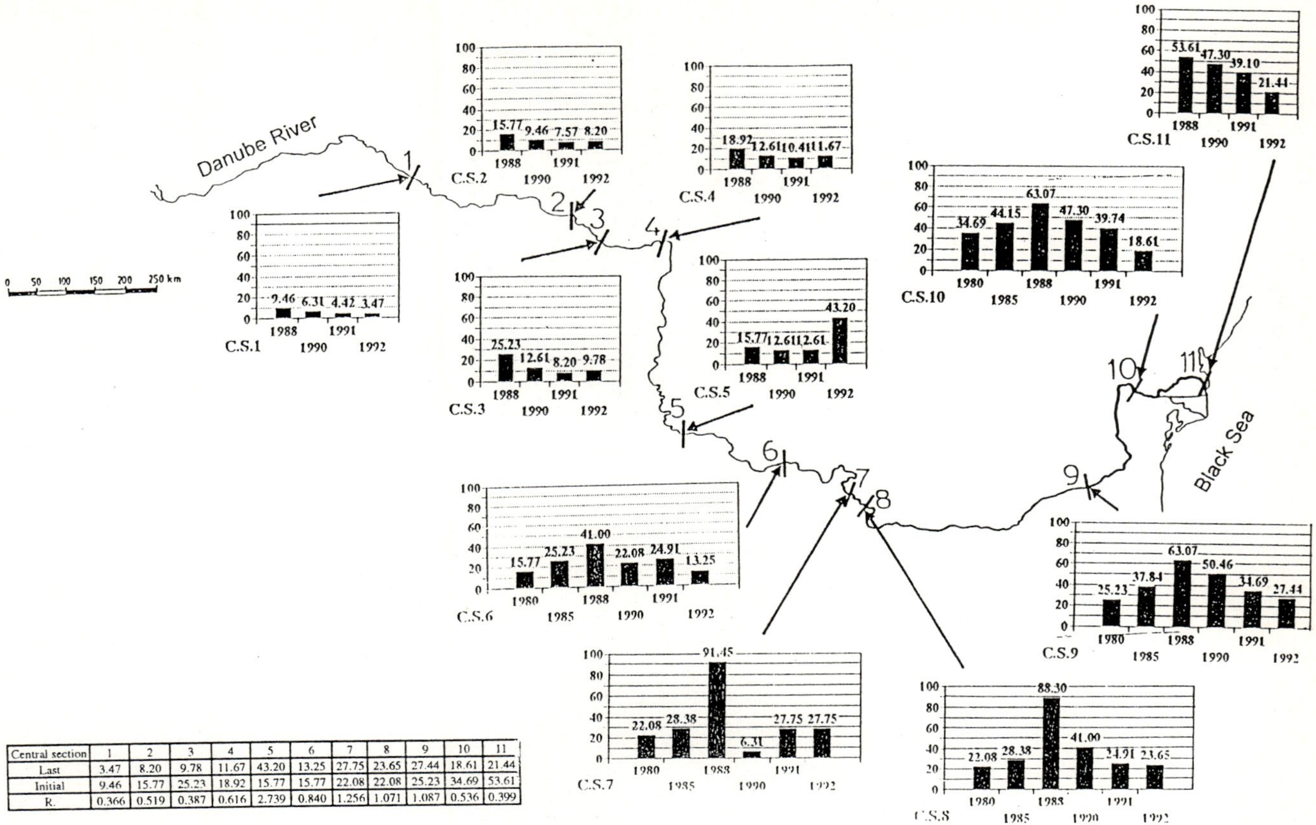

Central section	1	2	3	4	5	6	7	8	9	10	11
Last	3.47	8.20	9.78	11.67	43.20	13.25	27.75	23.65	27.44	18.61	21.44
Initial	9.46	15.77	25.23	18.92	15.77	15.77	22.08	22.08	25.23	34.69	53.61
R.	0.366	0.519	0.387	0.616	2.739	0.840	1.256	1.071	1.087	0.536	0.399

Figure 3 Time evolution of fertilizers loads (annual average values: P kt/year).

The same time and space distribution also apply to chloride, where the specific parameters are proportional to Cl^-, Na^+.

Parameter	Concentration	Load
a) ratio "R": last year (91) station (1) to the reference station (1) - 1988	1.97	6.96
b) "R"	1.41 (7, 10)	4.4-7.3 (7, 9, 10, 11)

Micropollutants

An overview of the heavy metals significant functions and parameters is presented in Table 1.

Table 1 Variation in heavy metals characteristics.

Metal	T*		R max**		Time trend	
	concent-ration	load	concent-ration	load	upstream (1-6)	downstream (6-11)
Cd	27	152.7	2.3 (S8)	3 (S8)	decrease	decrease
Hg	3	27.6	-	-	-	- (no date)
Pb	11.5	57.5	2.12 (S10)	11.4 (S10)	increase (S6)	increase
Cr	10	48	0.6-0.7 (S7/S8)	3.2-3.3 (S7/S8)	decrease	decrease

* report between concentrations (loads) in the entrance of the Danube Delta (1991) and those in reference station (1) (year of the beginning in data base).

** report between concentrations (loads) in the last year and those in the first one at the corresponding station number (time trend).

Load discharged into the Danube Delta (1991): - Cd 489 T/yr , Hg 88.3 T/yr (1985), Pb 4,793 T/yr, Cr 1,810 T/yr.

Cd and Pb especially point out the most significant pollution increase from the upstream countries to the Danube Delta and the time extension at least for load. There are some super positions of "hot spot" (local one) - (S8), which show the anthropogenic effect cause by tributaries heavy metals transport.

In the case of pesticides, there are a few data which point out the following observations: There is a general increasing pollution of the Danube River from the station 1 to station 9 (the concentration ratio shows an increase of about 5 times - and for load 22.4 times). The most polluted area begins with area 5, but after that is decreasing. However, the remaining concentration is higher in the Danube Delta compared with station one. The approximate load discharged into the Danube Delta is around 4 T/yr (1992).

Diffuse / Point Sources Based on the Immission Data

The importance of diffuse sources is different, depending on the country, time and determinant (see Table 2, Figs. 4 and 5). Even the range of variation is very large: minimum % was found for chloride (12%) and the maximum one for phosphorous (89%).

Table 2 Relative Importance of Pollution from Diffuse Sources .

Determinant	Range%(station)	Predominant area	Time trend	Average %
NITROGEN	28.2(S3/91)-65.6(S5/92)	5-11	increase	55
PHOSPHOROUS	31(S1/92)-88.9(S11/90)	-	decrease	50
BOD_5	31(S7/90)-64.7(S1/90)	-	increase	50
COD-Mn	21(S10/90)-65(S1/91)	1-3	increase	45
TDS	26.6(S5/90)-53(S11/92)	5-11	increase	35
Cl-	12.1(S10/90)-61.6(S9/92)	6-11	increase	30

However, as can be seen from the above table, the N, P are the most significant as the "determinant" for diffuse sources, based on the immission data processing. The trend of the % is obviously to increase with some exception in the case of phosphorous.

Country Specific Problems

The most important problems for Romania are related to the following:

(i) focal areas where the "R" is inside the country showing an "export" of pollution through the Danube River into the Danube Delta and the Black Sea,

(ii) the natural retention capacity of pollutants in the Danube Delta,

(iii) ground waters nutrient trends.

The issues (i) and (ii) should be related to the "import" of pollutants.

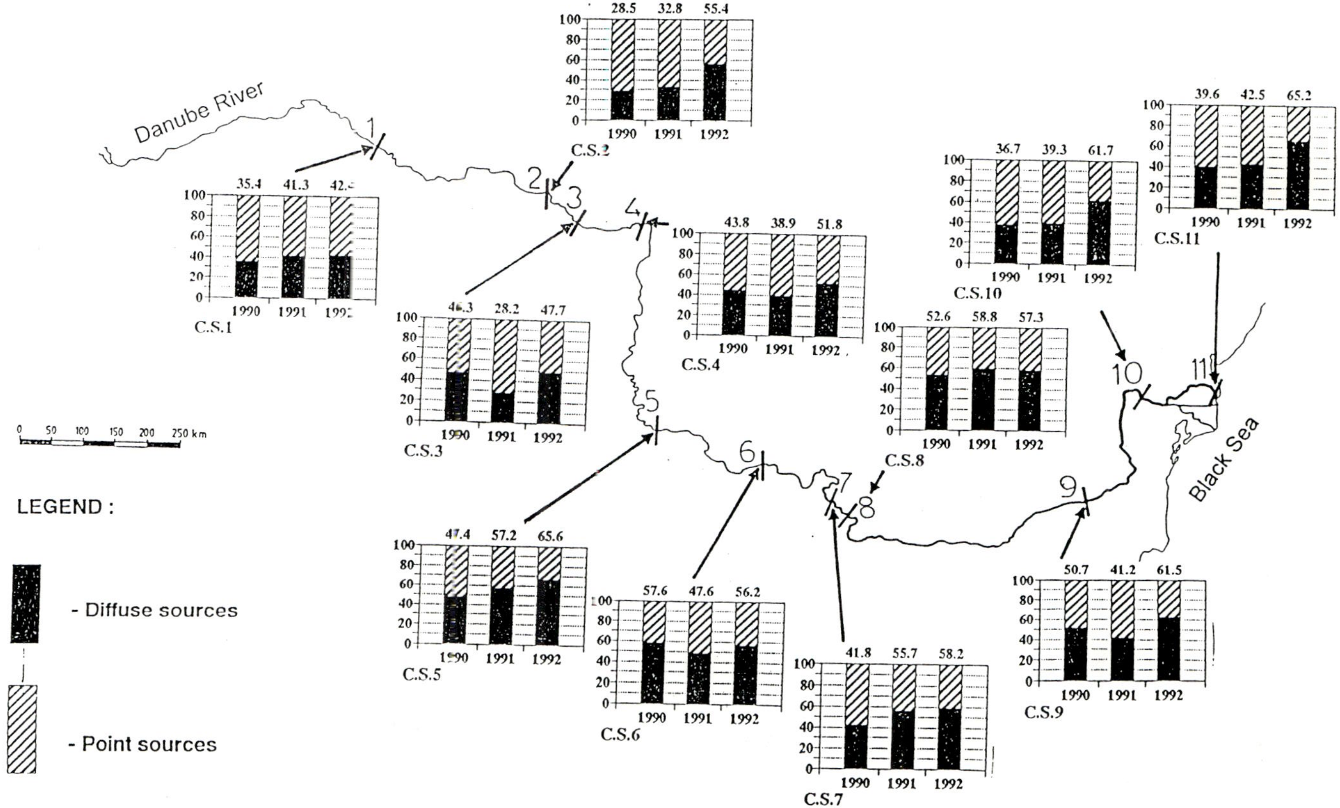

Figure 4 Time evolution of fertilizers diffuse sources percentage (annual average values: N %)

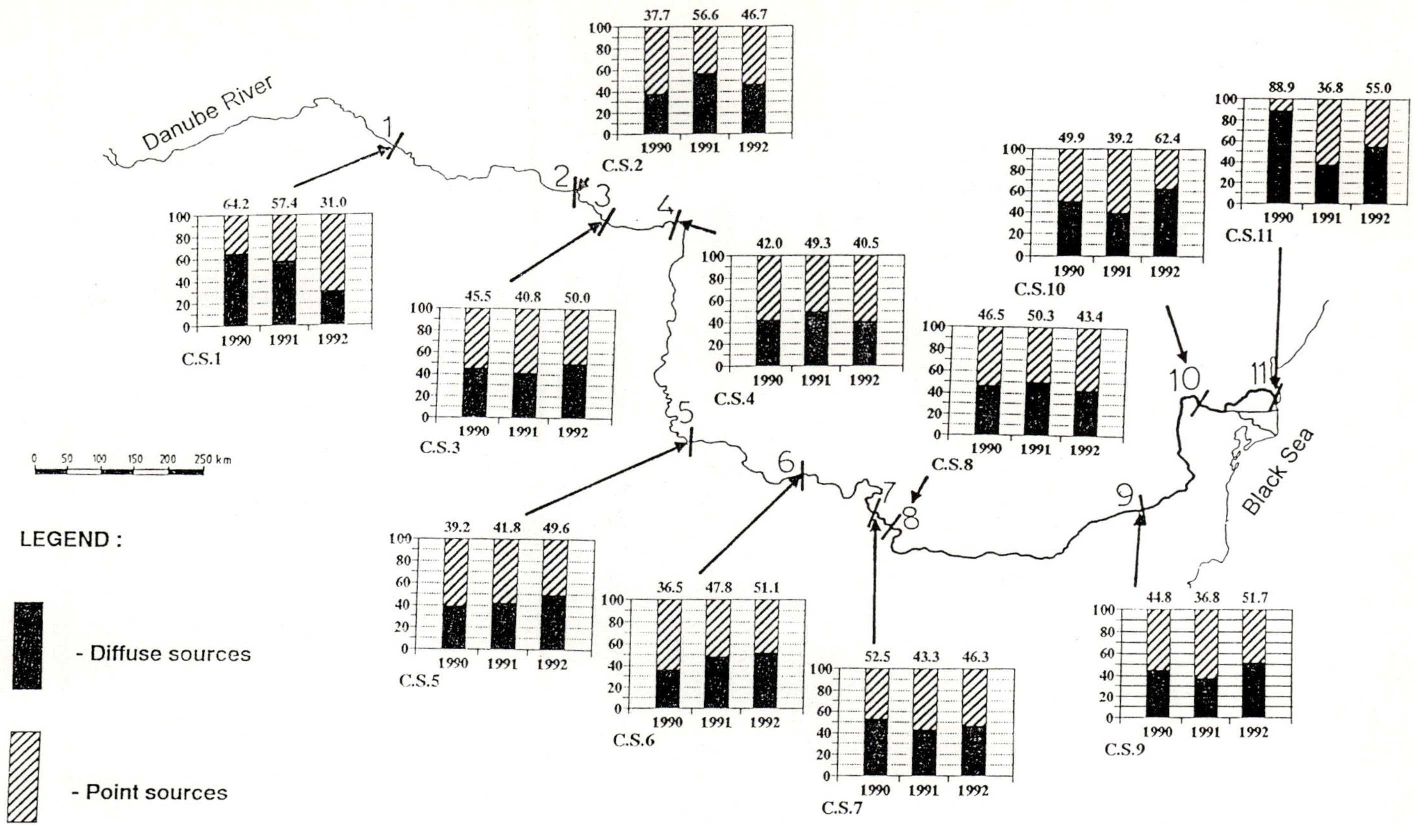

Figure 5 Time evolution of fertilizers diffuse sources percentage (annual average values: P %).

Space Distribution of the Ratio "R"

In accordance with the Figs. 2 and 3 the following remarks could be made:

Areas 7 and 8 make local significant contribution in the increase of nitrogen loads. Focal area 7 reflects a local contribution in the increase of phosphorous load. Data at point "6" (entrance of Danube River in Romania), indicate a maximum value of R for nitrogen loads (Varduca, 1994). On the other hand, the maximum R for COD-Cr (1.9 and 1.3) and for P (control station 5-2.7) are located in the upstream area (Towards Sustainability, 1992; United Nations, 1990). This is another problem related to the "import" of pollution and the state of the "Iron Gate 1" lake, which is almost affected.

Nutrient Budget on the Danube Delta and the Export to the Black Sea

As previously mentioned, the wetlands generally and most likely the Danube Delta have a natural capacity for removal of nutrients. Using the data concerning nitrogen and phosphorous concentrations (1991) upstream of the Danube Delta and at the discharges from each Danube arm into the Black Sea, the following remarks should be made:

(i) 14.4 % of the nitrogen load and 8.2% for the phosphorous are "filtered" by the Danube Delta

(ii) the retention coefficients are different for each Danube branches:

* for Chilia	1.6% (N) and 1% (P)
* for Sf. Gheorghe	53.8% (N) and 17.4% (P)
* for Sulina	31.6% (N) and 26% (P)

In comparison to the literature, these figures are smaller. One of the most important reasons is the regulations-dams and other hydrotechnical water works provided against flood conditions and for recovering arable lands. As a result, the Danube River arms water is not spread to the land but circulates mostly on the canals (especially Chilia). There are some positive exceptions (Sf. Gheorghe arm and Sulina arm) where the retention percentage for N and P are still good. It is obvious that there is no significant "filtering" effect of the Danube Delta for the non biodegradable organic compounds. The retention capacity of the Danube Delta is more effective than denitrification or sedimentation process which could occur in the Danube River.

Global Ecosystems Water Quality Trend

Despite the chemical and physical variables based on the biological determinants, the overall ecological water quality trend could be assessed.

Saprobic index trend

The Pantle Buck method was adopted for accounting the saprobic index (Bucharest Declaration). It is based on the following relation:

$$S = \frac{\sum(s \times h)}{\sum h}$$

where:

S = saprobic index for the station (1...11)

h = organism frequency: (1 - not usually, 2 - frequent, 3 - very often);

S = 1.0-1.5 for oligosaprobe condition, (I)

S = 1.5-2.5 for ß-mezosaprobe condition, (II)

S = 2.5-3.5 for -mezosaprobe condition, (III)

S = 3.5-4.0 for polisaprobe condition. (IV)

As an overall image, the saprobic index corresponds to the II class. The water presents a slight tendency of decreasing the quality from biological criteria point of view in time and space which reflects obviously the chemical biota quality.

Quality classes

As a guide line, the Draft ECE Standard Statistical Classification of Ecological Freshwater Quality (CES 668/1990) proposed for international water bodies (Rhine, Danube etc) has been adopted. In accordance with this, there are five Water Quality Classes:

I - Excellent (Blue)
II - Good (Green)
III - Fair (Yellow)
IV - Poor (Orange)
V - Bad (Red)

The variables which are considered in this classification are divided in seven groups as:

A - oxygen regime (DO%, BOD_5, COD-Mn);
B - eutrophication (total P, total N);
C - acidification (pH, alkalinity);
D1 - harmful substances (heavy metals and cyanide);
D2 - pesticides;
E - radioactivity;
F - microbial pollution.

Taking into consideration the available data, the following remarks could be raised:

Concerning the space distribution of the Danube water quality, there is a general trend of decreasing, but in different degree for each variables group.

The oxygen regime started from the excellent class, after which follows a stationary good class. There are 331 km in the upstream course of excellent class (15.3%) and 1873 km (84.7%) of the good class.

Referring to the eutrophication variables, as a consequence of high concentrations for N and P, even in the first segment there are 550 km (24.3%) of bad class. In the downstream there appears a slight improvement - 499 km (22%) of fair class, but the most significant weight belongs to the poor class - 1175 km (53.7).

From the buffering capacity point of view, the entire Danube River belongs to the excellent class (100%). Heavy metals, especially downstream of station 6 (Cd, Hg, Pb), have a very important effect on the quality class. Only 15.1% (331 km in the upstream countries) belong to the "fair class" and the rest, 1855 km (84.9%)m, to the "poor class". Nearly the same quality class corresponds to the pesticide group - 6.9% "fair class" (152 km) and 93.1% to the "poor class" (2.034 km). The bad conditions referring to the harmful substances are amplified to the downstream countries.

Referring to the time trend, there is an evident tendency of nutrient pollution and as a result, 100% of the Danube River belongs to the "poor class". For the heavy metals and pesticides, it is a standstill condition in 1992, as compared to 1988. As an overall image, despite the dissolved oxygen regime and buffering capacity, the global Danube water quality corresponds to the "poor class".

Comparative Analysis for Nutrient Reduction

At least two elements should be analyzed for the selection of the most advisable scenario:

(i) the balance between strategic objectives and technical/financial support needs, and

(ii) a holistic approach for the establishment of consequences which could occur by intersectorial relationships between environmental functions reported to environmental themes specific for each scenario.

Balance between Strategic Objectives and Technical/Financial Needs

The contribution of each target group for N, P emission loads are different and as a result, the measurements will give differential outputs. In other words supposing only a 10% reduction for each target group, the whole reduction could or could not cover the scenario objective as a function of proper weight of the respective group. The population group emission load represents 15.3% in N and 9.5% in P of the total emission. A 10% reduction will cover only 1.5% of N and 0.9% P. Therefore, considering the measures only for population, it will be possible to balance 2.7% N and 1.3% P from the agriculture emission or 5.6% N and 5.2% P from the industry . As a consequence, the scenario with emphasis on population only is fair to reach the target reduction. The agriculture target group is more representative in the total emissions: 57.6% N and 72.3 P. A 10% reduction applied only to this group will cover 37.6% N and 76.6% P for the industry. The same 10% reduction only for the agriculture emission could cover nearly 6% N and 7% P for the total emission decrease. This is 3 times (N) and 8 times more (P) than if reduction to the population group only was applied or 2 times (N) and respectively 3.5 times (P) higher than the industry weight in this respect. As a whole, the investment in this "agriculture group" is more advisable. 10% reduction to the industry group only could cover 2.7% N and 1.8% P from the total emission, which is equivalent to about 18% N and 20% P reduction from population. As a general remark, the hierarchical measures should be taken in the following order: (a) agriculture, (b) industry, (c) population.

More details concerning alternative actions to reduce nutrient emissions, the related costs and proposed legislation measures can be found in Varduca, 1993 and 1994.

Conclusions

The actual nutrient load transfer by the Danube River into the Danube Delta and the Black Sea exceeded the self purification capacities and generated significant disturbances of aquatic ecosystems with all consequences even for environmental functions (recreation, fisheries, drinking water supply). In Romania, the agriculture target group is more representative in the total emission of N and P, followed by industry and population. Investment in this "target group" is more advisable. For Romania, the most acceptable scenarios are based on the transition of point discharges from agriculture large farms to the diffuse sources and charge to the organic farming, and the most expectable legislation will create new, strict obligations for point discharges quality conditions. One step to stop this process is the harmonization of national scenarios into a general framework strategy at the basin area. The approach presented in this paper has been applied also to micropollutants.

References

Varduca A. (1993) - Danube Integrated Environment Study - Country Report Phase 1 - Dec. 1993 - I.C.I.M.

Varduca A.(1994) - Danube Integrated Environment Study - Country Report Phase 2 - March. 1994 - I.C.I.M.

Towards Sustainability (1992) - A European Comunity Programme of Policy and Action in Relation to the Environment and Sustainable Development - Brussels, 27 March 1992, vol II.

United Nations (1990): Economic and Social Council Statistical Commission and Economic Commission for Europe - Statistics of Water Quality, CES/668, 15 March 1990

Surface Waters Categories and Quality Technical Requirements - STAS 4706/88

Drinking Water Quality Technical Requirements - STAS 1342/91

V.6 Management and Protection of the Black Sea Environment: An International Approach

Laurence D. Mee
Coordinator, GEF Black Sea Environmental Program
Florya, Instanbul
Turkey

Introduction

The European continent has three large semi-enclosed seas of similar area, which are heavily impacted by human activities: the North Sea (area 5.75 x 10^5 km^2), the Baltic Sea (3.73x10^5 km^2) and the Black Sea (4.2x10^5 km^2). The Black Sea is unique among these as it is deep (maximum 2212m) and virtually isolated, connected with the Mediterranean by a narrow and shallow channel (the Bosphorus). This situation has resulted in its being permanently anoxic below a depth of 150-200m. Indeed, 90% of its 5.37x10^5 km^3 is anoxic, the largest anoxic water mass on our planet. Not all of the Black Sea is deep, since about 25% of the Black Sea's area is occupied by its north-western continental shelf less than 200m deep.

This area receives the discharge of two of Europe's largest rivers, the Danube (alone contributing about 203 km^3 of water/year, more than the entire fresh water input to the North Sea) and the Dnieper (contributing an additional 54 km^3), as well as additional input from the Dniester (9.3 km^3/yr.) and smaller rivers. The eastern Black Sea also receives water from the Don (28 km^3/yr.) and the Kuban rivers (13 km^3/yr.), which discharge their waters through the shallow Sea of Azov, and there are numerous smaller rivers draining from the southern and eastern coasts (see Fig. 1).

The shallow, biologically productive layer of the Black Sea thus receives water from a vast drainage basin over five times its own area covering huge industrial and agricultural areas of 9 countries. At least 162 million people live in the Black Sea Basin (cf. 90 million for the Baltic). About 81 million of these live in the Danube basin alone (cf. 41 million for the Rhine). The Russian and Ukrainian portion of the Black Sea basin includes 68 million inhabitants (25% of the population) and the Turkish portion, 12 million inhabitants (20% of the population of Turkey). Additionally, one of the world’s major cities, Istanbul (pop. about 9 million), is situated on the narrow Bosphorus Strait, which is the critical connection between the Black Sea and the World Ocean via the Mediterranean.

The Black Sea drainage basin covers a vast area of the European Continent and drains a large part of Turkey, Bulgaria, Romania, Ukraine, Russian Federation, Georgia, Hungary, Belorussia, Croatia, Slovenia, Serbia, Slovakia, Austria, Moldova and Germany and even receives minor inputs from the Czech Republic, Switzerland, Poland, Italy and Albania. The names of the principal rivers are shown in italics.

The Black Sea itself has already been the victim of unmanaged fisheries (on an international level), of unrestricted intense shipping activities (with no formal controls on deballasting, etc.), of mineral exploitation and the dumping of toxic wastes. Clearly, the threat to the Black Sea from land-based sources of pollution is potentially greater than in any other marine sea on our planet. In this report, the author will describe some catastrophic environmental changes in the Black Sea which have already seriously reduced its value as a resource to the coastal states and will discuss how these problems might be tackled. Three priority environmental issues have been selected for detailed consideration. These are eutrophication, induced alterations in marine ecosystems and chemical and microbiological pollution. These problems will be dealt with from the point of view of scientific evidence for the problem (including causes and effects), socioeconomic consequences and possible remedies. Coastal development issues (including changes in coastal morphology and the loss of freshwater resources) will also be briefly reviewed.

Figure 1 Black Sea Drainage Basin.

Eutrophication

The evidence is overwhelming that a major part of the Black Sea (notably the northwestern shelf) is increasingly critically eutrophic. The Danube (as monitored within the framework of the Bucharest Declaration) currently introduces about 60,000 tons of total phosphorus/year, about the same as the total river input of P to the North Sea or to the Baltic, and some 340,000 tons of total inorganic nitrogen/year, more than double the load from the Rhine but less than half of the total river input to the North Sea. Some data is available from other rivers, but loads have yet to be compiled. Sufficient historical data exists, however, to demonstrate that there has been an enormous increase in the nutrient load to the Black Sea in the past 25 years, probably as a consequence of the intensification of agriculture (often using fertilizer but also increasing the leaching of nutrients from soils) and the more widespread use of phosphate detergents. Recent estimates for the Danube basin and its tributaries have revealed that about half the nutrient load comes from agriculture, a quarter from industry and a similar amount from domestic sources.

Fig. 2 illustrates the increase in nutrient discharge for the case of the Danube and the continental shelf waters of Romania, where a spectacular increase in dissolved phosphate was observed. Increases in concentration of threefold (nitrate) and sevenfold (phosphate) have also been measured since the 1950s in the Dniester. Additionally, some coastal settlements discharge their sewage and industrial wastes directly to the sea.

The immediate consequence of increased nutrient supply was the intensification of phytoplankton blooms, particularly on the shallow north-western continental shelf where nutrients are readily recycled. The fate of nutrients in the open sea where the oxic/anoxic boundary acts as a "liquid bottom" is less certain. Such blooms may have initially been beneficial to the marine food chain but the longer term effects were devastating. The effects can be summed up as follows:

(i) A gradual basin-wide shallowing of the euphotic zone (the surface layer in which the light level is sufficient for net biological primary production). Even in the open Black Sea, Secchi depth measurements have decreased from 50-60m in the early 1960's to present values of about 35m, or often less than 10m in coastal waters.

(ii) The decreased light penetration has resulted in the massive loss of shallow water macrophytes, formerly an important component of the Black Sea ecosystem and a major economic (commercially harvested) resource. For example, on the northwestern shelf, only 300,000 tons of Phyllophora remain, a 50 km^2 patch representing 5% of the original area.

(iii) The eutrophication has dramatically modified the base of the marine food chain, encouraging the development of almost monospecific dense blooms of nanoplankton (and occasionally algae). Higher trophic levels have also become modified with vast explosive blooms of zooplanktonic Noctiluca miliaris (with a very low food value for higher species) in the 1970s, spectacular basin-wide biomasses of jelly fish (Aurelia aurita) of up to 450 million tons in the mid 1980s and the present virtual total dominance of the opportunistic predatory ctenophore Mnemiopsis leidyi (with biomasses of up to 1 kg/m^2 in the open sea and 5 kg/m^2 on the shelf and a negligible food value to higher organisms).

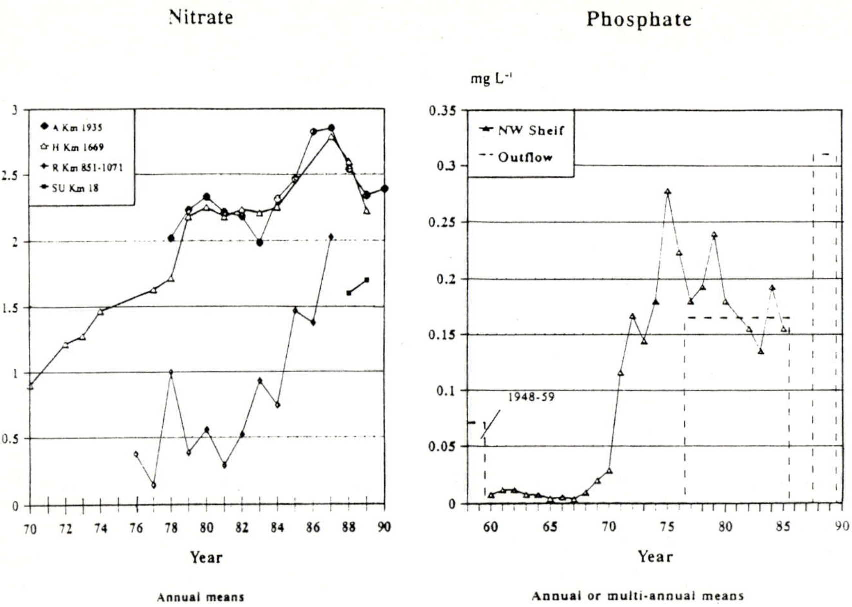

Figure 2 Changes in annual mean nitrate concentrations at various points in the Danube, phosphate concentrations in the Danube outflow averaged for 1948-59, 1977-85, and 1988-89 and the annual mean phosphate concentrations on the Black Sea continental shelf of Romania. Country codes are: A = Austria, H = Hungary, R = Romania, and SU = Ukraine.

(iv) The enormous increase in organic matter falling to the shelf floor from blooming and decaying organisms has resulted in widespread hypoxia. Thus, 95% of the Soviet NW Black Sea shelf and the entire Sea of Azov are now prone to hypoxia and the occasional formation of an anoxic benthic layer of hydrogen sulfide rich water (distinct from the permanent anoxic layer in the main basin). This has led to the complete elimination of a large proportion of macrobenthic organisms and the death of a formerly rich fishery (e. g., of bivalves).

On the Romanian coast alone, a single event of anoxia in 1991 eliminated an estimated 50% of the remaining benthic fish. The loss of benthos is not only restricted to the shelf, but, to a greater or lesser degree, affects the entire northern Black Sea coast. Recent evidence (Fig. 3) suggests that macrobenthic communities are disappearing from formerly pristine sites on the central and eastern Black Sea coasts.

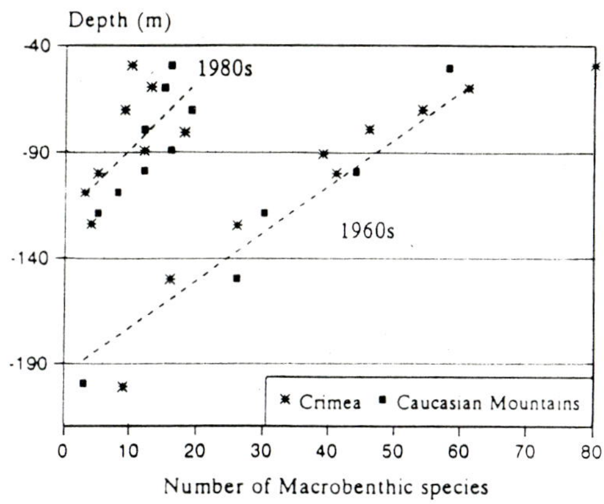

Figure 3 Change in the number of macrobenthic species as a function of depth from the 60s to the 80s at the same sampling sites in the coastal waters adjacent to the Crimea and the Caucasian mountains (after Zaika).

(v) The profound alteration of the marine ecosystem of the Black Sea primarily as a result of eutrophication has contributed to the demise of its fisheries. Sensitive, high-value species disappeared to be replaced ephemerally by large quantities of small omnivores such as anchovies, the fisheries of which has now also collapsed (see section 3.3) for reasons which are not entirely clear. Of 26 species of commercial fish in the 1960's, only 6 remain in significant exploitable quantities.

Socioeconomic Consequences

In a matter of 25-30 years, the Northwestern Black Sea has been converted from a diverse ecosystem supporting a highly productive fisheries industry to a eutrophic plankton culture, which leads to environmental conditions unsuitable for most higher organisms. Furthermore, the amenity value of these areas has diminished; the sea is now green or brown, and beaches tend to smell of decaying organic matter.

In the case of countries such as Romania, with only some 240 km of coastline, the loss of fisheries and amenities is an important socioeconomic problem. The Romanian fishing fleet is currently laid-up (except for a limited Atlantic fishery), but 1.5-2 million national tourists still visit the beaches annually, as there are no alternatives available. Certainly the potential for attracting foreign tourists is not strong. The economic consequences to Ukrainian fisheries and tourism is probably proportionally greater, though the Crimean coastal area is still not visibly affected by severe eutrophication. On the shelf area, however, there is a loss of biodiversity, which is not merely an academic matter since the disappearing macroalgal communities are proven resources for the production of natural products including agar, natural dyes and raw materials for pharmaceuticals.

There is a clear need for a study of the environmental economics of eutrophication in the Black Sea if this issue is to be given a higher place on the political agenda of riparian states. This is an important missing element from the otherwise conclusive scientific evidence for urgent action to be taken to control nutrient sources to the Black Sea.

Remedies

The fate of the Black Sea will be closely linked to the success of the Environmental Management Plan for the Danube basin and future similar efforts for the other major rivers. With effective waste treatment and management, nutrient loads can be reduced substantially. This probably explains the large difference between nitrate/phosphate ratios of 10.8:1 in the Rhine and 4.1:1 in the Danube, given that 20 years of effort have been put into reducing the phosphate load of the Rhine. However, this matter has to be approached in a pragmatic manner, which takes into account the economic realities of Eastern European nations. Effective reduction of nutrient loads requires major changes in the industrial production of detergents, changes in agricultural techniques and the efficient treatment of domestic and industrial wastes. This is an onerous task. Current liquid waste discharges directly to the sea from Romania, for example, amount to over 1.7×10^8 tons annually and for Turkey 5×10^8 tons. Reducing phosphate loads to the Rhine by 50% has taken 15 years--this with the strong economies and the political commitment of all riparian states. With 5 times the human population of the Rhine basin and struggling economies, the clean-up of the Black Sea drainage basin may take many decades.

The immediate scenario is, unfortunately, a rather bleak one. The provision of adequate and relevant scientific information for taking necessary management steps is of vital importance. For example, eutrophication cannot merely be "switched off" by reducing nutrient loads. Nutrient recycling (particularly from the continental shelf areas) would fuel eutrophication for many years or even decades, of course with gradually diminishing intensity. Recycling rates for the limiting nutrients (nitrogen and phosphorus) are not equal and the gradually changing ratios of these two elements would modify the nature of the plankton community forming the base of the food chain. This phenomenon must be modeled in order to make economic assessments of the benefit of different management regimes (management of industrial and domestic waste discharges and of agricultural practices). Similarly, economic constraints will not permit a total shut-down of nutrient supplies in the foreseeable future. Environmental engineers should be given clear priorities to limit nutrients and to evaluate how the marine environment will respond to such measures. Scientists have been debating whether nitrogen or phosphorus are the critical limiting nutrients in the sea for half a century, and sadly this futile debate still continues in fora within the Black Sea region. The current more realistic view is now that communities gradually adjust to a wide range of N/P ratios and also to other chemical and physical constraints (often by changing species composition), but that a steady-state system is rarely (if ever) attained. It is the food value of these communities to higher organisms (and the trophic continuity to exploitable species of economic value) which underpins our current concept of "value" of the ecosystem. Clearly, though highly productive, a heavily eutrophic environment resulting in trophic "dead ends" is valueless, but it may not be necessary to restore the Black Sea to its pre-detergent nutrient levels in order to regenerate an ecosystem with an economic yield. Just how much the levels should be cut is a matter requiring urgent study. The view of the author is that current research tends towards a post-mortem approach rather than one leading to better environmental management.

As a final note on the topic of eutrophication, it is worth recording that in most Eastern European countries, as a result of economic hardships in the past five years, fertilizer application has dropped considerably. The decrease is typically four or five fold. In the case of Moldova, for example, application has dropped from about 150 kg/hectare to some 40 kg/ha, and agricultural production has been halved. This decreased productivity clearly has serious economic consequences for the riparian countries, but conjectural evidence suggests that the fisheries on the NW shelf of the Black Sea have recovered only slightly in the last two seasons. Much hard scientific evidence would have to be accumulated to establish a relationship between these two observations. If the two observations are linked, the response time of the system would appear to be rather fast. Clearly, this would give encouragement to the inclusion of measures to reduce nutrient runoff as part of agricultural reform. Specific measures to make such a reduction could include proper pricing of agrochemicals and regimentation to enforce a buffer zone between plowed arable land and rivers. The efficiency of such meadowland and wetland buffers has been shown to be considerable. They provide additional benefits, including recreational areas, pasturelands and protection of subjacent aquifers containing drinking water. The present period of economic transition and reform offers a unique window of opportunity to address these issues. Such policy recommendations, however, must be backed with scientific data and with research on environmental economics.

Induced Alterations in Marine Systems

Eutrophication is not the only culprit for catastrophic changes in the Black Sea environment. Other causative factors include the introduction of opportunistic species, changes in the hydrological balance, overfishing and chemical pollution. The first three of these will be considered in this section, and the chemical pollution issue will be treated separately.

Opportunistic Settlers

With its high primary productivity (both natural and by eutrophication), and low salinity (average about 22 $^{o}/oo$), the Black Sea presents ecological conditions that in many ways resemble those of the lower reaches of a large temperate estuary. As species diversity becomes reduced (often as a result of eutrophication), opportunistic settler species, brought in the ballast water of ships, can easily find an ecological niche in which to flourish. The first documented case was that of the predatory sea snail Rapana thomasiana, which probably settled from Japan in the late 1940s and is widely held as responsible for the demise of commercially harvested oyster populations and a general decrease in biodiversity. In recent years, it has itself become the target of heavy fishing (the harvest is exported to Japan!), and stocks are now declining.

A second notable case was the soft-shelled clam, which first appeared in the late 1960s and successfully competed against the local species Corbula Mediterranean to achieve densities of over 1,000 individuals/m^3 (about 1 kg/m^2) on the Romanian shelf. These large populations may actually improve the capacity of the ecosystems for self purification. However, large banks of Mya and mussels have now been eliminated by the effects of temporary hypoxia and anoxia.

There have been several other species introduced, but none has had the impact of the ctenophore Mnemiopsis leidyi, which appears to have been introduced from the estuaries on the eastern seaboard of the USA and, as mentioned earlier, has had an explosive development since 1988. Together with the jellyfish Aurelia aurita, it represents a trophic "dead end" and worse still, feeds on other zooplankton, including fish larvae. As yet, it appears to have no natural predators.

Changes in the Hydrological Balance

The construction of dams on major rivers for irrigation has resulted in a substantial net decrease in the runoff to the Black Sea (see Table 1). There may be an alteration in exchange through the Bosphorus and negative consequences, even in the distant Eastern Mediterranean. In the case of the Don and Kuban rivers feeding the sea of Azov, the reduction is already 27% and 49%, respectively. This has resulted in a net increase in salinity of more than 2% in the previously brackish Azov Sea (which is also catastrophically eutrophic) and a decrease in the exchange of water through the Kerch straits. As a result, the migratory patterns of many fish have been modified, valuable anadromous sturgeon have declined, and the fisheries production has dwindled.

Table 1 Reduction in annual discharge of some rivers draining into the Black Sea.

Natural % Reduction				
River	Flow ($km^3 yr^{-1}$)	1971-75	1981-85	1991-2000 (projected)
Don	27.9	19%	27%	43%
Kuban	13.4	39%	49%	65%
Dnieper	53.5	24%	52%	71%
Dniester	9.3	20%	40%	62%

The Case of Turkish Anchovies. Overfishing?

By 1985, the most significant fisheries of marine species was on the Turkish coast. The Turkish fisheries had become increasingly reliant on high technology, and the total capture of anchovy and horse mackerel reached 338,020 tons (1984), of which all but 15,262 tons was anchovy (Engraulis sp.). The Black Sea fishery supplied 80% of the national demand for fish and had grown from a total of 68,347 tons in 1976. Within the three year period from 1987-89, the anchovy fisheries collapsed to under 15,000 tons. Various factors were involved in this matter:

(i) The possibility of overfishing. Turkey invested very heavily in its fisheries and provided large incentives for the fishermen to improve their gear. When it became clear that smaller fish were being captured, a 9 cm minimum limit was imposed, but, following lobbying, this was reduced to 7.5 cm, permitting the capture of the 1-3 year class stock. Now that the fishery has collapsed, the 9 cm limit has been restored, and a two year moratorium to the fisheries was declared. There appear to have been no thorough stock assessments made (e.g., by the egg/larval method) in order to guide stock management.

(ii) The possibility of recruitment failure. The Turkish anchovy stock migrates from the northwestern Black Sea where reproduction occurs. This throws it to the mercy of the eutrophication and related problems, particularly to the predatory ctenophores. These may have inflicted heavy losses on the larval population and caused a failure in recruitment.

(iii) Alternative concepts. Turkish fishery scientists have suggested other possibilities. The first is that the migrating anchovies have been affected by anthropogenic pollutants. This seems unlikely, as there were no sudden changes in land-based activities over the time period of the collapse. The second, more worrisome theory is that the local (protected) dolphin population is getting more than its fair share of the stock. Formerly, dolphins were hunted in large quantities, particularly in the Soviet Union, but this practice has long ceased. Soviet fisheries scientists comment that there is "hardly a dolphin to be found" in the northern Black Sea and doubt that the population has simply migrated South. A stock assessment would quickly resolve this issue.

Other Factors Reducing Fisheries

The Bosphorus/Marmara Sea has suffered considerable ecological damage as a result of overfishing and pollution, and the catches of those fish that formerly migrated into the Black Sea are rapidly declining. If this process continues, the gateway to the Black Sea will become a barrier to its rehabilitation, and it will be deprived of major stocks of valuable migrating species (e.g., the Bonito).

Socioeconomic consequences

Unlike the adjacent Mediterranean Sea, the Black Sea was a traditionally rich fishing area. The fisheries supported a total human population of some 2 million people (fisherman and their dependents). This economic resource has now been practically destroyed. Biodiversity has also suffered, and the future options for resource development of the Black Sea have been reduced.

Remedies

A fisheries convention (the Varna Convention) was signed between Bulgaria, Romania and the USSR in 1960. Turkey remains outside this agreement. This convention did not contemplate regulation of fisheries, but it focused on data gathering. It thus had no significant impact on restraining the fisheries effort or managing stocks. The loss of fisheries resources in the Black Sea is an issue that transcends the usual boundaries of stock management, commonly a process of managing the activities of the fishermen themselves (managing people, rather than fish). The declining fisheries are a clear consequence of the degradation of the ecosystem itself, which is in turn intimately related to land-based human activities. These matters cannot be divorced, and the corresponding international institutions should work very closely together to assist the riparian states to incorporate the fisheries issue in any new legal regime for the Black Sea. In this context, it is important to note that a new broader-based fisheries convention is currently being finalized and should be ready for signature before the end of 1994. This convention will govern information exchange, stock assessment (of wild stocks), fisheries practice and effort, aquaculture, and will eventually provide a mechanism for allocating quotas.

What, if any, are the other options for management? Marine biologists (and particularly fisheries scientists) are loathe to suggest introducing new species to the marine environment (which may be conceived as meddling with nature). In the case of the Black Sea, the environment is already devastated, and serious consideration should be given to a search for predator species of Mnemiopsis, for example. The concept of environmental management is one which may offer the only alternative to an economically valueless ecosystem.

Aquaculture has been attempted on a pilot scale in some Black Sea locations. A Norwegian/Turkish joint venture is producing small quantities of salmon. The salmon, however, require feeding, and the potential source of food, anchovy fish flour, has disappeared, making the process costly (as the food must be imported). A small Crimean pilot plant produces mussels (mostly distributed to Soviet sanatoria), and Soviet scientists are experimenting with the culture of macroalgae to recede devastated populations. These endeavors are small, however. Aquaculture requires clean water and a cheap source of nutrients. Neither of these is easily available in the Black Sea. There are, however, newly emerging technologies. An example of these is sea ranching, where fish fry are artificially reared through the early life stages when they would be easy prey for Mnemiopsis. The juvenile fish are then released to the open sea where they develop to adult stages and can be harvested. The adults may even be able to outcompete Mnemiopsis for food and perhaps, species may be found which can prey upon the ctenophore.

If sea ranching is to be successfully employed, international cooperation will be vital, and special legislation to control the activity will have to be incorporated within a new fisheries convention. Questions arise such as which species will be authorized? How will the work be shared among the coastal states of stocking juveniles and harvesting the products? This innovative technology would require an entirely new approach to economic cooperation in stocking and harvesting the commons - an approach which transcends current thinking on free-market economic strategies.

Chemical and Microbiological Pollution

Data on chemical and microbiological pollutants are very scarce, partly because of the limited capacity of regional laboratories to conduct pollution assessments and partly because of the previous unwillingness of some governments to disclose such information. Clearly, this second problem is no longer applicable, and all riparian states have expressed their deep concern with the assessment of land and marine-based sources of pollution.

Land-based Sources of Pollution

A first clue to the seriousness of the problem for the Black Sea can be gleaned from the 1989 Bucharest Declaration monitoring data on the Danube outflow. Annually, the Danube alone discharges 1000 tons of chromium, 900 tons of copper, up to 60 tons of mercury, 4500 tons of lead, 6000 tons of zinc and a staggering 50,000 tons of oil. In all cases, these considerably exceed the entire river load to the North Sea (by 2100% in the case of oil, see Fig. 4).

Although comparative load data is not yet available on contaminants from rivers in the Soviet Union, it is clear that many of these rivers are seriously contaminated with industrial and mining wastes. Additionally, many coastal industries appear to discharge directly to the sea with little or no treatment, and loads transported through the atmosphere have yet to be quantified.

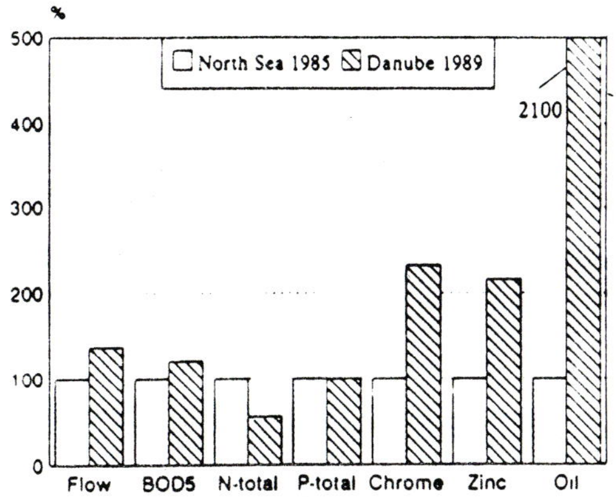

Figure 4 Total river loads of various chemical constituents to the North Sea compared with the Danube input to the Black Sea (from Dogterom by permission of the author).

Synthetic Organic Contaminants

The Bucharest Declaration monitoring program covers only a few of the more easily measured toxic contaminants and, as yet, excludes synthetic organic contaminants (such as pesticides and herbicides). There are no validated data (by international intercomparison exercises) on pesticides from the Black Sea. Notwithstanding the limitation, a group of scientists from the Romanian Marine Research Institute published data on chlorinated hydrocarbons in biota from the Danube estuary and Romanian coastline in the early 1980s. At that time, total DDT levels in fish ranged from about 1 g/g, dry wt. (Gobius sp.) to 4.6 g/g dry wt. (Sprattus sp.). This is close to the normal concentration limit for human consumption and is 5-10 times the comparable 1980 value in northern Baltic fish. Water measurements conducted in 1981-82 in the Sinoe lagoons of the Danube delta revealed total DDT levels of 32-486 ng/L and lindane concentrations in the range 20-550 ng/L, about 1000 times those observed in north-western Mediterranean coastal waters. Recent levels of DDTs from the Kerch Straits between the Azov and Black Seas are between 8-20 ng/L, two orders of magnitude higher than in the Mediterranean. It is, perhaps, unreasonable to draw too many conclusions on the basis of this scanty information. DDT production (but not lindane) is banned in Romania and the Commonwealth of Independent States, and its place has been taken by organophosphorus pesticides. In the CIS, however, there is considerable concern for apparent linkages between pesticide application and adverse human health effects. Certainly, water from the Dnieper is used to irrigate large areas of rice paddies where insecticides and herbicides are applied in considerable quantities. Throughout the region, the study of pesticides and PCBs is barely commencing, and much work will be needed to develop a regional capacity for assessment (which should also include an assessment of biological effects).

Heavy Metals and Radionuclides

In the case of heavy metals, there is a stronger data base but, in the case of sediment data, this is again rather difficult to interpret as "normalizing parameters" (sediment characteristics, aluminium and organic matter measurements) are rarely published, and there is usually no indication of whether the data are validated by participation in intercalibration exercises (although these were regularly conducted in COMECON countries). The preliminary evaluation of publications from Romania and Bulgaria suggests that trace metal contamination is generally limited to well defined source-related hot-spots and is not a basin-wide phenomenon. In the case of mercury, some seafood samples taken near the Dnieper estuary were above recommended limits for human consumption.

As a result of the Chernobyl accident, the concentration of artificial radionuclides in various compartments of the Black Sea is well studied (by scientists from CIS, Romania, Turkey and the USA, among others). In no other regional sea does nuclear contamination engender such widespread general concern. Understandably, Chernobyl has become the scapegoat of many people who blame all kinds of unrelated biological effects on this accident. In reality, radiation levels in fish (mainly from Cs-137, Cs-134 and Sr-90) are within acceptable limits for human intake, and it is estimated that the Cs-137 signal will be normalized in Black Sea surface waters by the year 2000. However, a more complete dose assessment is still required, and little information has been gathered on alpha-emitting radionuclides. Public concerns about radionuclides have recently been exacerbated further by reports of unsafe nuclear reactors (now shut down) on the Bulgarian banks of the Danube, activities of nuclear powered submarines and a recent (fortunately inconsequential) fire in Chernobyl. In addition to satisfying health concerns, measurements of natural and artificial radionuclides in the Black Sea have already proven valuable as a tool for studying particle fluxes and modeling the transport of other contaminants.

Microbial Pathogens

There is little published data on microbial pathogens in the Black Sea region. The main reason for this is that in some countries such data is obtained (if at all) exclusively by health authorities and considered to be a confidential matter. Some data have been published in Romania and Turkey but have not been subjected to a comparative analysis with internationally accepted norms (e. g., EEC water quality standards). Perhaps the clearest evidence that a problem exists is the reports of limited outbreaks of water borne cholera in the summer of 1970 (in the northern Black Sea estuarine areas and the sea of Azov) and again in 1991 (on the Romanian and Ukrainian sides of the Danube delta). Unsanitary conditions have led to the temporary closure of an increasing number of Ukrainian and Russian beaches during the holiday season.

Dumping and Toxic Waste

There are 16 official dump sites in the western Black Sea. Most of these are on the continental shelf and are used for the disposal of dredging spoils (sludge). The negative impact of this activity on benthic biota is said to be colossal. Since many of the dredged areas are also discharge sites for rivers (and industrial/domestic effluents), dumping may be a mechanism for bypassing the natural "estuarine filter" for contaminants and introducing toxic materials to the open sea. This activity requires urgent assessment if a future Convention for the Protection of the Black Sea is to be fully effective. Also worthy of mention is the recovery of about 364 drums of toxic waste (which originated in Italy) from the beaches and coastal waters of Turkey in 1988. This and later attempts at dumping demonstrate that even the virtually enclosed Black Sea does not escape from this insidious activity.

Oil Pollution

The author has found little published information on this subject despite the previously mentioned enormous loads of oil discharged via the Danube. Russian scientists published a report in 1985 showing extremely high hydrocarbon levels (n-alkanes measured by gas chromatography) along the navigation route from the Danube to the Bosphorus. Concentrations were typically around 0.3 mg/L in the shipping lane compared with typical total oil levels of 0.001 to 0.03 mg/L in the Gulf, 1983-1984. The origin of the pollution was clearly shipping related as a concentration gradient of 0.12 to 0.015 mg/L was associated with the Danube outflow itself (levels much lower than the shipping lanes). More recent data from coastal waters off Odessa reveal an average concentration (1988-91) of total petroleum hydrocarbons of 0.064 mg/L (range 0.005-0.18 mg/L). There are inadequate port reception facilities in the Black Sea, and deballasting activities are reported to occur, particularly in the Western region (where EEZ boundaries have not yet been agreed upon). Furthermore, concern about serious oil pollution from military installations and ships was publicly expressed in the former USSR in 1990, and remedial measures were agreed. Heavy tanker traffic in the Bosphorus has led to considerable public concern over the risk of oil spills and explosions in the vicinity of Istanbul. The recent accident of the oil tanker Nassia, which caught fire in the Bosphorus after a collision with a freighter, released several thousand tons of oil to the Black Sea, further heightening public concern. Turkey is introducing new regulations on passage through the Bosphorus, and there are calls for modifications to the 1936 Montreaux Treaty, which affords the status of international waters to this channel. Offshore oil and gas exploration (and limited production) is also underway in Romania and Bulgaria. A proper assessment of oil pollution is urgently required (particularly of the more toxic polycyclic aromatic hydrocarbon fraction).

A Black Sea Turnover: the ultimate threat?

The existence of a vast volume of anoxic water in the Black Sea rather close to the ocean surface has fascinated marine chemists for many years and has been the prime target of several oceanographic cruises (e. g., from the USSR, USA, Turkey) in the past 50 years. Autumn turnover in anoxic lakes has devastating consequences, often with the total elimination of aquatic animals, and it is a simple matter to extrapolate this to a Black Sea context. Of course, the Black Sea is protected by a very strong density gradient, and such a turnover is unlikely; even a reduction of 50% in the river discharge would not significantly alter the density structure within a 100 year period. The only risk of penetration of anoxic waters to the surface would be a major increase in the reduction of sulfate (fueled by increased loads of organic matter) and a correspondent change in the depth of the oxic/anoxic boundary. Following the 1988 Knorr expedition, some US scientists found that this boundary had become elevated by up to 30m with respect to their 1969 data and speculated that this could be a trend associated with decreased river flows. This idea was strongly contested, particularly by Soviet scientists. They have made a seventy-year series of observations in which large fluctuations were observed in the depth of the boundary (+ 20-30m variations), and they have further demonstrated that its depth typically varies from 95m to 190m within anticyclonic and cyclonic gyres, respectively. Of course, the situation could be modified in the future, particularly if global climate changes were to seriously alter the rainfall pattern in the Black Sea drainage basin. The debate continues. In the interim, however, the prospect of a sudden environmental cataclysm seems highly improbable.

Socioeconomic Consequences

The full socioeconomic consequences of pollution are impossible to evaluate without solid data and studies relating contamination with biological effects. Information on the levels of contaminants in foodstuffs and water supplies will eventually become an important factor within an emerging market economy. The possible use of water from the Dnieper river, for example, for domestic water supplies to Crimean towns and cities is already causing considerable controversy because of reports of pollution. As previously mentioned, reports of small cholera outbreaks may be a dramatic indication of a deeper problem affecting the human population.

Actions and Remedies

It is clear from the above review that proper assessments are urgently required of pollution in the Black Sea. Such assessments must include data on land-based (and marine) sources, levels in the environment and biological effects. The cost to the environment of marine pollution may also have to be estimated if this issue is to be given a higher place on the political agenda. The term "incalculable cost" is meaningless when the price of reducing the pollution loads of industry must be justified. Another term in the equation is the cost of wastage. As an example, the yearly cost of oil wasted down the Danube may be computed as US $7 million (at current market prices) - enough to fund an entire Action Plan for rehabilitating the Black Sea!

An urgent concerted action is required to enhance the capability of regional institutions to conduct marine pollution assessments. A broad-based pilot monitoring approach is required where a team of scientists addresses a specific priority action. For example, a preliminary land-based survey of a northern Black Sea river outflow should consider the type of upstream industrial and agricultural activity and adjust the initial parameter list to consider likely toxic effluents. Where significant contamination is discovered, seaward gradients must be measured and biological effects assessed along them. Full-scale region-wide monitoring is warranted where pollution problems are found to be common to all countries or basin-wide contamination is suspected. Once enough data is gathered to make initial assessments and take regulatory measures, a program for longer-term trend monitoring can be established.

Marine pollution assessments should lead to clear environmental quality objectives for the region. Achievement of these objectives depends upon the early linkage of the pollution control program with national planning bodies and with the objective of banks from which credits may eventually be obtained. There is also a need for improved public awareness and education, not only in the marine environment but as regards waste disposal, land use and agricultural practices.

If effective control of pollution is to be linked with the harmonious development and modernization of industry on a basin wide scale, innovative strategies will be required. The sweeping introduction of emission standards might result in the large-scale closure of plants, but is far more likely to render ineffective any enforcement.

Active consideration should be given to the concept of marketable load permits, which would be priced in such a way as to allow factories to finance emission control technology through the sale of part of their "load" to the newcomers. The total load should be set on the basis of environmental rather than wholly economic arguments, if the commons is to be a beneficiary.

Coastal Development Issues

The presence of a large human population living or vacationing on the Black Sea coast is placing heavy demands on the coastal environment, endangering the already limited freshwater supplies and placing an additional stress on the biological resources. The morphology of the coastline itself is changing as reduced river flows (and consequent transport of fresh sediments) lead to the erosion of beaches which are, in turn, a major attraction to tourism. Many Black Sea bathing beaches are already characterized by large numbers of unsightly groins constructed to limit erosion, which are also effective traps for solid waste and floating debris. This further exacerbates the pollution problem described earlier.

The amenity value of the Black Sea cannot be overstressed. It is the only marine area available to millions of Eastern European citizens. A successful transition to market economies and increased freedom of movement may result in a greater proportion of the population demanding access to Black Sea beach resorts. Unless a policy of integrated coastal zone management (ICZM) is applied together with heavy investments in services (repaid by appropriate direct or indirect user tariffs), the economic and social value of the Black Sea coastal zone will rapidly deteriorate in a similar manner to its fisheries resources and biodiversity.

First steps have already been taken by the Romanian government to protect the Danube Delta, one of the few remaining natural wetland areas in the region. The Danube Delta Biosphere Reserve was created in 1990 and is a critical habitat for many migrating birds and terrestrial species. Unfortunately, it may be the only area suitable for protection if the Black Sea environment continues to deteriorate at its present pace, and rehabilitation will be a necessary precursor to the creation of additional protected areas.

Towards a Regional Policy for Environmental Management

Legal and Policy Tools

Until recently, the Black Sea was unprotected by any common policy or legal regime. Efforts to negotiate a legal Convention for the Protection of the Black Sea, modeled along the lines of the Barcelona Convention, began in 1985, and it was finally signed in April, 1992 by all six riparian countries. The Convention has now been ratified by all six governments--a remarkable demonstration of commitment to managing the commons. The Convention entered into force in February, 1994. It contains a legal framework for the establishment of a Black Sea Commission and provides Protocols for protection against land-based sources of pollution, for regulating dumping and for emergency action in the case of spills of oil or other toxic substances. The Secretariat for the Commission, to be established in Istanbul, will have the task of elaborating detailed criteria for preventing, reducing and controlling pollution as well as general duties related to the implementation of the Convention. Amendments to the Convention shall require a consensus at a Diplomatic Conference.

All of the riparian governments also made a formal request to the Global Environment Facility (GEF) for a three year program for "Environmental Management and Protection of the Black Sea" and requested technical assistance in the design of an Action Plan for implementation of the Convention. This GEF Black Sea Environmental Program is designed to provide a framework for urgently needed support for environmental assessments, institutional capacity building, preinvestment feasibility studies and an "emergency investment portfolio" of actions designed to complete unfinished key pollution control projects (e.g., sewage treatment plants). This project should facilitate the establishment of a sustainable Secretariat for the Convention. It will also serve as a transition to the long-term Action Plan and provide basic data for the very large investments that will be needed for controlling pollution and rehabilitating the Black Sea. A similar project is already underway in the Danube basin, and the two projects together will provide a unique opportunity to link the concepts of river basin management, integrated coastal zone management and marine resource management, drawing upon all available expertise from throughout and beyond the UN system. It should promote closer coordination of the many independent multilateral initiatives presently planned or underway in the region.

The Convention itself is a legal and diplomatic tool for joint action and does not set out to establish environmental policy goals (e. g., targets for reducing the loads of specific pollutants, etc.). It also does not establish any regulatory mechanism for exploitation or development of the natural environment (e. g., straddled marine resources or specially protected areas). In order to develop a common policy framework, a clear "Declaration of Environmental Quality Objectives" was considered necessary. Following the initiative of the Government of Ukraine and employing the stewardship of the GEF partners led by UNEP, a Ministerial Declaration was formulated and signed by all six countries in Odessa in April, 1993 (the “Odessa Declaration”). This Declaration is a pragmatic and innovative policy statement which sets environmental goals and a time frame to guide management regimes and associated investments. It consists of a preamble underlining the seriousness of the environmental crisis, a general policy statement broadly reflecting the philosophy of UNCED, Agenda 21, and nineteen specific actions. These actions are designed to facilitate the rapid development of pragmatic measures for controlling pollution from land-based and marine sources (including the harmonization of environmental standards); to restore, conserve and manage natural resources; to respond to environmental emergencies; to improve the assessment of contaminants and their sources; to introduce integrated coastal zone management policies and compulsory environmental impact assessments; and to create a transparent and balanced mechanism for reviewing and updating the Declaration on a triennial basis. The implementation of the Odessa Declaration should be enhanced by its heavy emphasis on accountability, periodic review and public awareness.

The GEF Program and Sustainability

The GEF Black Sea Environmental Program became formally established in September, 1993 (Fig. 5.). By early 1994, a coordination unit (the PCU) had been established in Istanbul, and a workplan had been agreed by its National Coordinators (in many cases Ministers of the Environment or their deputies).

The program has three primary objectives:

* to strengthen and create regional capacities for managing the Black Sea ecosystem;
* to develop and implement an appropriate policy and legal framework for the assessment, control and prevention of pollution and maintenance and enhancement of biodiversity; and
* to facilitate the preparation of sound environmental investments.

Each of these objectives has enormous scope and involves the participation of a broad range of sectors. The program Steering Committee incorporates the National Coordinators, donor representatives and representatives of NGOs (including the business sector). It is clear that a mechanism is needed to share the burden of these program objectives if the Black Sea Environmental Program is to lead to a long-term Action Plan. Rather than to expect the coastal countries to contribute to a single centralized future program, it was decided to instigate a series of Working Parties based upon "Activity Centers", which are national institutions that already have the basic infrastructure and personnel to coordinate specific tasks in the region. Each Government has agreed to host one of these Centers. This enables costs to be shared with a minimal transfer of funds. The Working Parties themselves include at least one expert from each of the Black Sea countries, together with additional expertise where appropriate. These Working Parties will normally meet twice per year. The working parties will be action-oriented--at least one of the activities per year will consist of a seminar, training workshop or pilot study.

The Activity Centers and their Working Parties are as follows:

- Emergency Response (Varna Bulgaria),
- Routine Pollution Monitoring (i. e., assessment of land-based sources, bathing water quality, etc.) - Turkey,
- Special Monitoring Programs, Biological and Human Health Effects, and Environmental Quality Standards (Odessa, Ukraine),

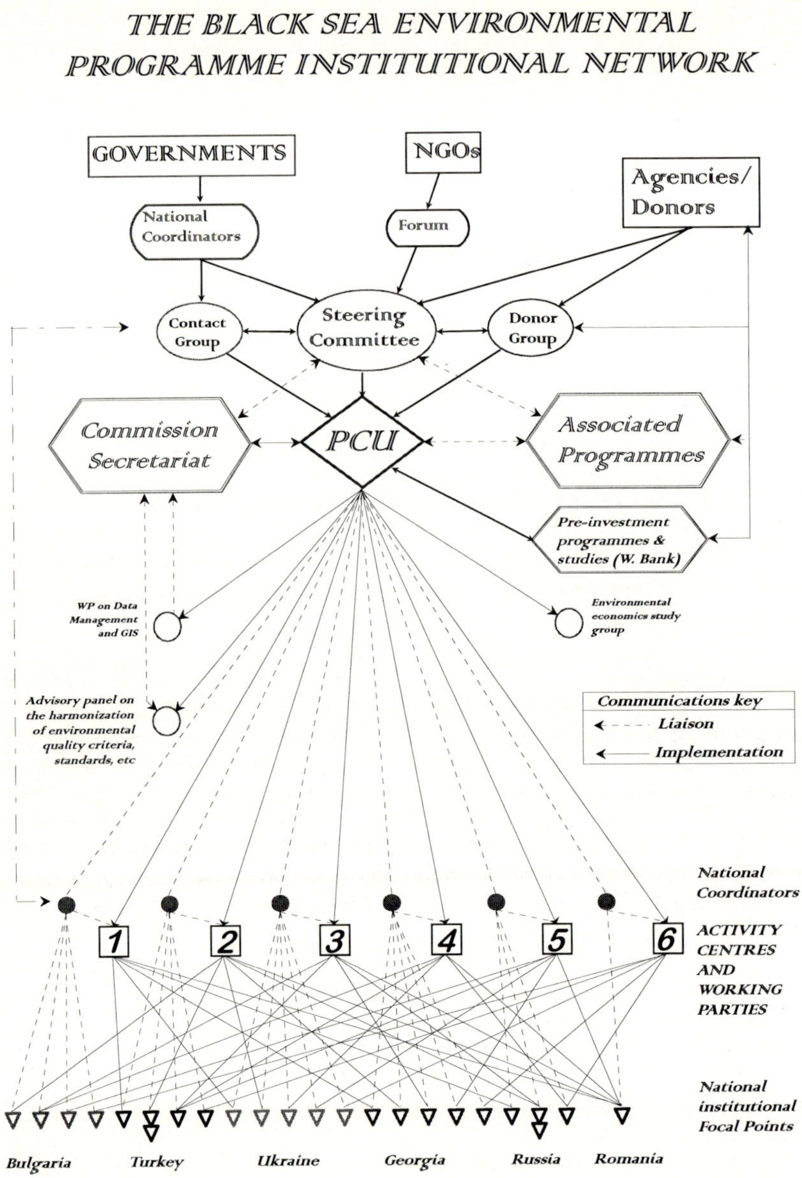

Figure 5 The Black Sea environmental programme institutional network.

- Protection of Biodiversity (Batumi, Georgia),
- Development of Common Methodologies for Integrated Coastal Zone Management (Novorossiysk, Russian Federation),
- Fisheries (Constantza, Romania)

In addition there are three Working Parties based upon the PCU:

- Data Management and GIS,
- Advisory Panel on the Harmonization of Environmental Quality Criteria, Standards, Legislation and Enforcement, and
- Environmental Economics Study Group.

The full structure of the network is shown in Fig. 5. It must be stressed, however, that the network is not the primary objective of the BSEP; it is simply a tool that should facilitate its implementation in a sustainable manner. The strength of the network will only be evidenced if the program results in the efficient implementation of the Odessa Declaration, the introduction of new environmental standards and legislation, improved economic instruments and the realization of effective environmental management through better planning, integration of sectors and the sensible use of investments.

The Final Challenge

Political good will, Declarations and Conventions can do little to rehabilitate the environment unless adequate financial support is available. Investments in pollution control and resource rehabilitation and management must be justified by solid environmental economics in order to win a position on the political agenda of the riparian states. The catastrophic degradation of the Black Sea will cost many millions of dollars to rectify, but the alternative would be to deprive a major part of eastern Europe of access to its renewable marine resources and amenities. This situation is elegantly summarized in the Bruntland report in a more general context, but is a fitting conclusion to the present article:

"....it is one thing to contribute a few million dollars for research, and quite another to incorporate the resulting findings into land-based development plans and to enforce strong pollution control programs. The massive US-Canadian clean-up of the Great Lakes over the past 15 years cost $8.85 billion for partial treatment of municipal and industrial wastes. Huge investments will also be required to roll back land-based pollution along UNEP's regional seas. Yet nowhere have the sums been committed under agreed schedules to construct the necessary urban and industrial pollution control systems and to underwrite policies to control agricultural run-off. The program now has to confront the regional seas challenge through the year 2000-- moving beyond general agreement on goals and research to a solid schedule of investment on a scale that will make a difference."

V.7 The Aral Sea Crisis: A Great Challenge in Transboundary Water Resources Management

L. Veiga da Cunha
Scientific Affairs Division[2]
NATO, 1110 Brussels
Belgium

"You cannot fill the Aral with tears"
Mukhammed Salikh, Uzbek poet

The Shrinking Sea

During the 1980s, the shrinking of the Aral Sea started to generate growing concern in public opinion in the West, mainly as a result of media coverage of the issue. Since then, numerous articles and scientific papers have been published, as well as a few monographs (such as UNEP 1992) and even entire books (such as Le Tolle and Maignuet,1993).

However, as early as the 1960s some articles about the risks of the worsening situation of the Aral Sea had started appearing in Soviet journals. Also, some research work done with respect to this concern was initiated in the same period, but the results were not generally made available. As Kuznetsov 1992 states in relation to the studies done on health conditions of the population most severely affected by the Aral Sea crisis, "unfortunately, secrecy over an entire series of research results in the 1970s, especially medical-epidemiological data, precluded their publication at that time and the prediction associated with them did not become available to the public in time."

In the early 1980s, Soviet authorities started to evaluate the consequences and the costs of the environmental problems in the USSR. In the period of Gorbachev, the discussion of environmental problems started to take place in open fora, the Chernobyl accident being considered as a catalyst for glasnost. During this period, the Aral Sea also became a leitmotiv in the Soviet media for those concerned about environmental issues.

Monin (1988), referring to the optimistic official projections made in the 1960s which predicted a doubling of irrigated land, said: "The total area of irrigated land in the Aral Basin has not doubled--it has only increased by a half; nor have the hypothetical yields of cotton and rice doubled. As far as foodstuffs are concerned, per-capita production in most prosperous Central Asia is 26 percent of the medical norm for meat, 42 percent for milk, 53 percent for fruits and vegetables ... What has increased? The use of chemicals. Cotton fields are inundated with ten times more pesticides than on average elsewhere in the Soviet Union or in the USA. What else has increased? Consumption of irrigation water."

---

[2]This paper represents the author's personal views and does not reflect an official NATO position.

A commission on the Aral Sea was officially created in the USSR in 1987, chaired by the head of the State Committee on Hydrometeorology and including the presidents of the Academies of Sciences of the republics of Central Asia. The Committee agreed on an action program, and its recommendations were considered at the highest political level in the USSR, finally leading to the approval in September 1988 of a resolution by the Central Committee and the Council of Ministers on "Measures for a Fundamental Improvement of the Ecological and Sanitary Conditions in the Aral Sea Region and for Increased Effectiveness in the Utilization, Strengthening and Safeguarding of Water and Land Resources in this Basin". However, the difficult economic situation which developed in the Soviet Union after 1989 made it impossible to allocate adequate financial resources to environmental problems.

Up until the late 1950s, the average level of the Aral Sea was relatively stable at around 53 meters above sea level. The table below shows the changes in level, surface, volume and salinity from 1960 to 1992, according to Glasovsky (1992). It can be seen that, between 1960 and 1992, the level of the Aral Sea decreased by more than 16 meters. At the same time, the surface of the lake decreased by about half, the volume decreased to less than 25%, and the salinity increased 3.5 times, reaching values identical to those of water in the oceans. At the same time as these physical and chemical transformations were taking place, biological changes were also occurring, both in the sea (as described, for example, by Aladin and Williams 1993) and in the delta (as described by Novikova 1992).

Change in Characteristics of the Aral Sea since 1960

Year	Level (m)	Area (km^2)	Volume (km^3)	Salinity (g/l)
1960	53.41	68,000	1,090	10.0
1971	51.05	60,200	925	11.2
1976	48.28	55,700	763	14.0
1987	40.50	41,000	404	26.8
1988	39.80	39,400	365	28.3
1989	38.60	36,900	330	30.1
1990	37.80	34,800	304	33.3
1991	37.00	34,000	248	34.0
1992	36.70	33,600	231	34.4
2000	33.00	23,400	162	42.0

The table above also indicates the prediction for the year 2000 which shows an aggravation of the present trend. A graphic view of the shrinking of the lake between 1960 and 1993 is shown in Fig. 1 according to Micklin (1993).

Before the 1960s, the Aral sea was about 450 km long and 300 km wide and was the fourth largest lake in the world in terms of surface area, after the Caspian Sea, Lake Superior and Lake Victoria. In 1993, the Aral Sea, due to its shrinking, dropped to sixth place, now also ranking below Lakes Huron and Michigan. However, the Aral Sea is a much shallower lake than the other five, with an average depth of only 16meters in 1960 and 7 meters in 1992.

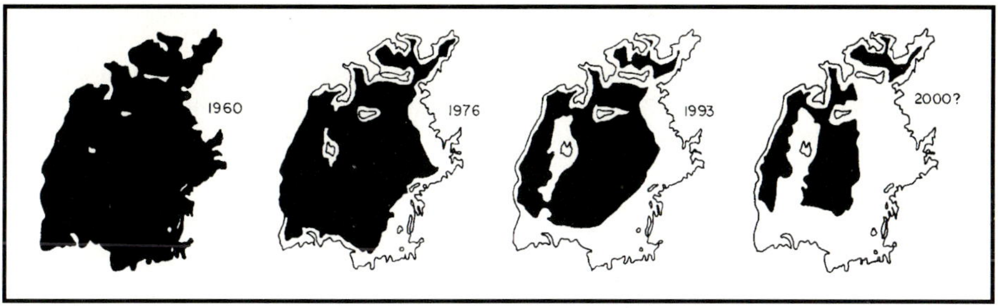

Figure 1 The Evolution of the Aral Sea Surface.

As Fig. 1 shows, in 1993 the Sea appears as two separate lakes, with the so-called Small Aral Sea to the north and the Large Aral Sea to the south. These two separate lakes are fed by two rivers, the Syr Darya and the Amu Darya (see Fig. 2). A connecting channel which existed between the small and large lakes was blocked in mid 1992 by a dyke, its top at 42 meters above sea level, preventing any water which flowed into the small lake from passing into the larger one.

These modifications to the Aral Sea have been due to the progressive reduction of flow reaching the lake from the two tributary rivers, which has caused an imbalance between inflow and evaporation from the lake. The main cause of reduction in the discharge of the two rivers has been the over-intensification of irrigation after 1960.

The most familiar world-wide image of the shrinking Aral Sea is that of abandoned ships lying on the dry bed of the Sea. These clearly symbolize the major economic losses resulting from the shrinking of the Aral Sea, namely the destruction of the fishing industry (all 24 native species of fish, once the source for an annual catch of 40,000 tons of fish, have been destroyed due to the increase in salinity) and the closing of the transportation routes across the Sea, mainly between the Kazakh port of Aralsk in the north and the Uzbek port of Muynak in the south. The water-front is now more than 100 km distant from these ports.

Another major consequence of the desiccation of the Sea has been the destruction of the ecosystems in the lake and in the deltas of the Syr Darya and the Amu Darya, and the wind-blowing of salts from 3.5 million hectares of exposed lake bed which are deleterious to crops and toxic to humans (Razakov and Kosnazarov 1994). By the action of windstorms on the exposed bed of the Aral Sea, tens of millions of tons of sand, salt and deposited agricultural chemicals are whipped up in the air. It has been reported that over half a ton of airborne salt has on average been deposited per hectare of agricultural soil. Moreover, the salts transported by wind can travel for thousands of kilometers, reaching the mountains, where the two rivers feeding the Aral Sea have their sources, and changing the flow of the rivers by accelerating the melting of snow in the mountains with consequent adverse effects on the water management of irrigation.

These very serious ecological, economic and social impacts have lead UNEP to consider the Aral Sea crisis as "one of the most staggering disasters of the twentieth century" (UNEP 1992). And this is not only because of the magnitude of the changes, but also because of the great speed at which they have occurred. However, as discussed below, the disastrous conditions directly associated with the drop in water level in the Sea appear now to be of lesser importance compared to the conditions in the Aral Sea basin as a whole.

During the Soviet era, some ideas for projects intended to ensure the refilling of the Aral Sea basin and restoration to its initial conditions have been put forward. These have included reversing the course of Siberian rivers flowing to the Arctic, transferring water from the Volga river or the Ural river and pumping water from the Caspian Sea. All these projects aimed at solving a mega-problem by resorting to mega-solutions have proven to be utopian, in view of the large expenses involved and the possible negative economic and environmental impacts of the solutions envisaged.

In fact, there is general agreement today that such grandiose solutions are not realistic and that restoring the Aral Sea to what it was in 1960 is not a viable option. The realistic proposal would then be to stabilize the Aral Sea at some sustainable level (see, for example, WB 1993).

It appears today that the only reasonable way to achieve a long-term solution to the problem is to define and implement effective land-use planning and water resources management policies in the Aral Sea basin, without obviously excluding short-term emergency measures, which are clearly needed to cope with the environmental problems that have developed and with the urgent human health problems. The figures are depressing. The Republic of Karakalpakistan, a territory with large autonomy inside Uzbekistan bordering the Aral Sea, is indeed the region where the problems are most serious. In the delta regions, infant mortality now stands at 180 per 1,000 births, compared to 35 per 1,000 in 1935 and a western average of 7 per 1,000. More than 80% of women and children suffer from anemia. Incidence of stomach cancer is 15 to 20 times higher than the average rate in former Soviet cities, with other types of cancer on the increase. Incidence of typhoid fever and hepatitis has increased dramatically. The average age for death fell from 64 in 1987 to 57 in 1991.

Transboundary Water Resources Management is the Real Issue

The situation in the Aral Sea basin was already very difficult before the collapse of the Soviet Union and before the attempt to evolve from one centrally-planned economy to five market economies. Indeed, water resources management of the Aral Sea basin was a great challenge even when the whole basin was largely within a single country. Superimposing the complicated network of borders of five new Central Asian countries on the already complex problems at stake has certainly raised the degree of difficulty in finding the most appropriate solutions. All of a sudden, a very complex water management problem has become a very complex transboundary water management problem. Indeed, this situation differs radically from that where there would have been a gradual development of international water resources problems between neighboring countries due to progressively increasing water scarcity. It is obvious that conflicts will tend to develop among the new countries in Central Asia, unless conditions are quickly created to implement reasonable solutions for effective transboundary water resources management in the region.

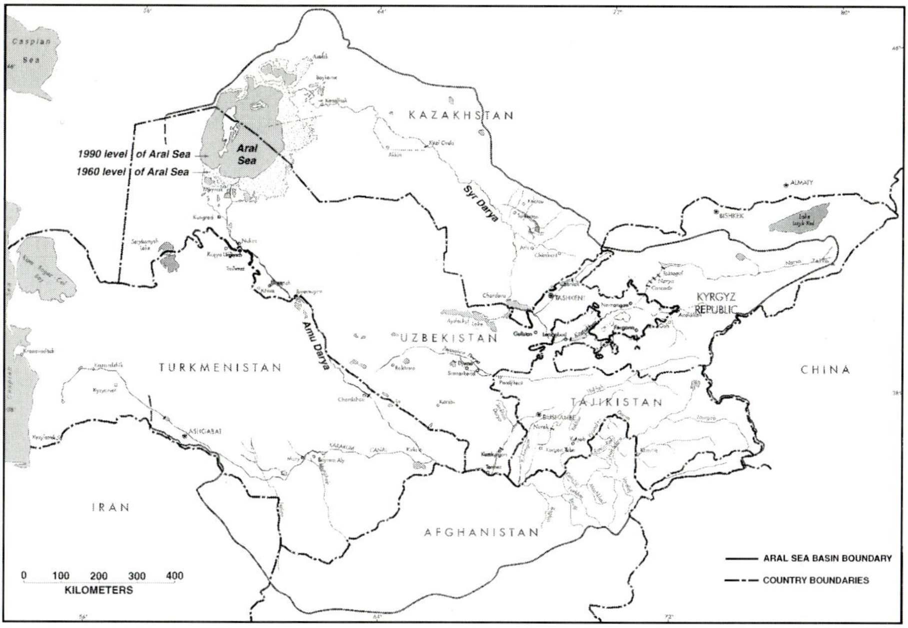

Figure 2 Aral Sea Basin and Countries Sharing the Basin.

Fig. 2, adapted from WB, 1992, presents the Aral Sea basin, showing the two rivers flowing into the Sea, the Syr Darya in the North and the Amu Darya in the south. The figure also shows the limits of the Aral Sea basin and the borders of the various countries. There are the five Central Asian Republics (Kazakhstan, Kyrgyzstan, Tajikistan, Turkmenistan and Uzbekistan) and also, in the southern part of the basin covering only 15% of its area, Afghanistan and Iran.

The area of the Aral Sea basin is 1.8 million km^2. The upstream part of the basin, which roughly coincides with the whole territory of Tajikistan and a large part of the territory of Kyrgyzstan, corresponds to the mountainous area where the flow is generated. The source of the Syr Darya is in the Tian Shan mountains and that of the Amu Darya in the Pamir mountains, both at altitudes of more than 7000 m. In the upstream mountainous areas of the two rivers, the annual precipitation ranges from 800 to 1600 mm.

After the rivers leave the mountainous areas they flow into the desert where annual precipitation falls abruptly to about 200 mm in the foothills region and less than 100 mm per year in the downstream regions. As a consequence, the natural discharges of the rivers reach their highest value as they enter the desert region, after which their discharge gradually decreases as they progress downstream, due to infiltration and evaporation.

However, the heavy reduction in the discharge of the Syr Darya and Amu Darya was, after 1960, mainly due to the over-intensification of irrigation, although it must be stated that, contrary to a rather well-established idea, irrigation in the Aral Sea Basin was not initiated in the 1960s, but has been present for many centuries in the Aral Sea basin in a sustainable condition.

According to data from Chub (1994), the irrigated area in Central Asia was already 3 million hectares in 1913, the major part of this being in the Aral Sea basin. In 1960, when the imbalance of the Aral Sea started, the irrigated area had grown to about 5 million ha. After 1960, in the period when the lake started to shrink drastically, the irrigated area increased by 50% to around 7.5 million ha. Fig. 3, adapted from WB, 1993, shows the distribution of the irrigated areas in the Aral Sea Basin.

The expansion of irrigation in the Aral Sea basin after 1960 has, indeed, induced non-sustainable conditions. In reality, in this period an increase of 50% in the irrigated area has corresponded to a 100% increase in water consumption for agriculture, changing from approximately 50 to 100 km^3 per year.

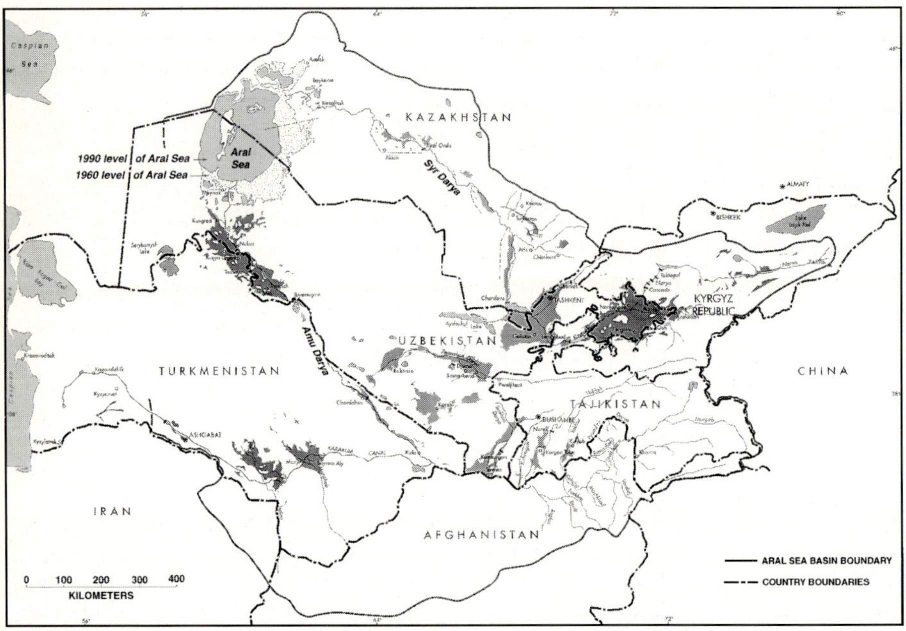

Figure 3 Irrigated Areas in the Aral Sea Basin

This increase in water consumption per irrigated hectare was due to the development of highly water-demanding crops, mainly cotton and rice, to keep pace with Soviet demand for food and fibre. Attaining self-sufficiency in cotton production was a particular obsession in the USSR, cotton being dubbed "white gold" by the planners. A number of factors have caused the abrupt increase in water consumption, such as: the building of large dams, creating reservoirs with large evaporation and infiltration losses; the building of long, unlined canals crossing the desert; the development of inefficient irrigation practices; the irrigation of poor soils requiring more water to irrigate crops and to flush salts; and the development of inadequate drainage systems discharging the drained water into depressions in the desert instead of draining them into the Aral Sea. In addition to the decrease in quantity of water available due to escalating consumption, reference should also be made to the decline in water quality.

Peterson, 1993 identifies three main reasons for this. Firstly, the demanding official plan for producing more cotton has prevented farmers from cultivating other crops which could build up the soil, resulting in cotton plantations becoming more and more dependent on fertilizers. Secondly, intensive cotton cropping has also forced farmers to increase the use of pesticides. It has been reported that in Turkmenistan, for example, the use of pesticides was 20-25 times the Soviet average. According to Smith, 1991, pesticide use in Uzbekistan was about 10 times the level recommended in the United States. Thirdly, the use of heavy doses of defoliants in the autumn has been developed to facilitate the harvesting of cotton. This excessive use of chemicals in irrigation and the lack of adequate drainage has contributed considerably to the pollution of the surface waters and contamination of groundwater with resulting damage to human health conditions, as drinking water in the region of the deltas is withdrawn and used without any treatment.

The discharge of the two rivers emptying into the Sea as they leave the mountains corresponds to 110 km^3/yr (or about 3500 m^3/s), two thirds relating to the Amu Darya and one third to the Syr Darya. More precisely, this represents 73 and 37 km^3/yr, respectively (Voropayev et al., 1987).

About half of the annual discharge generated by the two rivers, i.e. 55 km^3/yr reached the Aral Sea before 1960, when the lake was still in equilibrium. This amount of water decreased dramatically in subsequent years to an average value of about 25 km^3/yr in the seventies, 15 km^3/yr in the eighties, and 7 km^3/yr in the nineties.

The evolution of the annual discharge of the Amur Darya and Syr Darya rivers since 1989, with projections for the years 2005 and 2010, respectively, is shown in Figs. 4 and 5, according to Micklin, 1993. A stabilisation of the inflows at 4 km^3/yr for the Amur Darya and 3 km^3/yr for the Syr Darya is assumed, which is a somewhat arbitrary assumption and probably optimistic one. But even if this assumption were correct, Fig. 4 shows that the large Aral Sea, which is fed by the Amu Darya, would continue to shrink rapidly after the year 2005, when the water level would become 23 m below the level in 1960, and the area would be approximately 20% of the original. In the case of the Syr Darya, as Fig. 5 shows, because of the construction in 1992 of the dyke at the level separating the small and large lakes, there would be a gradual increase in the water level, stabilising at approximately 42 metres. The combined areas of the small and large seas would by the year 2005 amount to 18,000 km^2, i.e. 25% of the 1960 value, assuming that the annual flow of 7 km^3/yr were to remain stable.

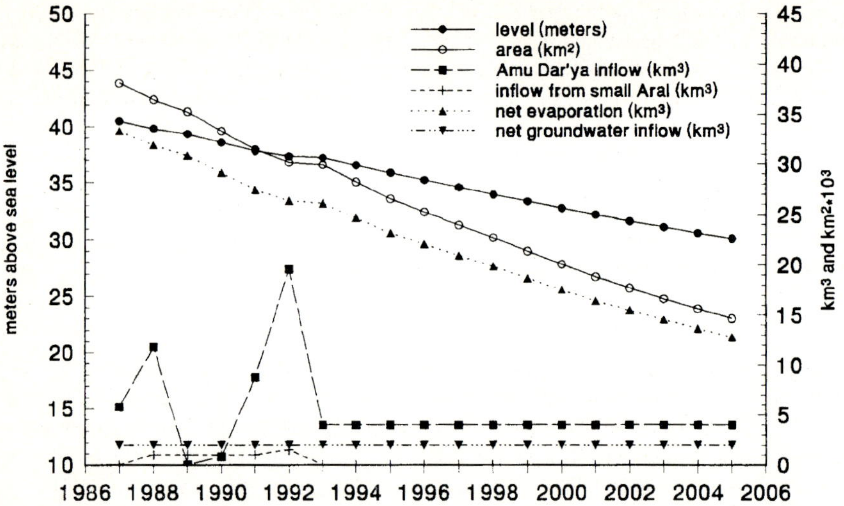

Figure 4 Evolution of the Discharge of the Amu Darya and Water Balance of the Large Aral Sea.

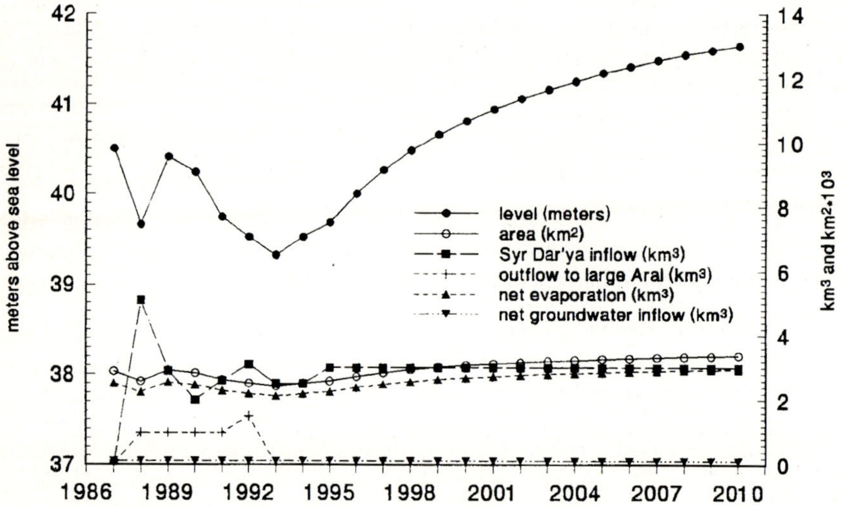

Figure 5 - Evolution of the discharge of the Syr Darya andWater Balance of the Small Aral Sea.

In conclusion, it seems clear that the Aral Sea as it was known for centuries is now condemned to vanish completely within a generation, or at least be reduced to one of the many insignificant lakes which will by then exist in the Aral Sea Basin. In fact, the Aral Sea is disappearing not only because of the water consumed by evapotranspiration and infiltration, but also because multiple lakes are being formed when floods occur and as a consequence of inefficient drainage of irrigated areas. One relevant example of this is Lake Aydar located 200 km east of Tashkent. This lake, which was virtually non-existent in 1969, has experienced a steady increase in level since then. For instance, when visiting the region in 1994, the author was informed that, in that year alone, its level rose by 2 m, with the discharge of the Syr Darya deviated to Lake Aydar being 9 km^3, a volume equivalent to that reaching the Aral Sea in the same year.

The New Central Asian Republics and the Aral Sea Problems

The independence of the new republics in Central Asia occurred on various dates during the latter half of 1991. The rulers of the republics have quickly acquired a consciousness of the relevance of water resources to the economy of the new countries, and have fully realized the importance of establishing sound agreements for managing the water resources of the Aral Sea basin to ensure the development and political stability of the region. Making clear that environmental security related to the Aral Sea basin issues is indeed a major issue in this part of the world, the heads of state of the five republics of the Aral Sea basin signed, on 26th March, 1993 in Kyzl Orda, in the delta of the Syr Darya, an agreement on joint actions aimed at coping with the economic, social and environmental problems of the Aral Sea basin. In particular, the states agreed to draft a joint statement addressing the Aral Sea crisis and the environmental rehabilitation of the region. They also agreed to establish a coordinated research program and to create a common information system of environmental monitoring. The states have established an Interstate Council for Addressing the Aral Sea Crisis and an International Fund for the Aral Sea. The structure of these two bodies is depicted in Fig. 6, according to ICWC, 1994.

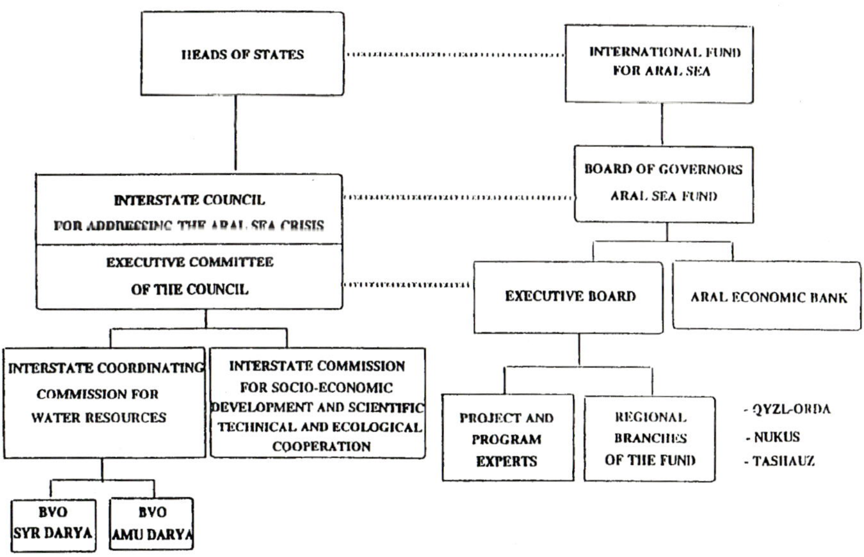

Figure 6 Institutional Framework in Central Asia for Addressing the Aral Sea Basin Problems.

The Interstate Council for Addressing the Aral Sea Crisis (ICAS) is comprised of the five republics of the basin, as well as Russia, which has observer status. Within the framework of the ICAS, which has an Executive Committee based in Tashkent, two committees have been created, the Interstate Coordinating Commission for Water Resources and the Interstate Commission for Socio-economic Development and Scientific, Technical and Ecological Cooperation. There are also two river basin agencies (BVOs) which report to the Commission for Water Resources, one for the Syr Darya and the other for the Amu Darya.

The International Fund for the Aral Sea (IFAS), has been set up to finance joint actions to improve the present situation in the Aral Sea Basin and especially in the region close to the Aral Sea. In particular, the IFAS is used to finance the following: the interstate ecological monitoring systems in the Aral Sea region; implementation of joint measures for protection of water, soil, air and the biosphere; development of projects aimed at coping with the Aral Sea crisis; fundamental and applied research; participation in international projects for rehabilitating the Aral Sea environment.

The Fund has a board of governors which is chaired by the head of one of the states, designated in rotation by the five republics. There is an Executive Board, as well as an Aral Economic Bank reporting to the board of governors. The Executive Board supervises project and program experts and three regional branches of the Fund. The Fund relies on contributions from the five countries (with initial recommended contributions of 1% of their national incomes) and donations from other countries or from international organizations.

On 11th January, 1994 the Heads of State met again in Nukus, in the delta of the Amu Darya, and approved a framework for the definition of action programs to be developed by the Interstate Council, taking into consideration three horizons: urgent (3-5 years), mid-term (5-10 years) and long-term (more than 10 years).

The initial steps taken by the Aral Sea basin states described above are certainly important, but in view of the gravity of the present situation, it is essential that conditions be created for immediate action to take place in the form of practical measures. In fact, enormous efforts will have to be made to face the difficulties which developed during the period of Soviet rule and which were only made more difficult by the sudden appearance of the borders of 5 new countries in the region.

The gravity of the problems differs from country to country. The two upstream countries, Kyrgyzstan and Tajikistan are basically water suppliers and do not exert great pressures on water use since, in view of their physical characteristics, they are not candidates for using great amounts of water for irrigation. The three downstream countries are the great users of the water for irrigation. They badly need water to cope with the very serious human health and environmental situation, particularly in the deltas of Amu Darya and Syr Darya. Uzbekistan and Kazakhstan withdraw water at several sites along the two rivers for supplying several areas of irrigation and human settlement along the rivers. The case of Turkmenistan is special, as it depends on water derived from the Amu Darya and conveyed by the 1,300 km long Karakum Canal (see Fig. 2). This canal, when opened in the mid fifties, was designed to divert up to 15% of the then current discharge of the Amu Darya, but was in fact initially diverting only a very small part of this discharge. However, due to an increase in the irrigated area, to an increase in water loss because of notorious leakage from the canal and to a decrease in the discharge of the Amu Darya, the proportion of diverted water has been gradually increasing, representing at present a value higher than that reaching the Aral Sea. Water diversion through the Karakum Canal also had undesirable secondary consequences. Rozengurt (1989), for example, states that in Ashkabad, the capital of Turkmenistan, engineers have been struggling to prevent the city from being inundated with seepage from the Karakum Canal by drilling 150 relief wells to pump the water out.

The amount of water withdrawn from the Syr Darya and the Amu Darya, at one time the result of political decisions taken centrally in Moscow to satisfy the needs of the Soviet economy, is nowadays mainly dictated by the needs of the economies of the three downstream countries for which cotton is a significant hard currency earner. However, the evolution of the economies of these countries is strongly related to their demographic evolution. Of the 51 million inhabitants of Central Asia, almost 70% live in the Aral Sea basin. The following table gives the demographic data in the five republics of Central Asia, according to the WB 1994a.

Two of the five countries, Kazakhstan and Turkmenistan, have quite low population densities, the latter having, however, a much greater birth rate than the former. The more critical demographic situation appears to be in Uzbekistan, where there is a very high population density and birth rate. This situation will lead to an extremely rapid increase in population, which may cause very strong pressure on the water resources of the two rivers but also tremendous economic and social pressures.

Demographic Data in Central Asian Countries.

Population in 1992	Population Density	Growth Rate 1985-92	(million) (p/Km^2) (%)
Uzbekistan	21.3	47.5	2.4
Kazakhstan	17.0	6.3	1.0
Tajikistan	5.6	38.8	3.0
Kyrgyzstan	4.5	22.8	1.7
Turkmenistan	3.9	9.2	2.5

The escalating growth in population will clearly determine a great increase in water demand. Moreover, any water possibly saved in cotton production, due to an increase in efficiency and/or decrease in the irrigated area, will be directly transferred to the production of food to cope with the needs of the increasing population. Large investments will be necessary to face the escalating water demand, but these will be difficult as governments will obviously be forced to invest in other areas to meet rising social needs and unemployment. Unemployment is not likely to be solved by resorting to further development of irrigation, as was the case in the past.

Global Climate Change and Water Resources in the Aral Sea Basin

As indicated, future changes in population may drastically condition the evolution of the use of water resources in the Aral Sea basin. However, population is not the only variable expected to change rapidly. Climate change should also be referred to when analyzing the present situation and may have to be considered when analyzing scenarios for the future.

The reduced surface area of the Aral Sea has modified the weather in the region, the climate becoming more continental and, as a consequence, reducing the growing season (Peterson 1993). Moreover, man-induced climate changes have recently become a major cause for concern, and their impact on future water resources availability should be taken into consideration in an analysis of the situation in a region such as the Aral Sea basin. The increase in atmospheric concentration of CO_2 and other greenhouse gases is likely to cause a global warming, of still uncertain magnitude, which is expected to result in significant changes in the regional distribution of precipitation.

The changes in the precipitation regime in the Aral Sea basin would normally have important consequences for water resources availability, with significant social, economic and environmental impacts. In fact, through the workings of the hydrologic cycle, climate plays an essential role in the quantity and quality of the available water and thus, climatic changes will have direct consequences on the vegetation, the soil and the environment in general. Water demand will also be affected by climate change, since the water requirements for municipal, industrial and agricultural uses are conditioned by prevailing climate conditions and particularly the temperature regimes.

From a practical standpoint, it is important to try and evaluate the magnitude of the consequences of climate changes on the sustainable development of water resources, to identify the possible difficulties that they will raise for water resources managers and, finally, to establish if and how the water resources management and planning strategies should be modified. The objective would be to assess the impact of the expected climate changes on water resources, in order to help water resources managers and planners to take adequate measures in due time to alleviate the adverse impacts, or to take the best advantage of favourable impacts.

The possible impacts of climate changes on inland waters would include changes in the global amount of water resources available, in the spatial and temporal distribution of these resources, in soil moisture, in the characteristics of extreme phenomena such as floods and droughts, in water quality, in sedimentation processes, and in water demand. The author has provided elsewhere a detailed discussion of the several possible consequences of climate change for water resources (Cunha 1989b and 1992).

The best existing tools for producing the effects of global climate change are the global circulation models (GCMs). However, for a relatively limited region such as the Aral Sea basin, these models do not ensure sufficient spatial resolution. Also, to specify adequately the effects of climate change, reliable predictions of transient changes would be required and these are not currently available.

The results of the various GCM experiments often lead to quite different conclusions. In order to avoid this difficulty, it is usual to consider composite GCMs, which result from averaging standardized GCM results. Hulme et al. (1992), in a study sponsored by the Climate Research Unit of the University of East Anglia, have prepared GCM composite scenarios for China and South-East Asia in order to analyze the implications of climate change for this part of the world.

Since the region studied by Hulme et al. (1992) is in the immediate vicinity of the regions where waters of the Aral Sea basin are generated, i. e., the mountains of Kyrgyzistan and Tajikistan, it was considered to be of interest to analyze what the results of Hulme's study could show for these regions. Fig. 7 represents a map adapted from the one prepared by the team of the Climatic Research Unit, which shows for the year 2050 the percentage variation of precipitation due to global climate change in the period December-January-February and June- July-August. The original map has been slightly extended to the West in order to cover the whole of Kyrgyzstan and Tajikistan.

According to Fig. 7, the precipitation in the upstream part of the Aral Sea basin, shown in the left-hand part of the maps in the figure, would increase by the year 2050 from 5 to 10% in winter and decrease in summer to between 0 and -5%. This would indicate an overall increase in precipitation in the Aral Sea basin. It is interesting to note that Budyko and Sedunov (1990), using completely different methods, also predict for the Aral Sea region an increase in rainfall due to global climate change.

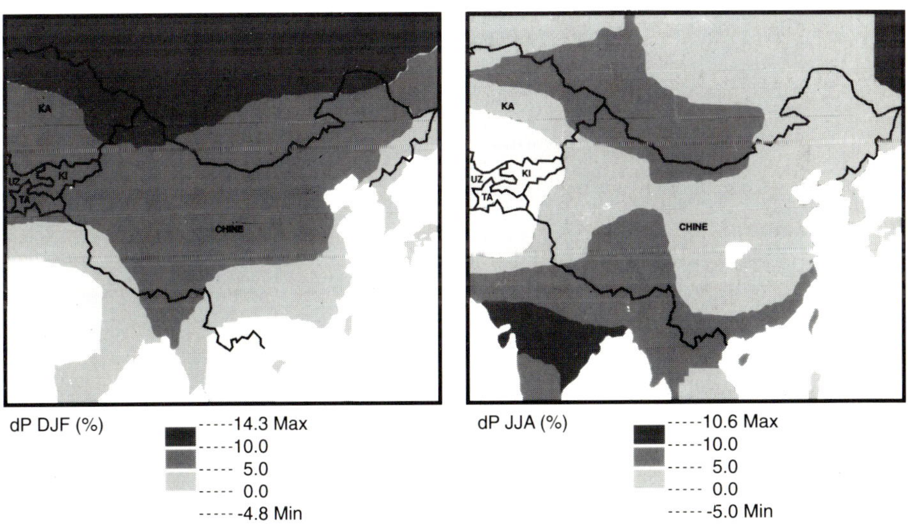

Figure 7 Percentage Variation of Precipitation in Central and South-East Asia due to Global Climate Change for the year 2050

Obviously, these predictions should be considered with caution, as many uncertainties remain about climate change and its impact on water resources. The predictions would need careful consideration and interpretation before they can be fully confirmed. The overall consequences of precipitation change should be carefully analyzed. The analysis shows, however, that considerations of climate change should not be neglected while studying the future water resources situation in the Aral Sea, and that it would probably be interesting to initiate a study of the Aral Sea basin of the type done by Hulme et al. (1992) for China.

Identification of Research Needs for the Study of the Problems of the Aral Sea Basin

An Advanced Research Workshop supported by the Scientific Affairs Division of NATO on "Critical Scientific Issues of the Aral Sea Basin: State of Knowledge and Future Research Needs" was held in Tashkent in May (1994). This workshop provided the first occasion after the collapse of the Soviet Union for scientists of the five republics of Central Asia and Russia to meet for discussing Aral Sea problems.

The NATO workshop made clear that, in solving the Aral Sea basin problems, the role of science and technology is of paramount importance. As a result of the NATO Advanced Research Workshop, a number of recommendations for further scientific research have been prepared, dealing with six areas of major interest for the problems of the Aral Sea basin.

The research needs identified are the following:

1. *Hydrologic processes and phenomena*
   - Studies of the hydrologic effects of the environmental changes in the basin of the Aral Sea;
   - Assessment of micro, meso and macro scale climatic change owing to desiccation of the Aral Sea;
   - Identification and evaluation of the intensity and range of salt/dust transfer from the dry bottom of the Aral Sea;
   - Computer modeling of key hydrodynamic processes in the Aral Sea;
   - More intense investigation of groundwater and its role in the water balance of the Aral Sea basin and sea proper;
   - Study of the water balances and hydrology of the large and small Aral Seas as separate water bodies;
   - Determination of the minimum amount of surface and groundwater that needs to be reserved (from consumptive and polluting uses) for ecological sustainability in the Aral Sea basin.

2. *Ecosystems and their changes*
   - Continued investigation of biotic (floral and faunal) changes in the Aral Sea and deltas of the Amu Darya and Syr Darya brought about by drying the Aral Sea, with better integration of studies of different aspects of the region's ecology and stress on the employment of contemporary methods of understanding ecosystem dynamics in a holistic framework;
   - Development of computer models of ecosystem changes as a means of integrating and understanding the dynamics of very complicated systems;

- Attention to issues of biodiversity and loss of endangered species, particularly in the deltas of the Amu Darya and Syr Darya.

3. *Agricultural production and management*

- Studies of land tenure and use in the Aral Sea basin and how these relate to water use and ecological degradation there;
- Investigation of the extent and nature of water use in the Aral Sea basin and of means to implement effectively water-saving technologies, particularly in irrigated agriculture;
- Evaluation of presently unutilized or under-utilized sources of water to augment currently fully- or over-utilized sources.

4. *Medical, health, social, economic, cultural, and demographic issues*

- Studies of demographic dynamics in the Aral Sea basin, of how these relate to environmental and other regional problems, and of means of alleviation;
- Investigations of the economic structure of the Aral Sea region and of means for its improvement;
- Study of the medical and health situation in the Aral Sea region and of means for its improvement;
- Investigation of the legal structure in the Aral Sea basin, both within nations and among nations, and its relationship to ameliorating the most serious environmental problems;
- Studies devoted to developing effective means of monitoring the health of human populations in the Aral Sea region.

5. *Toxic contaminants (biocides, metals, other organic and inorganic compounds)*

- More intensive study and monitoring of toxic contaminants, including their sources, amounts, environmental pathways, persistence and biological effects, and sinks in the Aral Sea region;
- Development of less harmful substitutes for toxic contaminants and alternative means of controlling pest species of plants and animals.

6. *Application of modern technologies*

- Studies and monitoring of hydrology, landscape and ecosystem change, irrigation and of other appropriate subjects in the Aral Sea basin employing contemporary computerized and remote sensing (e.g., satellite) technologies for data gathering, analysis, organization, and storage; of particular importance is the development of a GIS (Geographic Information System) for critical parts of the region;
- Investigations should be undertaken to determine the optimal means for introducing modern technologies on a broad scale into the Aral Sea basin and for training local scientists and technicians in their use.

This identification of appropriate research needs is an essential step in coping with the Aral Sea basin problems. In the same way, certain international initiatives, such as the diagnostic study made by UNEP (UNEP 1992) or the identification of key programs and 19 projects proposed for implementation by the World Bank (WB 1994b), are indeed very important steps. However, there are certainly real risks that a successful recovery of the Aral Sea basin may not be achieved. WB 1993 identifies three major types of risks:

(a) Failure of the Central Asian republics to cooperate and make the required sacrifices;
(b) An approach to the Aral Sea programs as normal operations rather than as formidable and challenging programs that require extraordinary efforts and long-term commitment;
(c) Possible unrealistic expectations that the Aral Sea crisis could be resolved in a short time.

It must be realized that international initiatives, such as those referred to above, or other ongoing multilateral and bilateral initiatives can only have results and be useful with the effective commitment of the governments of the Republics of Central Asia to cope with the crisis.

The governments should have sufficient political will to cooperate in order to take the measures that are needed to reverse the current course of events. Also, the governments of the countries of the region should effectively be willing to provide sufficiently-qualified staff to enable the recently created institutions to function and to allocate timely national financial resources to achieving a solution to the existing problems, as well as to the development of long-term policies on integrated water resources management, land-use, and social and demographic policies.

If this is not achieved quickly, the Aral Sea basin crisis will certainly be singled out as the most relevant example in the world of non-sustainable development.

References

Aladin, N. V.; Williams, W. D. (1993). Recent Changes in the Biota of the Aral Sea, Central Asia. Proceedings of the International Association for Theoretical and Applied Limnology, Barcelona, 1992.

Budyko, M.; Sedunov, Y. - Antropogenic Climatic Changes. In "Karpe, J. J, Otten, D; Trindade, S. C. (eds) - Climate and Development, Springer-Verlag, Berlin, p 270-284.

Chub, V. 1994 - Review of the Scientific and Environmental Issues of the Aral Sea Basin. NATO Advanced Research Workshop on "Critical Scientific Issues of the Aral Sea Basin : State of Knowledge and Future Research Needs", Tashkent.

Cunha, L. V. da 1992 - Mundanças Climaticas e Recursos Hidricos (Climate Change and Water Resources). 1st Water Congress, Lisbon.

Glasovsky, N. F. 1992 - Salt Balance of Aral Sea. Geo Journal.

Hulme, M.; Wiglex, T.; Jiang, T. - Climate Change due to the Greenhouse effect and its Implications for China.

ICWC 1994 - Aral Sea Problems: Review and Decisions, Tashkent.

Kuznetsov, N.T. 1992 - Geographical and Ecological Aspects of Aral Sea Hydrological Functions. Post-Soviet Geography 33 (5).

Le Tolle, R.; Mainguet, M. 1993 - Aral. Springer-Verlag, Paris.

Micklin, P. (1993). The Aral Sea Problem. Civil Engineering: Proceedings of the Institution of Civil Engineering, London.

Micklin, P. (Eds) 1995 - Critical Scientific Issues of the Aral Sea Basin: State of Knowledge and Future Research Needs, NATO Advanced Science Institutes, Springer-Verlag, Berlin.

Monin, A. 1988 - Zastoinye Zony (Stagnant Zones), Novyi Mir, 7, July.

Novikova, N. M. 1992 - La Dégradation de la Végétation dans l'Actuel Delta de L'Amu Darya. Secheresse, 3, September.

Peterson, A. J. 1993 - Troubled Lands. The Legacy of Soviet Environmental Destruction. Westview Press, Boulder.

Razakov, R.; Kosnazarov, 1994 - Investigations of Dust-Salt Transfer from Dried Bed of Aral Sea and Working Out Measures to Decrease its Environmental Impact. NATO Advanced Research Workshop on "Critical Scientific Issues of the Aral Sea Basin : State of Knowledge and Future Research Needs", Tashkent.

Rozengurt, M. A. 1989 - Water Policy Mismanagement in Southern USSR; The Ecological and Economical Impacts. National Council for Soviet and East European Studies, November.

UNEP 1992 - The Aral Sea: Diagnostic Study for the Development of an Action Plan for the Conservation of the Aral Sea. United Nations Environmental Programme, Nairobi.

Voropayev, G.; Blagoverov, B.; Ismayylov 1987 - Ekonomiko - Geograficheskiye Aspekty Formirovaniya Territorialriykh Edinits v Vodnon Khozyaystve Strany, Nauka, Moscow.

WB 1993 - The Aral Sea Crisis. Proposed Framework of Activities. The World Bank, Washington, 1993.

WB 1994a - The World Bank Atlas 1994. The World Bank, Washington D.C.

WB 1994b - Aral Sea program - Phase I Aide Memoir. Volumes I and II. Europe and Central Asia Region. World Bank.

Chapter VI: Transboundary Water Management in Regional River Basins

VI.1 Interbalkan Cooperation in Solving Interstate Water Resources Problems: the Case of the Vardar / Axios River Basin

Vujica Yevjevich
Professor Emeritus of Civil Engineering
Colorado State University
Fort Collins, Colorado 80523
USA

Introductory Statements and Goals, Purposes and Objectives in the Vardar River Basin Interstate Cooperation

Introductory statements

The Balkan Peninsula faces various water resources problems of interstate character, which require solutions acceptable to the states involved. Many similar regions worldwide face the same problems. Apart from Greece, Turkey (in its European Region), Bulgaria, Albania and Rumania, the final break up of Yugoslavia will likely result in five new independent states, namely: Slovenia, Croatia, Bosnia and Hercegovina, Slavic Macedonia and Yugoslavia. This latter state will likely be composed of Serbia, Montenegro and maybe the Serbian Republic (a part of the ex-Yugoslav Republic of Bosnia and Hercegovina). What all this means is that on the relatively limited surface of the Balkan Peninsula there will be ten independent states in the future instead of six prior to 1991. The more independent states a region has, the more numerous and complex the interstate water resources problems and solutions. They will likely require many international, interbalkan bilateral or multilateral agreements.

The ex-Yugoslav Republic of Macedonia is called Slavic Macedonia in this text This adjective 'Slavic' is used here only as a temporary term, to distinguish it from the Greek province of Macedonia, because of disagreement with Greece for the use of the name Macedonia only, without an adjective. The resolution is pending on its name, flag and the other items contested by Greece.

The Danube, Drava and Tisza rivers enter the Balkan Peninsula from Central Europe. The Central European states, with interest in interstate water resources problems in basins of these three rivers, likely are also interested in multilateral solutions of interbalkan water resources problems of these river basins. An interesting case is the Save (Sava) River Basin of the former Yugoslavia. Before the separation of Yugoslavia into independent states, the Sava River basin completely belonged to Yugoslavia, as a large national river basin. The split of Yugoslavia makes now the Save River an international river, belonging to several independent states.

In some way, a similar case is with the Vardar (Axios) river basin. Until 1992 only two countries shared its river basin and the river itself, Yugoslavia and Greece. By splitting up of

Yugoslavia into five new states, now this river basin will be shared by Slavic Macedonia (the largest part of the basin), Greece (the lowest parts of the river and its basin) and the new Yugoslavia (basically Republic of Serbia, with the upper river basins of the two tributaries of the Vardar River, the Pcinja and Lepenac rivers, belonging to Serbia).

Three types of borders divide states which have some territory in a river basin (Fig.l):

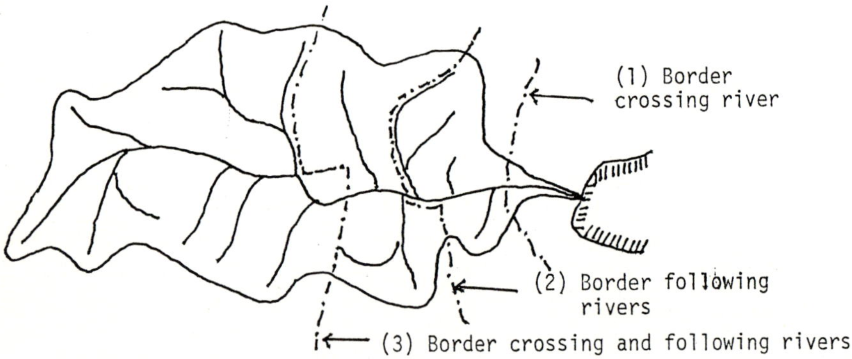

Figure 1 Schematic presentation of the three types on how the interstate borders cross or follow the rivers.

(1) Rivers serve as borders between states, like the Danube River of its lower course serving as the border between Bulgaria and Rumania;

(2) Borders cross the river at a point, namely there is no joint share of a river section by the two states, like the border between Greece and Slavic Macedonia crossing the Vardar (Axios) River near the city of Gevgelia, or the border between Hungary and Yugoslavia crossed by the Danube and Tisza Rivers at points, and

(3) Borders which follow but also cross the international rivers.

How the interstate borders follow or cross, or both, the international rivers and how they divide rivers and river basins, will determine what type of water resources problems are or will likely be posed for the bilateral or multilateral interstate solutions. A large number of international agreements in solving various types of interstate water resources problems is available in the world for reference and as already existing international precedents.

Goals of the states in water resources developments in the Vardar River Basin

It can be safely stated that the Slavic Macedonia, to which most of the Vardar River basin belongs, has nearly all the social, economic and political goals which may occur with water resources development, conservation, control and protection within a relatively large river basin. Greece is mainly interested in increasing food production by by obtaining new water for irrigation in its Macedonia, or in the lower Axios River basin, then in the water pollution control, defense from floods, drought control, with all these goals likely to be accomplished by the cooperative building of large reservoirs on the tributaries of the Vardar River in the Slavic

Macedonia. Yugoslavia will likely have as the goals in water resources developments in this basin: the new transportation capacity by the construction of the navigational route Danube-Morava-Vardar-Gulf of Salonika, and the diversion of water from the Upper Lepenac River (a tributary of the Vardar River) to Kosovo, to offset the lack of water for food production by irrigation. The accomplishment of these goals will require a good cooperation and solution of inevitable controversies by the three states.

Purposes in accomplishing goals. Goals are accomplished by various water resources developments, transfers of water from the water-surplus adjacent river basins, water conservation, control and protection. Each particular goal will require that some water resources purposes are satisfied, such as irrigation, drainage, hydropower production, navigation, water supply, water pollution control, flood defense, drought control, and similar (up to about 35 small, median and large likely purposes of water resources developments).

Objectives in accomplishing purposes and goals. Finally, to accomplish the purposes required by the state goals in development of water resources one must define, and then maximize or minimize the objectives of economical, social, monetary, political or fairness characters. This classical trinity: goals, purposes, objectives should be strictly applied in the future cooperation and resolutions of controversies between three states in water resources development of the Vardar River basin.

Geomorphology, Hydrology, and Hydroelectric Power Potential of the Vardar River Basin

Geomorphology

Four major mountain catchment divides separate the Vardar River basin from the adjacent river basins, with these divides mostly located between the Slavic Macedonia and the bordering countries (Fig. 2). To the South, the mountain divides separate this basin mainly from Greece, with the part of the Greek province of Macedonia belonging to the Vardar River basin. To the East, the mountains divide the northeastern part of the Vardar River basin from Bulgaria, and the southeastern basin from the Struma River basin. To the West, the maintains divide the Vardar River basin from the Black (Crni) Drim River basin and Albania. To the North, the lower mountains separate the Vardar River basin from the Morava River basin and Serbia. Nearly of a quadratic shape, the center of the Vardar River basin is a geomorphological bowl, experiencing a kind of rain shadow, with special patterns of abundant precipitation of humid mountainous regions and low precipitation of the dry lowland valleys in the center.

Precipitation and Evaporation

The Vardar River basin has a typical Mediterranean precipitation and evaporation regimes. Figure 2 shows isolines of the average annual precipitation in the ex-Yugoslavian part of the Vardar River basin, with annual precipitation in the Greek Macedonia being similar to that of the lowest part of the southeastern area of the Slavic Macedonia. The lowest average annual precipitation is about 400- 450 mm in the center and southeast parts of the basin, while the maximum average values reach to more than 1500 mm along some of the western mountainous catchments divides. The annual evaporation is relatively large, thus leaving only a small part of precipitation to become runoff. The Mediterranean type of precipitation and evaporation is usually characterized by a much larger precipitation in the winter half of the year, and relatively much smaller part in the summer.

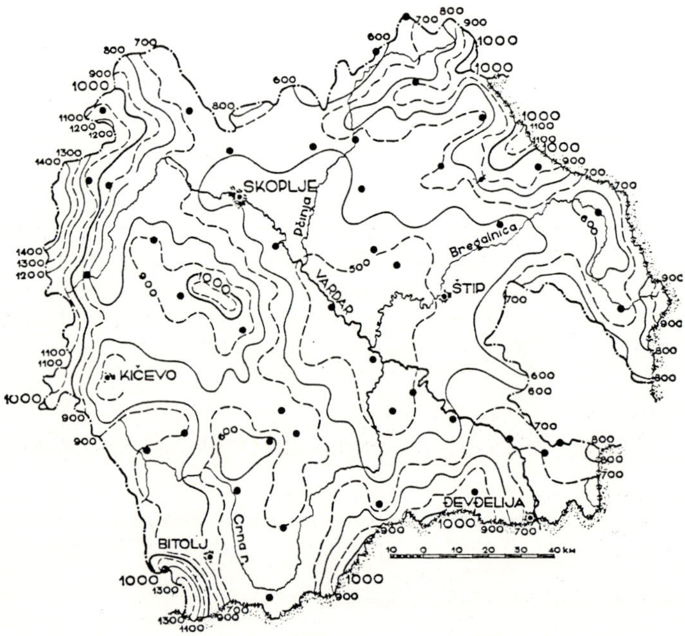

Figure 2 Isolines of the average annual precipitation of the Vardar River basin, in the ex-Yugoslavian part of the basin (Taken from the book: Water Power Resources of Yugoslavia, 1956).

Basic consequences of the Mediterranean precipitation and evaporation regimes of the Vardar River basin are the following:

(1) Limited moisture in vegetation season requires irrigation of the lowland and valley parts of the river basin, or a transfer of water from the winter to the summer season; and

(2) Because the western part of the Balkan Peninsula has much larger precipitation than the center and the lower parts of the Vardar River basin, the transmountain diversion of water from the West to the East should be considered as the interbasin water exchange or diversion, whichever of the two alternatives may come out to be the most attractive solution to the interested parties of this diversion strategy.

Runoff

Fig. 3 presents the daily flow hydrograph of the Vardar River at the border between the Slavic Macedonia and the Greek province of Macedonia, for a period of 16 years only. The purpose of this display is only to properly demonstrate and emphasize the Mediterranean regime of runoff. It shows a high variation in the total annual runoff, demonstrating that a very wet year may produce 5-7 times as much runoff as a very dry year. Fig. 4 presents the sequence of the twelve mean monthly flows of the Vardar River at the border between Slavic Macedonia and Greece. This figure demonstrates also the typical Mediterranean runoff regime of the Vardar River. It has the major winter-spring rainfall and snowmelt season, the major low flow summer-fall season, the minor fall rainy season, and the minor early winter low flow season. The fitted continuous line implies that the monthly runoff is periodic, likely only with the 12-month and 6-month significant harmonics of periodicity.

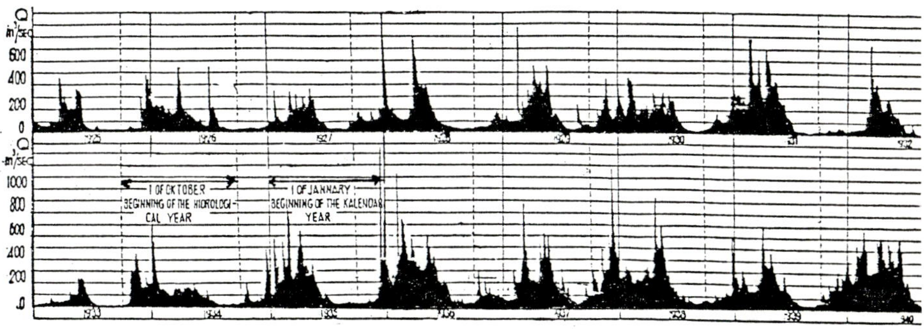

Figure 3 Daily runoff hydrograph of the Vardar River at Gevgelia (same source as for Fig. 2).

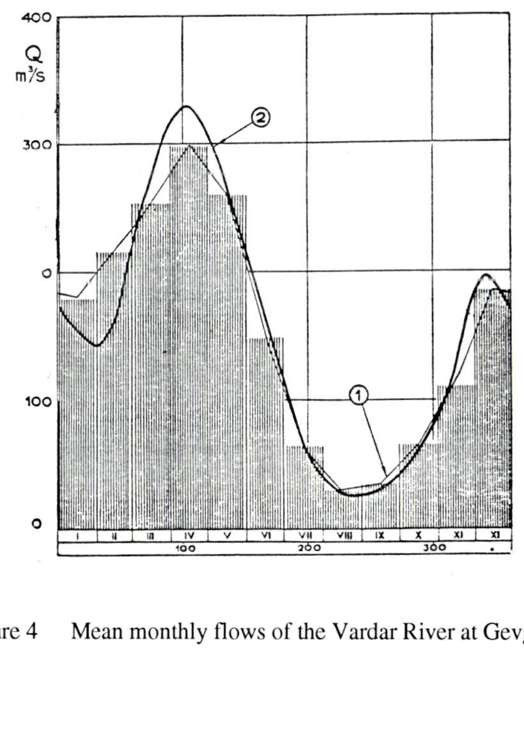

Figure 4 Mean monthly flows of the Vardar River at Gevgelia.

Fig. 5 shows the mean discharge along the streams in the Vardar River basin. It shows that the five major tributaries of the Vardar River: Lepenac, Treska, Pcinja, Bregalnica and Crna Reka (Black River) rivers, have relatively modest mean discharges, and that only the main stem of the Vardar River gets the sizable and concentrated mean discharges.

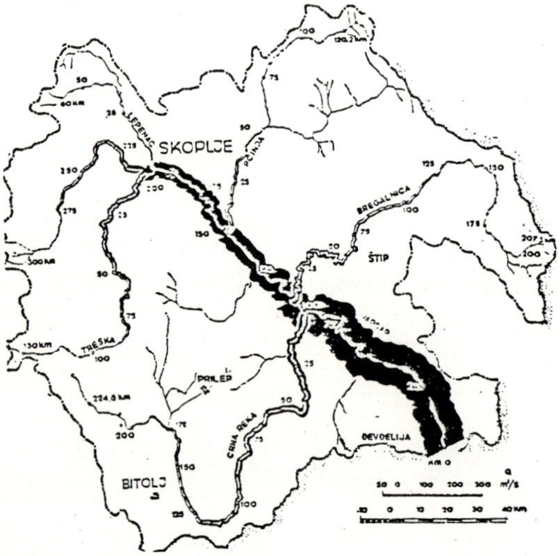

Figure 5. Mean discharge of major rivers in the Vardar River basin, proportional to the widths of the stripes along the streams.

Floods

Floods occur both during the longer rainy and snowmelt winter-spring season, but also in the short rainy fall season. Very large fall floods are of a special mechanism. When a large snow falls in the mountains and hills, and is followed by a large warm rain storm, which also melts the accumulated snow, the exceptional flood occurs. Figure 6 shows schematically how such a flood is generated by the two sources of water, the rainfall and the snowmelt. One may conclude that the regular annual floods occur nearly every late winter - early spring season, but large or exceptional floods occur mostly in the fall season.

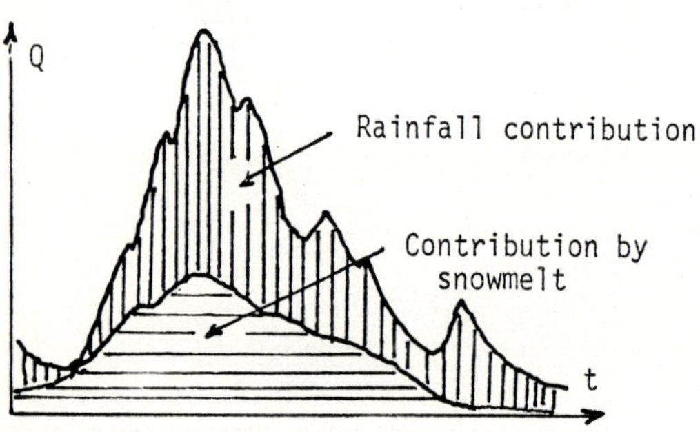

Figure 6 Composition by the source of water of the fall floods.

Droughts

Droughts in the Vardar River basin are frequent, not only because of the precipitation and evaporation regimes, but also because the allocated or needed water, basically for irrigation, makes even the median runoff year nearly as a drought year. Coping with droughts is one of the major water resources problems of this river basin.

Gross hydroelectric power potential

Fig. 7 presents the elevation versus the mean discharge for the streams of the Vardar River basin, with the shaded areas being proportional to the gross hydroelectric power potential of a stream. Fig. 8 shows the gross hydroelectric power potential along the major streams of the Vardar River basin, in kilowatts per kilometer of the streams, demonstrating the river sections with the significant gross power potential.

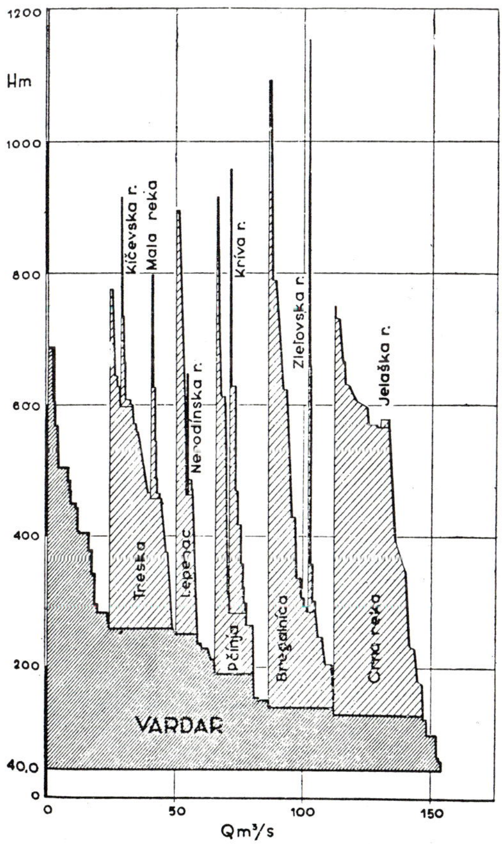

Figure 7 The elevation versus the mean discharge graphs of the major streams in the Vardar River basin.

General Interstate Cooperation in Developing Water Resources of the Vardar River Basin

The following are the major water resources problems in the Vardar River basin, which are, or will be, of mutual interest of the four countries: Slavic Macedonia, Greece, Yugoslavia (Serbia) and Albania:

(1) Transfer of water from humid to dry regions of the Balkan. These transfers in the case of the Vardar River basin mean taking water from the water abundant West to the water deficient East, basically from the Drim River basin to the Vardar River basin. An example of this strategy and policy already accomplished is the Mavrovo-Vardar River hydroelectric power scheme, by which the part of waters of the Radika River (tributary of the Crni Drim) is transferred via the Mavrovo Reservoir and hydroelectric power plant into the Vardar River.

(2) Development of water resources in the Vardar River basin. Storage of water for irrigation, increase of low flows, control of water pollution, coping with floods and droughts, water supply, etc., are the main purposes of development to be accomplished. Storage of water is only feasible on some tributaries of the Vardar River, or outside of its basin by water transfer into the Vardar Basin, by sharing benefits and costs among the interested states in building reservoirs and operating the released water in vegetation season according to interstate agreements.

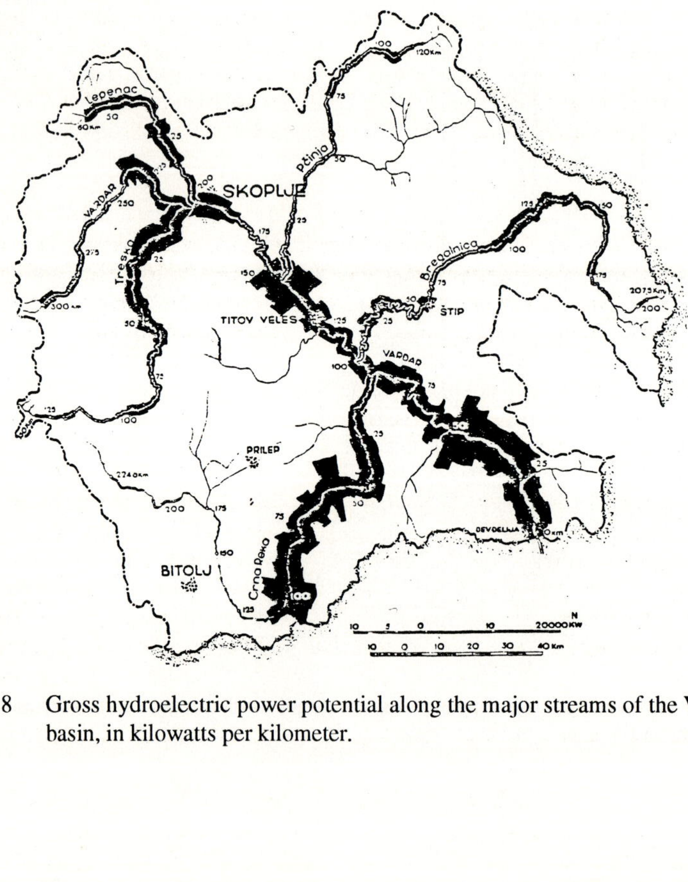

Figure 8 Gross hydroelectric power potential along the major streams of the Vardar River basin, in kilowatts per kilometer.

(3) Pollution control of the Vardar River waters. One of the major water resources problems, especially sensitive between Greece and Slavic Macedonia, is the pollution of the Vardar River waters, especially during low flows. Pollution abatement will be one of the most delicate problems requiring the proper interstate solution and cooperation. With time that pollution will likely become an issue if not a controversy between these two states, especially because of the resulting pollution of the Gulf of Salonika and the Region Sea.

(4) Step-by-step constructing of the Danube-Morava-Vardar-Gulf of Salonika navigational route. This is an old idea, generally known in the Balkan as the Morava-Vardar Canal. At present, this idea may finally become of interest as a long range goal and water resources purpose, not only to Yugoslavia (Serbia), Slavic Macedonia and Greece, but also to all European countries having the navigational access to the recently completed international transeuropean navigational route Rhine-Danube. These three countries may not have the necessary capital to construct this navigational route without the wide participation of European or even general international capital.

Transmountain Water Diversions

The only transmountain water diversion into the Vardar River basin at present is diversion of the Upper Radika River water, which is transferred via the Mavrovo Reservoir and hydroelectric power scheme into the Vardar River. To compensate Albania for the decrease of the mean discharge of the Drim River, the former Yugoslavia permitted Albania to extend the flooding by a reservoir on the Drim River into the White Drim in Metohija, which is the part of the Kosovo autonomous region of Serbia in Yugoslavia. This mean that there are feasible solutions for water transfers, by which the beneficiaries of transfers compensate the losers of water in some adequate ways.

A project already exists for construction of a reservoir on the Upper Lepenac River (a left bank tributary of the Vardar River) into the Kosovo Region, with the water transfer for irrigation, with drainage water from that irrigation scheme either returning to the Lepenac River or via the bifurcation to the Sitnica River into the Ibar River and then to the West Morava and Great Morava Rivers. For this transfer the agreements by the Slavic Macedonia and Greece would be necessary, since it would decrease by several cubic meters per second, on the average, the mean flow of the Vardar River.

Similarly, ideas exist for further transfer of the Radika River waters into the Mavrovo Reservoir and the Vardar River by pumping water from a point of Radika River 100-150 meters below the highest level of the Mavrovo Reservoir. This transfer would increase by several cubic meters per second the flow of the Vardar River and for that amount decrease the mean flow of the lower Radika River, the Black Drim and the Drim Rivers. For that transfer it will be necessary to obtain the agreement by the Western Slavic Macedonia and Albania. The trade-off between the transfer of water of the Lepenac River out of the Vardar River and the the transfer of water from the Radika River into the Vardar River may be a practical solution.

The major interstate water resources problems between Slavic Macedonia, Greece and Albania would be created by the utilization of waters of the Prespa Lake (Fig. 9). Waters from this

Lake spill over in the underground into the karst channels, leading them to the Biljana Springs at the southeastern shore of the Ohrid Lake. Fig. 9 shows the mean discharge along the underground and surface streams of the hydrologic system Prespa Lake - Ohrid Lake - Black Drim (left map), and the subsurface and surface gross hydroelectric power potentials of this system (right map).

Hydroelectric power of the Black (Crni) Drim between the Ohrid Lake and the border with Albania has been already developed. Both water and hydroelectric power of this system would be decreased if either irrigation of the land around the Prespa Lake is undertaken, or the Prespa Lake water diverted either to Greece for hydroelectric power development, or to the Bitola Plain (Pelagonia) for its irrigation and for irrigation of the Salonika Plain, all with the hydroelectric power development.

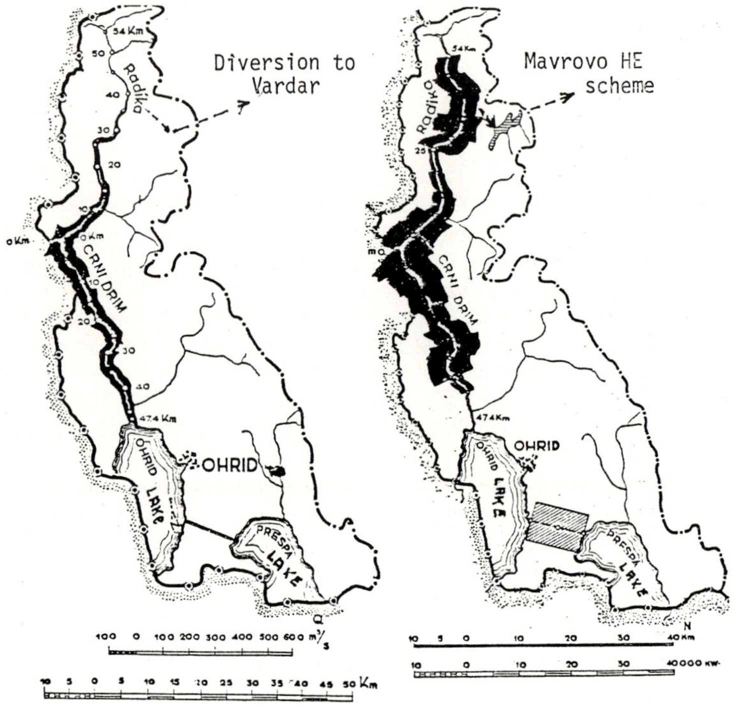

Figure 9 Mean discharge (left map) and gross hydroelectric power potential (right map) of the Prespa Lake-Ohrid Lake-Crni Drim basin.

Any scheme to utilize waters of the Prespa Lake would likely involve not only Greece, Albania and Slavic Macedonia, but also Europe and the international environmental movements, since this Lake is an important rest station for migratory birds. If a scheme for irrigation of land around the Lake is developed, so that the lake level is decreased in fluctuation and stabilized, and all three states received the benefit of it, that would be easier to agree upon by all interested parties. If a stable Lake level is planned with the transfer of water to Greece, the approval of Albania and Slavic Macedonia would be necessary and their losses compensated, because of the effects on the Ohrid Lake and the loss of production of hydroelectric power along both the Black Drim and the Drim rivers.

Similarly, if waters of the Prespa Lake are diverted toward Pelagonia in Slavic Macedonia, and used also to irrigate land in Salonika Plain, the compensation to Albania for losses and partly to Western Slavic Macedonia would be necessary.

Water Resources Development of the Vardar River

Any storage of water in the Vardar River basin, any diversion of water into the Vardar River basin or out of it, and the operation of these schemes would be of high interest to Slavic Macedonia but also of high interest to Greece as the downstream state. Of special interest to both of them will be a joint development of storage capacities which would supply water to both countries for irrigation, development of hydroelectric power (in order to partially offset the cost to irrigation by selling electricity), coping with floods and droughts, as well as pollution control of rivers.

No significant storage capacities are economical along the Vardar River itself, because of the impermissible flooding of important communities, communication lines, industries and productive land. The storage on the Lepenac River will be used for irrigation in Kosovo. Some potential storage capacities on the Pcinja River should likely be reserved for water supply of the future navigational Morava-Vardar canal. The Bregalnica River has the storage reservoir Kalimanci, as well as a small storage reservoir on the Kocane River, all used for irrigation of rice paddies in the Kocane Plain. Any potential storage site will likely be used for irrigation in this river basin. The Crna Reka (Black River) has only one reservoir at present, Tikves. Its stored water is used for multiple purposes, including irrigation, hydroelectric power, water supply to industry, and flood control.

Good storage potential exists on this river between the Bitola Plain (Pelagonia) and the Tikves Reservoir. It should be considered as a very attractive river section for joint development of reservoirs between Slavic Macedonia and Greece. The Treska River has significant potential for storage reservoirs upstream of the small Matka hydroelectric power plant, also attractive for joint developments. The Doyran Lake at the border between Slavic Macedonia and Greece is an ideal storage capacity for a joint development between these two countries, because it is easy to divert the available water from the Vardar River into this Lake.

Figure 10 shows the river sections available for development of storage capacities. The major reservoirs may be created on the Treska and Crna Reka Rivers, and minor reservoirs of local significance on the other rivers indicated in Fig.10.

In any eventual negotiations of jointly developing large reservoirs in the Vardar River basin between Slavic Macedonia and Greece, the first attention should be paid to the Doyran Lake storage capacity, then to the Crna Reka and after that to the Treska River potential reservoirs.

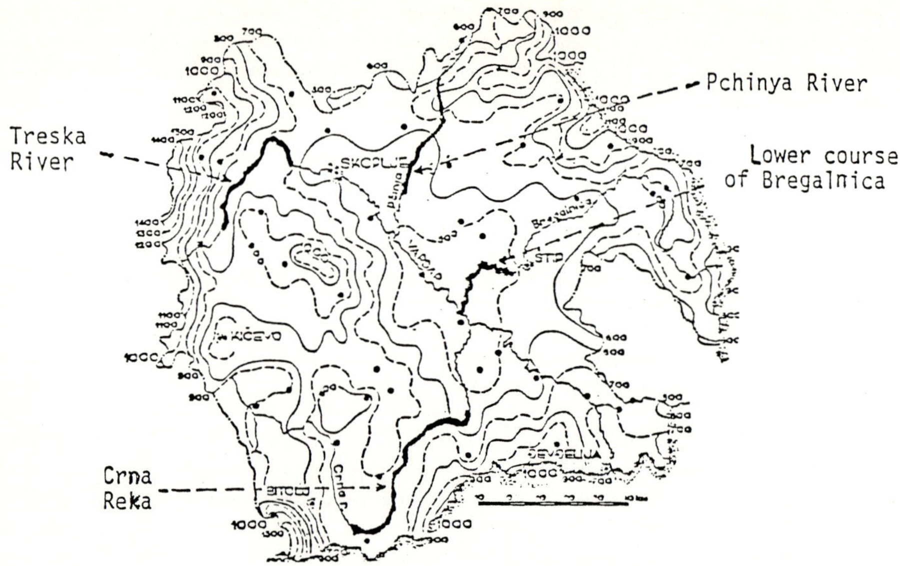

Figure 10 River sections on streams in the Vardar River basin for development of the future major and minor storage capacities

Pollution Control of Waters in the Vardar River Basin

As in many other countries of rapid development, Slavic Macedonia did not invest parallelly with industrialization the necessary funds also into the corresponding environmental protection, especially in cleaning urban and industrial wastewaters before they are discharged into the rivers. To meet the standards of water quality in the Vardar River and in its tributaries, all urban wastewaters should be cleaned at least up to the secondary treatment. The same is valid for the industrial wastewaters and waters from the presently active mines and smelters, or drained waters from the old mines. The downstream water users have all rights to insist that water quality standards of international recognition should be enforced. This aspect is also an important factor for the Mediterranean Sea, since the River Vardar brings all its pollutants into the Gulf of Salonika and the Region Sea. A lot of patient negotiations and cost sharing will be necessary in the near future to clean the waters of many sections of streams in the Vardar River basin.

Morava - Vardar Navigational Route

Figure 11 shows the hydrographic map of the rivers Danube, Sava, Morava, South Morava, Pcinja and Vardar, as the streams in the former Yugoslavia, connected with the future navigational Morava - Vardar route .

Seven particular and specific sections of that route are:

(1) Section of the Great Morava River between its confluence into the Danube River and Stalac (the confluence of the South and West Morava) of the navigational route will likely come out to be most economically developed by constructing the cascade of weirs with hydroelectric power plants and navigational locks, and likely with irrigation of land along this river (Energoprojekt, Beograd, 1993).

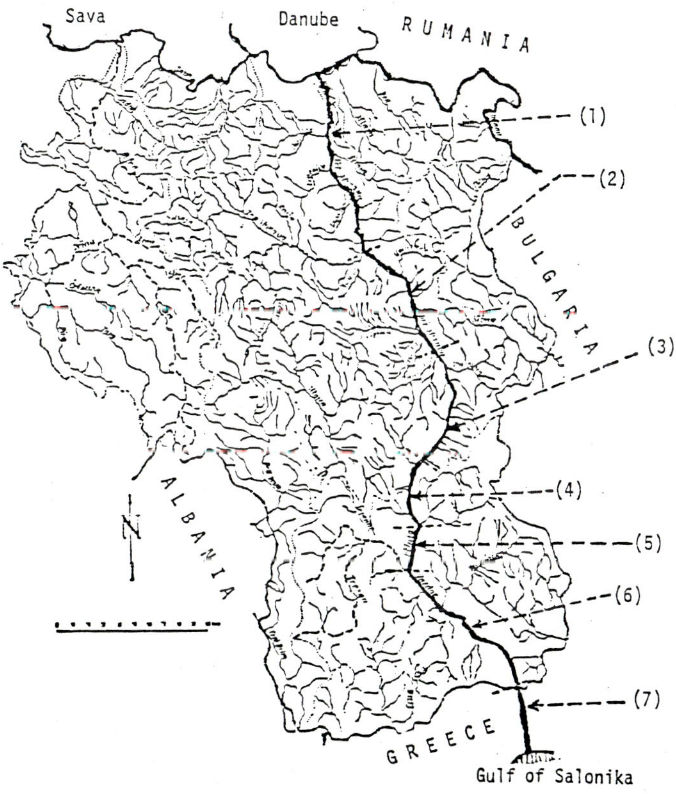

Figure 11 The Morava-Vardar navigational route, with seven specific sections.

(2) Section of the South Morava from Stalac to the Grdelica Narrows will likely be studied in two main alternatives, namely either as a cascade of weirs along the river channel, with navigational locks and hydroelectric power plants, or as a cascade of locks along the separate canal built along but outside of the river channel, or as a combination of the two.

(3) Section of the South Morava River along the Grdelica Narrows will likely be planned as the cascade of weirs with hydroelectric power plants and navigational locks, for the simple reason that these Narrows would not have sufficient space for a navigational canal apart from the existing and important communication lines of railroad and international highway. This section poses also the problem of sediments coming into the river, though the soil conservation measures in the last 50 years have been very successful by a substantial decrease of sediments reaching this river section.

(4) Section of the catchment divide between the South Morava River and the Pcinja River at Presevo (or between the Morava and Vardar river basins) will be likely planned as the navigational canal with locks. For water supply to this section of the navigational route, likely reservoirs on the Upper Pcinja River, or any other smaller tributary, close to that section, will be built.

(5) Section along the lower course of the Pcinja River down to its confluence into the Vardar River will be likely solved by construction of one or a couple of weirs with navigational locks. Narrows of the Lower Pcinja River will likely preclude the construction of a canal outside of the river channel.

(6) Section of the Vardar River between the confluence of the Pcinja River and the border with Greece will likely be made navigable by constructing a cascade of weirs or dams, with power plants and navigational locks. Though the valley of the Vardar River along this section has several enlargements, they are followed with narrow valleys which would nor favor a parallel navigational canal with the river channel.

(7) Section between the border of Greece with Slavic Macedonia and the Gulf of Salonika will likely be composed of a channel weir with power plant and the navigational lock in the upstream most part of this section, and the navigational canal with locks and gates for water release for irrigation down to the sea.

The fact that the Rhine-Danube navigational route is recently completed, thus connecting the North Sea and the Black Sea, traversing the Europe and passing through or along eleven countries, will likely stimulate the interest in construction of additional navigational routes which will feed the traffic into this central navigational route. The Morava-Vardar navigational route will cut significantly the length of navigation between the Central Europe and the ports in the Mediterranean Sea. It may take a century for this navigational route to be built, but it should be started as soon as a feasibility project becomes available. In this endeavor not only Greece, Slavic Macedonia and Yugoslavia should be interested, but also many countries of the Central, West and East Europe.

For various water resources projects, on rivers along the line of the Morava-Vardar navigational route, each of them should be planned and designed in such a way as to permit its future most economical incorporation into the navigational route system. Basically, it will be done by adding later the navigational locks. Since no significant dam and no weir has been yet built along the above mentioned river sections, it is the proper time to start planning this route by creating the International Commission for the Morava-Vardar Navigational Route. The purpose of this commission would be to advise the interested governments in financing the work on the general plan and the feasibility project of this route and to participate in its approval when finished. Any future project along the route should take into account the Morava-Vardar navigational plan and feasibility project at the most efficient and economical way. When a large number of such projects have been developed in the future, then the finishing of the route may be ready to be fully implemented. This may follow the similar patterns as the Rhine-Danube navigational route has had.

Interstate Water Resources Commission for the Vardar River Basin

When the process of separation of the former Yugoslavia into the independent states has been finally terminated, it may be the proper time to create an interstate water resources commission for the Vardar River basin. It should be composed of Slavic Macedonia, Greece and Yugoslavia (mainly Serbia). Its objectives should be to discuss and search for the mutually acceptable solutions of any pending or new water resources problem in the river basin involving the interest of two states at least, if not all three. This commission may be tailored according to some other, already existing, international river commissions. It may also initiate the planning of the Morava-Vardar navigational route until the interests develop in the Balkan and in the European countries for that route for a special Morava-Vardar Navigation Route Commission to be established. That commission then will take over the tasks of building that navigational route, and separate it from the Vardar River Commission.

Conclusions

1. Water resources issues, controversies and conflicts between states often occur, requiring mutually acceptable solutions, so they should be expected also in the Vardar River basin.

2. River Vardar offers several potential water resources projects of mutual interest of Slavic and Greek Macedonias, for future joint developments

3. The Danube-Morava-Vardar-Gulf of Salonika future navigational route should be conceived as a long-range, implemented step-by-step, strategic objective of Greece, Yugoslavia (Serbia) and Slavic Macedonia.

4. When interests are mutual, it is always feasible to find solutions by fairly dividing benefits and costs of international, bilateral or multilateral, water resources projects.

5. Division of water quantities of international rivers, preservation of water quality, jointly built and operated water storage capacities, copings with floods and droughts, all of them invite fair cooperation among the Balkan states.

References.

Energoprojekt, Beograd, 1993, Water Resources Scheme of the Great Morava between the confluence of the South and West Morava and the Danube River.

Yevjevich V., 1956, Water Power Resources of Yugoslavia (and Methodology of Investigating Water Power Resources), Beograd, 1956.

Note. The writer of the above text has worked as young hydraulic engineer on various water resources projects in Slavic Macedonia in the period November 1937 to April 1941, or prior to the Second World War.

Between 1945 and 1958 he was involved with many projects in Slavic Macedonia, particularly with the Mavrovo hydroelectric power scheme, as well as in planning water resources developments of the Vardar River and its major tributaries. After that period he has followed water resources developments in that river basin.

VI.2 Management Problems of Transboundary Rivers between Bulgaria and Greece

Tzovaridis S.N.
Ministry of the Environment, Land Planning and Public Works
General Secreteriat of Public Works, Hydrology Section, Athens
Greece

Moutafis N.J.
Public Power Corporation, Hydroelectric Development, Athens
Greece

Cavadias G.S.
Consultant in Water Resources Development, Athens
Greece

Introduction

The purpose of this paper is to describe briefly the current approaches to the management of transboundary rivers between Bulgaria and Greece. It is apparent that Southern Bulgaria shares all its river basins with northern Greece, with the result that Greece receives from Bulgaria an important proportion of its water resources.

The waters of these rivers are used in both countries mainly for irrigation, water supply and hydroelectric power production. In recent years the water quality and environmental problems have acquired special importance, due to the increase of population and the existence of biotopes protected by the Ramsar Convention.
In addition to the above the paper raises a number of questions of more general interest such as:

(a) The effect of a possible climatic change on the formulation of international agreements concerning transboundary rivers.

(b) The role that can be played by international Organizations and, for our Region, the European Community for the creation of conditions that facilitate the solution of problems related to the management of transboundary rivers.

Agreements between the Two Countries.

The basic agreement of cooperation between the two countries concerning the management of transboundary waters was signed in November 1964. The basic points of this agreement were the following:

a) The parties must avoid causing damages to the other party by the construction and operation of water resources projects in their transboundary rivers.

b) The parties will exchange useful information on water resources and hydrometeorological, topographical, geotechnical, geological, agricultural and economic data.

c) The parties will exchange the necessary information in order to avoid inundations caused by floods in transboundary rivers.

d) The contacts between the two countries concerning the problems of transboundary rivers will be made directly by the authorities responsible for water resources in each country.

e) Within five months from the signing of the agreement Bulgaria will carry out technical and economic studies on the possibility of construction of structures on Bulgarian territory for the retention of sediment and the irrigation of agricultural lands within Greece. On the basis of these studies the benefits to both countries will be determined and a special agreement may be prepared.

Following the agreement of 1964 a Joint Commitee and a working Group were formed to deal with specific issues such as the flooding in Bulgaria due to the operation of the reservoir in the Arda river, the channel adjustments of the river Evros (Maritza) and the use of transboundary rivers for fluvial transportation.

Recently at a meeting between the two countries (April 1992), a common proposal to the European Community was made for the establishment of observation stations for monitoring the quantity and quality of the flows of the transboundary rivers Strymon, Nestos and Evros and other subjects to be discussed below.

Although the basic agreement between the two countries was signed 30 years ago, the contacts between them concerning transboundary waters have not been as regular as they might otherwise have been. Thus, a number of problems still exist, which might be resolved by a closer cooperation between the two countries.

Problems in the Management of the Transboundary Rivers

The Nestos Basin

The Nestos (Mesta) river rises in central Bulgaria; flows in a general south-east direction and finally discharges into the Thracian Sea, 35 km east of the town of Kavala (see Fig.1). The river has a total length of about 250 km, out of which 100 km are in Bulgaria and 150 km in Greece. The total catchment area of the river is 5751 km2. Although the river course is longer in Greece than in Bulgaria, the catchment area in Bulgaria is larger than that in Greece, (see Tab. 1).

The flow of the Nestos river is used by both countries for municipal water supply,irrigation and hydroelectric power production. It must be noted that water from the Dospat (Dhespotis) river (one of the tributaries of the Nestos river in Bulgaria) is diverted to the Evros river within Bulgaria. The hydrology of both the Bulgarian and the Greek sections of the Nestos river has been summarized in the 1979 Engineering Report on the Thissavros Power Plant.(PPC/GMPT 1979)

Hydrology

The Greek and Bulgarian basins of the Nestos river are shown on Fig.1. The Greek basin is conveniently divided into six sub-basins, related to flow measurement stations and dam locations along the river. The sub-basins and their corresponding catchment areas are shown on Table 1.

Table 1 NESTOS RIVER CATCHMENT AREA

BASIN	GREEK SECTION	TOTAL	TOTAL minus DHESPOTIS DIVERSION
-----	km^2	km^2	km^2
Bulgarian Section	-	3,437	2,872
Delta Gauge Station	47	3,484	2,919
Papades Gauge Station	406	3,843	3,278
Thissavros Dam	821	4,258	3,693
Platanovryssi Dam	1,215	4,652	4,087
Temenos Dam	1,225	4,662	4,097
Temenos Gauge Station	1,521	4,958	4,393
TOTAL	2,314	5,751	5,186

Runoff from the Nestos Basin in Bulgaria

The hydrological study performed in 1972 (Moutafis 1991), estimated a mean annual runoff of 433 mm for the Nestos basin in Bulgaria. This corresponds to a mean annual inflow at the border with Greece, of 47.2 m3/s or 1488 x 106 m3. Part of that water is diverted from the Dospat reservoir to the river Vucha. The area of the Nestos basin drained by this diversion project was determined at 565 km2 (Moutafis 1991), thus resulting in the diversion of 245 x 106 m3 of water per year (7.76 m3/s). This figure is in agreement with the data presented in an article by Tzehova et al published in the February 1963 issue of the Bulgarian journal "Electric Energy".

In addition, in the 1970 Protocol that was signed between Greece and Bulgaria, is stated that in an average year 147 x 106 m3 of water are taken directly from the Dospat river, 79 x 106 m3 are diverted into the Dospat reservoir from the Canina and Vistcheritsa rivers and 24 x 106 m3 from the Bistrista river. So in total 250 x 106 m3 of water are diverted away from the Nestos basin during a year of average flow. Thus from three different sources it is concluded that the diversion from the river Dhespatis has caused a mean annual loss of 245 x 106 m3 of water to the flow of Nestos into Greece. Consequently the mean annual flow at the border had been reduced by the year 1964, to 39.43 m3/s or 1243 x 106 m3.

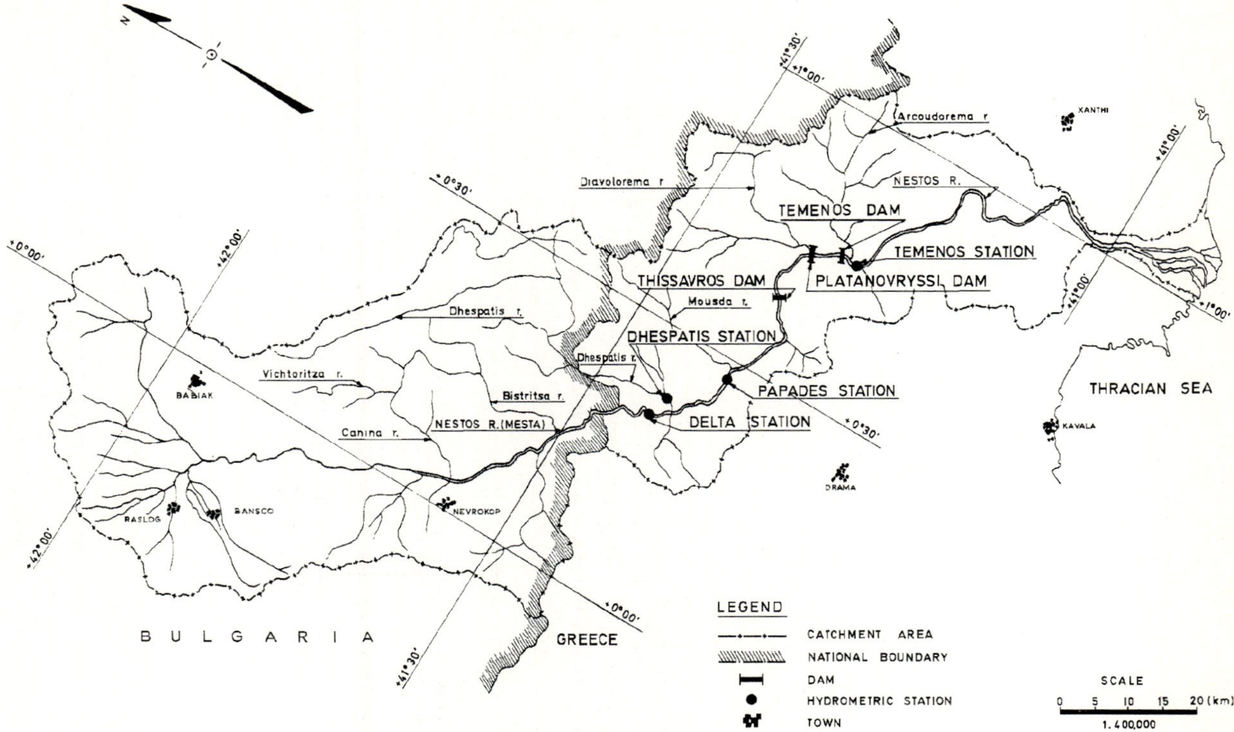

Figure 1 The Nestos Basin as shared between Bulgaria and Greece.

Runoff from the Nestos Basin in Greece

In the Greek basin of the Nestos river there are two types of hydrological measurements:

a dense network of rain gauge stations measuring precipitation
a set of hydrometric stations measuring the flows of the river.

There are three reliable hydrometric stations on the Nestos river and one station on the tributary Dhespatis (the location of the stations is shown on Fig. 1). These stations have been in operation for a number of years, as shown on Table 2.

Table 2 HYDROMETRIC STATIONS

STATION	YEARS OF OPERATION
Temenos	1964 -
Papades	1965 -
Dhespatis	1970 -
Delta	1978 -

It has been verified that the available data of the flow measurements are reliable and it has been proved that a good correlation exists between the observed precipitations and runoff measuring systems (PPC/J.V. COBA - CONSULPRESA 1970). The mean annual flow of Nestos river at Papades and Temenos measuring stations, for the period 1965~1990, is shown on Fig.2.

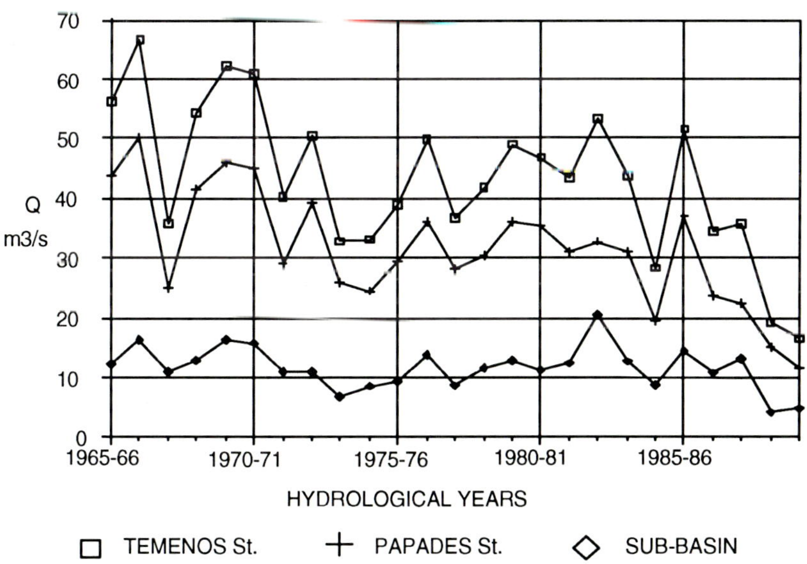

Figure 2 Flow of R. Nestos at Papades and Temenos Hydrometric Stations and their difference.

Since the location of Papades hydrometric station is fairly close to the border, the graph for the Nestos flow at Papades station is a good indication of the inflow from Bulgaria. On the same figure, the difference between the corresponding flows at these two stations is also shown. This difference is actually the runoff from the intermediate sub-basin between the Papades and Temenos stations and consequently is independent of the inflow from Bulgaria.

The sub-basin, between the Greek-Bulgarian border and the Papades station, has an area of 406 km2 and it represents 26% of the Greek basin upstream of the Temenos station.

General Remarks on Nestos Runoff

An indication of the trends and/or changes in water flow over the period 1965-1990 may be obtained by considering the mean annual flows , as shown on Table 3.

Table3 NESTOS RIVER WATER FLOW

PERIOD	PAPADES STATION		PAPADES-TEMENOS SUB-BASIN	
	FLOW m^3/s	STANDARD DEVIATION	FLOW m^3/s	STANDARD DEVIATION
1965-74	37.02	9.37	12.22	3.07
1965-84	34.02	8.11	12.20	3.00
1965-90	31.61	9.46	11.67	3.58
1965-74	37.02	9.37	12.22	3.07
1975-84	31.02	4.70	12.18	3.23
1985-90	21.99	8.70	9.53	4.29

The mean annual runoff of 345 mm has been calculated for the sub-basin between the Papades and Temenos measuring stations, for a mean annual flow of 12.2 m3/s (see Table 3), which has remained virtually unchanged for the period 1965 - 1988. From the data shown on Fig. 2 and Table 3, it can be stated that:

- It is evident from the graphs that there is good correlation of peaks between the inflow from Bulgaria and the flow from the intermediate sub-basin in Greece, indicating a similarity in precipitation between the two catchment areas.
- There is a marked reduction of flow in the last two years (1988~1990) for both basins, which is related to the generally dry period that extended to the middle of 1993.
- For the period 1965~1990, there is a tendency for reduction of the mean annual flow at Papades station. Comparing the values for the two decades (1965-1974 and 1975~1984) it is noticed that the mean annual flow is reduced by 6 m^3/s. It is not clear when this

reduction occurred, but it appears that it mainly occurred in the first half of the 70s, followed by a gradual reduction since then.

- For the period 1965-1990, the runoff of the Papades-Temenos sub-basin, i.e. the difference between Papades and Temenos flows, remains more or less constant.
- The reduction in water flow observed at Papades measuring station, for the period 1965-1990, may be related to other uses of water within Bulgaria, mainly the diversion of Nestos tributaries to other basins.

Assuming that a mean annual runoff of 345 mm (270 mm for the period 1985~1990) over the Papades ~ Temenos sub-basin, applies to the sub-basin upstream of the Papades station, the total inflow of Nestos and Dhespatis rivers at the border can be computed by subtracting 4.5 m3/s (3.5 m3/s for the period 1985~1990) from the corresponding Papades station measurements. The results are shown on Table 4.

Table 4 INFLOW FROM BULGARIA

PERIOD	MEAN FLOW m^3/s	MEAN VOLUME X 10^6 m^3
1965~1974	32.52	1025
1975~1984	26.52	836
1985~1990	18.49	583

An examination of tables 3 and 4 and figure 2 shows that the flows entering Greece from Bulgaria have a negative trend during the entire period 1968-1986 whereas the intermediate flows between Papades and Temenos stations show such a trend only for the period 1985-1990.

This difference in the time-pattern of the two flows indicates that the flow entering Greece may have been influenced by management decisions within Bulgaria such as the diversion of the Dospat river, whereas both flows may have been affected by the drought of 1985-1990 which may be either a temporary phenomenon or the result of a more permanent climatic change. This latter possibility raises important problems for the Management of Water Resources projects and particularly those on transboundary rivers. This problem is discussed in more detail later in this paper.

Strymon - Kerkini

1. The Strymon is a transbourdary river with relatively large catchment area (11.000 Km2 in Bulgaria and mean runoff 2500x106 m3/y. at the border) and of great importance for the irrigation of the Serres plain in Greece (100.000 ha). Irrigation is done through the reservoir of Kerkini lake which was built in 1933 in order to regulate the discharges of the Strymon river and protect the area from floods; it developed into an important wetland which is protected by the RAMSAR Convention. The course of the Strymon through Greek territory until it reaches Kerkini is 24km.

2. The subject we shall deal with is the transportation of suspended solids from the upper and middle course in Bulgaria to the Greek portion of the river and especially to the Kerkini lake. Indeed, upstream from the border the slope of the bed of the Strymon in the flat areas is superior to 2o/oo; when entering Greece (Roupel), the slope becomes 1,4o/oo and the river deposits the suspended solids. From measurements and observations (Psilovikos et al., 1992) it resulted that when there are floods, the suspended solids which have been deposited at Roupel are eroded and transported downstream. While flowing downstream the suspended solids are deposited at a very quick rate. Thus, part of them are deposited on the bed upstream from Kerkini and form islets while the biggest part is deposited in the lake.

3. When the Kerkini lake was created (1933) its maximum useful volume was 200x106 m3 and its surface 51 km2. According to the estimations of the project study the reservoir should have a time of operation of 80 years. In fact the deposition rate was much quicker and sedimentation of the lake was much greater (1984). The Greek state had to build new gates, new embankments and raise the old ones. The new works caused negative changes to the ecosystem of the lake area, causing reactions from Greek and international environmental organizations.

4. In order to face the increasing water demand and stem big floods, raising the embankments even higher so as to increase the stored volume was thought to be the solution. The responsible Greek services ordered a study and research on the environmental consequences of such works. That study has not been completed yet but the first indications seem to forbid the accomplishment of the works. Therefore the following measures are counterproposal:

a) Anti-erosion protection works in Bulgaria

b) Deviation of the river from the delta formed within the lake and partial removal of the deposit material from chosen points of the lake.

5. During a recent Greek-Bulgarian meeting the Greek side stressed the need for the anti-erosion protection works to be undertaken upstream of the borders in the catchment of Strymon river and further referred to solid material in suspension, entering and being deposited in the reservoir of lake Kerkini.

The Greek side underlined the importance of this issue and expressed its desire for appropriate studies to be elaborated by the Bulgarian side in 1994, so that pertinent projects could be identified and financed under the 1995 PHARE/INTERREG initiative. The Bulgarian side will consider the Greek request.

Ardas Basin

1. On the basis of the agreement of 1964 for the disposal of the waters of the Arda river by the power plant of Ivailovgrad, Greece has built a dam on the river by the Bulgarian border for the daily regulation of the discharge from the power plant for the irrigation of the downstream region. The dam was functioning in the early 70's while in the late 60's modern irrigation networks started being built, 1/3 of the budget being a loan from the European Bank. Networks have already been constructed on 25.000 ha while, after the works are completed, they will reach 30.000 ha.

2. Every year about mid-April, there is a conference of the Joint Greek-Bulgarian Technical Work Committee which decides on the flow-rate which will be given by Bulgaria to Greece through the Ivailovgrad dam, per month and per day, according to article 1, minute 2 of the agreement of 1964.

3. When the water level at the Arda dam approaches the overflow level, there are places where the river inundates Bulgarian and Greek areas. To solve this problem each side has done a survey and a technical study is already being elaborated by the Joint Committee in order to raise the embankments to avoid the flooding of the areas and to store a larger quantity of regulated water at the dam. The study is to be completed by the end of May 1994. To finance the work, about 0.4 M ECU, a proposal was made and it was included in the program PHARE/INTERREG 1994. The work the cost of which is to be tendered within 1994.

Erythropotamos Basin

1. The Erythropotamos river springs in Bulgaria, and after running for about 10 km on and alongside with the Greek-Bulgaria border, it enters Greece and flows into the Evros river at the level of the town of Didymotichon.

2. In the area where the river enters Greece, the cultivated lands-about 300 ha - are often flooded and eroded because of the continuous changes on its bed; therefore the Joint Committe proposed and undertook topographical works on the Erythropotamos, on the Greek-Bulgarian border, on a 11.250 m. length (1985). Both sides agreed on the necessity to construct flood-control works so that the border line would be the axis of the river, wherever possible. The Geographical Military Services of both countries have agreed on the border line and on the basis of that agreement a study has been elaborated for the flood-control works on the river and the protection of the cultivated land. That study has been approved by both sides. Whereas for the cost of the works, it amounts to about 3 MECU and the allocation is 50% for each country.

3. Like the works on the Arda river, the EU initiative PHARE/INTERREG, which concerns the development of actions of a transboundary character, will finance the works. The elaboration of the tender documents is expected to be completed in the course of this month and the works tendered by both Bulgarian and Greek sides. Those decisions were taken during the recent Greek-Bulgarian meeting of experts (21-22/4/94) in Athens .

General Remarks on the Transboundary Management Problems

Effects of Climatic Change

The design and operation of water resources projects is presently based in the assumption that the natural river flows are generated by a periodic-stationary process and therefore their statistical properties (e.g. means, variances, autocorrelations etc.) do not have long-term trends. The validity of this assumption has been questioned in recent years on the basis of indications of climatic change, caused by CO2 and other anthropogenic gazes, which affect the processes of the hydrologic cycle. The detection and modeling of changes in river flows caused by climatic change has been the subject of extensive investigations by various researchers and

United Nations organizations (e.g. Cavadias, 1992) and of the previous NATO Institute (Stakhiv, 1993). Under conditions of climatic change the basic assumption of stationarity must be replaced by scenaria of changes in the statistical characteristics of the river flows.

The best approach for taking climatic change into account in the design and operation of water resources projects is not yet clear.

Views expressed in the literature (e.g. Stachiv, 1993) range form adaptive modification of design and operation criteria according to current changes of hydrometeorological variables to basing the design parameters and operating rules on scenaria of change derived from the results of General Circulation models (GCM).

In the case of a water resources development of a basin within a single country it is possible to adopt flexible approaches to this problem. However, if the river basin belongs to more than one country, the water is shared on the basis of international agreements which include clauses assuming stationarity of the river flow process. In case of change in the statistical properties of the flows the modification of such agreements is very difficult. If conditions of climatic change are anticipated it will be necessary to include provisions for this eventuality in the international agreements. The nature and formulation of such provisions is an important and difficult area of study for Water Resources Engineers. This type of study may be carried out within the frameworks of many programs of International Organizations currently in progress.

The Role of International Organizations

The management problems of transboundary rivers will become more difficult in the future due to an increase in demand and a possible reduction of the flows caused by climatic change. On the other hand, the increase in awareness of these problems and the existence of the cooperative framework provided by the European Community are positive factors which will make possible a more detailed study of these problems and improved conditions for their solution.

Recommendations

It is apparent from the above that although the existing agreements between the two countries form a good basis for the management of transboundary rivers the current state of cooperation between Bulgaria and Greece is not completely satisfactory.

We would like therefore to suggest the creation, within the framework of the 1964 agreement, of a joint technical committee which would have the following tasks:

a) Interchange of hydrologic information on a regular basis

b) Formulation of common research and development projects on the basis of the programs PHARE - INTERREG

c) Coordination of the work of ad hoc committees formed to deal with specific problems (e.g. the management of the discharges of the Nestos river).

References

Cavadias, G. S (1992). A survey of current approaches to Modelling of Hydrological Time Series with respect to Climate Variability and Change. WMO/TD-No 534 WCASP-30 World Meteorological Organization.

Moutafis, N. J. (1991). Possibilities for further Development of Hydroelectric Projects in Greece. 2-day Conference:"Crisis on the Country's Electrical Energy Production?" Technical Chamber of Greece, June 18-19 1992.

PPC/J.V. COBA-CONSULPRESA (1970). Nestos river Development. Platanovryssi H.E.Project. Interim report.

PPC/J.V COBA-CONSULPRESA (1972). Nestos river Development Thissavros, Platanovryssi and Temenos HEPs. Preliminary report, Vol.1, 3 and 4.

PPC/GMPT (1979). Nestos River Hydroelectric Development. Thissavros HEP. Engineering Report, Vol.1 and 3.

PPC/GMF (1984). The Thissavros - Temenos Multi-Purpose Hydroelectric Projects Report on Economic Evaluation and Financing.

PPC/DAYE (1987). Thissavros HEP, Energy Study Summary).

PPC/DAYE (1988). Economic Evaluation of the Master Plan projects on Nestos River, (in Greek).

PPC/DAYE/TSMYS (1991). Evaluation of nestos River Flows, (in Greek).

PPC/DAYE/TMEPM (1991). Nestos River HEPs - Energy Study (Draft, in Greek).

Psilovikos et al. (1992) Study of the problem of Sedimentation of Kerkini reservoir and of the Strymon watercourse.(In Greek). Research Commitee, Aristotle University of Thessaloniki Er.Pr.2343.

Stakhiv, E. Z. (1993). Water Resources Planning and Management under Climate Uncertainty. (Paper proposed to the NATO Institute, Deauville May 1993.

VI.3 Current Status of the Anthropogenic Impact on the Bulgarian Part of the Mesta (Nestos) Transboundary River

G. Mihailov and R. Arsov
Department of Water Supply, Sewage, Water and Wastewater Treatment
Faculty of Hydrotechnics, Higher Institute of Architecture and Civil Engineering
1 Chr. Smirnenski Blvd., 1421 Sofia
Bulgaria

Introduction

Mesta (Nestos) is one of the transborder rivers which originate in Bulgaria and, crossing the Greek territory, flow into the Aegean Sea. It runs through a region with comparatively well developed industry and agriculture. While the industry is centralized mainly at the upper and middle part of the river basin the agriculture is more developed at the upper and lower part. The quantity of municipal wastewaters is usually much less than the industrial wastewaters because of the low population density in the region. Under these circumstances the entire course of the river in the territory of the Republic of Bulgaria is of special concern with primary respect given to the main indicators for contamination with inorganic and organic substances of industrial origin. This is to keep the water quality in the standards and eventually to take all due measures to prevent transboundary contamination as well.

The great political and economical changes in Bulgaria during the last four years reflected negatively on the national industrial productivity which decreased drastically. While the gross national product was decreasing in this situation, the river's water quality was improving because of the reduced quantity of the industrial wastewaters and pollutants. Most of the industrial enterprises in the Mesta (Nestos) River basin reduced by up to ten times their wastewater emission.

Some of these industrial enterprises have been closed for economic reasons or due to privatization problems, and the Yeast plant at Raslog has been closed due to ecological reasons. It is impossible to predict when the industry in this region will recover it's structure and capacity so the current status has to be considered as a starting point for eventual development in the future. In this sense a brief review of the existing and planned technical structures and measures for wastewater treatment would be useful in order to assess their contribution to the current status of the river water quality and as a real base to match the needs of a full capacity wastewater treatment, if it will be necessary after some time.

On the other hand actual information about the Mesta River water quality, regarding the physical, chemical, hydrobiological and saprobiological indicators, is of a great importance for predicting the direction of the expected changes in this sense after the increasing of the wastewater load in the near future.

Obviously, all these activities are of a mutual interest for both sides - the Bulgarian and the Greek. Consequently, joint efforts have to be made to keep the water quality good enough for water supply and irrigation and to save the ecological system of the region, a main part of which is the River Mesta itself.

General Hydrological Characteristics of the Mesta (Nestos) River Basin

The valley of the Mesta (Nestos) River is located between the Pirin Mountain in the West, the Rila Mountain in the North and the Rhodopi in the East. The confluence of the Bjala Mesta River and the Cherna Mesta River near the town of Jakoruda is accepted as the origin of the Mesta (Nestos) River. A spring with an altitude of 2240m near the Granchar Lake is accepted as the origin of the Bjala Mesta River.

In the territory of the Republic of Bulgaria the river runs from the North to the South. At the Bulgarian - Greek border the river's length is 126km with a catchment area of 2767 km^2. The River Mesta has 25 tributaries of which the Dospat River is the biggest. The total catchment area of the Mesta and Dospat Rivers in Bulgaria is 3111,4 km^2. The Dospat River flows into the Mesta River in Greek territory.

The Mesta River is surrounded by the highest Bulgarian mountains, and this determines a high average altitude for the whole river valley, namely 1318 meters above sea level. This is the highest river valley in Bulgaria.

There is a well-developed hydrometric network along the Mesta (Nestos) River Valley (Fig. 1). The number and the arrangement of the tide battens is adequate for reliable determination of the hydrological parameters with the required accuracy. Most of the tide battens have been put into operation during the period from 1949 till 1951. All of them are provided with limnographs for continuous water level recording.

The Mesta (Nestos) River receives its tributaries from the surrounding mountains Rila, Pirin and the Rhodopi. The annual hydrograph of Rila's tributaries is characterized by a clearly expressed phase of increasing, spanning from April till June. The runoff during this period is 70% of the annual one. During the phase of increasing (at the end of May) the daily average flowrate is about 15 times the annual average flowrate.

There are two periods of low water level - winter minimum (January and February), and summer minimum (September and October).

After the influx of the Istok River the regime of the Mesta (Nestos) River changes considerably. Its mountain type regime gradually modifies. The annual hydrograph of this section of the river indicates a considerably flattening flow with a longer high - water period which in some years extends till February.

After putting in operation the Granchar Channel (with 67 water intakes) and the Vatcha Cascade certain disturbances of the river flow have been observed.

The average annual turbidity and the average annual silt volume R are measured along the Mesta (Nestos) River at the tide batten posts at Jakoruda, Mamina Kula and Hadjidimovo. Generally for the whole period of monitoring (natural and disturbed regime) at Jakoruda R is equal to 1,30kg/s, water discharge Q = 4,7m^3/s and suspended solids concentration ρ = 575g/m^3.

Figure 1 Hydrometric and water quality measurements stations

Anthropogenic Impacts on the Mesta (Nestos) River Basin

Irrigation

The potential for irrigation in the valleys of the Mesta (Nestos) and Dospat Rivers is 25000ha. In 1990 the irrigated areas were 18000ha. According to the three water quality measurement stations - Jakoruda, Bukovo and Hadjidimovo (Fig.1.), the location of the irrigated fields is as follows (Fig. 2):

- Jakoruda area: 950 ha are irrigated with water from the Mesta (Nestos) River;
- Jakoruda - Bukovo area: 840 ha are irrigated with water from the Belitza River;
- 6520 ha are irrigated with water from the Bjala, Bachevska and Glazna Rivers;
- at Bukovo 1600 ha are irrigated with water from the Mesta (Nestos) River and its tributaries;
- there are no irrigated fields between Bukovo and Gospodintzi;
- Hadjidimovo area: 6800 ha are irrigated with water from Mesta (Nestos) River and its tributaries;
- after Hadjidimovo towards the Bulgarian - Greek border 1650 ha are irrigated with water from the Dospat, Kochan, Matniza and Bistritza Rivers.

The total area under irrigation is 18600 ha with a water consumption of 50 million cubic meters in an average year. The water intakes are in operation during the irrigation period. Their number is about 50. There are also about 17 small reservoirs for irrigation at Hadjidimovo.

One of the biggest hydro-power systems in Bulgaria, the Dospat cascade, is located in this region. It includes two big reservoirs - "Dospat", with a volume of 445 million cubic meters, and "Shiroka Poljana".

The soil types determine the method of construction of the irrigated areas, the irrigation type and the optimal time for irrigation.

The existing irrigation fields are mainly of gravity type. This type may cause non-point pollution in a case of weak control on fertilization.

The suitability of water for irrigation is evaluated through quantitative indexes. It depends on the geographic location, climate, terrain, flora, fauna, anthropogenic activity, etc. The dissolved and suspended substances change the qualities of water. They influence the quality of the water stored in reservoirs as well. The quality of water for irrigation is analyzed during the stage of design of the irrigation field using statistical data and taking control samples for each water intake as well. The control over water quality is exercised by the local Irrigation Department and the Regional Environmental Inspectorates. In the case of small reservoirs there are certain requirements related to safety against pollution. The control samples are taken in the beginning of the irrigation season.

Hydro-Power Systems

At the present moment the Granchar Canal takes 88 million cubic meters of water in an average year and carries it into the Belmeken reservoir in the Rila Mountain (Fig. 2). Other 105 million cubic meters of water are taken by the Dospat reservoir for power utilization through the Vatcha cascade. Because of the complex character of both cascades - "Belmeken" and "Vatcha", this amount of water is used also for irrigation and water supply.

The hydro-power capacity of the Mesta (Nestos) River is in a stage of feasibility study.

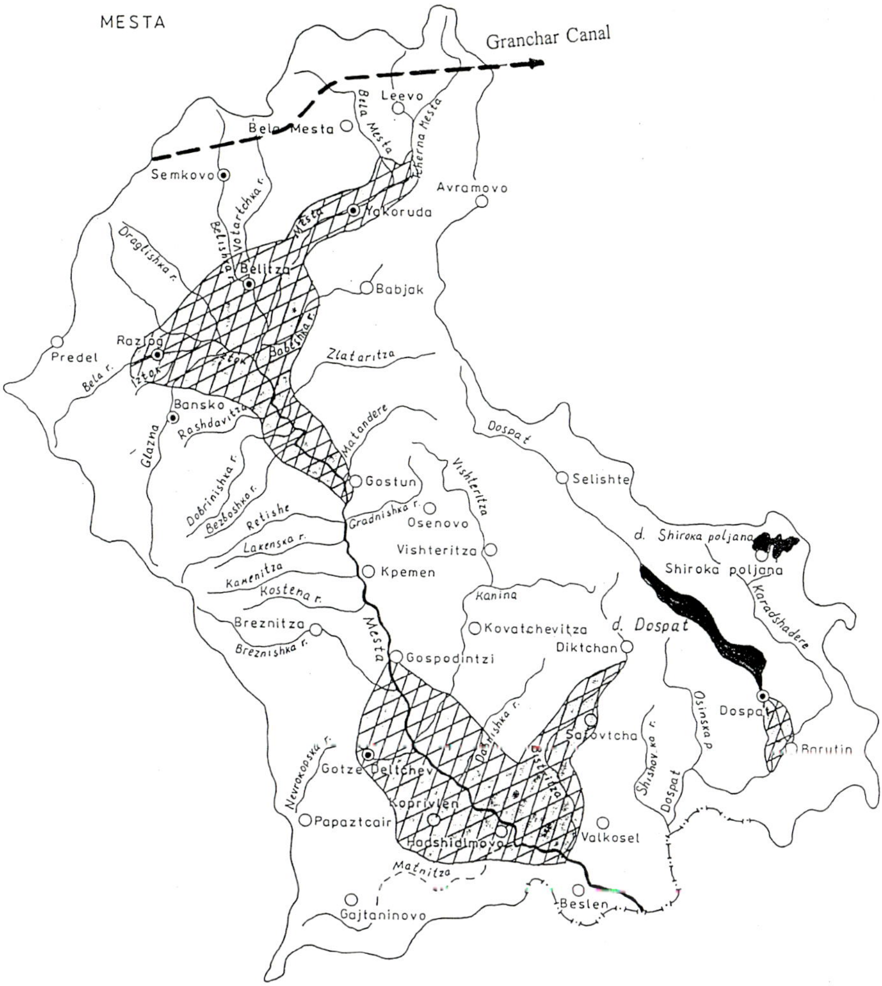

Figure 2 Irrigation fields.

Impacts from Industrial and Urbanized Areas on River Quality

The industry in the Mesta (Nestos) valley is concentrated in the bigger towns and villages - Razlog, Bansko, Jakoruda, Gotze Deltchev, Dospat, Hadjidimovo, Phillipovo, Baroutin, etc. The types of industries and their locations are shown in Fig. 3.

The industrial wastewaters are the main source of river pollution along with the untreated domestic wastewaters from the towns and villages. The most dangerous sources of river pollution were the Yeast plant (which was closed in 1988) and the Paper and Mill plant in Razlog (operating now at about 1/10 of it's capacity), at the upper river branch.

There are three permanent stations for the Mesta (Nestos) river quality monitoring (shown in Fig. 1), belonging to the state's Regional Environmental Inspectorates with Headquarters in Blagoevgrad and Smoljan. They are part of the state net for surface water quality monitoring.

The data collected and processed is the base for establishing the current state and planning of the river's water quality, according to the accepted national water quality rank scale. This scale takes into account 87 parameters, separated in 5 groups: Physical and inorganic; Organic; Inorganic with an industry origin; Organic with an industry origin; Biological.

According to the values of these parameters and river's water usage, three river quality ranks have been established in 1986 - I, II and III. Table 1 is an illustration of this.

At 1992 the total annual volume of wastewaters being released into the river basin was 75708 m^3, broken down as shown in Fig. 4.

The wastewater sources include several main points - small towns, villages, resorts, tourist bases and industrial areas as follows:

- *Towns* - Raslog, Bansko, Yakoruda, Belitza, Gotze Deltchev, Dospat;
- *Villages* - Banja, Eleshnitza, Hadjidimovo, Baroutin;
- *Tourist bases* - Vihren, Banderitza, Shiligarnika, Damjanitza, Bezbog, Popina Laka (Papazchair);
- *Resorts* - Semkovo, Dospat Lake;
- *Industrial areas* - the towns and villages given above plus Uranium Mining and the Ore Enrichment Plant at the Eleshnitza village on Zlataritza river (in process of closing).

The information cited above indicate wastewater sources have been considered with regard to availability, current condition and planned construction of the sewage and treatment plants.

The location of the wastewater treatment plants of different kinds at the settlements, industrial areas and resorts is shown in Fig. 5. The river water quality monitoring is a real base for the current wastewater load magnitude and the treatment plants' performance efficiency.

LEGEND:

- 1 Building materials and coal;
- 2 Chemical industry;
- 3 Machinery;
- 4 Electronic and electrotechnical industry;
- 5 Garages;
- 6 Building industry;
- 7 Wood industry;
- 8 Paper and mill industry;
- 9 Plant growing;
- 10 Food industry;
- 11 Sugar industry;
- 12 Textile industry;
- 13 Cattle and sheep;
- 14 Meat products industry;
- 15 Milk industry;
- 16 Public services;
- 17 Uranium mining

Figure 3 Type of Industries releasing waste waters into the river at 1992.

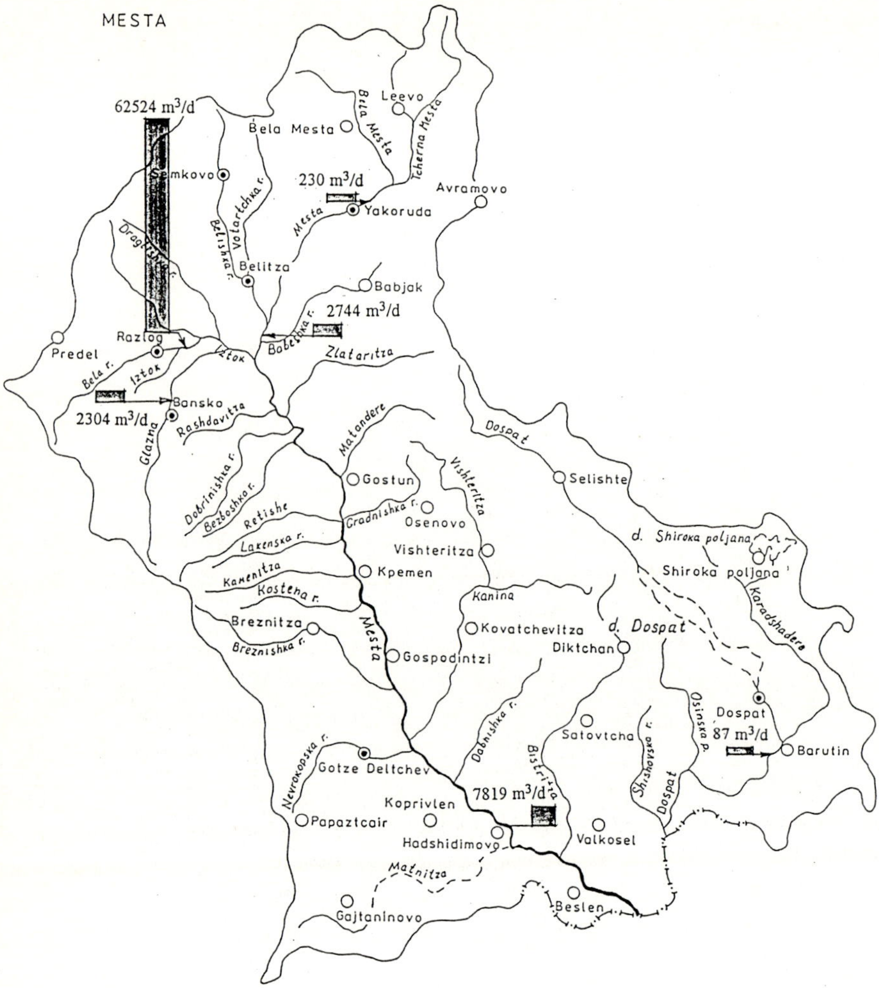

Figure 4 Wastewater load at 1992.

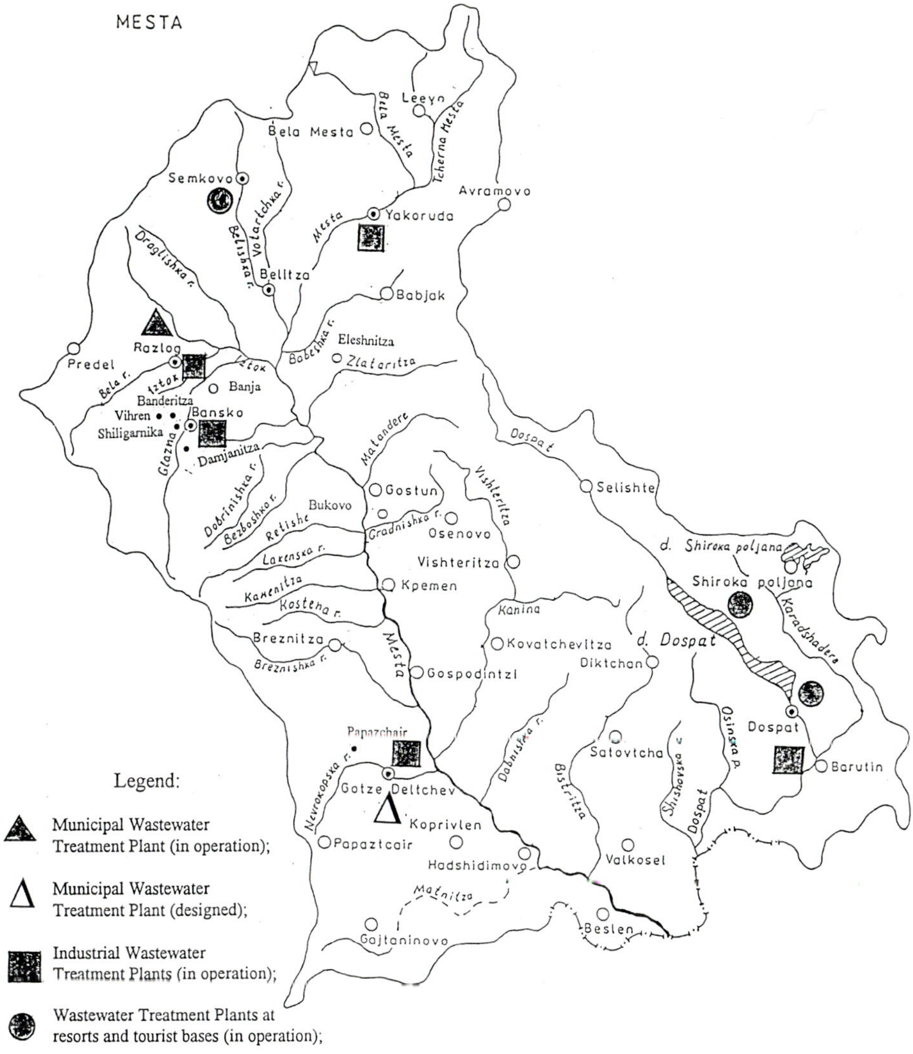

Figure 5 Location of Wastewater Treatment Plants.

Table 1 Limits of surface water quality standards.

Indicator / Rank	River water usage	pH	t° C	DO mg/l	NH_4^+ mg/l	NO_3^- mg/l	NO_2^- mg/l	org.N mg/l	PO_4^{3-} mg/l	Pb mg/l	Ni mg/l	Zn mg/l	Cu mg/l	COD mgO_2/l	BOD_5 mgO_2/l
I	Water supply	6.5 - 8.5	average season temp.+3°	> 6	0.1	5	0.002	1	0.2	0.05	0.1	3.0	0.2	25	5
II	Industry	6.0 - 8.5	average season temp.+3°	> 4	2.0	10	0.04	5	1.0	0.10	1.0	5.0	0.5	70	15
III	Irrigatio n	6.0 - 9.0	average season temp.+3°	> 2	5.0	20	0.06	10	2.0	0.10	1.0	10.0	3.0	100	25

The State of River Water Quality

Water Quality Chemical Analysis

Methods

The water quality of the Mesta (Nestos) River has been investigated during a period of one year starting from October 1992. The following indicators were measured: active reaction (pH), temperature (t^0, C), dissolved oxygen (DO), ammonia (NH_4^+), nitrates (NO_3^-), nitrites (NO_2^-), organic nitrogen (org.N), phosphates (PO_4^{3-}), lead (Pb), nickel (Ni), zinc (Zn), copper (Cu), chemical oxygen demand (COD) and biochemical oxygen demand -5 (BOD_5).

Standard methods and procedures described in "Standard methods for the examination of water and wastewater" - 1985, APHA, AWWA, WPCF were used.

The analyses were carried out at the laboratory of the Department of "Water Supply, Sewage, Water and Wastewater Treatment" at the Faculty of Hydrotechnics, UACEG, with the exception of the atomic absorption analysis of the heavy metals which was conducted at the Department of Chemistry at Sofia University.

The places where the samples were gathered are called in the text *water quality measuring stations* and coincide with the places of the hydrometric stations. They have the names of the settlements Hadjidimovo, Bukovo and Yakoruda, which are situated at a distance of several kilometres from the points of sampling (Fig. 1).

Results

The corresponding data are given in Table 2. The figures in it represent the average values of the measurements of three parallel samples.

The values of pH varied between 6.5 and 8.5 are in the range of the standard limits during the whole period with the single and negligible exception for the months of September and October at the Bukovo measuring station only. As it can be seen from Table 2 there is a general tendency of increasing the pH values.

As it has to be expected that the relation between t^0 and DO may have a contrary tendency. DO in the river is generally much more than the standard limit of the first rank class and it's lowest value in the three points of sampling in July can be explained by the unusually high temperatures at that period and not by an accidental contamination of organic matter, which would become obvious by COD and BOD_5 data.

The values NH_4^+ in the Yakoruda measuring station are usually low because it lies in the upper part of the river. According to the Bulgarian standards the river water can be used for water supply. As far as the Bukovo and Hadjidimovo measuring stations are concerned, remarkable deviations of the NH_4^+ values from the I rank limits are observed. It is valid especially during the winter months due to the low velocity of the aerobic biochemical reactions of the river self purification processes.

Table 2 Mesta (Nestos) river water quality data.

Water quality measuring station	Date	pH	t° C	DO mg/l	NH_4^+ mg/l	NO_3^- mg/l	NO_2^- mg/l	org.N mg/l	PO_4^{3-} mg/l	Pb mg/l	Ni mg/l	Zn mg/l	Cu mg/l	COD mgO_2/l	BOD_5 $mgO_2/$
1	2	3	4	5	6	7	8	9	10	11	12	13	14	15	16
Hadjidimovo	09.10.92	7.1	16.7	12.0	0.04	0.00	0.000	0.000	0.020	0.000	0.00	0.020	0.010	19.9	7.6
Buckovo		7.9	14.4	11.4	0.08	0.08	0.000	0.000	0.090	0.000	0.00	0.010	0.020	27.9	7.2
Jacoruda		6.8	11.2	10.1	0.04	0.00	0.000	0.000	0.030	0.000	0.00	0.080	0.020	39.8	6.6
Hadjidimovo	06.11.92	8.2	11.8	14.9	0.35	1.56	0.070	0.000	0.080	0.000	0.00	0.030	0.020	11.8	4.2
Buckovo		8.2	8.8	13.4	0.00	4.10	0.000	0.000	0.070	0.000	0.00	0.030	0.030	6.7	4.2
Jacoruda		7.2	5.0	12.4	0.08	0.68	0.000	0.000	0.000	0.000	0.00	0.080	0.020	6.7	2.6
Hadjidimovo	17.12.92	6.7	2.0	12.5	0.43	6.80	0.050	0.070	0.070	0.000	0.00	0.010	0.000	6.4	2.0
Buckovo		6.6	1.3	13.3	0.40	6.20	0.050	0.060	0.000	0.000	0.00	0.010	0.020	22.5	1.6
Jacoruda		6.5	0.1	13.5	0.00	3.40	0.000	0.020	0.000	0.000	0.00	0.060	0.020	12.8	0.8
Hadjidimovo	04.02.93	7.6	3.4	16.1	0.50	1.25	0.050	0.000	0.060	0.000	0.00	0.070	0.038	22.0	1.8
Buckovo		8.4	3.2	16.1	0.00	2.92	0.050	0.000	0.050	0.000	0.00	0.075	0.030	11.9	3.4
Jacoruda		7.1	0.2	13.7	0.00	1.16	0.000	0.000	0.010	0.000	0.00	0.085	0.030	10.2	3.0
Hadjidimovo	24.03.93	7.9	12.1	10.3	0.05	0.55	0.098	0.000	0.080	0.000	0.00	0.050	0.030	15.7	3.0
Buckovo		7.9	8.7	11.4	0.00	1.27	0.070	0.000	0.050	0.000	0.00	0.030	0.020	6.3	1.8
Jacoruda		7.3	1.8	12.5	0.02	1.72	0.000	0.000	0.020	0.000	0.00	0.060	0.030	6.3	1.6
Hadjidimovo	22.04.93	7.4	12.6	12.6	0.00	1.00	0.090	0.000	0.070	0.000	0.00	0.020	0.020	6.0	1.8
Buckovo		7.6	12.2	12.0	0.00	0.00	0.060	0.000	0.040	0.000	0.00	0.040	0.020	2.0	1.2
Jacoruda		7.5	11.0	11.1	0.40	0.32	0.000	0.000	0.010	0.000	0.00	0.060	0.030	2.1	1.2

With regards to the NO_3^- data it is clear that the river water belongs to the I rank and during the prevailing part of the period the values are well below the limits. The concentrations of NO_2^- at the measuring points at Bukovo and Hadjidimovo exceed significantly the standards during May, July, and September (at Hadjidimovo) and during October (at Bukovo and Hadjidimovo). Taking into consideration the low values of NH_4^+ and NO_3^- at the same time a conclusion can be made that some inhibition of the second stage of the oxidation process has taken place:

$$2NH_4^+ + 2OH^- + 3O_2 \rightarrow 2NO_2^- + 2H^+ + 4H_2O$$
$$2NO_2^- + O_2 \rightarrow 2NO_3^-$$

On the other hand the comparatively high value of org. N at Hadjidimovo only during September is a signal for disturbances in the N - containing substances' transformation process.

The content of PO_4^{3-} in the water is under the limits of the I rank with one exception only during the month of September at the Bukovo measuring station where at this time the water can be classified between I and II rank. Obviously it is an exception which could has been provoked by different factors of accidental origin. That is why there is no need to change the conclusion for the PO_4^{3-} content, already made.

The atomic absorption analysis of the river water for Pb, Ni, Zn and Cu demonstrate a presence of these metals, generally in traces. For example the Pb concentration is 10 times lower than the I rank limit, the Zn concentration - 30 times lower, the Cu concentration - average 5 times lower and the Ni concentration is zero or approximately zero with an exception for the months of September and October when the value exceeds the limits of the I rank but does not reach the II rank.

The COD and BOD_5 measurements show a tendency of decreasing from October till December '92 and after that the values stabilize at an average of 12 mgO_2/l for COD and 1.6 mgO_2/l for BOD_5. The river is classified as I rank without any doubt.

Hydrobiological and Saprobiological Analysis[1]

Methodological Approach

While the hydrochemical analysis of the river water quality gives an idea of the state of the water flowing in the river at a certain moment, the biological methods provide a more reliable and lasting characterization of the quality, based on the respective resistance of the bioindicators to its temporary or steadier alteration.

The classical method of Zelinka and Marvan (1961) known as the method of the saprobic valences, further developed by Rothschein (1962) has been applied to evaluate the purity of the Mesta river water. The method is based on the knowledge of the percentage distribution of the different bioindicators in variously polluted zones. These zones, called "saprobic", are distinguished according to the quantities of dissolved oxygen and organic matter. The purest zones are the xeno- and oligosaprobic (I category according to the Bulgarian State Standards

---

[1] The analysis is prepared by Prof.Dr.Sci.B.Rousev and Assoc.Prof.Dr.Iv.Janeva, Institute of Zoology, Bulgarian Academy of Science, Sofia.

BSS), followed by the β - mesosaprobic zone (II category according to the BSS) and α - mesosaprobic zone (III category). Most polluted is the polysaprobic zone (IV category).

In the analysis of the communities (benthic zoonecoses), developed under the respective saprobic conditions, in order to characterize the quality of the Mesta river water, besides the general relative and the total number, also 4 basic structural indicators are introduced: individual species diversity - "H" (Shannon & Weaver, 1963), general species diversity - "d" (Margaleff, 1958), dominance - "c" (Simpson, 1949) and eveness "e" (Pielou, 1966).

The number of species (S) and the relative number (N) of the biocenoses are basic characteristics which allow the following of the constancy of the species composition and its numbers, the occurring changes, the seasonal pattern in the development of the various species. It is also the initial data for establishing the other structural indicators.

Results

In the period from October 1992 till October 1993 the in-situ investigations envisaged in the work program were carried out in October, November and December 1992 and March, April and September 1993 at the three permanent stations - upstream of Yakoruda, downstream of Bukovo and downstream of Hadjidimovo. All 21 samples collected were processed in a laboratory. Their relative number was established. The organisms found belong to 17 benthic groups (Table 3).

The species of all groups were determined, with the exception of Heteroptera and Hydracarina (Table 3). The following species were found: from Hydrozoa - 1, from Turbellaria - 2, Oligochaeta - 23, Hirudinea - 3, Gastropoda - 3, Lamellibranchia - 2, Isopoda - 1, Ephemeroptera - 23, Plecoptera - 10, Odonata - 4, Coleoptera - 8, Trichoptera - 35, Chironomidae (Diptera) - 17, Simuliidae (Diptera) - 9 and Diptera - 16.

The values of the saprobiological index of S_R, expressing the integral evaluation of the pollution at the respective station, are presented in Table 3.

At this stage no structural parameters are calculated and no analysis of the benthic zoocenoses structure is made, since, due to methodological difficulties, the numerous samples of the Oligochaeta class and the Chironomidae family are still not determined quantitatively.

Mesta River, Upstream of Yakoruda

The values of *S* at this station varied between 69.10 and 74.33 (oligosaprobity, I category). Here organisms of 106 species from 13 benthic groups were found (Table 3). The sample taken in December 1992 was most scanty - 2 groups with 5 specimen, which is natural for the water temperatures measured at that time - 0^{0}C, since most of the water insects go into diapause. In March the benthic groups were 9, and in the remaining cases 10 (October 1992 and April 1993) and 11 (November 1992 and June and September 1993). The relative sample numbers varied between 5 specimen and 1341 specimen (March 1993). Their average value was about 800 specimen. Dominant and subdominant during the various periods of sampling were species of the Ephemeroptera and Plecoptera orders - typical inhabitants of the upper

reaches of the rivers and oligosaprobic bioindicators Brachyptera Seticornis and Leuctra Sp. (Plecoptera), Ecdyonurus gr. venosus, Epeorus sylvicola, Ephemerella ignita and Baetis rhodani (Ephemeroptera).
All these data correspond fully to the conditions of life offered in the oligosaprobic zone.

Mesta River, Upstream of Bukovo

The river at the village of Bukovo is in its middle reach. The station is located several kilometres after the inflow of considerable quantities of sewage.

The saprobiological index here varied between 63.92 (oligosaprobity, March 1993) and 36.18 (α - mesosaprobity in April 1993). During all the remaining months the saprobity was within the range of β - mesosaprobity. The heavier pollution in April was due to the massive development of the α - mesosaprobic indicator Nais elinguis (Oligochaeta), while all remaining indicators were found in a few numbers.

Representatives of benthic groups with a total of 59 species were found at this station. No Turbellaria and Lamellibranchia species were established; these were found in the previous station. Plecoptera and Gastropoda were represented only by one species. Dominants or subdominants at this station were the β - mesosaprobic indicators Baetis fuscatus and B.lutheri (Ephemeroptera) and α - mesosaprobic indicators Asellus aquaticus (Isopoda), Nais elinguis and Limnodrilus profundicola (Oligochaeta).

In principle the β - mesosaprobic zone is a climax one and offers relatively quite favorable conditions for the development of the biocenoses. It is featured by sufficient quantities of dissolved in the water oxygen and basic biogenic elements (inorganic and organic), necessary for the feeding and development of the hydrobionts. However, bearing in mind the above given data for the biocenoses at this station, it should be pointed out that the β - mesosaprobity in this case is unstable and temporary and in a comparatively short period goes over from β - to β - α - and even α - mesosaprobity. For this basic reason the welfare of the biocenoses typical for the β - mesosaprobity is not observed.

Mesta River at Hadjidimovo

The saprobiological index at this point varies from 41.53 (March 1993) to 50.17 (November 1992) (with the exception of April 1993 when it was 39.10), which demonstrates a permanent β - mesosaprobity.

Only representatives of Plecoptera were not found here. This can be explained by the biology of the species of this order, which inhabit mostly the upper and more rarely the middle reaches of the rivers. By contrast with the previous station, the biocenoses found at the village of Hadjidimovo consist of a large number of species; 66. On the other hand the dominants at this station are only β - mesosaprobic indicators - Stylaria lacustris (Oligochaeta) and Baetis vernus (Ephemeroptera). That is why the results gathered so far justify the location of the Mesta River in the II category.

Water quality measuring station	Date	Hydrozoa	Turbellaria	Oligochaeta	Hirudinea	Gastropoda	Lamellibranchia	Hydracarina	Isopoda	Ephemeroptera	Plecoptera	Coleoptera	Trichoptera	Odonata	Heteroptera	Diptera	Chironomidae	Simuliidae	S_R
1	2	3	4	5	6	7	8	9	10	11	12	13	14	15	16	17	18	19	20
Hadjidimovo	October '92	-	-	197	34	2	-	1	1	44	-	-	10	-	-	-	17	-	43.56
Buckovo		-	-	195	6	7	-	4	41	58	-	1	14	-	1	-	66	-	40.23
Jacoruda		-	21	11	-	3	-	2	-	340	141	4	128	-	-	5	58	-	71.78
Hadjidimovo	November '92	-	18	500	1	2	-	-	-	10	-	-	2	-	-	3	500	-	50.17
Buckovo		-	-	319	6	6	-	8	13	66	-	1	24	-	-	1	60	-	44.29
Jacoruda		-	4	500	-	6	8	1	-	193	58	-	65	-	-	9	156	3	71.08
Hadjidimovo	December '92	-	-	198	2	2	1	-	1	37	-	-	7	-	-	-	11	-	46.80
Buckovo		-	-	82	1	-	-	3	1	53	-	1	49	2	-	5	22	-	51.06
Jacoruda		-	-	-	-	-	-	-	-	1	-	-	-	-	-	-	4	-	-
Hadjidimovo	March '93	-	1	500	3	1	-	1	2	6	-	-	1	5	1	2	15	-	39.10
Buckovo		-	-	80	-	-	-	-	-	32	-	-	1	-	-	-	17	-	60.52
Jacoruda		-	11	293	-	6	-	-	-	262	184	-	36	-	-	8	500	9	73.96
Hadjidimovo	April '93	-	-	2000	3	-	-	4	-	13	-	-	1	3	1	-	91	-	41.53
Buckovo		-	-	440	4	-	-	4	2	28	-	-	7	-	-	2	53	-	36.18
Jacoruda		-	1	141	-	-	14	-	-	197	145	1	36	-	-	7	184	7	74.33
Hadjidimovo	June '93	2	-	500	7	2	-	12	2	80	-	3	-	3	7	3	500	1	43.09
Buckovo		-	-	152	11	-	-	-	5	414	-	-	18	3	-	2	153	4	45.41
Jacoruda		-	29	18	1	-	-	5	-	392	15	14	125	-	-	20	47	3	69.10
Hadjidimovo	September '93	-	-	25	6	2	-	-	7	30	-	1	10	4	1	-	23	2	43.52
Buckovo		-	-	28	4	1	-	3	22	232	1	-	248	-	-	6	71	106	46.91
Jacoruda		-	33	58	-	1	-	9	-	525	65	5	25	-	-	2	48	2	72.67

Table 3 Relative number of the gathered benthos animals in Mesta (Nestos) river.

According to the hydrobiological investigations the river at Yakoruda is of I category and at Bukovo and Hadjidimovo of II category. The divergence from the results of the hydrochemical investigations is due to methodological reasons - the hydrochemical analysis take into account the qualities of the water flowing at the moment of investigations, while the benthic hydrobiological analysis consider the permanent state of the river bed which is under the effect of previous pollutants.

Summary and Conclusions

The Mesta (Nestos) River water quality analysis in the territory of Bulgaria, performed in the course of one year (October '92 - October '93), shows an improvement of it's qualitative indicators. Several reasons for this can be pointed out:

First of all is the drastic reduction of industrial activities in the region. The second reason is the availability of the wastewater treatment plants at the main wastewater flows entering the Mesta (Nestos) River, as well as the fact that all of them are presently underloaded. The third reason is the new environmental policy of the Government with a more precise and stronger control over the contamination with domestic and industrial wastewaters.

According to the investigated chemical indicators the Mesta River is of a I rank under the Bulgarian classification of surface waters quality and according to the hydrobiological and saprobiological indicators it is classified between I and II rank. This has to be confirmed more definitely after an integral exploration of the water quality for a longer period, including the foreseen in the standard indicators - physical, chemical, microbiological, hydrobiological, etc.

Concluding, it must be pointed out that the remarkable improvement of the Mesta (Nestos) River water quality, observed during the period under consideration, as well as during the last few years, is to a great extent the result of the drastic decrease in industrial activities in this region and has to be considered as an exception which hardly could be maintained for a long time. In order to stabilize the current status of the rivers' quality in a normal economic situation, definite technical measures and investments have to be undertaken immediately. Most of the bigger settlements still don't have wastewater treatment plants and the existing ones need reconstruction and technological improvement. In this sense we hope that the joint research, carried out by working groups from UACEG - Sofia and Aristotle University-Thessaloniki, will contribute in the effort to better understand the situation, and for planing the necessary technical measures for rivers' water quality stabilization and improvement in the near future.

VI.4 Water Quality Assessment of the Greek Part of Nestos (Mesta) River

D. Argiropoulos, J. Ganoulis, and E. Papachristou
School of Engineering
Aristotle University of Thessaloniki
54006 Thessaloniki
Greece

Introduction

The assessment of the water quality in rivers is very often performed without using adequate methods and tools in order to quantify the state of pollution. It is quite usual to see in some reports that estimations about the pollution of a river, the state of water quality and the causes of pollution are based on evaluation of the results of costly monitoring campaigns by the use of simplified graphical representation of data or sometimes by means of elementary statistics like mean values and variances.

Nevertheless, these estimations may sometimes support decisions for measures for the polluted river or water bodies that might have a high cost. Yet the results of these measures and the money spent might not be efficient or even useful because inaccurate initial estimations lead to non-adequate remediation, trying to address the wrong problem. Thus the need for reliable estimations of data is beyond any doubt very important.

Within the European Union (EU) project INTERREG, a scientific team from the Engineering School of Aristotle University of Thessaloniki, under the responsibility of Prof. Papachristou, has conducted an investigation for the pollution of Nestos river (Papachristou, 1994). The fact that this river is transboundary and flows to the Mediterranean sea as well, gives to this investigation a special importance. Besides the extended sampling and laboratory analysis performed for the project, special care was given to the appropriate statistical analysis of data. The investigators have tried to avoid problems that were identified in the beginning of this introduction so that valid conclusions could be drawn to help appropriate management of the water quality of the river which could be done in a second stage. In this paper the main points of the statistical analyses are presented which have been used for the assessment of the water quality of Nestos river and might have a broader interest for different uses of water.

Description of Nestos Area

The river Nestos originates from the Rila mountain in South Bulgaria and after a distance of about 250 km in Bulgarian and Greek territory it flows into the North Aegean sea (Fig. 1). The length of the river in Bulgaria is about 120 km, while in Greece it is about 130 km. The total watershed area of the river is 5749 km^2, of which 2312 km^2 belong to Greece. The total area is mountainous with the exception of the delta region near the last part of the river. The mean monthly flow rate close to the Potami station, according to measurements from 1966 up to 1990 is 32.7 m^3/s.

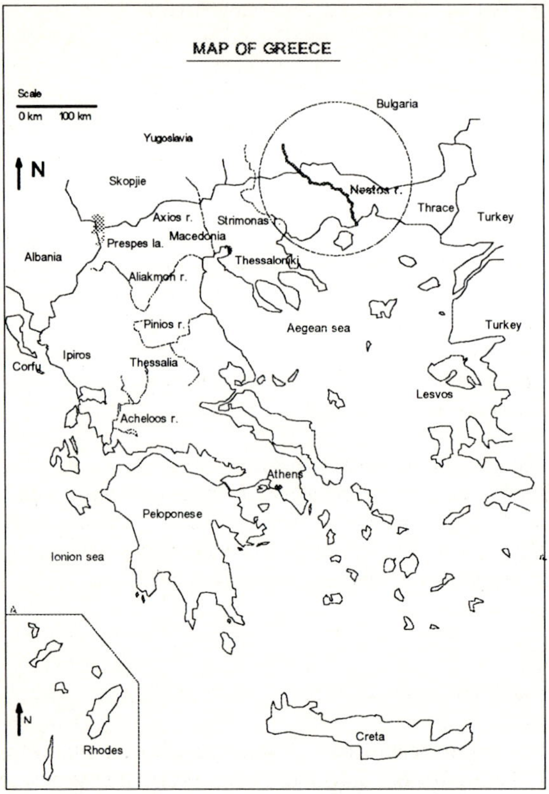

Figure 1 Location of the river Nestos in Greece.

During this work, sampling has been executed at three stations in the Greek territory, that is at the Potami, Stavroupoli and Delta station. The station Potami is the first monitoring station in Greek territory downstream the Greek-Bulgarian border, in a distance of about 7 km. Due to the fact that there are no pollution sources between the border and that point it is estimated that water quality at that point reflects incoming pollution from Bulgaria. Stavroupoli is the second station in a distance of 47 km from the border, near the middle of the Greek part of Nestos. The Delta station is located about 1 km upstream the mouth of the river.

Methodology

The statistical analysis followed in this project may be divided in Exploratory Data Analysis (EDA) and Confirmatory Data Analysis (CDA). In the first stage, a preliminary step is accomplished in which missing values of time series are found followed by the presentation of plots of the parameters and their characteristics as they are derived by the use of different graphical and statistical methods.

In the second stage, confirmation of the characteristics of the parameters presented in the first stage is made with the aid of graphical and statistical methods as well. Nevertheless, the distinction of the statistical techniques used in the above two mentioned stages is not always clear. The graphical methods that have been used in the project include amongst others the presentation of data with plots, the frequency histogram, the simulation of data sets using curves fitted to different statistical distributions and the autocorrelation function plots of data time series.

The statistical methods that are used are both parametric and nonparametric. As it is known, environmental data usually do not follow the normal distribution. In this case the use of non parametric statistics is preferred as more appropriate for the data sets, since the use of parametric tests, although more powerful in general, requires that the data follow the normal distribution. The latter assumption seems not to be valid for certain data sets, thus the need arises for nonparametric statistics. Among the nonparametric statistics used in this project are the Sign test and the Wilcoxon test for testing the hypothesis of medians equality.

The classical statistical techniques used in this work include the calculation of summary statistics, the statistical hypothesis tests, the suitability of simulated distribution functions according to the x^2 test as well as the Smyrnov-Kolmogorov test, and finally the data set quartile analysis of standards exceedence. Finally certain conclusions are drawn referring to the main variables and their characteristics.

Data Analysis

Data Preparation

The data time series present some missing values. These gaps have been filled, so that statistical methods could work better for the analysis of time series. In this project the linear interpolation is used to complete the time series. The final time series, which result after the completion of the actual data, covers the period from June 1992 to November 1993 with 33 fortnightly observations (28 real measurements plus 5 missing values estimated): The samples were taken *approximately* every fifteen days. It should be pointed out also that observations in raw data recorded as "not detected" or "below detection limit" were substituted in the working time series with the half value of detection limit, according to international practice in this field.

Graphical plots

Time series plots of basic variables are presented in Fig. 2. In the x axis the time of measurements is denoted by n=1,...,33. In the y axis values of time series variables are represented (usually concentrations in mg/l, except if it is marked differently). In Fig. 2 the plots of basic water quality variables indicating organic pollution of the Nestos River at Delta station are shown (NH4, BOD5, NO3, PO4, DO and DOsaturation). From these plots some preliminary indications may be pointed out. There is a high variance in the values of all variables. Dissolved oxygen takes values around the saturation point with a lot of supersaturated values. First observations, occurring in summer period, have indicated higher values for almost all variables. Similar remarks were also derived from the plots of the other two stations not presented here.

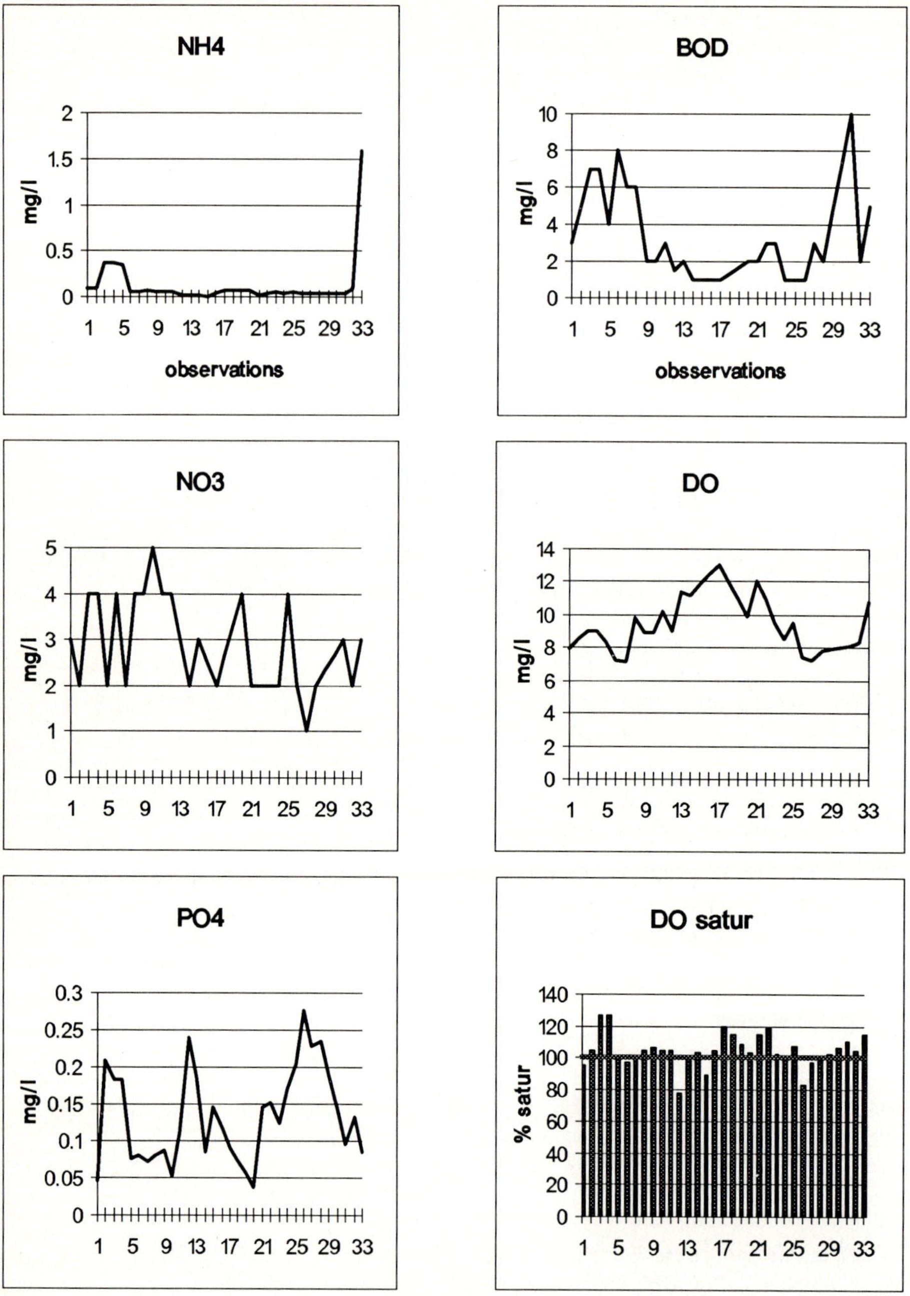

Figure 2 Plots of basic water quality variables in River Nestos at Delta station (6/92-10/93).

Summary Statistics

Information about some parameters at Delta station is presented In Tab. 1 in the form of summary statistics showing the central tendency, the variation and shape. From these statistics a first estimation is that many differences exist among the various parameters. They differ in the variance (some parameters have high variances relative to their mean value, like ammonia and cadmium with a coefficient of variation 226.2 and 109.7, respectively), while most of them have differences between the mean and median and present significant values of kurtosis and skewness. According to the above remarks almost all variables appear to follow non-normal distributions, with existence of outliers. Further, more details about the form of statistical distributions will be investigated in the next section.

Tab. 1 Summary statistics of water quality parameters of Nestos river at Delta station.

	DO	SS	BOD	NH4	NO2	NO3	CU	Pb	Cd
Number	33	33	33	33	33	33	33	33	33
Mean	9.46	23.19	3.34	0.12	0.03	2.86	0.0084	0.0019	0.001
Median	9	15	2	0.05	0.016	2.7	0.005	0.001	0.0005
Min	7.1	2	1	0.008	0.006	1	0.002	0.001	0.0005
Max	13	100	10	1.58	0.13	5	0.028	0.008	0.006
St. Dev.	1.68	23.46	2.45	0.28	0.029	0.95	0.0078	0.0018	0.0011
Skewness	0.46	2.39	1.07	4.82	1.82	0.32	1.321	1.93	3.17
Kurtosis	-0.89	5.23	0.26	25.11	3.02	-0.85	0.357	2.94	10.78
Coef. Variation	17.8	101.14	73.2	226.2	92.95	33.47	92.66	94.54	109.78

Frequency Distributions of Water Quality Parameters

In this section, the frequency distributions of the water quality variables at Delta station are investigated. First, the values of each variable are grouped in classes and its frequency histogram is constructed. A visual check of the shape of the histogram gives a first impression about the distribution of the parameter and whether this distribution could be simulated with the normal one. Next follows a quantitative control with the test of the statistical hypothesis whether that distribution function fits adequately with one of the known theoretical distribution functions, like the normal (N), the lognormal (LN), the gamma (Γ) or the Weibul distribution (W). The x^2 test as well as the Kolmogorov-Smyrnov test have been used to check the null hypothesis of good fitting, which may be rejected at a significance level of α=0.05. In Tab. 2 the results of the application of the aforementioned tests are shown for most of the basic water quality parameters at Delta station. From this table it can be seen that most variables fit best to lognormal and gamma distribution.

In Fig. 2 are also shown the histograms of distribution functions of some specific parameters as well as the curves of the theoretical distributions that fit best to them after the aforementioned tests. The tests at the two other stations give relevant results, revealing some similarities and differences with data at Delta station as well.

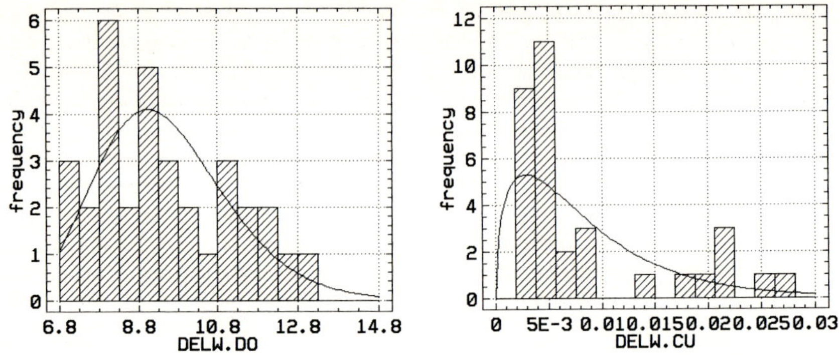

Figure 2 Histograms and best fitting frequency distribution curve for DO and Cu at Delta station.

Tab. 2 Distribution functions of choice of parameters at Delta station.

parameters	distribution function	p value of x^2 test	p value of K-S test
DO	LN	0.27	0.67
DOSAT	N	0.08	0.69
BOD5	Γ	0.12	0.18
COD	Γ	0.41	0.77
NH4	-	-	-
NO3	-	-	-
PO4	Γ	0.76	0.82
CU	Γ	0.07	0.06
PB	-	-	-
CUSED	LN	0.14	0.64
PBSED	-	-	-

Autocorrelation-Seasonality

It is known that time series of water quality parameters often present lack of independence. This depends on the sampling frequency as well as on the characteristics of the water body and pollution sources. The presence of seasonality or trend in data affects the aurocorrelation function. Unfortunately the short length of the time series can give only an *indication* for seasonality and no information about trends. In the present project, the existence of autocorrelation is examined, while only exploratory analysis for seasonality is performed. The check includes the use of graphical and statistical methods (exploratory and confirmatory data analysis). The autocorrelation function (ACF) of each parameter is calculated and the time series with terms 1, 2, 3, ..., 12 lags apart are compared. The autocorrelation coefficients are checked statistically at a significance level of α=0.05. The results of this check for Delta station is presented in Tab. 3 while the plots of ACF of some parameters and the curves of 5% significance level of their coefficients are shown in Fig. 3. As it is shown *most time series at Delta station present autocorrelation of terms one lag apart* (first order autocorrelation). In the same table, the terms with partial autocorrelation which are also shown, confirm the previous conclusion. The shape of each plot of autocorrelation function gives also an indication about the possible existence of seasonality in the time series, which is also indicated in the table. The check for the other two stations gave similar results as at Delta station.

It should be pointed out however, that these indications could be affected by the possible presence of trends (which also cannot be detected and then eliminated because of the short length of time series). Moreover, the estimations for seasonality for the above time series cannot be confirmed with statistical methods due to their short length.

Tab. 3 Autocorrelation function of time series of water quality parameters at Delta station.

parameter	terms with significant autocorrelation (lag of ACF)	terms with significant partial autocorrelation (lag of PACF)	indication of seasonality
TEM	1, 2	1	yes
PH	1	1	no
CON	1, 2	1	yes
DO	1, 2	1	yes
DOSAT	1	1	no
SS	-	-	yes
BOD5	1	1	yes
NH4	1, 2	1,3	no
NO2	-	-	no
NO3	-	-	-
PO4	1	1	yes
COD	-	-	no
CU	1	1	yes
PB	1, 2	1	yes
CD	1, 2	1	yes
CUSED	1, 2	1	yes
PBSED	1, 2	1	yes
CDSED	1	1	no

The results of the above analysis include a degree of uncertainty, due to the possible existence of some autocorrelation because of seasonality and trend. Anyway, if we accept the above results, we can conclude that, due to the first order autocorrelation, the fortnightly measurement frequency does not add information with regard to the long or moderate term behavior of most time series. Thus, if one is interested in the general water quality of the river or in the existence of seasonality or trends, the monthly measurement frequency seems to be sufficient. Naturally this conclusion is not useful, if one is interested in the detection of possible pollution accidents and sudden peaks in the water quality of the Nestos River, in order to take immediate measures to protect the river.

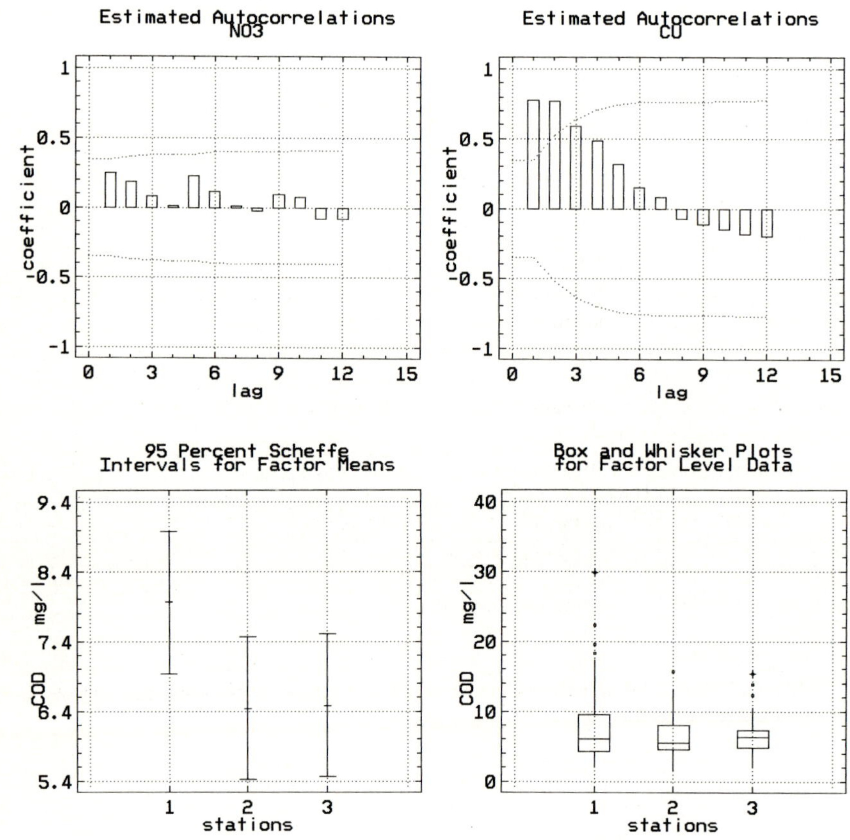

Figure 3 Autocorrelation function plots of NO3 and Cu for Delta station; plots of means and medians for COD for the three stations.

Statistical Evaluation of Data at Three Stations

In this section an overall statistical evaluation of data at three stations is accomplished aiming at conclusions that can be useful in the evaluation of the water quality of the river and the water quality management.

A question that is often posed in such studies is whether each value of water quality variable differs from one station to another. An answer can be given based on the three samples of time series that have been measured at three stations. If the central tendency statistics are used, the means or the medians of the samples are compared. Yet conclusions concerning the entire population of each parameter at each station may be valid only after some statistical elaboration of the samples. Thus, in order to draw the conclusion that the population of a variable at station 1 is higher than the population of the same variable at station 2, a feature of central tendency of the relevant sample at station 1 and 2 is compared with the use of an appropriate statistical test. Given the fact that most variables proved not to follow the normal distribution, it was decided to use the median instead of the mean. An indication about the different conclusions that can be drawn according to the used statistic is depicted in Fig. 3. As it is shown, the multivariate analysis of variance for the equality of means between the variables of COD for the three stations gives the impression that station 1 has greater values than the two others. Instead of that, in the next graph, the comparison of medians gives an impression of equality of these variables.

It was chosen finally, non-parametric tests, which are distribution-free tests, to be used. Thus, both the Sign test and the Wilcoxon rank-signed test were used to test the null hypothesis that the medians of the two populations are equals (Conover, 1980). Rejection of null hypothesis is done at 0.05 significance level for one-side test with the alternative hypothesis being that the median of the variable of one station (whose sample median is higher than the other's) is greater than the median of the same variable of the other station

In addition to the relations among the variables at three stations, the estimation of the absolute level of values is very interesting compared with water quality standards of various uses of the river. According to EU legislation (and Greek legislation as well) there exist imperative and guide values which the measurements of a period of one year should meet. According to the legislation the 90th and 95th percentiles are of interest in the aforementioned control (and in a few cases the 5th and 10th as well). These percentiles are shown in Tab. 4.

Tab. 4 95th and 90th percentiles of observations of time series at stations 1 and 3.

	Station 1		Station 3	
Parameter	90 %	95 %	90 %	95 %
SS	59	39	95	47
BOD	8	8	8	7
NH4	0.994	0.55	0.373	0.35
NO2	0.398	0.093	0.09	0.081
NO3	7	6	4	4
PO4	0.324	0.301	0.239	0.229
COD	22.31	14.26	13.85	10.47
CU	0.032	0.029	0.026	0.022
PB	0.021	0.018	0.006	0.005
CD	0.007	0.007	0.004	0.002
DO	8.5 (5 %)	8.9 (10%)	7.2 (5 %)	7.4 (10%)
DOSAT	83 (5 %)	88 (10 %)	83 (5 %)	95 (10 %)
CU	0.028	0.031	0.024	0.025
PB	0.018	0.021	0.015	0.019
CD	0.007	0.007	0.002	0.002

From the preceding analysis at the three stations in Greek territory the following conclusions concerning the water quality parameters at each station as well as across the river may be drawn:

Temperature of water (TEM) increases from the border up to the mouth of the river, varying within normal levels for the area. This is quite normal and has no connection with thermal pollution. The existence of periodicity seems very strong. Also there is a clear autocorrelation of the time series (which might be due to the periodicity or other reasons as well).
pH ranges in natural levels without statistically significant changes from station to station. The values at three stations meet the standards for water designated for drinking as well as for fish propagation.

Conductivity (CON) increases statistically from upstream to downstream, with much higher values at Delta, because of the sea water intrusion. The time series show a clear autocorrelation and possible seasonality. Violation of standards for water designated to be used for drinking happens at Delta station due to salt water intrusion (natural cause).

Dissolved Oxygen (DO) and saturation percentage of dissolved oxygen (DOSAT). DO along the river varies with frequent presentation of supersaturated values. This is indicative of good water quality with respect to organic pollution but also of the existence of phytoplankton. The water at station 2 seems to have the best values with higher DO concentrations than the other stations. DO presents autocorrelation and indications of seasonality at the last two stations, while DOSAT presents also autocorrelation but not seasonality. The values of both parameters meet the standards for drinking water and for fish propagation.

Suspended Solids (SS) in station 3 tend to have lower values than 2 and station 2 than 1. This could be explained either by imported pollution from Bulgaria or by natural causes like the decrease of water velocity as slope of river bed decreases gradually. There are no indications for seasonality or aurocorrelation at stations 1 and 2, while there are such for station 3. All stations have high variance, with station 2 being twice as high as station 1. Also the distribution at station 2 presents high values of kurtosis and skewness due probably to many extreme values. This change in "population" could be possibly explained by the contribution of the Despatis tributary which confluences downstream of station 1. Concerning legislation, the values at station 1 and 3 violate the guide values for drinking water class A1 as well as for salmonides and cyprinides propagation.

Biochemical Oxygen Demand (BOD5) has a moderate mean value and a high variance. All three stations exhibit indications of autocorrelation and seasonality. The fact that the population at station 1 has statistically higher values than at station 2, while stations 2 and 3 do not have statistically significant differences is attributed to the imported pollution from Bulgaria and its gradual assimilation along the river. Nevertheless, the values of stations 1 and 3 violate the guide values for drinking water classes A1 and A2 and A3, as well as the guide values for fish propagation (no imperative values assigned).

Chemical Oxygen Demand (COD) values present moderate variance, with no statistical differences among the three stations. Station 1 exhibits greater variance than the others, with higher extreme values (which might be pollution incidents in Bulgaria). On the other hand the stability in mean values and the existence of extreme values (although with differences among stations) show that the water section downstream Stavroupoli accepts also a pollution load coming from the surrounding watershed in Greece. Stations 1 and 3 show autocorrelation, while there is seasonality only at station 3. The values of COD along the river are lower than the guide values for drinking water of A3 class, meeting in this way the standards of legislation.

Ammonium (NH4) is characterized by the high variance at all stations and the great differences of the features (like distribution, seasonality, etc.) among the three populations. This variability can not be interpreted with the existing data and information. Concerning the comparisons of location parameters between stations, ammonium tend to have higher values at station 1 and 3 than station 2. The above as well as the fact that station 3 includes extreme values that occur in different times and have different magnitude from station 2 (max at station 3 is 1.57, while at station 2 is 0.45) can be attributed both to imported pollution (could be industrial, domestic or agricultural) and to pollution added downstream of Stavroupoli (domestic or agricultural). The measured values of NH4 present violations of water quality standards at all three stations concerning the guide values for cyprinides and salmonides. *From the above, it is obvious that management measures are needed for this parameter.*

Nitrites (NO2). The statistical comparison of parameters of central tendency show higher values at station 1, while the other stations do not differ significantly. Imported industrial or urban pollution might give an explanation of the above. Moreover there is clear violation of the standards. There is a violation of guide values for cyprinides and salmonides at all three stations. Especially at station 1 the measured values are about three times higher than the guide values for cyprinides.

Nitrates (NO3). The means range at low or relatively moderate levels, with station 1 presenting higher values than station 2 probably due to imported pollution (industrial, urban or agricultural). Moreover, station 3 has higher values than 2, probably due to the Greek contribution to pollution from the plains downstream Stavroupoli. Anyway, the parameter meets all the standards for potable water.

Phosphates (PO4). There is the same pattern as for nitrates, indicating the same pollution sources in the two countries. Aurocorrelation and seasonality exist at all stations. The measured values meet the guide values of standards for potable water of all classes.

Heavy metals in water and in sediment. Although there were certain difficulties in the statistical analysis of all six parameters because of the existence of many records with not detected notation, some conclusions were drawn for these parameters, yet with more uncertainty than for the other parameters. Generally, all metals presented clear indications for seasonality (with the exception of cadmium) and as a rule strong autocorrelation, usually of first, second and third order. It should be pointed out that there was a clear increasing trend from station 1 to station 3, indicative of pollution coming from Bulgaria. Moreover,

- Dissolved Copper exceeds guide values for drinking water A1 at stations 1 and 3, while the values are very close to the limits for fish propagation.
- Dissolved Lead meets the standards, yet very close to them.

- Dissolved cadmium exceeds the standards (imperative values) for drinking water A1, A2 and A3 at station 1.
- The three metals in the sediment present more or less the same behavior, especially copper and lead, with autocorrelation and strong indications for seasonality at all three stations. Unfortunately there are no standards in legislation related to the water uses and subsequently no comparison can be performed with the measured values.

Conclusions

The findings of this investigation, despite its rather limited duration in time, has shown the actual state of water quality of the Nestos River for the period of the research. The specific features of water quality parameters at all three stations have been analyzed, while differences among values of parameters at these stations have been recorded. Moreover, comparison with standards set by EU and Greece for various water uses has been made.

Differences in water quality among different sites of the river have been clearly distinguished and documented. The different properties of water quality parameters that were found can be related to different kinds of pollution sources. Thus, potential causes for pollution may be further investigated, since pollution can be now attributed to certain geographical areas, so their identification might be easier. The latter may further be promoted by combination of other tools, such as the use of information systems. Results from the statistical assessment may also help the application of risk analysis approach, which can be applied as a second stage of an environmental investigation.

According to the preceding results, it can be concluded that statistical analysis proved to be a useful tool for the assessment of rivers water quality. With this kind of analysis arbitrary estimations about real water quality may be avoided. Decisions can be made about the exceedence of certain standards in water quality and thus corrective measures may be provided. Consequently the planning and management of water quality in a river basin, even in transboundary rivers, may be supported with data and an advanced statistical analysis of the data.

References

Conover W. (1980) "Practical nonparametric statistics", second edition, John Wiley and sons

Papachristou E. (1994) "Investigation of the pollution of transboundary between Greece and Bulgaria Nestos river". EU, INTERREG program, project final report, Thessaloniki

Wonnacott T., Wonnacott R. (1990) "Introductory statistics", fifth edition, John Wiley and sons

VI.5 Transboundary Water Issues Between Greece-Bulgaria and Turkey: The Case of the Meriç / Evros River

Ahmet Samsunlu, Derya Maktav
Sedat Kapdaslii
Istanbul Tech. Univ., Environmental Eng. Department
Remote Sensing Department, Hydraulic Department, Instanbul
Turkey

The UN Convention on the Protection and Use of Transboundary Water Courses and International Lakes

An international convention was held on March 17, 1992 in Helsinki under the auspices of the United Nations. According to this convention; the parties involved shall, in particular, take all appropriate measures to prevent, control and reduce any transboundary impact to reduce pollution of waters causing or likely to cause transboundary impact; and to ensure that transboundary waters are used with the aim of ecologically sound and rational water management, conservation of water resources and environmental protection.

They should also ensure that transboundary waters are used in a reasonable and equitable way, taking into particular account their transboundary character, in case of activities which cause or are likely to cause transboundary impact.

Measures for prevention, control and reduction of water pollution, shall be taken, where possible, at the source. These measures shall not directly or indirectly result in a transfer of pollution to other parts of the environment. The parties should be guided by general principles such as the precautionary principle and polluter-pays in a way that needs of the present generation are met without compromising the ability of future generations to meet their own needs. They may cooperate in order to develop harmonized policies, programs and strategies covering relevant catchment areas. The application of this Convention shall not lead to the deterioration of environmental conditions nor lead to increased transboundary impact.

Water Resources and Water Quality

The concept of "Water Quality" defined as the combination of physical, chemical and biological features and the technical dimensions of which is determined by selecting appropriate parameters, is an important factor in the evaluation of water resources.

If multipurpose usage of transboundary waters is in question, it is necessary to consider the purpose needing the highest "water quality". But the priorities may be different for every country and this is quite a difficult problem to be solved by international authorities.

For each type of use there must be a quality dimension to be protected in the water resource (receiving water). This dimension named as "quality criteria" is defined as standards or legislation which change in different countries depending on their level of technical and scientific developments.

These differences between the quality criteria and water quality demands depending on the main goal of the usage in different countries can cause some dispute that can be finished by making an international management organization. This kind of organization is difficult to be made. Therefore, a communication should be constituted first in order to the discuss problems and to transfer the data obtained in national base and to exchange the knowledge.

The River Meriç is an example of transboundary waters due to the water quality problems resulting from the usage of river water for different purposes in Turkey, Greece and Bulgaria. Meetings between the relevant countries continue, but scientists must take part in these meetings, too.

The Meriç Basin

Meriç Basin which was coded as 01 in basin planning by General Directorate of State Water Works (DS) in Turkey, is located in Trakya. The boundaries of the basin, as shown in Figure 1, extend to Bulgaria and the water line of Istranca mountains in the north; Vize, Saray and Erkezky in the east, and Saroz Bay passing Orlu and Tekirda in the south.

Meriç River, with its five tributaries, including Arda, Tunca, Ergene and Doanca, Eskenderky streams is the most important river in the basin Ergene River and its branches are the main water dicharging river of the basin. Although the Arda and the Tunca Rivers join the Meriç River in Turkey, resources of these rivers are in different countries: Arda in Greece, and Tunca in Bulgaria.

Food and textile industries are mostly widespread in the basin. Ergene River and its branches are most polluted in the basin. Most of the industries have designed or constructed their treatment systems in recent years. The present treatment systems mostly were designed and operated by using physical and chemical treatment.

Water Quality of the Meriç River

In the basin, experimental measurements have been made and conducted at 27 stations located on the Meriç, Arda, Tunca and Ergene Rivers by General Directorate of State Water Works since 1979. [D.S.I. 1985] [Karpuzcu, et al. 1992]. The location and codes of these stations are shown in Figure 1. At these stations the parameters given in the Water Pollution and Control Regulations have been measured particularly. In this paper, instead of the detailed evaluation of the whole stations, the stations placed on the main changes and on the joint points were selected and the water quality was investigated. For this purpose, 001, 007, 008, 009, 010 (Meriç River), 002 (Tunca River), 003 (Arda River) and 0.17 (Ergene River) coded stations were selected. For evaluation purposes, parameters important from water pollution and control points of view, BOD5, dissolved oxygen, NO2-N, total dissolved solid matter parameters were considered.

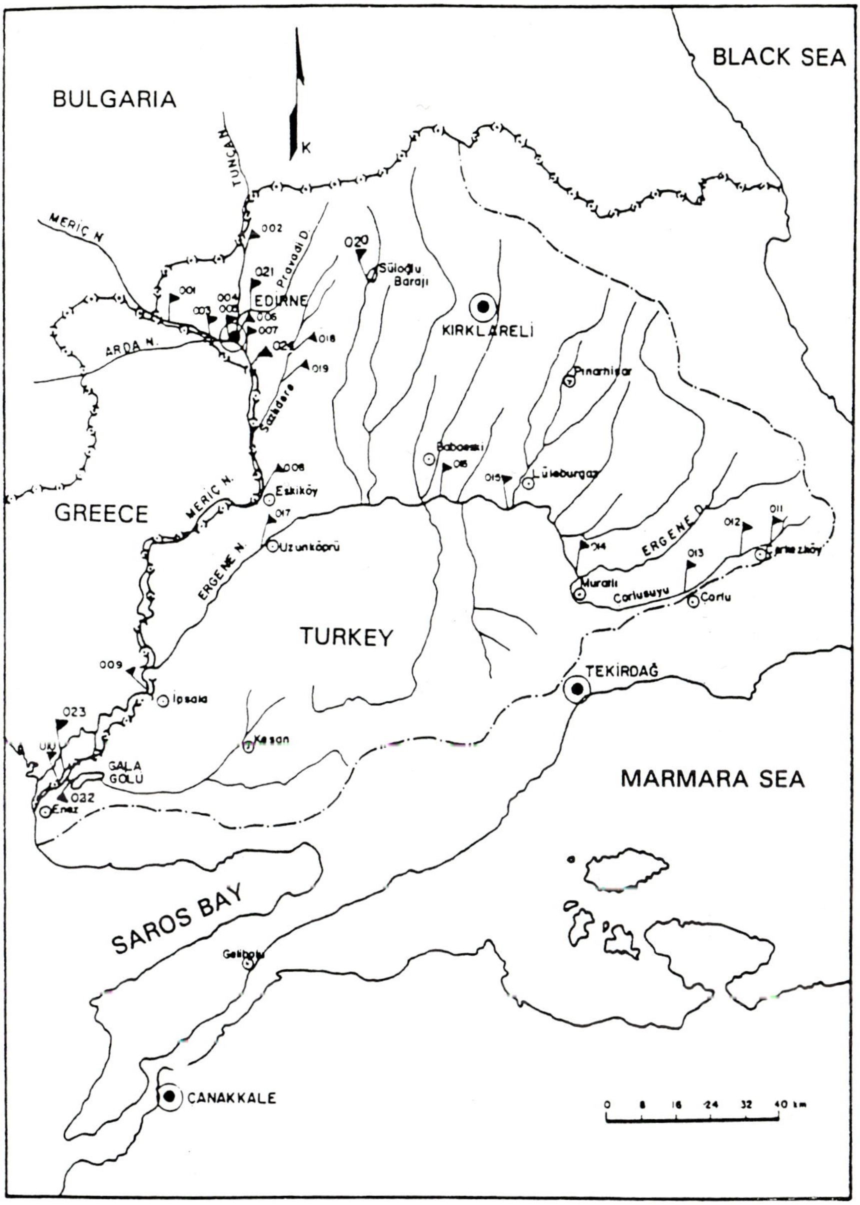

Figure 1 Meriç Basin.

The results obtained by using yearly averaged values of 1980, 1985 and 1990 are shown in Figs. 2, 3, 4, and 5. On the same figures the values used for the classification of water are also shown. If the BOD5 parameter is taken into account, no change is seen in water quality except 002 station in Fig. 2.

The river can be classified as second class at the 001-002 stations but for the year 1990, 002 station is in the third class and 003 cooled stations is in the first class. Water quality of 007-010 cooled stations were dropped to second class in 1985 and 1990 while being first class in 1980. During the monitoring period, the dissolved oxygen values which represent the first class water quality along the river, do not change significantly (Fig.3).

Fig. 4 shows the changes in nitrite concentration in the basin for the years 1980-1990. At the 003 station, in 1980 and 1985 third class and in 1990 second class water quality were observed while it was fourth class at 001 and 002 for all years. At 007-010 stations generally third class water quality was determined. Water quality class of 009 station increased to third class in 1990. Evaluating the TDS values, it can be concluded that water quality of the river is between first and second class except the measurement at 010 stations at which third class water quality was determined in 1980. Also Turkish Ministry of Environment stated that water quality of the Meriç Basin represents the first and second class water quality except few measurements of pH, NH3-N, NO2-N, BOD5, O-PO4, Fe, Pb and Cu at some stations. Annual mean values of chosen parameters, nitrite, TDS, DO, BOD5 were calculated and the graphic representing the yearly changes in between 1980 and 1990 are seen in Figs. 6 to 9.

As it can be seen from these Figures;

The average BOD5 value is around 6 mg/l and refers to 2nd class water quality.
The average DO value is around 10 mg/l and refers to 1st class water quality.
The average nitrite value is around 0.04 mg/l and refers to 2nd class water quality.
The average TDS value is around 0.9 mg/l and refers to 2nd class water quality.

The Meriç River can be classsified into second class according to the results mentioned above. In order to improve the water quality of the Basin, the pollution loads carried by Tunca River from Bulgaria and Ergene River from Turkey should be controlled and pollution level should be decreased. To sustain the pollution of Meriç River at its present level, Bulgaria, Turkey and Greece should take necessary precautions by being in communication as closely as possible.

Water Quality Problems in the Meriç Delta

In the Meriç Delta, the saltwater intrusion into the river through the river mouth at the coast consitutes an important water quality problem. The MeriçRiver arrives at the Aegean Sea at the Saroz Bay after flowing 187 km along the borderline between the countries of Turkey and Greece. Just before arriving at the sea, it forms a large delta with very interesting properties, such as lagoons and their channels. This delta with its lagoons is seen on the satellite image in Fig. 10. This image is formed using Landsat Thematic Mapper data [Maktav, D., 1994].

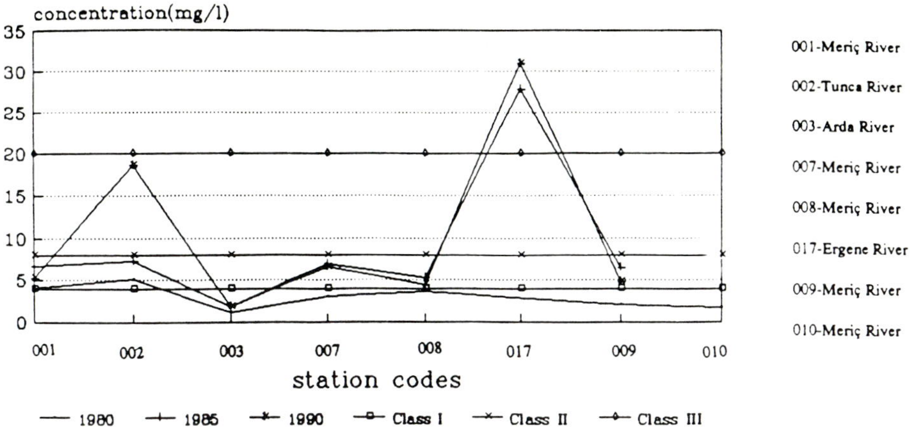

Figure 2 BOD_5 values for 1980, 1985 and 1990.

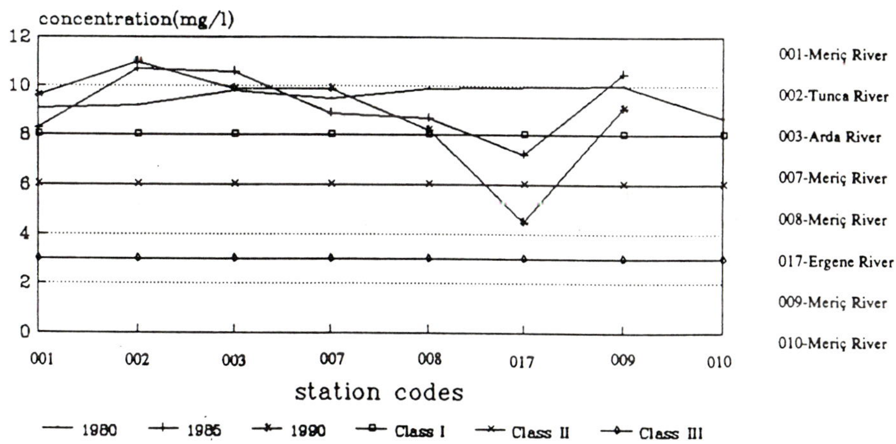

Figure 3 DO values for 1980, 1985 and 1990.

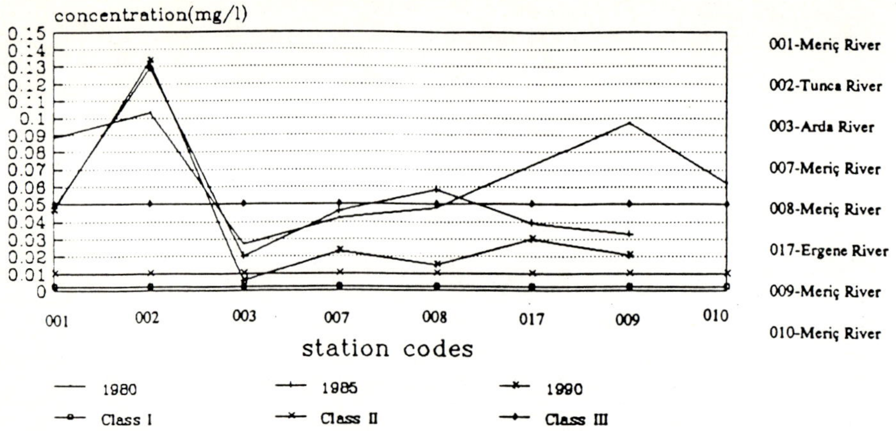

Figure 4 NO_2-N values for 1980, 1985 and 1990.

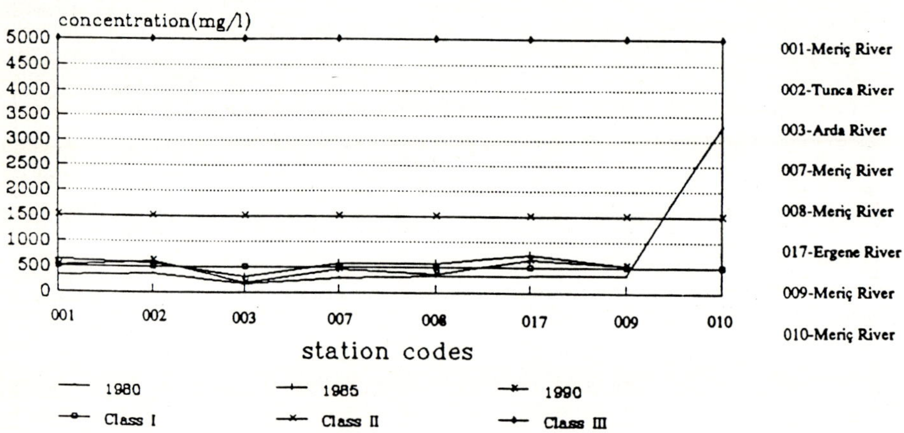

Figure 5 TDS values for 1980, 1985 and 1990.

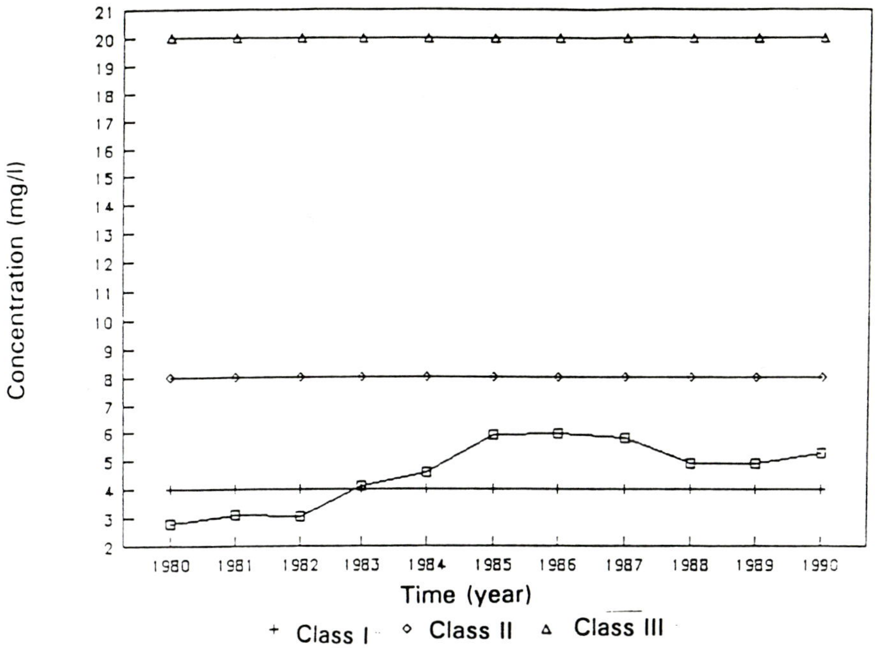

Figure 6 BOD$_5$ values from 1980 to 1990 in Meriç river.

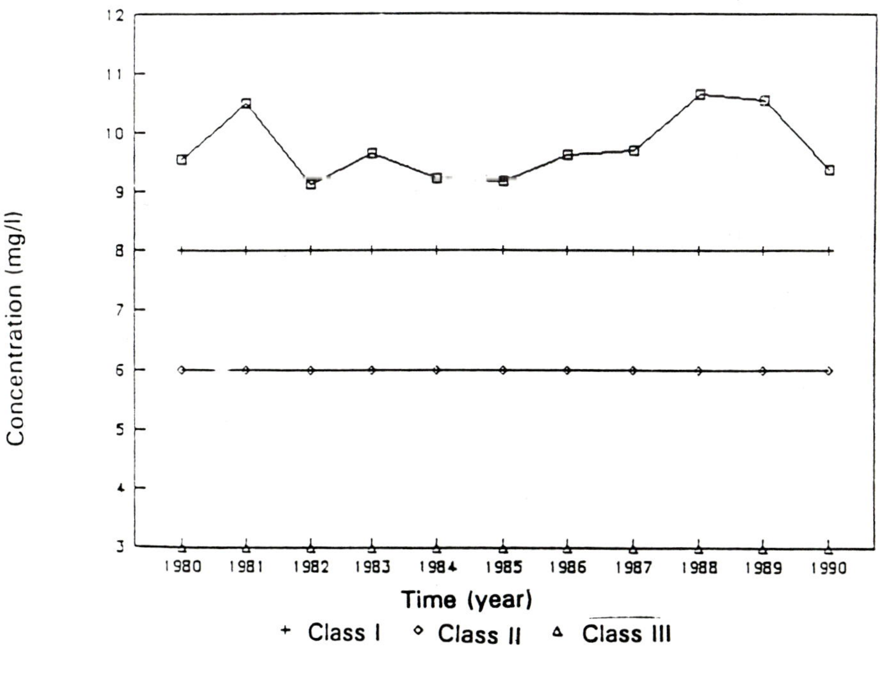

Figure 7 DO values from 1980 to 1990 in Meriç river.

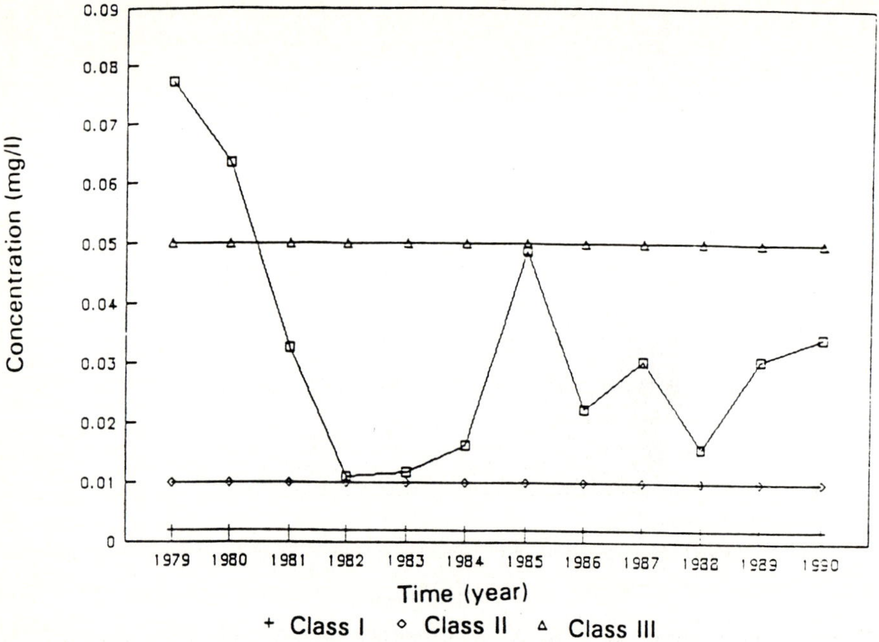

Figure 8 NO_2-N values from 1980 to 1990 in Meriç river.

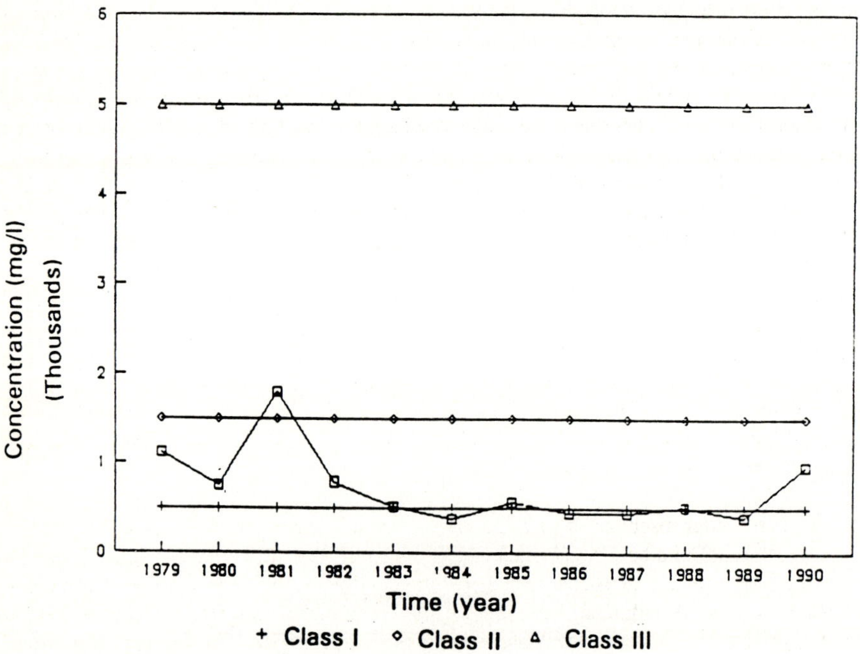

Figure 9 TDS values from 1980 to 1990 in Meriç river.

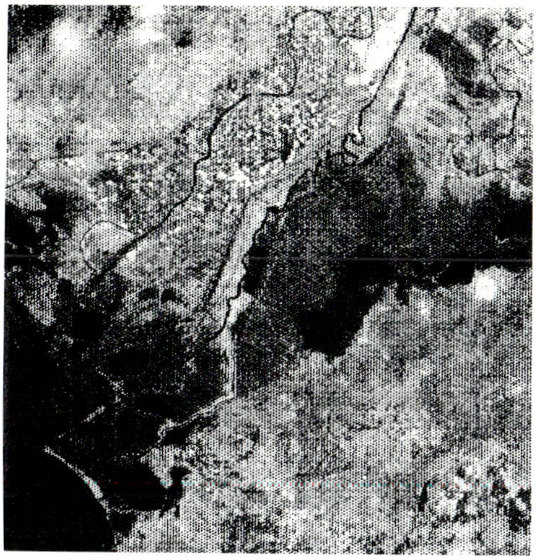

Figure 10 Landsat image of the Meriç Delta in the year 1992.

Generally, hydraulic processes in a river delta are mostly dominated by water discharge, water level, mean elevation of the bottom, water surface slope, mean flow velocity, cross section area, mean depth, sediment discharge, mean sediment concentration and hydrodynamic conditions of the coast in the vicinity of the mouth (Mikhailov, V.N., 1969). Also, the offing is a component of the river mouth area, and delta front is the most dynamic type of the shore as a result of interaction between river and sea (Skriptunav, N.A., 1969). In light of these considerations, it is obvious that Meriç Delta is under effect of highly dynamic conditions, and the saltwater intrusion occurs when the water discharge is less than 30 m3/sec. Because, the mean elevation of the river bottom is -2.50 m and inverse hydraulic slope can be seen when the mean sea level is higher than the still water level due to the wave motion. Additionally, the river bed slope is also inverse in the delta and tidal range is about 0.6 m in the Saroz Bay. As a result of all these, the saltwater can reach approximately 35 kms upstream the Meriç River.

In recent years, many large dams with large storage capacities in Bulgaria were constructed and water discharge in Meriç River has been reduced. Especially in summer periods, water discharge dropped to 4.9 m3/sec. On the other hand, in the delta, at both sides of the river rise farming were made. The rise farming covers an area of about 65000 ha in Turkey. For irrigation purposes Turkey purchases about 14 000 000 m3 of water from Bulgaria in the months of August and September. But, this still does not solve the saltwater intrusion problem in the delta.

The decrease of water discharge is also important for the river mouth, delta morphology and ecology. The Meriç Delta between Turkey and Greece has quite important ecologic properties because of the lagoons and their channels. In natural conditions, the river changes its route due to the yearly change of hydraulic conditions. Some erosion and accumulation can be seen at both sides of the river, even some small islands are formed. But, for the very low water discharge, quite a large sedimentation problem arises around the river mouth. The reason for this is, that the coast of Saroz Bay consists of very fine sandy beaches at which very strong this is, that the coast of Saroz Bay consists of very fine sandy beaches at which very strong alongshore sediment transport occurs. That is why at the Meriç River mouth the sand accumulation is a very big problem due to the significant waves in direction of SE. This problem will be chronic unless the river water discharges are high enough.

Conclusion

Turkey has great potential from water resources and quality points of view compared to other countries. But in order to prevent water resources and quality, the required precautions should be taken as much as possible. If the domestic wastewater caused by rapid population growth rate, and unsound urbanization, industrial and agricultural wastewaters are discharged to receiving media without any treatment and control, the water resources will be polluted or the present pollution level will reach a dangerous level.

Therefore, to prevent and increase the water resources potential a new study has been started.

In the first stage, it was decided to construct the treatment plants of whole big settlements for their domestic wastewaters. In addition, required regulations to force the industries to construct their treatment plants were established. And in order to support them financially, a fund was established.

Furthermore, the construction of the treatment plant of the erkezk^y Organized Industry Region, which is discharging its wastewaters to Orlu stream, tributary of Ergene River, was started.

It can be concluded that the water quality of the Meriç River, and morphologic and ecologic properties of its delta strongly depend on the regulation of water discharge by the Bulgarian dams. Therefore, cooperation between Turkey, Greece and Bulgaria is needed. It is also necessary, according to the Helsinki Convention, that all related countries have to take appropriate measures for conservation of ecological structures of transboundary waters. At that point, the difficulty is to evaluate the problems related to the changes in natural environment. Generally, in order to obtain a sufficient solution, an extensive monitoring through field measurements is necessary.

The application of remote sensing technology in addition to the field measurements can give a spatial view for transboundary waters. Besides, the integration of remote sensing technology and field measurements is essential for transboundary waters because of the following reasons:

a. A transboundary coastal zone covers generally large coastal areas. It is difficult to obtain spatial views for large areas.

b. In transboundary coasts, in-situ measurements should be carried out by transboundary states. This can raise some problems in obtaining global results as a whole.

If an integrated monitoring program may be applied with a scheme given in Fig. 11, all the results obtained by field measurements in national bases can be gathered in a central station, in

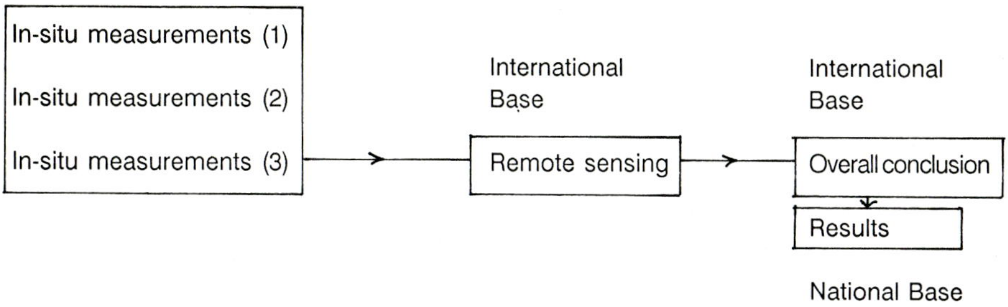

Figure 11 Integration of monitoring measurements in a transboundary framework.

References

DS, State Water Works, Water Quality, Data(1979-1982), Ankara,1985.

DS, State Water Works, Water Quality, Data(1983-1984), Ankara, 1985.

Karpuzcu, M., et.al. Ministery of Environment of Turkey, An Investigation Project on Pollution Levels and Water Qualities of Water Catchment Areas, Meriç Basin, Bosphorus University, Environmental Sciences Institute, 1992.

Maktav, D., Kapda*fl*ì, S., Landsat Thematic Mapper Monitoring of the Lake Salda in Turkey, 1994 ACSM/ASPRS Annual Convention, Reno, USA, 1994.

Mikhailov, V.N. Hydrologic-morphometric characteristics of delta branches, Hydrology of deltas, Proceedings of the Bucharest Symp. May 1969, pp.146-158.

Skriptunov, N.A., Hydrological processes in offings and their role in formation of a delta front, Hydrology of deltas, Proceedings of the Bucharest Symp. May 1969, pp.164-172.

VI.6 Regional Water Transfers and Drought Management Strategies

Christos A. Karavitis
Research Associate
Water Resources Planning and Management, Colorado State University,
Fort Collins, Colorado, U.S.A.

Introduction

Drought constitutes a rather severe hazard to all human activities, and especially to water supply. There are numerous efforts for planning and management actions against droughts, nevertheless, deficiencies still pertain in such attempts. The resulting major challenge concentrates on the development of comprehensive and effective drought management schemes. In this context, the primary problem of this research is composed of the following two key elements:

1. The overall difficulty in understanding the complex environment of drought. The phenomenon per se is an extreme event, a widespread and elusive natural hazard, having a multiplicity of impacts and interactions with the whole human system.

2. The additional dimension of managing drought under extreme and adverse conditions. Such conditions include the physical, structural, socio-economic and institutional environment. The metropolitan Athens area, where the water supply is accomplished by a series of reservoirs and conveyance structures approximately 190 km to the west of the region, reflects such a case.

Then, the present paper is trying to identify appropriate solutions for the above problematic areas. The resulting options may be used as a means to avoid undesirable effects or major future crises in meeting the continuous challenge of drought and in fortifying a risk management commitment.

Drought and Drought Management

Drought Definition

The existing tendency among professionals, politicians, managers or even common citizens is to view and consider drought as a natural hazard, a traditional disaster causing emergency mobilization (Wilhite, et al 1987; Grigg, N.S and Vlachos, E.C. 1990). However, cumulative experience from scientific investigations of the recent decades is indicating that the recurrence of drought is unavoidable, as the phenomenon seems to be an inevitable and permanent part of the world's climate, especially with the recent indications of potentially increasing instability in the environment (greenhouse effect, ozone depletion etc.).

In this aspect, drought and its ramifications must be recognized and anticipated for all planning and management efforts in water resources. The first step in studying and responding to drought is the examination of its dimensions. Drought dimensions ideally incorporate the definition, the causes and the impacts of drought. The causes and the impacts of drought will not be examined, but a drought definition is essential for the scope of this paper. Nevertheless, a precise, unambiguous definition of drought remains elusive (Yevjevich, et al 1983; and Grigg, N.S. and Vlachos, E.C. 1990). Difficulties in devising an objective definition of drought are probably threefold. The first source of confusion may be that drought implies a variety of things to various professionals according to the specialized field of study (meteorology, hydrology, water resources, agriculture etc.). A second problem is eliciting because the definition of drought is strongly related to the geographical, hydrological, geological, historical and cultural traits of a given locale. A third factor is the difficulty to modify existing drought terminology according to updated techniques and practices.

However, a general definition of drought may be the state of adverse and wide spread hydrological, environmental, social and economic impacts due to less than generally anticipated water quantities (Karavitis, C.A., 1992). Such water deficiencies may originate from precipitation decreases, physical and/or operational inefficiencies in water supply and distribution systems, as well as from incompetent water management. Then, drought is not only the lack of precipitation as it is generally believed, but a complex phenomenon influencing the whole human system.

Drought Management Approaches

A review of the drought literature may indicate that most water agencies or authorities are not implementing drought contingency measures, responses or plans until a drought is imminent (Yevjevich, V.et al., 1983; Wilhite, D.A. et al., 1987; Grigg, N.S. and Vlachos, E.C. 1990; U.S.A.C. of E., 1991, Karavitis, C.A. 1992).

Even then, the applied management actions are usually short range, piecemeal and segmented, in other words typical of crisis management. It is exactly this spirit of crisis management that should be avoided, and to move instead into a risk management posture, in other words into anticipatory and proactive planning.

Thus, this paper describes the application of integrated management strategies in order to confront droughts. The proposed responses may be described as a synthesis of existing potential alternatives (i.e. administrative mobilization, economic incentives, supply and demand measures, etc.), with the incorporation of the latest in computer technology.

Physical Dimensions: Regional Water Transfers and Drought

Water Resources in the Greater Athens Area

In describing the water resources of the Greater Athens area the key element seems to be the uneven distribution, both in time and space, of precipitation, activities and population. The continuing increase of water demand as well as environmental considerations pose additional constraints. The Attica sector, where Athens is located, is the poorest in water resources in

Greece, having about 400 x 10^6 m^3/yr (Karavitis, C.A. 1992). Western Greece is by far richer in water than the eastern part of the country, creating a "water crescent", from the north through the south. Thus, water has to be transferred to the east where population and activities are concentrated.

In Attica sector urban water supply exceeds by far any other use reaching approximately 300 x $10^6 m^3$/yr. The current water supply is based upon two surface water sources: the reservoir on Mornos river in Western Greece, and the system of river Voiotikos Kefissos - lake Ylike to the Northwest of Athens. Water from Mornos is carried to Athens via a gravity conveyance system with a length of 189 km.

The lake Ylike - V. Kefissos branch supplies water through a 75 km conveyance scheme boosted by a series of pumping stations. Existing water resources management practices, the inefficient wastewater and storm water systems are also compounding to the problems.

In this regard, the context of the Metropolitan Athens area may be summarized in the following components: - the area is semi-arid and subject to shocks due to the water scarcity - there has been a rapid and high concentration of population and activities with marginal and decaying infrastructure, and - the management experience of complex systems seems to be limited. Hence, Athens is under stresses with very little resilience to natural hazards.

Drought and Water Supply in Greater Athens

The detailed examination of the 1990 drought in Athens as well as the analysis of the collected data and management conditions may lead to the following key evaluation points (C.A. Karavitis, 1992): - the existing data bases are not complete nor reliable - the water allocation plan is at best haphazard- the 1990 drought was the worst on record - updated information technology was not applied for management efforts. To illustrate these points the impacts of the 1990 drought in the water reserves of Athens are presented in Fig. 1. In order to give an indication of the magnitude of the problem, in October of 1990 the Athens metropolitan area had water reserves for only 56 days.

Drought Management Responses in 1990 and 1992

The complete presentation and analysis of the 1990 drought responses, was essential before any attempt to evaluate them, and has lead to the following conclusions (C.A. Karavitis, 1992): the framework within which the responses were launched is characterized by a centralized decision making process with probable political ties. In addition, water resources planning and management policies are not well organized and operation and maintenance of water schemes seem to be sluggish and inefficient. At the same time, the initial analysis of the 1992 drought has accentuated the above concerns. Overall, for both the droughts incidents, drought risk and impact assessment, drought alert mechanisms and drought contingency planning were absent in the drought management responses.

Responses were implemented in a piecemeal and segmented approach with an emphasis on short range actions. Measures were applied well after the onset of the event, with little consideration to the crucial timing factor. Therefore, the 1990 and 1992 drought management responses in Greater Athens are typical of crisis management.

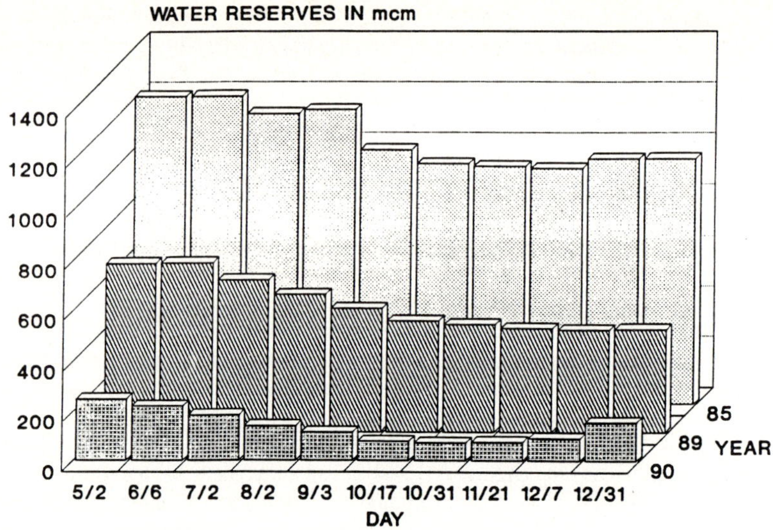

Figure 1 The 1990 drought impacts on the water reserves of Greater Athens.

Towards Integrated Drought Management Strategies

Drought management scheme

The demarcation, assessment and evaluation of the 1990 drought, as well as of the applied drought responses in the urban environment of Greater Athens, resulted in accentuating the need for holistic drought management strategies.

The research has concluded, that such comprehensive strategies should comprised of the following components (Karavitis, C.A. 1992): - pre-drought preparation and planning (drought forecasting, drought alert, drought contingency master plan etc.) - drought responses plan (short and long - term measures - post - drought measures (evaluation, revisions in planning etc.). In applying such a procedure, the incorporation of computer technology becomes crucial.

Development of a drought management knowledge based system (KBS)

It was noted that, drought management strategies should be a continuing endeavor and not a process initiated during a drought crisis, and ending when the hazard dissipates. Drought responses should lead towards getting the optimum use out of the available limited supplies. In this quest, the advantages of computer technology and specifically of the Knowledge Base Systems (KBS) are unambiguous in confronting drought. Such a KBS could be of use in achieving: - the delineation of a wide range of alternatives for drought measures - the evaluation and selection of the main areas where time and resources should be focused - the development of a consensus among various groups (politicians, managers, experts, etc.) in confronting drought, particularly under adverse conditions such as in the case of Athens. Thus, the KBS becomes ''the arena of focused disagreement''.

Since 1988 several programming approaches were used in developing an integrated drought management KBS, with the characteristics previously described. Finally, the program was created using the PROLOG language, designed to run in an IBM environment and it was named XERASIA.4 (Karavitis, C.A., 1992). Briefly, it examines three basic categories of drought responses:

· supply augmentation presenting different methods to enlarge water supply (i.e. construction of new facilities, rehabilitation, mixed type development, etc.).

· demand reduction including proactive and reactive measures, technological adjustments such as reuse, recycling, etc.

· impact minimization focusing drought loss absorption, loss reduction, anticipatory measures (forecasting system, consumption regulation etc.).

The final output of the program reflects the above main categories and their subcategories (new, mixed types, reuse, anticipatory measures etc.).

Application of the Xerasia - KBS for the 1990 Athenian Drought

The developed DSS (XERASIA 4) was applied and the results were analyzed and evaluated in the context of the metropolitan Athens droughts (Karavitis, C.A., 1992). Briefly, from the output of the KBS, (Figure 2). reactive demand reduction measures and anticipatory impact minimization responses have scored the highest. Demand reduction measures have to be concentrated on maintenance efforts of the water supply system, water reuse, reduction of non- essential uses and long - term water saving programs. Anticipatory measures should include a drought forecasting system, regulation of consumption and public information programs.

These categories have more or less been neglected in the 1990 - 92 drought responses. The applied measures have concentrated on new reservoir construction, groundwater utilization and a sporadic maintenance program without an overall drought contingency plan. Concluding, the application of the DDSs illuminated a complete spectrum of potential drought responses demonstrating its use as a mnemonic device. At the same time, it has indicated the main areas of drought policy options.

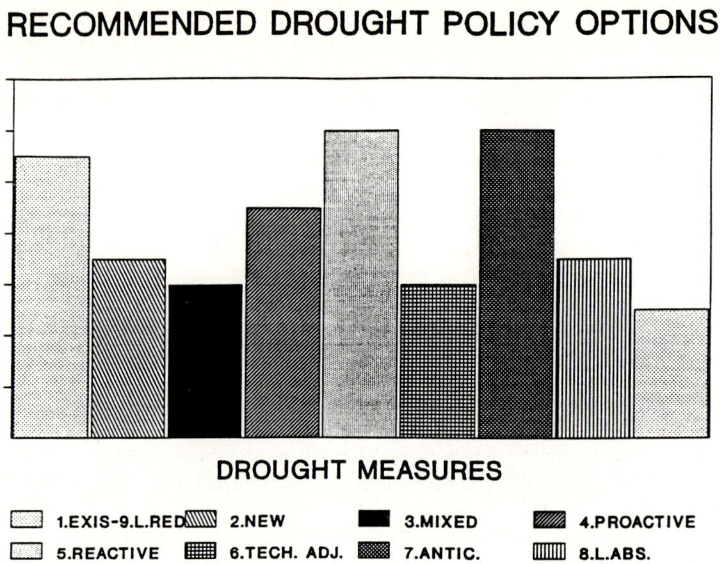

Figure 2 XERASIA 4. Program output with the various drought responses categories.

Drought and Policy Options

In 1991, as soon as the drought hazard dissipated for the Athens metropolitan area, it seemed that drought management efforts have also dwindled. Plans for a water resources policy, although announced, seem to be still pending. The incorporation to the water supply of the aquifers adjacent to the Ylike scheme, as well as the construction of a new reservoir and diversion in the Euenos river, seem to be so far the only long- range measures. Improvements in the Athenian maintenance practices, although announced, have not yet been possible to follow. The imposed pricing on water has not been lifted and more or less existing water consumption practices are still in use. Overall, it would seem that the 1990 drought has not changed the management practices. In this regard, it may be safe to predict that if another drought is set, crisis management will be again applied (Karavitis, C.A. 1992). This has been ruefully proved as data on the 1992 drought have started to be collected. Therefore, the need for the application of holistic drought management strategies in the Athens area is more than essential. In this context, the character of drought policy options for a complex spatial and organizational system such as the megalopolis of Athens should have a two prong emphasis: towards the development of a water resources policy with a long - range time table, and towards the incorporation of systems engineering techniques for managerial efforts in drought responses.

Conclusions

In order to assess future droughts in the metropolitan Athens the most pressing issue regarding drought management strategies should be an effort to incorporate the whole system in a regional or even a national basis. Therefore, two main areas of potential recommendations have been identified:

1. Actions concentrating on the water infrastructure of greater Athens (maintenance and operation programs, assessment of water supply and use, development of accurate and complete data bases, application of the information technology).

2. Actions centered around comprehensive drought management strategies (development of a complete water resources policy not only for Athens, but for the whole of Greece, implementation of drought management responses in a timely and coordinated fashion, improvement of the decision making process particularly through the uses of DSS, the conduct of impact assessments and the incorporation of risk management).

References

Grigg, N.S., and Vlachos, E.C., 1990. Editors, Drought water management, Dep. of Civil Engineering, Colorado State University, Fort Collins, Colorado.

Karavitis, C.A., 1992. Drought management strategies for urban water supplies: The case of Metropolitan Athens. Ph.D. Dissertation Dep. of Civil Engineering, Colorado State University, Fort Collins, Colorado.

U.S. Army Corps of Engineers, 1991. The National Study of Water Management During Drought, Virginia, Vol 1 and 2.

Vlachos E.C. 1982. "Drought Management Interfaces" ASCE Annual Meeting, Las Vegas, Nevada.

Vlachos, E.C. 1988. Water Resources Planning for Droughts. In Grigg and Vlachos op cit.

Vlachos, E.C. and Karavitis, C.A. 1992. (Personal communication). Assessing the 1990 Drought in Athens. Colorado State University, Fort Collins, Co.

Wilhite W.A., Easterling, W.E. and Wood, D.A., 1987. Planning for drought, editors, Westview Press, London.

Yevjevich, V. da Cunha, L., and Vlachos, E.C., Eds. 1983. Coping with Droughts, Water Resources Publications, Littleton, Co.

VI.7 Transregional Environmental Problems in the Tyumen Region and the Bashkortostan Republic, Russia

A.A. Gubaidullin
Tyumen Institute of Mechanics of Multiphase Systems,
Siberian Branch of the Russian Academy of Science,
74 Taymirskaya St., 625000 Tyumen,
Russia

R. Khamitov
Institute on Applied Ecological Problems
12/1,8 Mazta Str.
Ufa, 450005, Bashkortostan
Russia

Introduction

The Tyumen region occupies a vast territory on the West - Siberian Plain. Its area makes up as much as 1.44 million square kilometers. This territory is home to the multinational population of 2.7 million people. The population of the region is predominantly urban; about 72% of the people live in 16 towns including Tyumen which is the center of the area. The Tyumen region includes the Jamalo - Nenetski and Khanti - Mansijsky autonomous districts.

The average January termperature is from minus 17 to minus 29 degrees Celsius. The average July temperature is from 14 to 18 degrees Celsius. Total average annual precipitation measures from 200 to 600 millimeters per year.

The Tyumen region is Russia's basic source of raw materials such as oil, gas, and timber. The largest oil-field is called Samotlorskoye ; the largest gas deposits are called Urengoyskoye and Medvezhje. The oil and gas produced in the region are transferred through pipelines to the European part of Russia and abroad.

The basic environmental problem of the Tyumen Region is oil pollution of soil, water and atmosphere including the pollution of minor rivers and lakes. The problem of drinking water is also very urgent. The northern rivers are especially vulnerable to contamination. While in the midland areas the organic river pollutants are neutralized by the "natural filters", in the northern areas because of low water temperatures and lack of oxygen the rivers can't be cleansed of such waste. The water ecosystem filters which would be effective in another type of climate don't work there and the northern rivers are virtually defenseless against the pressure of non-purified waste.

The Problem of the River Ob

The Ob River is the major water «artery» of the region. The problem of the Ob River can be put as a priority on the agenda because it's out ground republican and state borders and turned into an international one, for it's of deep concern to all the circumpolar countries. The ecological expedition «Ob - 90» and the discussion of its results at the inter-regional science and practice conference held under the motto : «Saving Ob is Saving Siberia», confirmed the gravest fears that the River Ob is turning into the dump of all Western Siberia and that the pollution of the Ob - Irtish River basin has a harmful effect on a global scale.

The ecologists' estimate indicated that the River Ob annually brings into the Arctic Ocean over a hundred thousand tons of hydrocarbons. This is the consequence of water pollution during oil pipeline breakdowns and mud sump leakages. Moreover, tank filling is usually done with violations and large quantities of oil products are spilled over board. It's worth mentioning that one ton of oil on the water surface turns into about 12 -15 square kilometers of oil film. Not only do we face the decrease in fishing but a considerable part of the fish catch has a distinctive kerosene smell and is not edible. But it's not a secret that the Ob - Irtish River basin is a unique fish farming water body. Here are the world's largest resources of lake whitefish and extremely rich variety of its species. Within the basin one can find the most numerous herds of white salmon and a considerable stock of sturgeon.

In the Tyumen Region, which currently produces 80% of the total fish catch in the basin, there has been a serious decline in the output which dropped down to 17 thousand tons in 1992. This was the lowest level of fish catch for the past 23 years.

This situation is the result of the river basin being under the strongest anthropogenic pressure. Because of the hydro-construction performed on the River Ob, about 40% of all fish have been cut off. The most annoying thing is that no one really cares about building water purification facilities while in the meantime 42 thousand cubic meters of waste is dumped each day into the River Ob. In general the river basin of the region takes 300 000 cubic meters of non-refined sewage daily.

Impact of the Oil and Gas Complex

Most disastrous is the development of the "oil and gas complex" in Western Siberia, accompanied by the increasing pollution of water bodies with oil products which eventually will be harmful not only to the fish but to Man as well. For decades many perilous substances spilled into the rivers with the waste from metallurgical and mining plants located in the Urals have been passing through our region. It's obvious now that of negative consequence for the local environment are air masses moving from vast territory of Kazakhstan and also the fact that the sources of major rivers lie in the areas of environmental pollution. The proximity of the Novaya Zemlya island and the use of nuclear explosive technologies for intensive oil production have been of serious concern to the environment, too.

The development of the oil and gas complex caused destruction of the environmental balance, disintegration of landscapes and violation of the hydrological conditions to the rivers of the Ob

Basin as well as contamination of soil, water, open reservoirs, ground waters and atmosphere. It led to the reduction in the number of industrial fish and fur-bearing animals, to considerable worsening in conditions of their reproduction, shortening graze fields for the deer, decrease of timber resources, irrational use of hydrocarbonic raw materials and mineral resources.

According to very approximate calculations, during the period of the oil and gas boom (which lasted for 25 years) the total damage to nature is estimated to be billions of dollars. Geologists alone destroy 10 thousand hectares of land every year and cut down 1.7 million cubic meters of wood. Nowadays billions of cubic meters of gas, millions of tons of unstable gasoline are being burnt in flares per year. About 5 million tons of pollutants are spit out into the atmosphere. Today's size of Tyumen Taiga can't make up for the amount of oxygen burnt in flares.

The length of corrosively-dangerous parts of the pipelines is constantly increasing causing frequent pipeline breakdowns and more oil spills into the environment. The pipelines currently under production still do not have nature-protection facilities. Throughout the vast territory the pollution of soil, surface and ground waters with oil products and various drilling muds is going on. Besides, there are the processes of surface deformation inundation and marsh formation. The uncontrolled water injection for sustaining the stratum pressure has already led to seismic phenomena and earthquakes (at Nefteyugansk). The destruction of permafrost cover has caused formation of a new landscape : the polar desert, and to the Tundra's melting which will add to its slipping down into the Arctic Ocean.

The contamination of rivers and other water bodies also causes irreparable harm to the fauna. There are about a hundred rivers that have completely lost their role for the fish industry. The total annual loss makes up 25 % of all the fish catch and this tendency puts in danger the existence of the regional fish industry.

Water Ecosystems in Tyumen

In Tyumen there are two major water ecosystems : the ecosystem of the Tura River and the ecosystem of Lake Andreyevskoye. The Tura River is the right - bank tributary of the Tobol River, and the basic water reservoir. The water from the river is used as water supply for a number of localities and for the needs of the local industry. In particular in its middle flow the Tura is the source of water supply for Tyumen. The river streams through two regions : Sverdlovsk Region and Tyumen Region. Along with its tributaries it occupies the area of 80 000 sq. kilometers and the total length of 1015 kilometers. There are several major tributaries flowing into the Tura River, such as the Salda, the Tagil, the Nitsa and the Pishma.

The Tura River is a type of river with prevailing spring floods. On the banks of the river within the city limits and outside Tyumen there are several enterprises, agricultural plants which use water from the Tura and serve as the source of pollution. Widely-developed navigation adds to the river's contamination. Within the city, along the banks of the river there are 25 sewage outlets, with the total amount of dumped waste 314 million cubic meters per year, including 59 million cubic meters of purified waste. At the factories discharging waste into the Tura there are 13 refineries of which 10 have sufficient level of purification. The refineries at the shipbuilding plant and at the sheepskin and fur processing factory are inadequate. The water in the Tura River comes to Tyumen from the Ural Mountains already polluted. There is an increased content of various metal salts, iron, copper, lead, etc. The inconsistency of their presence in the water is the evidence of waste discharges from the industrial enterprises. The content of oil products is twice as high as the highest allowed concentration and there is a tendency for its further growth because of the waste from urban sewage and the river transport. So, the water in the Tura River may be qualified as a contaminated type in summer as well as in winter periods. Only during the period of snow melting and rains does it become moderately-polluted.

Lake Andreyevskoye is located 11 kilometers to the South-West off Tyumen and belongs to the group of lakes lying in between the Tura and Pishma Rivers. The lake has an outlet via the system of lakes (Gryaznoye, Bolshoi Duvan, Duvan) into the Duvan River flowing into the Pishma. The area of Andreyevskoye's water surface is 16.2 square kilometers and it belongs to the type of medium size lakes in the world classification. The average depth is 1.65 meters, the maximum is 2 meters. There are 43 species of plankton. The total water-related area is 189 square kilometers consisting of podzoled black soil and loams.

Within the area there are several lakes, the largest of which is called Kirtchim. All the neighboring lakes through the system of marshes have surface water drainage into lake Andreyevskoye. The lake has an oval shape and sloping shore. Here one can find different types of fish. On the territory of the lake, bottom deepening is being done in order to reach the depth of 7 down to 20 meters. It's been discovered that deepening lower than 5 meters negatively affects the ecosystem. However this work is being done for the purposes of sand extraction.

Ten lakes located in the south of the Tyumen Region including Lake Andreyevskoye were examined for the presence of nitrogen and the content of ammonium, nitrate and nitrite forms of nitrogen was found. It was also noted that there are annual, seasonal and even daily changes in the concentration of mineral forms of nitrogen in the lakes. However in a number of cases they trace the tendency for its further growth. It is confirmed by the fact that the maximum indices for the nitric type of nitrogen exceed the highest allowable concentration by 20 times. As nitrates are the evidence of current contamination, the anthropogenic influence upon the lake becomes obvious. Biogenic forms of nitrogen not only pollute the water they worsen the destruction by other pollutants for they increase the consumption of oxygen and lead to hydrogen sulphide poisoning.

The lake water is used as the water supply for the people. According to the data of the World Health Organization, use of water containing high concentration of nitrates and nitrites may cause child mortality.

Ecological Situation in Tyumen

In Tyumen and all around it there are vivid examples of a relentlessly destructive attitude towards Nature. For instance, the beautiful Gilyouskaya grove is damaged by a stream of sewage cutting through it. There are piles of garbage dumped around the neighbouring woods. Tyumen has a population of 540 000 people.

In Tyumen there are about 25 machine-building plants; around 20 objects of building complex; the plastic-producing plant, 6 factories of light industry and 12 of food industry; 2 medical plants, a printing-house, 2 powerful Heating Stations, working on natural gas and about 160 minor boiler-rooms which use solid fuel. There are such transport enterprises as : a depot about 500 motor-fleets of which only 36 have control stations. In the city there are 100 000 vehicles.

Tyumen is characterized by the absence of large monosources of pollution. Some enterprises like, for example, the autotractor electric equipment plant, the plastic-producing plant, the storage cell plant, the machine - tool plant and others are located within the residential area. Others are grouped together into industrial complexes. Provided with water, energy and other services from centralized sources they are economically efficient but at the same time, obvious is the real and potential harm as a result of excessive power and damaging materials concentrated within a relatively small area. There are 10 such industrial centers and the degree of danger to environment carried by each of them is determined depending on different factors.

Tyumen has a very high level of air-pollution. The average annual concentration of formaldehyde and benzopyrene is 4 - 5 times higher than sanitary norms. On some days there have been occasional concentrations of dust, carbon oxide, nitrogen dioxide and phenol which exceeded the highest allowed concentration by 3 up to 9 times. The emissions of suspended substances; hydride sulphide, carbon oxide, nitrogen dioxide, hydrocarbons make up as much as 99 % of total waste. About 32 tons of formaldehyde are emitted into the lower layers of the atmosphere by the local wood-processing plants. During the period from 1985 to 1989 the content of formaldehyde and benzopyrene increased. Thirty seven per cent of all the refineries are out of use or low - efficient. Seventy per cent of all emission are the exhaust gases from motor transport.

Air pollution in Tyumen in 1991

Pollutants	Average annual concentr. (mg/m^3)	Maximum accepted concentr. (mg/m^3)	Maximum annual concentr. (mg/m^3)	Maximum accepted concentr. (mg/m^3)
Dust	0.2	0.15	3.6	0.5
CO	1.0	3.0	20.0	5.0
Formaldehyde	0.015	0.003	0.288	0.035
Phenol	0.003	0.003	0.044	0.01
Sulphur	0.009	0.05	0.312	0.5

Ecological Situation in the Tyumen Region

The average annual emission from motor transport and enterprises into the atmosphere of the region is over 3 million tons of ecologically harmful substances. Only 1 % of these pollutants are caught. The use of motor transport makes up 88 % of all urban pollution.

The major danger to the people of northern territories is the toxic substrances dumped into the environment by the oil and gas industrial enterprises. For example the share of such emissions in the pollution rate of Sprout is 70 % and in the city of Nizhnevartovsk it even soars up to 96 %. At the same time the average annual concentration of some substances is within the limits of the highest allowed concentration. However their continuous action and their interaction with other pollutants in severe climate conditions may cause the damaging effect. The problem is intensified by the use of invaded river fish for food. Some fish also contain pesticides and other contagious substances. At oil and gas fields people often drink water from open reservoirs. This water is rated as «muddy water» and contains oil products, phenols in concentrations exceeding the highest allowed ones. The ground waters used in the North for centralized water supply contain various mineral admixtures, reducing their taste quality.

Nearly every substance ranked toxic which is found in the atmosphere of the Tyumen Region, combined with the influence of low temperatures, promotes respiratory ailments. Such ailments are more frequent in the North where they are accompanied by severe chronic forms. The complicated and combined influence of ecological pollutants and epidemic factors upon the people destroys their immune system, deforms the immune response of the organism, and complicates the treatment of infections and parasitic diseases. In the North the number of patients suffering from such ailments is 4 times higher than in Russia in general. The number of patients who have active forms of tuberculosis in the region of extreme North is over three times higher than Russia's average. All these facts are evidence that improper an environmental situation may promote the appearance and the complication of chronic diseases.

The industry of the southern areas has been developed according to the needs of the "Fuel and Energy Complex". The environmental situation in this part of Tyumen Region has a specific character. There are wood-processing and furniture factories emitting wood dust, the source of allergies; machine-building in which various nitrodyes, lead, and non-ferros metals are widely used; the construction that discharges cement dust, causing silicosis; the petrochemical industry polluting the environment with oil products, phenols, synthetic resin, etc. The Southrern part of the region, especially Tyumen, has always been a base of construction materials, equipment for the oil and gas complex, food and various goods sent to the North. That's why over 80% of the toxic substances are emitted into the atmosphere as a result of engine output. There are also a lot of oil and gas pipelines which, when they break, lead to water and soil contamination.

Of deep concern to ecologists is the problem of the polluted lake and river water being used for centralized water supply. Practically all major farms providing all the region with food products are located along the banks of polluted rivers which don't have refinery facilities. That is why, for example, the polluted area within the Irtish River basin cases of dermatitis and eczema are more frequent than throughout the rest of Russia. Sometimes the disease rate is 830% higher.

The Southern territories of the Tyumen Region located next to the industrial parts of the Urals perform the functions of the filter absorbing the pollutants discharged during the runoffs from the enterprises. Evidently, that's why in the populated area situated along the river Iset there are more cases of malignant tumors and hematogenic diseases.

All the facts mentioned above support the idea that in solving the problem of environmental pollution it is hard to have any hopes for steady improvement in the health of a native as well as a migratory. It's impossible to solve all these grave problems on the local scale within the structure of the regional government. The natural resources of this land have been relentlessly exploited by the whole country for 30 years while the heavy load of environmental disaster is upon the local inhabitants involved in this tremendous and harmful machine, directly or indirectly.

However, there is a lot of optimism in the activity of the regional Nature protection movement. Thanks to common efforts there have been a few bans set upon the construction of potentially environmentally harmful petrochemical plants and on building a channel supposed to drain water from the Irtish to other regions as well as closing down the construction of some towns in Yalutorovsk area, which could not guarantee environmental safety. Ongoing is the process of forming the service of environmental monitoring in the Tyumen Region.

Ecological Situation in the Bashkortostan Republic

The Bashkortostan Republic refers to one of the most developed regions of Russia. It is located in the Pre Urals, junction of the Urals with the Russian Plain and Western Siberia valley, occupying almost the whole South Ural and adjoining plains. Such a geographic position of the Republic causes large variety of its landscape, climatic conditions contrast, flora, fauna and mineral wealth. There are great reserves of oil, coal, manganese, copper-ironores, non-ferrous metals.

The territory of the Bashkortostan Republic is 143.6 thousand sq. km. with a population of more than 4 million people, among whom more than 60% live in 20 cities (including Ufa with 1 mil. inhabitants). There are 2720 lakes and 1120 rivers in the territory of the Bashkortostan with total length of more than 20 000 km. The availability of rich mineral resources and suitable geographic position favor the republic to become a center of oil-refining, petroleum-chemical industry of the Russian Federation with total capacity of 55 mil. tons of oil per year and the center of highly developed machine building. It is observed the deterioration of environmental conditions and the contamination of air, soil, surface-ground waters as a result of constant recovery growing up, the application of power resources and mineral raws, foundation of powerful chemical productions, which are not supplied with modern treatment equipment and other fields of economic activities.

Bashkortostan is the region of Russia with the most unfavorable ecological situation and Ufa, Salavat, Sterlitamak cities are at present among 60 cities where the greatest amount of hazardous pollutants are discharged into the environment. The complicated ecological situation in the Republic requires real picture revealing of environmental pollution, most of all, surface-ground waters. The main sources of contamination of waterbearing horizonts in the cities are the industrial sewage containing hazardous materials including phenol, chlororganic compounds, dioxin, salts of heavy metal and so on. In 1993 the total volume of water intake from natural water reservoirs was 1210 mil. m^3 which is equal to 4.7% of annual run-off.

Sewage water discharge to the surface water objects is 717.7 mil. m^3 including contaminated ones - 555.5 mil. m^3.

Discharged pollutants in water reserves of the Republic in 1993 were 841.2 thousand tons, the most of which came from Ufa, Sterlitamak and Salavat cities. Thus, only during the last year sewage waters brought to the environment 560 tons of oil products, 1.5 tons of phenol, 82 tons chlororganic compounds, 9 tons of plumbum, 15 tons of vanadium, 51 tons of zinc. About half of polluted sewage in the Bashkortostan Republic is discharged in Ufa area. The run-off from Sterlitamak city is equal to 15% of the total run-off, but it comprises 67% of all pollutants. The share of oil-chemical enterprises is 10% of total amount of discharge and 17% of sewage water volume.

Generally, pollutants occurrence in water is due to the absence of treatment equipment or due to their poor operation caused by backward technology and more out equipment. Only more than 5% of contaminated sewage is treated to the required level. It is necessary to note that treatment technology which is applied on the majority of industrial enterprises of the Republic does not meet the up-to-date demands and does not provide the complete treatment of industrial and urban run-off. More particularly hazardous substances are discharged into the largest river of the Republic - Belaya River (see table).

Around the area of big industrial cities is contamination by oil products (up to 38 norms), chlororganical and nitride compounds, heavy metals (to 8 norms) including mercury, phenol, phosphate and others. In 1993 alone 70 cases were registered of dangerous pollution of water objects and 4 cases of the extremely high level of contamination of the Belaya and Dyoma Rivers. Within the total amount of industrial run-off discharged into the rivers only 39% correspond to the Russia-wide norms.

Mean pollutants content in river water, mg/l

Pollutant	rivers				
	Belaya	Urshak	Dyoma	Ufa	Ashkadar
Oilproducts	0.5-1.9	0.2-0.7	0.7	0.4	0.1
Phenols	0.003	0.003	0.002	0.002	-
Copper	0.006	0.004	-	0.006	0.003
Zink	0.01	0.015	0.019	0.01	0.01
Mercury	0.003	-	-	-	0.002
Hexachlorbenzol	0.15	-	-	-	0.17
Nitrites	0.5	-	0.05	0.04	0.05
Ammonium	0.6	0.8	0.6	0.7	0.4

Along with industrial enterprises a tremendous amount of pollutants is sewed into the rivers and lakes from agricultural regions of the Bashkortostan. In 1993 204.3 thousand tons of fertilizers were applied, 2342 tons of pesticides were used, among them 1545 tons of herbicides, 2.4-D. On the water protected areas are located 34 large stock-raising complexes and 61 farms use the hydraulic system of dung removal, besides the latter is not utilised and are piled into the rivers and fields. The majority of storage facilities for fertilizers and pesticides do not meet the ecological norms and they are wrecked.

Technogenez becomes the main cause of ground water contamination. During the last decade this process has been transferred from a local to a regional one. The real scales of technogenious ground water pollution is still under future assessment but the existing facts are seriously impressive. So, to enhance the oil recovery there was widely used in the Republic the method of salt fluid injection and industrial run-off that led to the groundwater pollution of a large area. Of great danger are the pipeline breaks through which the salt waters are injected into the oil formation. Only the Tuymaza oil field within one year has registered 2277 cases of accidental breaks in the system of formation pressure maintenance. Some portion of them is discharged into the rivers and tributaries, but the majority (to 90-95%) enters the ground water through the ground. In some oilproducing regions the area of fresh water salting achieves 500 sp. km and even more. As a result of waste storage at some metallurgic and mining industry enterprises the ground waters were contaminated by copper, plumbum and by other heavy metals. The ground water contamination by the enterprises of chemical oil-chemical industry causes great anxiety. So, it is discovered that the ground water pollution by toxic components contained in waters is from the fertilizer production plant of Meleuse city located on Belaya river bank. Practically all the groundwater sources of Ufa city contain heavy metals, phenol and chlorphenoles. The most contaminated groundwaters are at a depth of 6-10 meters. The abnormally high concentrations of chlororganic substances (chlorphenols, pesticide 2.4-D, 2,4,5-T) are recovered in groundwater on the territory of production corporation "Khimprom" (Ufa). The phenol content ranges from 0.4 to 2500 mg/l; of 2.4-D from 0.2 to 425 mg/l. Even at a depth of 75 metres phenol has been detected in ground water.

The complicated and multicomponents picture of groundwater and subsurface hydrospheres contamination, it's mutual tie with environmental pollution arise the necessity of complex solving of ecological problems in the Bashkortostan Republic. Under the condition of increasing of technogenious loads it is very important to reveal options for production structures installation, to maintain the priority of environment protection arrangements, to develop measures on rational use of natural resources. The problem becomes more complicated due to the fact that the majority of the industrial enterprises in the Republic were founded in the 1950-s and 60-s (80%) and there is an obsolescence of machines. Most of them utilize old technologies which create a lot of waste, and modern treatment equipment is not available.

It is also necessary to create the complex monitoring system for water resources of Bashkortostan. The existing control posts are oriented to the observation and assessment of some separate parts by their own independent programs and are not linked by any organization or methods, which creates some difficulties in the evaluation of obtained data. Besides, these posts are equipped with old analytical equipment and an outdated apparatus and it is not possible to effectively receive the information on water contamination.

References

Gureeva Z.P. , Ignatova V.A., Kosterina L.P. (1993) Principles of Ecological Knowledge, Tyumen, 164 p. (in Russian).

Istok (1994) Regional Ecological Bulletin, Tyumen, No 1-4 (in Russian).

Kutsev G.F. (1987) The Man in North Town, Sverdlovsk, 224 p. (in Russian).

Index

A

B

C

D

F

G

H

Q

R

S

T

Y

Z

The Partnership Sub-Series incorporates activities undertaken in collaboration with NATO's Cooperation Partners, the countries of the CIS and Central and Eastern Europe, in Priority Areas of concern to those countries.

The volumes published as a result of these activities are:

Vol. 1: **Clean-up of Former Soviet Military Installations.** Edited by R. C. Herndon, P. I. Richter, J. E. Moerlins, J. M. Kuperberg, and I. L. Biczó. 1995

Vol. 2: **Cleaner Technologies and Cleaner Products for Sustainable Development.** Edited by H. M. Freeman, Z. Puskas, and R. Olbina. 1995

Vol. 3: **Remediation and Management of Degraded River Basins.** Edited by V. Novotny and L. Somlyódy. 1995

Vol. 4: **Earthquakes Induced by Underground Nuclear Explosions.** Edited by R. Console and A. Nikolaev. 1995

Vol. 5: **Transportation Infrastructure.** Edited by R. M. Gutkowski and J. Kmita. 1996

Vol. 6: **Sustainable Development of the Lake Baikal Region.** Edited by V. A. Koptyug and M. Uppenbrink. 1996

Vol. 7: **Transboundary Water Resources Management.** Edited by J. Ganoulis, L. Duckstein, P. Literathy, and I. Bogardi. 1996

Vol. 8: **Urban Air Pollution.** Edited by I. Allegrini and F. De Santis. 1996

Springer-Verlag and the Environment

We at Springer-Verlag firmly believe that an international science publisher has a special obligation to the environment, and our corporate policies consistently reflect this conviction.

We also expect our business partners – paper mills, printers, packaging manufacturers, etc. – to commit themselves to using environmentally friendly materials and production processes.

The paper in this book is made from low- or no-chlorine pulp and is acid free, in conformance with international standards for paper permanency.